Microbial Interactions and Communities

Volume 1

Microbial Interactions and Communities

Volume 1

Edited by

A. T. Bull

Biological Laboratory
University of Kent
Canterbury
England

J. H. Slater

Department of Environmental Sciences
University of Warwick
Coventry
England

1982

ACADEMIC PRESS

A Subsidiary of Harcourt Brace Jovanovich, Publishers

London · New York
Paris · San Diego · San Francisco · São Paulo
Sydney · Tokyo · Toronto

ACADEMIC PRESS INC. (LONDON) LTD.
24/28 Oval Road
London NW1

United States Edition published by
ACADEMIC PRESS INC.
111 Fifth Avenue
New York, New York 10003

British Library Cataloguing in Publication Data
Microbial interactions and communities.
Vol. 1
1. Microbial ecology 2. Biotec communities
I. Bull, A. T. II. Slater, J. H.
576'.15 QR100

ISBN 0-12-140301-7

LCCCN 81-71576

Photoset in Great Britain by Paston Press, Norwich and
printed by St. Edmundsbury Press, Bury St. Edmunds

Contributors

Alan T. Bull, Biological Laboratory, University of Kent, Canterbury, Kent CT2 7NJ, England

Richard G. Burns, Biological Laboratory, University of Kent, Canterbury, Kent CT2 7NJ, England

John Cairns, Jr., Center for Environmental Studies, Virginia Polytechnic Institute, Blacksburg, Virginia 24061, USA

Jan W. Drozd, Shell Biosciences Laboratory, Shell Research Centre, Sittingbourne, Kent ME9 8AG, England

John C. Fry, Department of Applied Biology, University of Wales Institute of Science and Technology, King Edward VII Avenue, Cardiff CF1 3NU, Wales

Alvin K. Jones, Department of Botany and Microbiology, University College of Wales, Penglais, Aberystwyth SY23 3DA, Wales

Jan C. Gottschal, Department of Microbiology, Biology Centre, University of Groningen, Kerklaan 30, 9751 NN Haren (GR), Netherlands

Don P. Kelly, Department of Environmental Sciences, University of Warwick, Coventry CV4 7AL, England

William J. Kelly, Department of Biochemistry, Lincoln College, Canterbury, New Zealand

J. G. Kuenen, Department of Microbiology, Biology Centre, University of Groningen, Kerklaan 30, 9751 NN Haren (GR), Holland

John D. Linton, Shell Research Ltd., Sittingbourne Research Centre, Shell Biosciences Laboratory, Sittingbourne, Kent ME9 8AG, England

Paul R. Norris, Department of Environmental Sciences, University of Warwick, Coventry CV4 7AL, England

R. John Parkes, Scottish Marine Biological Association, Dunstaffnage Marine Research Laboratory, P.O. Box 3, Oban, Argyll PA34 4AD, Scotland

Darryl C. Reanney, Department of Microbiology, La Trobe University, Bundoora, Victoria, Australia 3083

William P. Roberts, Department of Microbiology, La Trobe University, Bundoora, Victoria, Australia 3083

J. Howard Slater, Department of Environmental Sciences, University of Warwick, Coventry CV4 7AL, England
Keith M. Steinkraus, Institute of Food Science, Cornell University, Geneva, New York 14456, USA
Meyer J. Wolin, Division of Laboratories and Research, Environmental Health Institute, Tower Building, Empire State Plaza, Albany 12201, USA

Preface

In 1881 Robert Koch published the account of the methods he had devised to obtain pure cultures of microbes. This single paper, now one hundred years old, laid the foundations for the rise of modern bacteriology and microbiology but at the same time distracted attention from one important fact: namely, that most microorganisms naturally inhabit environments containing more than one species of organism and, indeed, in most instances containing very many different types. Thus, the expression of the microbial genome, through the growth and physiological capabilities of a particular organism, is normally modified, attenuated or enhanced by the activities of associated populations of microbes occupying the same environmental niche. By comparison with our understanding of topics such as microbial genetics, molecular biology or biochemistry, our knowledge of the effects of mixed microbial existence and growth is rudimentary. However, it is quite apparent that in many cases the sum of the individual microbial population activities, capabilities and behaviour is often less than, and sometimes markedly different from, those expressed by mixed microbial systems.

In the light of this general perception, this treatise has been started to reflect the growing interest in researches on the behaviour and properties of mixtures of microorganisms. Not only is there considerable interest in the intrinsic properties and features of interacting microbial communities—at various levels, including ecological, physiological, biochemical and genetic—but there is also substantial value in studying mixed cultures from more applied standpoints including medical, nutritional, and biotechnological applications. The treatise will be an occasional one designed to highlight the various topics relevant to each area, containing reviews written to provide a comprehensive coverage. Necessarily, in this fast-growing field of microbiology, some of the study areas are comparatively under-developed and it is hoped that their discussion will stimulate and encourge future developments. Indeed some facets of mixed culture interaction studies are in their infancy and one aim of this treatise is to foster their progress through comparison with the approaches already adopted in the more established fields. Inevitably the reviews will reflect the particular biases of the authors but it is hoped that interested students and researchers will be able to develop their own interests through the cited literature associated with each chapter. In this first volume a number of major topics have been covered and succeeding volumes will seek to treat those areas, such as competition, prey–predator interactions, and mixed culture biodegradation, not dealt with in the present volume.

It is with considerable pleasure that we acknowledge the initial contribution made by Anthony Watkinson of Academic Press who encouraged us throughout the gestation period of the project and whose enthusiasm was clearly evident from the discussions which we had with him and from which this treatise was evolved. We are grateful also to Peter Brown of Academic Press who has guided the project to maturity along with the assistance of June Nelson, Deborah Sanderson, and the other staff of Academic Press. We also appreciate the good-natured assistance of our secretaries, Ann Branch and Vicki Wade at Warwick and Marian Williams at UWIST, whose help in the preparation of the manuscripts was indispensable.

We hope that these collaborators, as well as the readers, will be interested in the final product for which we, of course, shoulder any responsibilities for its failings or shortcomings.

Alan T. Bull
Biological Laboratory
University of Kent
Canterbury CT2 7NJ
Kent, England

J. Howard Slater
Dept. of Environmental Sciences
University of Warwick
Coventry CV4 7AL
Warwick, England

Contents

1

Historical Perspectives on Mixed Cultures and Microbial Communities

A. T. Bull and J. H. Slater

1. INTRODUCTION

The vast array and variety of activities that constitute contemporary microbiology are dominated by one quintessential practice, that of the study of pure, or monospecies, cultures. Indeed, as Slater (1981) has remarked, mixed species cultures are normally considered to be the antithesis of good experimental technique and students of microbiology are nurtured in the pure culture doctrine from their earliest acquaintance with the subject. Undeniably most of our knowledge of the properties and behaviour of microorganisms derives from the examination of isolated single species, and such practice will persist as a mainstay of experimental microbiology. The reasons for the pervasive position of the pure culture tradition in microbiology are not difficult to define and will be considered below. Less readily comprehended is the relative neglect and avoidance of mixed cultures by the majority of microbiologists, especially when one considers that the monospecies is strikingly atypical of nearly all habitats that are colonized by microorganisms. This introductory chapter will trace the development of the pure culture tradition and suggest how interest in mixed culture systems has been stimulated and quickened in recent years.

2. THE ROBERT KOCH TRADITION

The existence of mixtures of microorganisms in natural habitats was demonstrated conclusively by van Leeuwenhoek, and his skill as a lens-maker and microscopist enabled him to distinguish different morphological types of bacteria. Van Leeuwenhoek's much-cited letter to the Royal Society of London concerning "animals in the scurf of the Teeth" (van Leeuwenhoek, 1684) contains particularly clear observations of bacteria. However, it was almost two centuries later before pure cultures of bacteria were unequivocally obtained. Meantime, the quest for pure culture methods was taken up by mycologists, among whom Micheli and Brefeld can be counted the pioneers.

Micheli was born in 1679 and from humble beginnings became the botanical director of the public gardens in Florence. Mycology was a particular interest of Micheli's and of the nearly two thousand plants decribed in his *Nova Genera Plantarum* of 1729, approximately half were fungi. Some of the new genera of microfungi characterized by Micheli included *Mucor*, *Botrytis*, and *Aspergillus*, and he developed simple but effective means of culturing them on the cut surfaces of fruits. On the separate faces of pyramids cut from melon, quince and pear, Micheli sowed the "seeds" of *Mucor*, *Botrytis*, and *Aspergillus*, and observed their germination and subsequent development of characteristic colonies, or "plants" as he described them. Micheli made serial subcultures of these moulds and recorded that he "always observed the same mode of growth in them, not in one trial only, but however often and whenever I attempted it" (Buller, 1915). Brefeld was born 102 years after Micheli's death in 1737, and his researches on fungi were made in a number of German universities where he held chairs of botany. Brefeld took Micheli's culture method several stages further and the influence which he must have had on Koch is obvious. He turned his attention to many different groups of fungi and was particularly interested in unravelling life cycles and tracing patterns of growth and morphological differentiation: the success of such studies was due very largely to his development of liquid cultures in glass flasks, in which he could observe fungal growth. An elegant example of Brefeld's careful experimentation, the elucidation of rhizomorph development in the honey fungus, *Armillaria mellea*, is recounted in E. C. Large's history, *The Advance of the Fungi* (Large, 1940). To Brefeld, success in achieving pure cultures was dependent on:

(i) the use of an appropriate nutrient solution that was both transparent and sterile;

(ii) inoculation of the nutrient solution with a single spore; and
(iii) the necessity of preventing contamination of the solution by other organisms.

The use of gelatin to solidify solutions is claimed to originate from Davaine in 1858 (Large, 1940) and Brefeld adopted the practice to prepare solid culture media for fungi. Brefeld's contributions to experimental microbiology have often gone unnoticed by students of the subject but his protocol for pure cultivation was simple, reliable, and innovative. Its one drawback was the necessity of manipulating single fungal spores with which to initiate the culture, an operation that was a practical impossibility with bacteria and similar sized organisms and which was resolved only later by Lister and Koch. But before considering bacterial culture methods, mention of one further contribution from the study of fungi is appropriate.

In 1869, Raulin, one of Pasteur's devotees at the *École Normal* in Paris, published his *opus magnum* on fungal growth, *Chemical Studies on Growth*. The object of Raulin's work was the development of a completely chemically defined nutrient medium for the growth of *Aspergillus niger*, so that:

(i) optimum conditions for growth could be defined;
(ii) effects of physical and chemical culture variables on growth could be assessed; and
(iii) the efficiency of nutrient assimilation into fungal biomass could be quantified.

Raulin certainly appreciated that pure culture was obligatory for his investigations and is at pains to point out that those cultures of *Aspergillus niger* that become contaminated by "foreign vegetation" (usually *Penicillium* species) had to be disregarded.

Attempts to resolve two closely interrelated controversies provided nineteenth-century microbiologists with the greatest impetus to develop pure cultures of bacteria: one was the microbial causation of diseases, particularly animal diseases, the other was pleomorphism. It was Henle who, in 1840, first recognized and advocated that isolation of the organism from the host was a *sine qua non* requirement in proving that microorganisms could induce disease. Unfortunately, Henle's perspicacity could not be translated into practice because the technology for pure cultivation of bacteria was a long way from being developed. During the nineteenth century, wide acceptance of a notion that came to be known as pleomorphism posed a serious challenge to the scientific progress of microbiology. Briefly stated, the advocates of pleomorphism believed that microorganisms were endowed with considerable capacities for morphological and functional variability. With the benefit of hindsight it is obvious that the confusion created by these claims of pleomorphism stemmed, unwittingly, from the study of mixed microbial populations and an inability to recognize succession in such populations. It took the development and exploitation of chemostat techniques a century later for the extent of true phenotypic variability in microbial species to be appreciated (Bull and Brown, 1979). Proponents of the alternative monomorphist thesis lead by Pasteur and Koch regarded study of pure cultures to be decisive in any analysis of microbial activity and morphology.

The epic investigations of Pasteur and Koch on such problems as alcoholic fermentation and anthrax were made without the advantage of pure cultures. In both instances the high reproducibility of their results almost certainly reflected the selectivity of the nutrient media being used to culture the organisms. During his work for the French wine industry, Pasteur analysed numerous different sources of "ferment" which he knew to contain a variety of microorganisms. Good ferments contained a preponderance of one particular organism, ineffective ferments were dominated by other organisms. Thus the practice developed of using only organisms from good ferments to ensure the production of good wine. Koch's successful investigation on anthrax also enjoyed a fair proportion of aleatory experimentation. He isolated the anthrax bacillus from the blood of infected mice and observed its growth in serum and aqueous humour, but his procedures were not designed to ensure the development of cultures from single organisms. Fortune favoured Koch because *Bacillus anthracis* grew rapidly in the bodies of experimental animals and outcompeted any contaminating bacteria. Not all microbiologists trying to follow Koch's lead were so fortunate. The method used by Pasteur for the cultivation of the anthrax bacillus was to inoculate sterile urine or wort with a drop of blood from a diseased sheep. Following serial cultivation of the growth in sterile urine, Pasteur assumed that a pure culture had been obtained of anthrax bacillus. However, neither Koch's nor Pasteur's procedures guaranteed pure cultures and it was Lister who reasoned that microscopic categorization of sterility or pure culture was apparent, not real, due to the very small size of the observed sample in comparison with the size of its source. Lister circumvented this difficulty by serial dilution of the source material to extinction, and argued that a drop from the dilution which produced growth contained but a single organism. Lister's paper "On the lactic fermentation and its bearing on pathology" was published in 1878 and contained the first reliable isolation method for pure cultures of bacteria. This dilution isolation method, elegant and simple as it was, had limited application and numerically minor bacteria in a mixed population would be lost by early dilution. By 1880, Koch was fully convinced of the importance of pure cultures for bacteriological research and aware that the necessary techniques were not yet available. One year later and Koch fulfilled that urgent requirement and thereby enabled a spectacular blossoming of medical microbiology.

The history of Robert Koch's career following his removal in July 1880 from Wollstein to the Imperial Health Office in Berlin will be familiar to most microbiologists, but a few recollections of his exploits are appropriate here. In Berlin Koch found himself part of a talented and active bacteriology section that included Gaffky, Löffler, and Proskauer. So productive was the Office that in 1881 it was able to sustain its own periodical, *Mitteilungen aus dem kaiserlichen Gesundheitsant*, and launching the first issue was Koch's "*Zur Untersuchung von pathogenen Organismen*", a paper described by Brock as the most significant for the rise of microbiology (Brock, 1961). Initially Koch followed Micheli's strategem of using natural materials on which to isolate and grow bacteria. The sterile slices of potato used by Koch, while facilitating the isolation of certain bacteria, was a very selective substrate and failed to support the growth of many species. Moreover, the wet surface and opaque nature of the material made colony development and identity a difficult matter. Subsequently, Koch experimented with gelatin as a setting agent for nutrient solutions, such as meat infusion, and his preliminary experiments, in which organisms were trapped in the medium as the gelatin cooled and solidified, laid the basis of the pour plate technique. Koch also prepared slabs of gelatin medium and with the aid of platinum wire streaked mixed populations of bacteria on their surfaces: isolated bacteria grew where they were dispersed and gave rise to colonies that were easily differentiated on the transparent substrate. A significant modification of Koch's streak plate technique involved replacement of gelatin by the algal polysaccharide agar–agar which was stable over a much wider temperature range and much less susceptible to microbial degradation. This revolutionary technique was demonstrated to an august assembly, including Pasteur and Lister, at the International Medical Congress at London, in 1881. The following year Koch isolated the causal bacterium of tuberculosis and for which the significance of his newly introduced pure culture technique was amply revealed. Thus was heralded in what has come to be known as the golden age of medical bacteriology. The years to the end of the century were dominated by the exploits of Koch's colleagues and others—among them Löffler, Fehleisen, Gaffky, Kitt, Fraenkel, Escherich, Weichselbaum, Kitasato, Yersin, Shiga—who were instrumental in isolating the bacteria responsible for most of the common infectious diseases. It was during this period, and in connection with his study of tuberculosis, that Koch finally established the practical test of Henle's much earlier ideas, a test which universally has come to be known as Koch's Postulates.

Robert Koch died in 1910. His postulates were the product of an eminently pragmatic approach to the understanding, cure, and prevention of infectious diseases, and of having the objective of achieving the unequivocal attribution of the causative organism to the disease. Koch has frequently been dubbed the ablest of technicians and his great achievements in medical microbiology accounted for by "his simplicity of outlook" (Stephenson, 1949). Marjory Stephenson wrote "No guesses as to the 'how' are hazarded by this great man; he possessed the empirical outlook and aimed at the perfect technique. He was the right man at the right time and medicine probably owes as much to his limitations as to his great gifts." And earlier Topley and Wilson (Wilson and Miles, 1955) had concluded that "the advances which he made in staining methods, in the use of the microscope for the observation of bacteriological preparations, and in the technique of cultivating bacteria, revolutionized this branch (bacteriology) of science". Whatever epithet history bestowes on Koch, there can be no question that the experimental tradition which he founded had the most profound influence on the course of all facets of microbiology during the following 100 years.

3. STIMULI FOR CHANGE

The use of nutrient gelatin as a culture medium for microorganisms was soon shown to have limitations. Soil fertility was a question addressed by several notable scientists in the late decades of the nineteenth century and nitrogen transformations, in particular, received much attention. But, although nitrification in soil was proven to be a biological phenomenon in the 1870s, it was not until 1891 that Winogradsky succeeded in demonstrating the process in soil-free media and isolating nitrifying bacteria. This he did by avoiding the use of complex organic materials like gelatin in his culture media: instead he used simple salts, later to be made into a solidified medium by the means of silica gel, on which to grow bacteria of this so-called autotrophic type. These early studies of autotrophic bacteria revealed some pitfalls associated with the isolation of microorganisms into pure culture. Stephenson (1949) recounts one such episode surrounding the work of Burri and Stutzer in 1895, who described the isolation of a *Nitrobacter*-like organism from soil on silica gel medium. This organism also grew in nutrient broth but lost its capacity to reduce nitrite even when subcultured back on to nitrite silica gel. Winogradsky's suspicions were roused by this report and he examined the culture himself. He showed that the *Nitrobacter* species was contaminated by three heterotrophs that grew in the nutrient broth but not subsequently on nitrite medium. Such associations of autotrophs and heterotrophs were repeatedly encountered by soil microbiologists in the first decades of the twentieth century, and the accompanying growth of heterotrophic bacteria with nitrifying species following serial subculture in inorganic media was not uncommon. Thus Sack in 1925 (cited in Stephenson, 1949) described mixed colonies of *Hyphomicrobium* species and *Nitrosomonas* species on silica gel medium which, if subcultured into nutrient broth, produced ammonia (due to growth of the heterotroph) and, if the amino acid concentration was not inhibitory, nitrite (from weak growth of the autotroph).

The study of distinct physiological groups of microorganisms, like the nitrifiers, lead Winogradsky and Beijerinck to the development of enrichment culture methods for their isolation. This universal microbiological method seeks to encourage the predominance of one type of organism by imposing selective growth conditions on a mixed microbial population: the type to be enriched can be varied at will by manipulating the physio-chemical conditions under which the isolation is made. Armed with such experimental techniques it is not surprising, therefore, that microbiology quickly blossomed to the extent where now we have a very detailed understanding of the physiology, biochemistry, and genetics of many diverse species. But, as in medical microbiology, this understanding has come from the study of pure cultures of organisms. The complementary studies of mixed populations have taken very much longer to develop and to gain acceptance. Thus, the effect of the all-important biological factor—species interactions—on microbial behaviour suffered neglect from all but a few microbiologists and in consequence remains the poorest understood facet of microbiology. There are several reasons why this situation arose and was perpetuated, not least being the established Koch tradition of experimental microbiology. The complexity of microbial communities in nature made their study a daunting prospect, added to which ecology was considered by most microbiologists to be something of a Cinderella subject until quite recent times. The predicament has been nicely construed by Brock (1966) in his *Principles of Microbial Ecology*: "ecology is physiology carried into the actual habitat; ecology is physiology under the worst possible conditions".

The analysis of microbial communities and the interactions occurring within them has been constrained by a number of conceptual and experimental dilemmas. Despite the ingenious work of pioneers, such as Gause, experimental studies of interacting microorganisms have only become satisfactorily feasible since the development of continuous-flow cultures such as the chemostat and its many elaborations. Most microbiologists now would recognize the chemostat as the most appropriate system for examining population interactions but it is pertinent to recall that the chemostat was introduced only 30 years ago and that its exploitation by those interested in mixed populations has occurred even more recently (Veldkamp and Jannasch, 1972). Among the conceptual constraints was that relating to the assumed instability of interacting populations (particularly those of a prey–predator type) in simple laboratory models of the sort used by Gause in the 1930s. On the one hand microbiologists appeared hesitant of accepting the Lotka-Volterra prediction of neutral stability and, on the other, animal ecologists were pointing to the necessity of spatial heterogeneity for the stability of prey–predator populations. In addition, a body of opinion grew up that, intuitively, was sceptical of the possibility of establishing stable microbial communities in the laboratory, irrespective of the interactions involved. Within the last few years these views have been emphatically refuted and many illustrations of the experimental proof will be found in this book. The recent work of Lewin and his colleagues on bacteria–bacteriophage dynamics in a chemostat reveals the extent of stable community complexity that can be established (Chao *et al.*, 1977): three populations of primary consumers in association with two populations of predators maintained in a habitat supported by a single limiting nutrient, glucose. In nature, microbial communities are exposed to a range of environmental perturbations and stresses but homeostatic responses generally enable them to retain their stability in the face of all but the most extreme stresses. The mechanistic bases for the various biological interactions that contribute to community homeostasis is reviewed later by Bull and Slater (pp. 13–44). Nevertheless, it is worth emphasizing at this point that stability within complex communities is usually affected by the simultaneous occurrence of different categories of interactions.

Stimuli for encouraging the study of microbial communities and the interactions upon which they are based have come from diverse sources. Quite naturally, the desire to treat microbial ecology in a synecological frame of reference has been a potent force for the development of mixed culture studies and laboratory model ecosystems. Systems of the latter type have seen increasing application in the area of environmental monitoring where the limitations of single species models have become recognized by researchers and protection agencies alike. In an analogous way we are fast appreciating that the pure culture approach to biodegradation, particularly of xenobiotic compounds and recalcitrant natural products, engenders serious drawbacks. Thus, not only may the biodegradative capacity of mixed populations be greater, both in quantitative and qualitative terms, than monospecies populations, but concentration on the latter denies the possibility of assessing the significance of genetic exchange in the evolution of novel degradative activities (Bull, 1980).

It may appear somewhat ironic in view of the earlier discussions in this chapter that medical microbiology has also provided a spur to mixed culture research. Community activities are now recognized as being instrumental in the establishment and virulence phases of many infectious diseases. Moreoever, the efficacy of antimicrobial agents *in vivo* can be altered dramatically by the indigenous mixed microflora, and continuous culture models are proving valuable in determining the pathogenic potential of organisms under simulated infection and therapy conditions (Onderdonk *et al.*, 1979).

Another area of microbiology that is dominated by the pure culture philosophy is the modern fermentation industry, a situation which runs quite contrary to those numerous traditional microbiological processes which evolved empirically from the activities of mixed populations. Indeed, traditional uses of microorganisms invariably entailed the deployment of mixed cultures and even a cursory glance at the literature will reveal a multitude of beverage, food, and dairy fermentations that have their origins in antiquity and derive from the fortuitous intervention of such mixtures (Steinkraus, pp. 407–444). Moreover, traditional and contemporary biotreatment processes for wastes also rely on the co-operative activities of mixed populations of microorganisms. As Harrison (1978) has rightly pointed out, the production of antibiotics, amino acids, vitamins, and many other microbial products could not have been contemplated in the absence of developed strains of appropriate organisms and the means of maintaining large-scale pure cultures. However, in more recent times the deliberate use of mixed cultures has been developed for a number of industrial processes and products, especially where the activities of a mono-species or mono-strain cultures have been found wanting. Under this latter heading come processes for the production of, for example, β-carotene, alcohol, and microbial protein; and perhaps the most dramatic application has been the development of mixed cultures for single-cell protein (SCP) from substrates such as methane, natural gas, and methanol. The incentive to develop a mixed culture process for SCP from gaseous *n*-alkanes was considerable because pure cultures of methylotrophs frequently showed erratic growth and biomass yields, while the mixture of substrates in natural gas required a mixed population to ensure their complete assimilation (see Linton and Drozd, pp. 357–406).

An understanding of microbial interactions and of the survival and activity of microorganisms deliberately introduced into specific environments, has also prompted research into mixed population dynamics. Three examples of such technology are found in seed inoculants, biocontrol systems, and microbiological agents in pollution control. The technology of seed inoculation has grown considerably since the pioneering studies of the 1950s and 1960s. It involves the inoculation of seeds (usually of staple crops such as beans) with specific nitrogen-fixing bacteria prior to planting either into low fertility soils or soils where the appropriate nitrogen-fixing microbial species is absent. Most success has been achieved with *Rhizobium* species inoculants. Here the host plant–*Rhizobium* symbiosis is highly strain specific and, in order to introduce a crop, such as soybean, into an area where it has not previously been cultivated the specific *Rhizobium* symbiont should be introduced simultaneously as a seed inoculant. Microorganisms as biocontrol agents have been researched and frequently commercialized for use against insect pests and phytopathogenic fungi. The antagonistic modes of action of these microbial insecticides, fungicides, and the like are diverse and may be highly specific (toxin production) or be of a more general nature (wide spectrum antibiosis, competitive colonization). Whatever the mechanism, "biocontrol must be viewed, not as a single-shot, all-purpose treatment, but as part of an integrated control programme and, obviously, it must work within the limits of biological balance, of which it is a specialized application" (Baker, 1980). Implicit in any action designed to manipulate successfully the microflora in this way is a need to understand its homeostatic basis and how it will respond to the input of alien species. These problems also attend attempts to detoxify xenobiotic compounds with microorganisms, both at sites of intentional or fortuitous discharge into the environment and in the more contained sites associated with their manufacture. Some success has been reported by seeding biodegrading microorganisms at polluted sites but this relatively new area of technology is unlikely to have a major impact unless the microbiology of interacting populations is appreciated, and practical protocols developed on the basis of such understanding.

4. PROSPECT

Earlier in this chapter we remarked on the conceptual and experimental problems that have constrained the study of microbial communities. We believe that we are now at a stage where many of those constraints have been removed entirely or can be overcome. The range of physical models now available for experimental research is such that most, if not all, degrees of habitat complexity can be explored in laboratory systems. The impact of the introduction of the chemostat into this field has been mentioned already. Much of mixed culture research relates to ecology *sensu stricto* and in this context the development of environmentally tolerant laboratory models, known generically as microcosms, has been of critical importance. The range of microcosm designs is large and includes simple closed (batch) systems of the sort used by Gause, and complex instrumented chambers and channels that comprise an array of trophic levels and spatial heterogeneities. Also it is now possible to model gradient systems by deployment of multistage chemostats, in which there is a counterflow of nutrients, inhibitors, oxygen and so on, or gel-stabilized microcosms. Similarly interface phenomena can be studied quite readily in fluidized-bed and microbial film reactors (Bull, 1980). Considerable advances have also been made in the study of microbial communities and interactions in their natural habitats and the opening discussions in this volume (Parkes, pp. 45–102; Fry, pp. 103–152) provide a state-of-the-art position on laboratory and field studies. From an ecological viewpoint, laboratory–field interaction studies, of course, are essential.

Not all the contributions to *Microbial Interactions and Communities* are primarily concerned with ecology in the usually accepted sense. For example, a number are concerned with microbial technology (Linton and Drozd, pp. 357–406; Norris and Kelly, pp. 443–474). However, the necessity, or desirability, of revealing the underlying principles of community behaviour is common to both types of pursuit and this will appear as a unifying theme throughout the work. Fredrickson (1977) concluded a stimulating review of mixed microbial cultures by considering what research was required to further our comprehension of such populations. Among the suggestions offered by Fredrickson for investigation were: coexistence of competing populations; space- and time-dependent phenomena; antagonistic interactions not primarily of a resource-limited nature; complex communities containing more than two or three populations and having more regard for positive interactions; mutation and selection; and development of new and improved mathematical models. Discussion of these issues will be made in the following chapters and in many instances the rate of research progress has been impressive and extensive.

REFERENCES

Baker, K. F. (1980). Microbial antagonism—the potential for biological control. *In* "Contemporary Microbial Ecology", pp. 327–347. (Ellwood, D. C., Hedger, J. N., Latham, M. J., Lynch, J. M. and Slater, J. H., eds.) Academic Press, London and New York.

Brock, T. D. (1961). "Milestones in Microbiology". Prentice-Hall, Inc., Englewood Cliffs, N.J.

Brock, T. D. (1966). "Principles of Microbial Ecology". Prentice-Hall, Inc., Englewood Cliffs, N.J.

Bull, A. T. (1980). Biodegradation: some attitudes and strategies of microorganisms and microbiologists. *In* "Contemporary Microbial Ecology", pp. 107–136. (Ellwood, D. C., Hedger, J. N., Latham, M. J., Lynch, J. M. and Slater, J. H., eds.) Academic Press, London and New York.

Bull, A. T. and Brown, C. M. (1979). Continuous culture applications to microbial biochemistry. *International Review of Biochemistry* **21**, 177–226.

Buller, A. H. R. (1915). Presidential Address. Micheli and the discovery of reproduction in fungi. *Transactions of the Royal Society of Canada* (3) **9**, 1–25.

Chao, L., Levin, B. R. and Stewart, F. M. (1977). A complex community in a simple habitat: an experimental study with bacteria and phage. *Ecology* **58**, 369–378.

Fredrickson, A. G. (1977). Behaviour of mixed cultures of microorganisms. *Annual Reviews of Microbiology* **31**, 63–87.

Harrison, D. E. F. (1978). Mixed cultures in industrial fermentation processes. *Advances in Applied Microbiology* **24**, 129–164.

Large, E. C. (1940). "The Advance of the Fungi". Jonathan Cape, London.

Onderdonk, A. B., Kasper, D. L., Mansheim, B. J., Louie, T. J., Gorbach, S. L. and Bartlett, J. G. (1979). Experimental animal models for anaerobic infections. *Reviews of Infectious Diseases* **1**, 291–301.

Slater, J. H. (1982). Mixed cultures and microbial communities. *In* "Mixed Culture Fermentations", pp. 1–24. (Bushell, M. E. and Slater, J. H., eds.) Academic Press, London and New York.

Stephenson, M. (1949). "Bacterial Metabolism". Longmans, Green & Co., London.

Veldkamp, H. and Jannasch, H. W. (1972). Mixed culture studies with the chemostat. *Journal of Applied Chemistry & Biotechnology* **22**, 105–123.

Wilson, G. S. and Miles, A. A. (1955). "Topley & Wilson's Principles of Bacteriology and Immunology". Edward Arnold (Publishers) Ltd., London.

2

Microbial Interactions and Community Structure

A. T. Bull and J. H. Slater

1. INTRODUCTION

There are relatively few environments in which conditions have selected for and preserved monospecies populations of micro-organisms. On the contrary, most environments are characterized by a diversity of microbial species and, moreover, this diversity frequently persists in laboratory culture systems when theory predicts otherwise (Slater and Bull, 1978; Bull, 1980). Changes in environmental conditions induce successional changes in population composition and, just as the case with plants and animals, long term seral shifts also occur in microbial populations. A not infrequent succession of change starts with a severe selection pressure and a low species diversity representing organisms that can survive such pressure, and culminates in a climax community which is in balance with the environment and which often has high species diversity. The capacity of communities to maintain their stability under circumstances of a varying environment is known as homeostasis.

Self-regulation of community composition and relationships is made possible by homeostatic mechanisms based on interactions between the community members. Numerous schemes have been proposed for defining biological interactions: none of them is perfect, not least because, as Alexander (1971) has remarked, they attempt to establish a more or less rigid boundary between interrelationships where none exists in nature. Nevertheless, attempts to categorize interactions are useful because they help to focus attention on mechanisms and to facilitate description of microbial communities. The terminology applied to community interactions and behaviour will be appraised later in this chapter. It must always be appreciated, however, that consideration of individual interactive mechanisms will be of little or no value in revealing community behaviour because each population within the community is likely to be involved in multiple, simultaneous interactions, and that it is their complexity of interactions that engenders community stability.

Many ecologists have argued that the crucial element in homeostatic systems is negative feedback, wherein an environmentally-induced change in one (or more) population effects responses in other populations such that the initial biological fluctuation is opposed, dampened or neutralized. In contrast, positive feedback, where a change is amplified by a community, appears infrequently in microbial communities. It is reasonable to suppose that the homeostatic effect of negative feedback becomes greater as the greater is the species diversity of the community, and that one accompanying effect will be an increased capacity to withstand invasion by other species. However, under certain circumstances dramatic increases in the population size of single species have been observed: such population changes are referred to as ecological explosions or ecological upsets. Ecological upset usually follows a perturbation of the environment, or, of a force holding the species in balance (Alexander, 1971), rather than an escape from homeostasis. A common cause of ecological upset is an increased input of carbon and energy substrates into a community as frequently occurs in soil, aquatic, and waste treatment systems. Marsh (1980) has described sudden, localized increases in populations of *Streptococcus mutans* in dental plaque communities following increases in sucrose consumption; while we (Senior *et al.*, 1976) have reported population explosions of *Pseudomonas putida* within Dalapon-metabolizing (2,2-dichloropropionic acid) communities following regulatory mutations in the enzyme aliphatic dehalogenase which enable the herbicide to be assimilated. Antibiotic and radio therapies provide very clear and dramatic instances of the breakdown of community homeostasis with, for example, ecological explosions of opportunistic fungi and yeasts in patients under medication. The basic concepts of community stability are shown in a simplified way in Fig. 1.

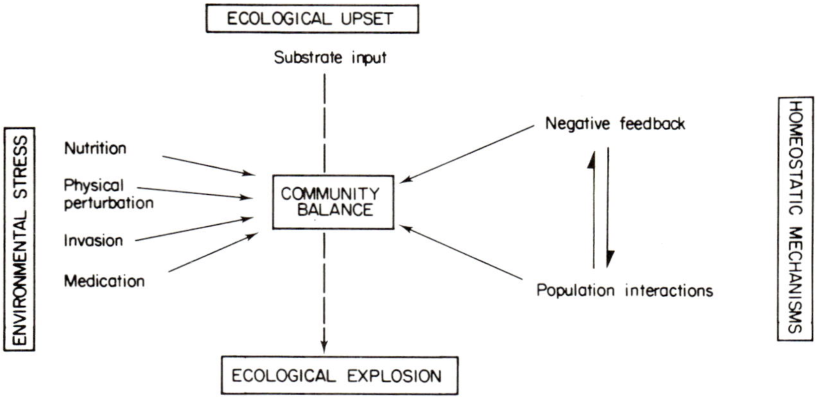

Fig. 1. The elements of community stability (adapted from Marsh, 1980).

2. THE CLASSIFYING OF MICROBIAL INTERACTIONS

2.1 Approaches

Attempts to describe the many types of interaction that occur between microorganisms—and between microorganisms and higher eukaryotes—have lead to extensive and sometimes perplexing systems of classification and nomenclature. In general, the approach to classifying biological interactions has been made either in terms of the effect of one population on another, or, in terms of the mechanisms involved. Williamson (1972), in addressing the problem of classifying interactions, has argued that systems based on the nature of biological processes are likely to produce ones which are simple but naive, or comprehensive but unwieldy: his resolution of the problem was to recommend classifications based on effects. Odum (1953) proposed a scheme which was among the originals of this type and has been adopted widely in all branches of biology. Essentially, interactions are defined by a community matrix A in which the elements, say a_{ij} and a_{ji}, describe the equilibrium effects of organism j upon organism i and *vice versa*. The interactions are classified in terms of the signs of the resulting effects, so that a_{ij} +, 0, or − would represent positive (beneficial), neutral, or negative (harmful) effects of organism j on organism i, depending on whether the population size of organism i increases, is unaffected, or decreases. Consequently, for our pair of matrix elements a_{ij} and a_{ji} we can construct a square matrix of all possible interactions (Fig. 2).

		Effects of population j on i (a_{ij})		
		+	0	−
Effects of population i on j (a_{ji})	+	++	+0	+−
	0	0+	00	0−
	−	−+	−0	−−

Fig. 2. Matrix of interactions between two microbial populations, i and j.

This basic classification has been elaborated in various ways. For example, Fredrickson (1977) used a "higher-order taxonomic level that has a mechanistic basis" to qualify Odum's scheme. Interactions that necessarily involved physical contact between members of populations i and j were termed direct interactions, whereas interactions which did not require physical contact between the populations were termed indirect. Fredrickson justified this qualifying definition to emphasize the role of the abiotic environment and to reiterate the impossibility of predicting behaviour of two directly interacting populations from monopopulation studies.

A number of terms, including commensalism, mutualism and parasitism, are used to describe the categories of interaction which result from matrix analysis (see Section 2.2). Pirt (1975) has defined the basic conditions for the maintenance of two microbial species in a chemostat (Table 1) and at the same time is cautionary about the use of terms like commensalism and symbiosis because they tend to be overlapping and do not match with any of his basic types.

Table 1. Basic conditions required to maintain two microbial species in chemostat culture (Pirt, 1975)

I. With the same growth-limiting substrate
(a) Specific growth rates coincide.
(b) Faster growing species inhibited by its own product(s).
(c) Faster growing species produces substances which activate the growth of the other species.
II. With different growth-limiting substrates
(d) Different growth-limiting substrates supplied to the culture.
(e) Product of one species is the growth-limiting substrate for the other.
(f) Predator–prey associations.

Organism interactions can be specific or nonspecific. Specificity implies attributes of the interacting organisms that result in an interaction having a degree of selectivity. In such interactions "each organism forms a more or less critical portion of the environment of its partner" (Dubos and Kessler, 1963). Not surprisingly, indirect interactions rarely show the degree of specificity that is characteristic of direct interactions, such as parasitism and symbiosis. The extreme examples of specificity are to be found in genetic interactions involving cell fusion and heterokaryosis. Finally, interactions can be additionally described in terms of interdependency, and facultative and obligate categories thereby identified.

Starr and his colleagues, and Lewis (Lewis, 1974) have attempted to devise a classification of organism interactions which is applicable to the full range of types. Thus, Starr's proposal of a set of interacting continua is founded on three criteria:

(a) locational–occupational (what is the relative location of interacting species and how do they sustain their existence?);
(b) valuational (does the interaction produce beneficial or harmful effects on the two or more populations involved?);
(c) dependency (is the interaction necessary or contingent?).

The components of these continua can be defined in terms of the Odum classification as follows:

(a) locational–occupational: commensalism, symbiosis, parasitism, predation;
(b) valuational: mutualism, neutralism, antagonism;
(c) dependency: obligate, facultative.

Lewis (1974) has appended a fourth continuum to the Starr scheme, *nutrition*, and has attempted to identify evolutionary relationships between the interacting components of the valuational and dependency continua. Thus, for example, five categories of interactions can be delineated on the basis of ecological and nutritional criteria: obligate saprotrophs, facultative necrotrophs (organic nutrients from dead organisms), obligate necrotrophs, facultative biotrophs (organic nutrients from living organisms), and obligate biotrophs. These ideas have not been adopted very extensively by microbiologists but they are important because they focus on mechanisms of interaction and emphasize that such interactions exist in a dynamic state and can be readily modulated by environmental factors. The interested reader is referred to the review by Lewis (1974).

Finally, it is necessary to refer briefly at this juncture to the existence and utility of verbal and mathematical models of microbial interactions. One of the pioneers of microbial interaction studies (Gause, 1934) clearly advocated mathematical modelling, in Gause's case, once sufficient observational data were obtained to enable analysis of population kinetics. Now it is more usual to construct models that have predictive value and to follow this with a portfolio of experiments designed to test the assumptions. Frequently, there is no shortage of explanations for the inadequacy of mathematical models, but whether such inadequacy is based on faulty simplifying assumptions or faulty anacalyptic (i.e. the hypotheses *per se*) assumptions, or both, is not always easy to decide. As Williams (1980) rather pithily remarked, in the context of modelling predator-prey interactions, often "we don't know whether we understand or not". The analysis of populations must be made in quantitative terms, consequently definitions of interactions tend to be of a mathematical nature. Williamson (1972) is of the opinion that it is those who analyse in terms of individuals who prefer verbal definitions. He cites Milne's (1961) definition of competition ("the endeavour of two animals to gain the same particular thing or to gain the measure each wants from the supply of a thing when that supply is not sufficient for both") as an example of a thoroughly unwieldy and imprecise definition in which one undefined term (competition) is replaced by seven.

Some discussion of particular models will be made below and will be found in several other chapters of this book.

2.2 Classification Based on Effects

We will begin our discussion by reference to interactions that can occur between two-membered populations. Inspection of Fig. 2 reveals that for such populations a maximum of six effects can be generated. These effects can be described verbally as follows.

(i) Matrix element 00: *neutralism*, lack of interaction between populations.
(ii) Matrix elements 0+, +0: *commensalism*, one population benefits from the presence/activity of a second population but the latter does not derive reciprocal benefit, or harm.
(iii) Matrix element ++: *Mutualism*, reciprocal benefit to the two interacting populations. *Synergism*, *protocooperation* and *symbiosis* are terms frequently used to describe particular types of mutualistic interactions that will be referred to below.
(iv) Matrix elements +−, −+: *parasitism*, *predation*, direct attack and

feeding of one population on the other. In parasitism the consumer organism is smaller than the one being consumed, while predation usually implies the ingestion of a smaller organism by a larger one. Lewis' term "biotrophy" is inclusive of interactions of these types.

(v) Matrix elements −0, 0−: *amensalism*, the antithesis of commensalism, in which the growth of one population is restricted by the presence of a second population but the latter is not affected.

(vi) Matrix element −−: *competition*, both populations are mutually restricted because of their common dependence on a limiting factor in the environment, usually a nutrient. In the absence of other types of interaction between the two populations, the less vigorous organism is displaced by the more vigorous one.

Parasitism, predation, amensalism and competition all represent domains within a continuous spectrum of antagonistic interactions. The use of these terms is frequently indiscrimate and imprecise. The difficulty has been resolved in part by Fredrickson (1977) who reserves parasitism and predation for positive–negative effect interactions that are direct and involve physical contacts. He has coined the term "indirect parasitism" to cover interactions that involve parasitism in the absence of cell contact, as happens in interspecific lysis. Some forms of amensalism and indirect parasitism have in common the fact that one population releases into the environment a material that induces a negative effect on the other; such materials include, respectively, antibiotics and lytic enzymes. The boundaries between amensalism, indirect parasitism and competition may be very difficult to draw because amensal and parasitic populations can benefit from the increased availability or release of nutrients consequent upon their activities, and thereby alleviate competition for nutrients. It would be interesting to know if true amensalism only occurs under conditions of nutrient excess. Similarly, the distinction between amensalism and competition needs to be emphasized. Competition as defined above is a two-way process but the term is sometimes used by ecologists for interactions in which only one population is affected.

2.2.1 Neutralism

Although Brock (1966) stated that it was rare for two populations not to interact, the significance of neutralism in community structure has been difficult to evaluate and is a poorly researched phenomenon. Brock offered some explanations for non-interaction including the populations being too far apart, and the environmental requirements being so different that neither population alters the qualities required by the other. Brock also considered that neutralism would be found most frequently in mixed populations of "extremely diverse organisms" under which circumstances competition would be minimal. However, one of the few experimental analyses that is relevant in this context revealed that a presumed case of neutralism occurred in chemostat populations of *Lactobacillus* and *Streptococcus* yoghurt starter cultures: individual population sizes in the mixed culture were the same as those of separate monocultures under identical growth conditions (Lewis, 1967). Similarly, a remarkably high incidence of neutralism (ca. 85%) has been detected among activated sludge bacteria (M. Davies, cited in Slater and Bull, 1978). It is clear that the bases of non-interactive phenomena deserve systematic investigation.

2.2.2 Commensalism

Schopfer in his classic treatise "Plants and Vitamins" (Schopfer, 1943) coined the term unilateral stimulation for situations where one of a pair of organisms is favoured in its development by the presence of the other; in contemporary parlance we substitute the term commensalism. Schopfer had considerable insight of the involvement of vitamins in unilateral stimulation (satellitism) and recognized the fact that prior modification of a substrate by one organism also often leads to a commensal interaction (metabiosis). Nitrification by *Nitrosomonas* and *Nitrobacter*, and the production of sake via the koji process were cited by Schopfer as instances of metabiosis.

In contrast to neutralism, commensalism is a frequently occurring interaction and most commentators are agreed that it is a significant factor in organism successions within microbial communities. Commensal interactions are known to be generated via a number of different mechanisms of which the unilateral provision of vitamins is particularly well known. An early report of this sort came from Jensen (1957) who identified vitamin B_{12} provision, by commensal *Streptomyces* species, for a trichloroacetic acid-metabolizing bacterium. More recently Miura *et al.* (1978) have observed that biotin was implicated in a commensal interaction that formed the basis of a model hydrocarbon waste treatment process: here *Pseudomonas oleovorans* grew on phenol and excreted biotin which was required by the *n*-tetradecane assimilating *Mycotorula japonica*. From Tsuchiya's laboratory was reported a careful study of yeast-bacterium commensalism in which riboflavin required for the growth of *Lactobacillus casei* was produced by *Saccharomyces cerevisiae* (Megee *et al.*, 1972).

In an analogous way commensalism may be based on the generation of carbon substrates:

(i) *Acetobacter suboxydans* oxidizes mannitol to fructose and the latter—but not the former—can be metabolized by *Saccharomyces carlsbergensis* (Chao and Reilly, 1972);

(ii) Swiss cheese manufacture in which competition for glucose is alleviated by *Lactobacillus plantarum* producing lactic acid which is the preferred carbon and energy substrate for *Propionobacterium shermanii* (Lee *et al.*, 1976).

Commensal interactions also are known in which one organism modifies the physical environment and thereby encourages growth of a second one. Such modifications encompass removal of oxygen to enable obligate anaerobes to grow, raising the water content or reducing the osmotic pressure to enable less tolerant organisms to grow, and the detoxification of environments with respect to antibiotics, metals, toxins and the like.

2.2.3 Mutualism

In Schopfer's (1943) terms mutualism is synonomous with the reciprocal or bilateral stimulation of interacting populations. A distinction between mutualism and protocooperation can be made by invoking Starr's notion of dependency: in mutualism the interaction is obligate, while in protocooperation the interaction is facultative and transitory. In symbiotic interactions both organisms again derive benefit but only certain organisms can symbiose with one another, i.e. symbiosis describes specific, direct interactions. Finally, whereas mutualism defines interactions that are necessary for survival or growth in a particular environment, symbiosis defines the development of some activity which neither organism alone can affect, or, colonization of an environment which would not be achieved by one or other organism alone.

Many of the investigated examples of mutualism *sensu stricto* are based on the cross-feeding of vitamins. Thus *Lactobacillus arabinosus* and *Streptococcus faecalis* can grow together in minimal media, despite the fact that they both have vitamin requirements, because of the cross-feeding of phenylalanine and folic acid respectively (Nurmikko, 1956). In an identical way *Bacillus polymyxa* (nicotinic acid requiring) and *Proteus vulgaris* (biotin requiring) interact mutualistically (Yeoh *et al.*, 1968). Mutualistic cross-feeding also is a feature of some fungal interactions. An early example was reported by Kögl and Friesa (1937) who observed that the reciprocal stimulation of *Nematospora gossypii* and *Polyporus adustus* was due respectively to the satisfying of biotin and thiamin (pyrimidine) requirements. The case of mutualism described by Müller and Schopfer (1937) revolves around the exchange of the two components of thiamin: *Mucor ramannianus* synthesizes pyrimidine but not thiazole while *Rhodotorula rubra* has the complementary synthetic capacity. Thus, dual culture in synthetic media produced growth that was equivalent to that of each mutualistic partner when furnished with thiamin and grown in monoculture.

Mutualism also may have other mechanistic bases. Complementation of lecithinase has been found between two closely related pseudomonads, for example (Bates and Liu, 1963). Again, the interaction may be related to the removal of a growth inhibitor and a clear illustration of this is to be found in Munnecke and Hseieh's (1974) study of parathion degradation by a bacterial community. Here *Pseudomonas stutzeri* brings about a cometabolic hydrolysis of the insecticide which produces *p*-nitrophenol as a product. Unfortunately, *p*-nitrophenol inhibits the growth of *P. stutzeri* but the inhibition is relieved by *P. aeruginosa*, a species which assimilates the phenol as a growth substrate. *Pseudomonas stutzeri* grows at the expense of metabolites and lytic products which, in part, are released by *P. aeruginosa*. An especially interesting interaction was detected by Gray *et al.* (1973) during their investigation of the photosynthetic bacterium *Chloropseudomonas ethylica*. The latter was shown to comprise a very intimate mutualistic association of *Chlorobium limicola* with a *Desulfovibrio* species, the coupling between them being made by cyclic oxidation and reduction of sulphur compounds which provides an electron donor and acceptor respectively. Other, subsidiary, reactions also may occur between these two bacteria but until further researches have been made it is impossible to decide whether the association is one of mutualism or protocooperation.

Protocooperation is evident in a stable bacterial community enriched to grow on methane (Wilkinson *et al.*, 1974). The community contained just one species, a pseudomonad, which could utilize methane and in doing so produced methanol which partially accumulated in the culture medium. A second member of the community was a *Hyphomicrobium* species which had very high affinity for methanol and prevented it from accumulating to concentrations that would be growth inhibitory to the pseudomonad. When a pulse of methanol was fed to the culture the numbers of *Hyphomicrobium* rose from the one or two percent characteristic of the stable community, to as high as 70% while the transitory high concentration persisted. Pseudomonad and *Hyphomicrobium* populations returned to equilibrium sizes as the methanol concentration declined.

The most extensively studied examples of symbiosis are those involving microbial interactions with animals and plants and a short survey of types is made in Table 2. The best example of a symbiosis involving two

Table 2. Common symbioses involving either animals or plants with microorganisms

A. *ANIMALS*

Animal partner		*Microbial partner*
Invertebrates:	protozoa	cyanobacteria, dinoflagellates, *Chlorella*
	coelenterates	dinoflagellates, *Chlorella*
	porifera	cyanobacteria
	platyhelminthes	dinoflagellates, *Chlorella*
	molluscs	luminous bacteria, dinoflagellates, *Chlorella*
	annelids	heterotrophic bacteria
Tunicates		luminous bacteria
Insects		bacteria, yeasts, filamentous fungi, protozoa
Vertebrates:	fish	luminous bacteria
	herbivores	bacteria, protozoa

B. *PLANTS*

Plant partner		*Microbial partner*
Legumes (root nodules)		*Rhizobium*
Non-legumes (root nodules):		
	mosses (*Sphagnum*)	*Haplosiphon*[a)]
	liverworts	*Nostoc*[a)]
	ferns (*Azolla*)	*Anabaena*[a)]
	cycads (*Cycas*)	*Nostoc, Anabaena*
	angiosperms)	
	(*Alnus*)	*Frankia*[b)]
	(*Gunnera*)	*Nostoc*
Angiosperms (mycorrhizae):		
	orchids	*Rhizoctonia*
	trees	*Agarics*
	various	*Endogone*

(a) cyanobacteria; (b) actinomycete

Further information: Brock (1966), Lynch and Poole (1979), Smith (1978).

microbial partners is lichens, in which fungi (mainly ascomycetes but some basidiomycetes and fungi imperfecti) and algae (mainly Chlorophyceae of the genus *Trebouxia*), or cyanobacteria (frequently of the genus *Nostoc*), are the interacting organisms. The reader is referred to the text of Hale (1974) for a succinct but comprehensive review of lichenology. Marine and freshwater protozoa (radiolarians, foraminifera, and ciliates) are known to establish symbioses with unicellular algae (most frequently Chlorophyceae but also dinoflagellates and occasionally diatoms) but the biology of these interactions has been rather neglected.

Finally, it is worth recalling that the term symbiosis as first used by de Bary (1866) was taken to describe microorganisms inhabiting other organisms and thus, parasitic interactions were embraced by this definition. However, such a definition is no longer accepted and, in order to avoid exacerbating an admittedly confusing terminology, we believe that the generally agreed definitions summarized at the start of this section should be adopted. For example, the situation wherein the interaction of *Bdellovibrio* with Gram-negative bacteria is described as symbiosis (p. 172) and as parasitism (p. 199) in the same work (Lynch and Poole, 1979) is unfortunate.

2.2.4 Parasitism

We have referred earlier to the distinction between parasitism *per se* and indirect parasitism. Considerable insight has been gained into the mechanisms of microbe–microbe parasite–host interactions and much information is available on parasite attraction and attachment to, penetration of, multiplication within, and killing and lysis of the host species. In this brief discussion it suffices to record that a wide range of microbial groups contain parasitic members—viruses (bacterial, fungal, algal, and protozoal hosts), bacteria (bacterial hosts, e.g. *Bdellovibrio*), fungi (fungal, i.e. mycoparasitism and algal hosts), and amoebae (fungal hosts).

Parasitic interactions illustrate a range of host specificities and degrees of obligation. Thus bacteriophages are obligate parasites *sensu stricto* and their host specificity may have significant practical consequences. For example, *Rhizobium* species are susceptible to bacteriophage attack, and resistant and sensitive strains can be isolated: unfortunately, resistant strains are often those which are least efficient in nodulation and nitrogen fixation. Fungi, such as Aspergilli and Penicillia and the cultivated mushroom, are also parasitized by viruses and in turn such infection can impair the pathogenicity of the host organism. Thus, viral infection of the cereal pathogen *Gaeumannomyces graminis* is thought to contribute to its loss of virulence.

In general, indirect parasitism has a lower specificity than direct parasitic interactions of the types referred to above; in addition it has been less exhaustively investigated. Most research in this field has focused on bacterial lysis of fungi. Such lysis is widespread in Nature and many reports are available on the biochemical mechanisms involved, the bases of susceptibility and resistance and its significance in natural ecosystems (see Alexander (1971) for a review). Fungi whose walls are largely composed of β-1,3 glucan, cellulose or mannans usually are susceptible to lysis but the effect may be reduced or entirely prevented by the presence of materials such as melanin (Bull, 1970). Indirect parasitism is readily demonstrated in the laboratory but a substantial body of data has accumulated which indicates its ecological relevance. In particular, work in Alexander's laboratory has strongly suggested that the control of plant pathogens, such as *Fusarium oxysporum*, in soil is contributed to by lysis (Mitchell, 1963). Similarly, damping-off of bean, tomato, and egg plant seedlings by *Rhizoctonia solani* has been controlled by indirect parasitism, in this case by another fungus, *Trichoderma harzianum* (Hadar *et al.*, 1979). Again, eradication of *Blastomyces dermatitidis* (causal agent of human blastomycosis) from soil is affected in part by lysis (McDonough *et al.*, 1965). In natural environments antagonistic interactions undoubtedly will be complex and involve communities of organisms of the type reported by Austin *et al.* (1977). The latter authors found that reductions in disease symptoms produced by *Drechslera dictyoides* (causal agent of net blotch of ryegrass) were effected by phylloplane bacteria including *Pseudomonas fluorescens*, *Listeria denitrificans* and *Xanthomonas campestris*, and that among the antagonistic reactions was hyphal lysis.

2.2.5 Predation

Microbial predation is brought about by endocytosis and thereby restricted to certain protozoa and those other microorganisms, like the Acrasiales, that have amoebal phases in their life histories. Although predation can lead to elimination of the prey, the usual endpoint of the interaction in laboratory cultures is some form of oscillating state between prey and predator populations which eventually may become damped.

The literature is repleat with mathematical models describing predator–prey population dynamics and laboratory experiments attempting to confirm them. This is not the occasion to appraise the various models. However, it is appropriate to recall that the Lotka–Volterra equations originally formulated in the 1920s and based on the logistic equation, have been the starting point for much modelling work; and, that the extensive predator–prey experiments of Gause (1934) have been repeatedly examined in attempts to validate one or other model. Both of these factors have dominated quantitative thinking about the predator–prey interaction to an extent that may have seriously retarded our understanding. Two recent discussions highlight how conceptual errors, both in the common predator–prey models of the Lotka–Volterra type and Gause's experimental systems, probably have compounded misunderstanding (Williamson, 1972; Williams, 1980). Salt (1974) remarked that it is not a question of Gause's observations giving an incorrect representation of predation but that they are incomplete and only reveal some of the features of predator–prey interactions. Thus, Williamson (1972) pointed out that Gause worked with too small a volume for his *Didinium–Paramecium* system to have a reasonable probability of persisting and that one of only four times larger would have given satisfactory results. Realistic models will probably predict damped oscillations of predator and prey but, as Williams (1980) has emphasized, the assumptions implicit in the Lotka–Volterra equations have been too readily accepted. In a rigorous examination of the Lotka–Volterra equations Williams recognized three major classes of assumption: simplifying environmental, simplifying biological, and explanatory (anacalyptic) assumptions. Failure of these assumptions means:

(a) that the design of observations lacks the required level of control,
(b) that the model is unsophisticated in biological terms, and
(c) that the nature of the phenomenon has been misunderstood.

Failures of types (a) and (b) have nuisance value and simply delay understanding; failures of anacalyptic types are calamitous and necessitate a complete re-evaluation of all assumptions. Williams concluded that the Lotka–Volterra equations are invalid models of predator–prey interactions and, much more disturbingly, that it is unclear whether simplifying or anacalyptic assumptions are at fault. So the position rests.

2.2.6 Amensalism

Amensal effects occur when microorganisms produce major metabolites, or antibiotics (including bacteriocins), which antagonize the growth of other species. Our original definition of amensalism implied that the antagonizing population derives no benefit from the interaction but there could be indirect advantages accruing from the release of growth substrates that otherwise would be assimilated by the antagonized population. Non-specific amensalism may result from a lowering of environmental pH, usually by the excretion of organic acids (e.g. lactic acid production which controls the invasion of dairy products and sauerkraut, and prevents fish spoilage), or by the production of oxygen or hydrogen peroxide.

Antibiosis as a mechanism of amensalism is well known but its significance in nature has frequently been questioned (Brock, 1966). However, in recent years the development of microbiological control methods for phytopathogens has reawakened interest in this subject and several reasonably well-substantiated instances of antibiotics being produced by amensals have been published (Baker, 1980). Thus, cereal grain inoculated with *Bacillus subtilis* or *Chaetomium globosum* protected against seedling blight; antibiotic production in the cereal rhizosphere during the period of greatest susceptibility to *Fusarium roseum* infection was considered responsible for the protection. Similar explanation is given for the "possession principle" by which, for example, protection of pine from *Heterobasidion annosum* infection by the low-grade pathogen *Peniophora gigantea* is effected. Equally interesting is the commercial process of "bacterization" whereby seed or propagule inoculation with bacteria increases plant growth. Bacteria are usually selected on the basis of their broad-spectrum antibiotic effects *in vitro* and growth stimulation is thought to be due to the inhibition of non-parasitic rhizosphere microorganisms that adversely effect the host plants. It is significant that non-antibiotic producing mutants of antibiotic-producing, growth-stimulating *Pseudomonas* species did not have the capacity to improve plant growth (Kloepper and Schroth, 1978).

An interesting case of amensalism occurs in mixed cultures of *Lactobacillus casei* and nisin-producing *Streptococcus lactis*, and has practical relevance in the development of continuous milk fermentations. Nisin is a narrow-spectrum, polypeptide antibiotic active against Gram-positive bacteria including *L. casei*. Thus, amensal interactions between component strains has to be understood when designing fermentation processes (Lawrence and Thomas, 1979). The latter authors go so far as to suggest that, for the above reason, lactic streptocci are unlikely to be used in continuous milk fermentations. However, attention to mixed culture physiology can alleviate problems resulting from amensalism. Oberman and Libudzisz (1973) found that although *L. casei* did not grow in the presence of 50 units nisin ml^{-1}, satisfactory mixed, batch cultures were established if the inoculum contained numbers of lactobacilli and streptococci in a 10 to 1 ratio. Moreover, stable continuous cultures were obtained when inocula of the above proportions were used and the dilution rate was $0.12\ h^{-1}$ or lower. Under the latter growth conditions nisin was not detected and amensalism was prevented.

Arguably the most dramatic exploitation of amensalism in plant protection involves crown gall control (Moore and Warren, 1979). Seeds or nursery plants of the crop concerned are inoculated with an appropriate strain of *Agrobacterium radiobacter* and this has the effect of preventing the pathogen (*A. tumefaciens*) transferring its T_i plasmid to the host plant and hence causing disease symptoms. It appears that the amensal produces a bacteriocin that either prevents attachment of the pathogen to receptor sites on the host roots or kills the pathogen outright.

Finally, it should be mentioned that amensalistic interactions can induce dormancy or morphogenesis in the antagonized population. As Baker (1980) has remarked "the progression through the stages of the life cycle of a soil microorganism is determined at least as much by the associated microflora . . . as by the genes of the microorganism". The available evidence suggests that antagonistic interactions such as amensalism are particularly potent in this regard.

2.2.7 Competition

Competition ensues when two or more populations are limited, in terms of growth rate or population size, by a common dependence on a growth substrate or other environmental factor. Competition underpins much ecological thought and, by providing the selective mechanism for evolution, can be considered the single most significant interaction in nature. Paradoxically, therefore, competition is very difficult to analyse under natural conditions. Williamson (1972) makes an interesting comparison of predator–prey and competitive interactions: the former, which is relatively easy to observe in the field, has not produced many definitive laboratory experiments, while the latter is relatively easy to examine experimentally in the laboratory, but its occurrence in nature is subject to dispute. It has been argued (Bowen, 1980) that microbiologists should adopt some of the experimental approaches developed in other fields, such as agronomy, and use replacement series and diallel analyses (Williamson, 1972) with which to probe competition. The reader naturally will be aware of the importance of competition in the design and operation of many industrial and agricultural practices such as algal biomass production (Goldman and Ryther, 1976), activated sludge processing (Krul, 1977) and *Rhizobium* inoculant development (Johnston and Beringer, 1976).

Several recent publications have detailed the theory of simple competition in closed and open laboratory systems, and the reader is directed to Fredrickson (1977) and Slater and Bull (1978) for kinetic analyses and examples. At this juncture we will restrict the discussion to just a few salient points. Chemostat theory predicts that when a number of organisms are competing for a single, growth rate limiting nutrient, a single (most competitive) organism will become dominant to the exclusion of all others after sufficient time has elapsed. However, numerous experiments have demonstrated that some or even all the less competitive organisms do not become eliminated but, instead, stable coexistence of competitors can occur. A notable example of the coexistence of competitors can be seen in the work of Senior *et al.* (1976) on Dalapon degradation where four distinct species of herbicide utilizers coexisted in a very stable chemostat community for over 2 years. Several factors can account for coexistence including the stabilizing influences of commensalism and mutualism (Meyer *et al.*, 1975; Miura *et al.*, 1980); and the presence of a population that preys on the competing populations (the interaction of *Tetrahymena pyriformis* with *Azotobacter vinelandii* and *Escherichia coli* competing for glucose is a classic illustration of such stability; Jost *et al.*, 1973). It was noted above that amensalism could be prevented in mixed cultures of lactic acid bacteria by appropriate manipulation of the inoculum ratio. In an analogous way competitive elimination of *Bacillus* species from mixed cultures with *Torula* (*Candida*) *utilis* was dependent on the inoculum sizes (Meers and Tempest, 1968). In the latter system the limiting nutrient was magnesium and its assimilation by *Bacillus* was more dependent on the secretion of a growth-promoting substance(s) than was magnesium assimilation by the yeast. Thus, at low population densities the *Bacillus* species were unable to maintain themselves against proportionately low yeast populations.

Competing populations can also be stabilized artificially by means of feedback control systems and the reader is referred to Wilder *et al.* (1980), for example, for a discussion of control options.

Before leaving the topic of competition it should be emphasized that it, like other microbial interactions, is subject to modulation by environmental factors. One of the first thorough demonstrations of such an effect was reported by Harder and Veldkamp (1971). These authors examined the effect of dilution rate on chemostat populations of a marine pseudomonad and a marine spirillum, and showed that dilution rate effects on competition could be completely reversed by temperature. Similar modulating effects of temperature can be seen in the outcome of competition within phytoplankton communities. The work of Goldman and Ryther (1976) is particularly interesting because

(a) it indicated that temperature optima defined in laboratory cultures were obscured by interactions of other environmental factors in field experiments (e.g. *Phaeodactylum tricornutum* was dominant in the field at 10°C but not in laboratory systems); and

(b) it showed that competition in the field differed depending on whether the temperature within a prescribed range was falling or rising.

Not surprisingly the nature of the growth limiting factor also has a strong influence on the outcome of competition: reference to recent work by Tilman (1977) will illustrate this point. In dual mixed cultures of diatoms, whereas *Asterionella formosa* was competitively dominant under conditions of phosphate limitation and *Cyclotella meneghiniana* became dominant under silicate limited conditions, stable coexistence was achieved when each population was growth limited by a different nutrient.

2.3 Final Thoughts on Classifications Based on Effects

"There may be a tendency to view the interaction between two organisms as if it were the same under all conditions" (Brock, 1966). Such an impression may be gained from a discussion of the *types of interaction that are possible* of the sort given above (Section 2.2). However, as Brock went on to emphasize, relationships between organisms are not permanent and interactions must be considered to be dynamic and to be subject to modification by the environment. A particularly striking illustration of the modifying effects of environment is the often-quoted work on bacteriophage–host interaction in saline sediments (Roper and Marshall, 1974). Protection against parasitism by the bacteriophage was afforded by an envelope of sorbed clay and sorption in turn was controlled by the degree of salinity in the environment.

We implied earlier (Section 1) that it is unreasonable to think in terms of strictly single interaction systems—even in two-membered laboratory populations. Perhaps the most useful way of describing such systems is to extend the existing classification and to refer to competition plus commensalism, or, less cumbersomely, to competitive commensalism, for example. Several instances of multiple interaction populations have been described and analysed. One of the first to be recognized was the *Bacillus polymyxa–Proteus vulgaris* system mentioned above (Yeoh *et al.*, 1968). The dominant interaction is one of mutualism but the production of extracellular protein by *P. vulgaris* had an inhibitory effect on the *Bacillus* and the interaction was more aptly described as antagonistic mutualism. In the same way the commensalistic interactions described by Chao and Reilly (1972) and Megee *et al.* (1972) are clear examples of antagonistic commensalism and competitive commensalism, respectively. More recently Miura *et al.* (1978) have studied a complex system in which a mixed population of *Pseudomonas oleovorans* and *Mycotorula japonica* was cultured on two carbon and energy sources, phenol and *n*-tetradecane: *P. oleovorans* assimilated phenol and excreted biotin, *M. japonica* assimilated both carbon substrates simultaneously and required an exogeneous supply of biotin. Thus, both commensalism and competitive commensalism occurred in the culture. Subsequently Miura *et al.* (1980) have constructed mathematical models (based on Monod's equations) in order to predict population stability when the interactions are competitive mutualism and competitive commensalism. In the case of competitive commensalism previously studied by this group (*P. oleovorans–M. japonica*), the model predicted damped oscillations in chemostat populations, a prediction that was borne out by the experimental data (Miura *et al.*, 1978).

Until now discussion has been almost entirely restricted to two-organism interactions. It is important not to forget, therefore, that interactions also occur in monospecies populations. Co-operative effects have long been recognized in relation to resistance to toxic materials, infectivity and the provision of organic growth factors (Brock, 1966). Chemotactic aggregation in the morphogenesis of cellular slime moulds and other microorganisms can also be considered as specific manifestation of positive interaction in a monoculture.

We remarked at the start of this chapter that most environments are characterized by a diversity of microorganisms. In discussing this diversity, both in nature and in laboratory models, we have implicitly adopted Fager's (1963) definition of "a group of species which are often found living together" to designate a community (Senior *et al.*, 1976; Slater and Bull, 1978). Other authors have used terms such as "consortium" and "association" to describe assemblages of microorganisms, particularly when investigated in laboratory cultures. Sampling and related difficulties usually determine that only a portion of "real" communities which occur in the environment can be analysed and Fry and Parkes discuss these issues in the two following chapters (pp. 103–152 and 45–102). As recently as 1977 Fredrickson was moved to conclude that "not much can be said about experimental studies involving more than two populations". The intervening years have witnessed a dramatic increase in community studies and the development of appropriate methods (microbiological, biochemical, genetic and mathematical) of analysis, as perusal of this book will reveal. Communities will exhibit a number of basic characteristics among which are successional development, homeostasis, defence against severe stress, and evolution. These matters, similarly, will be scrutinized in the remainder of the present chapter and in other chapters.

A number of terms have been coined to distinguish members of a microbial community on a functional basis. Thus, Brock (1966) refers to:

(a) dominants (species having a dominant influence in the community);
(b) associates (species dependent on the dominant(s) for their survival); and
(c) incidentals (species more or less indifferent to the actions of dominants and associates).

We have used somewhat analogous terms to differentiate microbial communities, particularly those which are involved in the biodegradation of xenobiotic chemicals:

(i) primary utilizers (species capable of assimilating or metabolizing the sole or major carbon and energy substrate in the system);
(ii) secondary organisms (species unable to assimilate or metabolize the major substrate but which, because of their ability to utilize substrate metabolites or lytic products, are stable members of the community).

Microorganisms are frequently distributed in discrete spatial arrays within the environment but rather few detailed studies have been attempted. Nevertheless, it is possible to define three-dimensional distributions of microbial populations and communities, and hence to extend understanding of the factors engendering stability and change. This approach is proving especially rewarding in the investgation of fungal communities reference to which is made below (Section 4).

Finally, we wish to emphasize the importance of making genetic analyses of microbial communities. Williamson (1972) pointed out a decade ago that geneticists usually studied selection acting on monospecies populations and have neglected the more difficult problem of selective interaction between several species. This is not to say that the genetics of interacting populations have been totally avoided; valuable insight has been gained of the genetics of phytopathogenic fungus–host, prey–predator, Rhizobium–legume,and virus–host interactions. However, little attention has been given to the genetics of mixed microbial populations and we agree with Fredrickson's (1977) prompting that much more needs to be paid to the effects of mutation and selection in all dynamic situations. Competition has been discussed as an important driving force for enzyme evolution in microorganisms (Slater and

Bull, 1978; Slater and Godwin, 1980) and the selection, in a complex community, of a strain of *Pseudomonas putida*, which can utilize chlorinated alkanoic acids (Senior *et al.*, 1976), is a clear illustration of such selection. Further consideration of the genetics of interacting populations is provided in the Chapter by Reanney and his colleagues (pp. 287–322).

2.4 Classification Based on Biological Mechanisms

During the last several years of studying microbial communities in our laboratories, we have suggested a system of classification based on the biological mechanisms involved (Slater, 1978; Slater and Bull, 1978; Slater and Somerville, 1979; Bull and Brown, 1979; Bull, 1980; Slater, 1981). Although analyses of most microbial communities that have been isolated are incomplete, the accumulated data enable a simple classification to be advanced (Table 3). Such a classification is obviously preliminary in nature and will be refined as further understanding of community interactions is achieved; at present the delimitation of Class 3 and Class 7 community structures is especially uncertain. In addition, a given community often will show the characteristics of more than class. The bacterial community isolated by Osman (Osman *et al.*, 1976; Bull and Brown, 1979) provides clear

Table 3. Classification of microbial communities based upon biological mechanisms involved (after Slater, 1981)

Class 1:	Structure based on provision of specific nutrients by different members of the community.
Class 2:	Structure based on the alleviation of growth inhibition, including removal of metabolites which are inhibitory to the producer species (see also hydrogen transfer communities).
Class 3:	Structure based on interactions which result in the alteration of individual population growth parameters thereby producing more competitive and/or efficient community performance.
Class 4:	Structure based on a combined metabolic activity not expressed by individual populations alone.
Class 5:	Structure based on cometabolism.
Class 6:	Structure based on hydrogen transfer reactions.
Class 7:	Structure based on the interaction of several primary species (Section 2.3).

illustration of this latter point: partial relief of substrate (3, 5-dihydroxytoluene) inhibition and alteration of substrate affinity and inhibitor constants.

Many examples of Class 1 to Class 7 communities are given in the review of Slater (1981) and in the context of specific discussions found in the subsequent Chapters of this book.

3. MATHEMATICAL MODELLING OF MICROBIAL INTERACTIONS

It is not our intention in this section to provide a critique of the various mathematical models developed in population biology. May (1973) has commented on the types of model-building which ecologists have adopted, ranging from empirical, pragmatic descriptions of specific systems to more abstract, strategic approaches that forego precision in an attempt to reveal universal principles. Rather, our wish is to point out certain difficulties that are inherent to model-building and which, during this explosive phase of analysing microbial communities, if disregarded are likely to impede understanding. The following observations will doubtless seem facile to the mathematically minded reader but we hope that they will serve as timely reminders to microbiologists of the dangers of adopting seemingly ideal models. The generally unquestioned acceptance of Lotka–Volterra equations as models of predator–prey systems has already been raised (Section 2.2).

The Monod model, relating specific growth rate, μ, to the concentration of growth limiting substrate, s, is frequently used as the starting point for the kinetic analysis of mixed cultures (Veldkamp and Jannasch, 1972; Slater and Bull, 1978) and appears to be a generally acceptable approach to describing the dynamics of competing populations. That is to say, although the Monod model is almost certainly an over-simplification, a "simple relationship reasonably expresses inter-relationships even though the physical meaning of the model parameters is unknown or perhaps does not exist" (Bailey and Ollis, 1977). Can the model be applied to cases of commensalism or mutualism? Miura *et al.* (1980) have modelled competitive commensalism (a) and competitive mutualism (b) on the basis of the Monod relationship:

where M_A and M_B represent the two interacting microorganisms; G_A and G_B growth factors produced by M_A and M_B respectively; and *s* the primary growth limiting substrate. In the case of competitive commensalism, the specific growth rates of M_A and M_B are given by:

$$\mu_A = \mu_{MA}\left(\frac{s}{K_{sA} + s}\right) \tag{1}$$

$$\mu_B = \mu_{MB}\left(\frac{s}{K_{sB} + s}\right)\left(\frac{G_A}{K_{GA} + G_A}\right) \tag{2}$$

where K_{sA}, K_{sB} and K_{GA} are the saturation constants of organism M_A for substrate *s* and organism M_B for substrates *s* and G_A, respectively. Depending on the magnitude of the maximum specific growth rates and saturation constants, four cases are possible (Slater and Bull, 1978, Miura *et al.*, 1980), three of which predict coexistence of the two organisms in chemostat culture, viz:

(a) $\mu_A > \mu_B$ and $K_{sA} > K_{sB}$

(b) $\mu_A < \mu_B$ and $K_{sA} > K_{sB}$

(c) $\mu_A < \mu_B$ and $K_{sA} < K_{sB}$

It is possible to predict the dynamic behaviour of the system for these various conditions and simulate the effect of dilution rate. When commensalism accompanies competition damped oscillations tend to develop when $\mu_A < \mu_B$; under conditions of competitive mutualism damped oscillations are predicted when $\mu_A < \mu_B$ or when $\mu_A > \mu_B$ (Miura *et al.*, 1980).

To date very few tests of predictions, such as those of Miura, have been made. The competitive commensal system of *Mycotorula japonica* and *Pseudomonas oleovorans* (Section 2.2) when examined in continuous culture (Miura *et al.*, 1978) was found to be unstable and show oscillations over a long period. Whether this unpredicted behaviour is system specific (i.e. low affinity of *Mycotorula* sp. for the water insoluble substrate, *n*-tetradecane; possible toxicity of phenol) or is based on faulty anacalyptic assumptions, as implied by Haas *et al.* (1980), as yet cannot be determined. The latter authors believe that the concept of growth-limiting substrate has been misused in the context of mixed culture kinetics. However, equation (2) has precedence in enzyme kinetics and several examples of dual substrate limitation (often carbon source and oxygen) have been observed with monocultures. In our opinion experimental testing of the Miura model merits considerable attention.

We stated above that the Monod model is most often used to predict the outcome of competition between microbial populations. However, alternative approaches to it and to the Lotka–Volterra equations have been developed for resource competition and one such is the variable internal stores model of Droop (1974). The latter assumes that internal nutrient concentration determines the specific growth rate and in turn is determined by the nutrient uptake and growth. Recently Tilman (1977) has used both models to predict the distribution of two algae, *Asterionella formosa* and *Cyclotella meneghiniana*, along a silicate–phosphate gradient. The two models made similar predictions and these agreed substantially with the observed experimental data. Again, steady-state coexistence was attained, when each species was limited by a different nutrient. Resource based models of the Monod and Droop types are mechanistic, whereas Lotka–Volterra equations are descriptive and their parameters are assumed to be constants. However, as Tilman points out, such parameters are constant only under steady-state conditions. Thus, although descriptive models can reveal features of community structure, mechanistic models are to be preferred because, in addition, they can be used to analyse dynamic situations.

The specific growth rate of an organism often responds to an environmental perturbation only after a time lag (see May (1973) for detailed discussion) and this may be modelled by introducing a first-order time constant into the Monod relationship. Relatively little use of time constants has been made in mixed culture modelling but a recent paper by Wilder *et al.* (1980) illustrates its application. Competition between *Candida utilis* and *Corynebacterium glutamicum* for glucose was modelled with the Monod relationship and first-order time constants added to simulate growth transients. Such simulations generated oscillations in population sizes with the largest oscillations appearing when the faster growing species was assigned the longer time constant.

Our discussion has been restricted to pair-wise populations. The modelling of multispecies microbial systems is in its infancy, but its development is required to gain an understanding of community stability and complexity. Stability refers to the capacity of a community to recover from a given perturbation to an equilibrium point, whereas complexity defines the number of species within a community, the number of interactions between the species, and/or the strength of the interactions. Mathematical analysis of model ecosystems suggests that increased complexity is associated with decreased stability and reduced ability to survive environmental stresses (May, 1979). Thus, community complexity may develop only in relatively stable environments, whereas relatively unstable environments require dynamically robust and simple communities. The testing of these generalizations against biological reality has been restricted almost entirely to populations of plants and animals. However, we are now beginning to see the emergence of microbial community studies and the description—at least in the laboratory—of complex, stable systems (Bull and Brown, 1979; Slater and Somerville, 1979). And with these studies will come the opportunity to pose appropriate questions about the structure and dynamics of multispecies communities.

4. COMMUNITY DEVELOPMENT

In concluding this chapter we wish to return to a theme which was introduced briefly in the Introduction, namely community development and succession.

Prior to their colonization, substrates provide an excess of some, if not most, nutrients, a situation that frequently encourages the growth of a diversity of microorganisms. As environmental conditions change, one community of organisms (usually) becomes replaced by another and such successions may occur over short and long time intervals. Environmental changes can be imposed from without either as climatic or anthropogenic disturbances, or as a consequence of colonization *per se*. In the majority of habitats successional changes in the microflora are not observed; instead studies become focused on the established or climax communities. Nevertheless, microbial succession is amenable to investigation and many descriptions have been given of colonization and succession in habitats such as soil, leaf litter, water, gastrointestinal and urogenitary tracts, the rumen, and food materials. Brock (1966) makes the important point that microbial ecosystems lack the great, long-term stability characteristic of macroecosystems in which such climax systems as forest so modify the environment that it can maintain itself even under circumstances of severe disturbance. Microbial ecosystems rarely create stable conditions in their environments; instead they adapt to changing environments by changing the community structure.

In recent years some particularly interesting research has been done into the colonization of decaying wood and this has become a most suitable model for analysing microbial succession. Thus, direct field observations of the three-dimensional organization of fungal mycelia into populations and communities can be combined with physiological and genetic analyses in the laboratory thereby enabling a rational understanding of community structure and dynamics to be made. The brief discussion of this system which follows relies very much on the recent, excellent review of the subject made by Rayner and Todd (1979).

Rayner and Todd (1979) argue that one difficulty of the succession concept is the encouragement which it gives to thinking in terms of "single, simple underlying causes, such as changes in nutrient status". Rather the central issue is the nature and development of fungal communities: what are they, how are they maintained and what forces cause them to change? Rayner and Todd's view of fungal community development within a spatially defined resource like wood is shown in Fig. 3. The mycelia of fungi A to K are primary colonizers and spread throughout the wood until contact is made and the available space is fully occupied. Species B and J are quickly replaced and deadlock interactions become established between species A and C, C and D, D and E, G and H, H and I, and I and K. Species F replaces species C, D, E and G; subsequently species L, M and N become competitive invaders and with F and H comprise the final community. The constituent

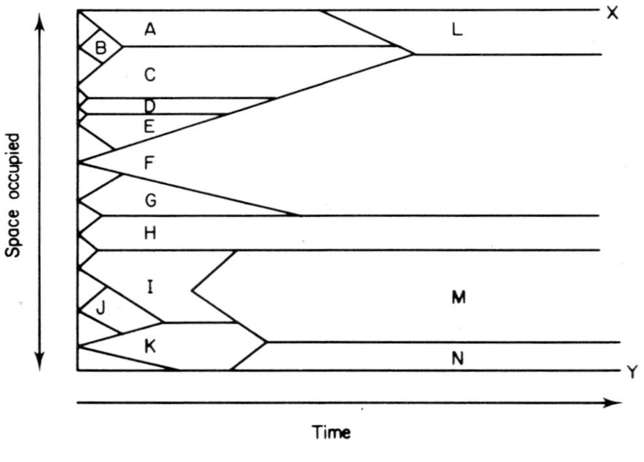

Fig. 3. Schematic representation of fungal community development in wood (from Rayner and Wood, 1979). Available colonizable space is encompassed by the region X–Y and the changing distribution of different fungal mycelia (A to N) within this space as a function of time is depicted as a sequence of tapering and expanding bands. Horizontal lines between the bands represent deadlock interactions; oblique lines represent replacement.

mycelia of the community interact in various ways, important among which are *intermingling*, *deadlock*, and *replacement*: deadlock interactions will tend to maintain the community structure while replacement reactions will lead to change or succession. Both deadlock and replacement reactions can be caused by direct and indirect interactions (see Section 2.1). Among the latter can be included: nutrient availability, antibiotic production, removal of toxic compounds and host resistance.

In the pioneer phase, wood typically is colonized by a wide range of fungi only some of which cause decay. The latter include certain microfungi (*Acremonium*, *Botrytis*, *Phialophora*, for example) and fast-growing basidiomycetes, most of which are replaced quite easily by other species which dominate in the established phase I. Gradually the deadlocked community structure of this second phase is disrupted by colonization of aggressive species (e.g. *Hypholoma fasciculare*, *Phanerochaete velutina*). A second series of deadlock interactions is likely to develop leading to an established phase II. Ultimately, the activity of decay species of the latter phase declines and a final replacement phase is set up which comprises non-decay microfungi such as *Trichoderma*.

Finally, it may be useful to make brief reference to classes of evolutionary strategies which plant ecologists have defined as responses to

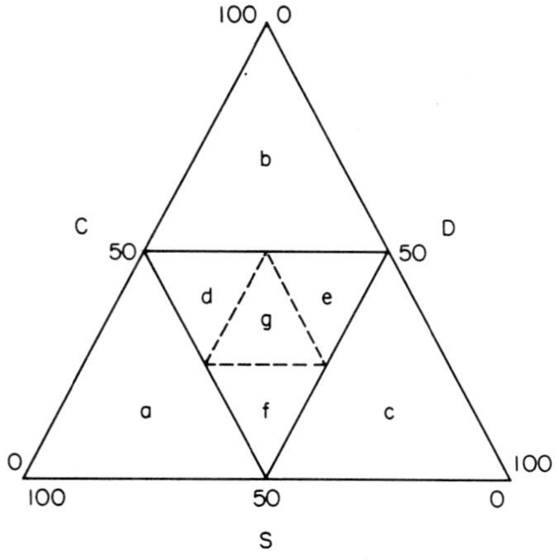

Fig. 4. Equilibrium positions between stress, disturbance, and competition, and strategies of survival (after Grime, 1979). C, D, and S indicate the relative importance of competition, disturbance, and stress, respectively; a = stress-tolerators; b = competitors; c = ruderals; d = stress-tolerant competitors; e = competitive ruderals; f = stress-tolerant ruderals; g = C-S-R strategists.

threats to existence in established communities, viz. to severe stress, frequent disturbance, and competitive exclusion (Grime, 1979). The types of primary and secondary strategies which occur can be defined by a triangular model of the type shown in Fig. 4. Plants have been referred to the three classes of primary strategists and their characteristics described in terms of morphology, life history, and physiology. It may be a rewarding exercise for microbial ecologists to adopt a similar system of classification but we are not aware of attempts to develop such an approach.

REFERENCES

Alexander, M. (1971). "Microbial Ecology". John Wiley, New York.

Austin, B., Dickinson, C. H. and Goodfellow, M. (1977). Antagonistic interactions of phylloplane bacteria with *Drechslera dictyoides* (Drechsler) Shoemaker. *Candian Journal of Microbiology* **23**, 710–715.

Bailey, J. E. and Ollis, D. F. (1977). "Biochemical Engineering Fundamentals". McGraw-Hill Book Company, New York.

Baker, K. F. (1980). Microbial Antagonism—the potential for biological control. *In* "Contemporary Microbial Ecology", pp. 327–347. (Ellwood, D. C., Hedger, J. N., Latham, M. J., Lynch, J. M., and Slater, J. H., eds.). Academic Press, London and New York.

Bary, A. de (1866). "Morpholgie und Physiologie der Pilze, Flechten und Myxomyceten". Engelmann, Leipzig.

Bates, J. L. and Liu, P. V. (1963). Complementation of lecithinase activities by closely related pseudomonads: its taxonomic implication. *Journal of Bacteriology* **86**, 585–592.

Bowen, G. D. (1980). Misconceptions, concepts and approaches in rhizosphere biology. *In* "Contemporary Microbial Ecology", pp. 283–304. (Ellwood, D. C., Hedger, J, N., Latham, M. J., Lynch, J. M., and Slater, J. H., eds.) Academic Press, London and New York.

Brock, T. D. (1966). Principles of Microbial Ecology. Prentice-Hall, Inc., Englewood Cliffs, N.J.

Bull, A. T. (1970). Inhibition of polysaccharases by melanin: enzyme inhibition in relation to mycolysis. *Archives of Biochemistry and Biophysics* **137**, 345–356.

Bull, A. T. (1980). Biodegradation: some attitudes and strategies of microorganisms and microbiologists. *In* "Contemporary Microbial Ecology", pp. 107–136. (Ellwood, D. C., Hedger, J. N., Latham, M. J., Lynch, J. M., and Slater, J. H., eds.) Academic Press, London and New York.

Bull, A. T. and Brown, C. M. (1979). Continuous culture applications to microbial biochemistry. *International Review of Biochemistry* **21**, 177–226.

Chao, C.-C. and Reilly, P. J. (1972). Symbiotic growth of *Acetobacter suboxydans* and *Saccharomyces carlsbergensis* in a chemostat. *Biotechnology and Bioengineering* **14**, 75–92.

Droop, M. R. (1974). The nutrient status of algal cells in continuous culture. *Journal of the Marine Biological Association of the United Kingdom* **54**, 825–855.
Dubos, R. and Kessler, A. (1963). Integrative and disintegrative factors in symbiotic associations. *In* "Symbiotic Associations", pp. 1–11. (Nutman, P. S. and Mosse, B., eds.) University Press, Cambridge.
Fager, E. W. (1963). Communities of organisms. *In* "The Sea", volume 2, pp. 415–437. (Hill, M. N., ed.) John Wiley, New York.
Fredrickson, A. G. (1977). Behaviour of mixed cultures of microorganisms. *Annual Review of Microbiology* **31**, 63–87.
Gause, G. F. (1934). "The Struggle for Existence". Williams & Wilkins, Baltimore.
Goldman, J. C. and Ryther, J. H. (1976). Temperature-influenced species competition in mass cultures of marine phytoplankton. *Biotechnology and Bioengineering* **18**, 1125–1144.
Gray, B. H., Fowler, C. F., Nugent, N. A., Rigopoulos, N. & Fuller, R. C. (1973). Re-evaluation of *Chloropseudomonas ethylica* 2–K. *International Journal of Systematic Bacteriology* **23**, 256–264.
Grime, J. P. (1979). Competition and struggle for existence. *In* "Population Dynamics", pp. 123–142. (Anderson, R. M., Turner, B. D.,and Taylor, L. R., eds.) Blackwell Scientific Publications, Oxford.
Haas, C. N., Bungay, H. R.,and Bungay, M. L. (1980). Practical mixed culture processes. *Annual Reports on Fermentation Processes* **4**, 1–29.
Hadar, Y., Chet, I. and Henis, Y. (1979). Biological control of *Rhizoctonia solani* damping-off with wheat bran culture of *Trichoderma harzianum*. *Phytopathology* **69**, 64–68.
Hale, M. E. (1974). "The Biology of Lichens", 2nd Edition. Edward Arnold, London.
Harder, W. and Veldkamp, H. (1971). Competition of marine psychrophilic bacteria at low temperatures. *Antonie van Leeuwenhoek* **37**, 51–63.
Jensen, H. L. (1963). Carbon nutrition of some micro-organisms decomposing halogen-substituted aliphatic acids. *Acta Agriculturae Scandinavica* **13**, 404–412.
Johnston, A. W. B. and Beringer, J. E. (1976). Mixed inoculations with effective and ineffective strains of *Rhizobium leguminosarum*. *Journal of Applied Bacteriology* **40**, 375–380.
Jost, J. L., Drake, J. F., Fredrickson, A. G.,and Tsuchiya, H. M. (1973). Interactions of *Tetrahymena pyriformis*, *Escherichia coli* and *Azotobacter vinelandii* and glucose in a minimal medium. *Journal of Bacteriology* **113**, 834–840.
Kloepper, J. W. and Schroth, M. N. (1978). Association of *in vitro* antibiosis with inducibility of increased plant growth by *Pseudomonas* spp. *Phytopathological News* **12**, 136.
Kögl, F. and Friesa, N. (1937). Ueber den Einfluss von Biotin, Aneurin und Mesoinosit auf das Wachstum verschiedener Pilzarten. *Hoppé Seyer's Zeitschrift für physiologische Chemie* **249**, 93.
Krul, J. M. (1977). Experiments with *Haliscomenobacter hydrossis* in continuous culture without and with *Zoogloea ramigera*. *Water Research* **11**, 197–204.

Lawrence, R. C. and Thomas, T. D. (1979). The fermentation of milk by lactic acid bacteria. *In* "Microbial Technology: Current State, Future Prospects", pp. 187–219. (Bull, A. T., Ellwood, D. C.,and Ratledge, C., eds.) University Press, Cambridge.

Lee, I. H., Fredrickson, A. G. and Tsuchiya, H. M. (1976). Dynamics of mixed cultures of *Lactobacillus plantarum* and *Propionobacterium shermanii*. *Biotechnology and Bioengineering* **18**, 513–526.

Lewis, D. H. (1974). Microorganisms and plants: the evolution of parasitism and mutualism. *In* "Evolution in the Microbial World", pp. 367–392. (Carlile, M. J. and Skehel, J. J., eds.) University Press, Cambridge.

Lewis, P. M. (1967). A note on the continuous flow culture of mixed populations of Lactobacilli and Streptococci. *Journal of Applied Bacteriology* **30**, 406–409.

Lynch, J. M. and Poole, N. J., Editors (1979). "Microbial Ecology. A Conceptual Approach". Blackwell Scientific Publications, Oxford.

Marsh, P. (1980). "Oral Microbiology". Thomas Nelson & Sons Ltd., Walton-on-Thames.

May, R. M. (1973). "Stability and Complexity in Model Ecosystems". University Press, Princeton, N.J.

May, R. M. (1979). The structure and dynamics of ecological communities. *In* "Population Dynamics", pp. 385–407. (Anderson, R. M., Turner, B. D.,and Taylor, L. R., eds.) Blackwell Scientific Publications, Oxford.

McDonough, E. S., Prooien, R. Van and Lewis, A. L. (1965). Lysis of *Blastomyces dermatitidis* yeast-phase cells in natural soil. *American Journal of Epidemiology* **81**, 86–94.

Meers, J. L. and Tempest, D. W. (1968). The influence of extracellular products on the behaviour of mixed microbial populations in magnesium-limited chemostat cultures. *Journal of General Microbiology* **52**, 309–317.

Megee, R. D., Drake, J. F., Fredrickson, A. G.,and Tsuchiya, H. M. (1972). Studies in intermicrobial symbiosis. *S. cerevisiae* and *L. casei*. *Canadian Journal of Microbiology* **18**, 1733–1742.

Meyer, J. S., Tsuchiya, H. M.,and Fredrickson, A. G. (1975). Dynamics of mixed populations having complementary metabolism. *Biotechnology and Bioengineering* **17**, 1065–1081.

Milne, A. (1961). Definition of competition among animals. *Symposia of the Society of Experimental Biology* **15**, 40–61.

Mitchell, R. (1963). Addition of fungal cell wall compounds to soil for biological disease control. *Phytopathology* **53**, 1068–1071.

Miura, Y., Sugiura, K., Yoh, M., Tanaka, H., Okazaki, M.,and Komemushi, S. (1978). Mixed culture of *Mycotorula japonica* and *Pseudomonas oleovorans* on two hydrocarbons. *Journal of Fermentation Technology* **56**, 339–344.

Miura, Y., Tanaka, H.,and Okazaki, M. (1980). Stability analysis of commensal and mutual relations with competitive assimilation in continuous mixed culture. *Biotechnology and Bioengineering* **22**, 929–948.

Moore, L. W. and Warren, G. (1979). *Agrobacterium radiobacter* strain 84 and biological control of crown gall. *Annual Review of Phytopathology* **17**, 163–179.
Müller, W. F. and Schopfer, W. H. (1937). L'action de l'aneurine et de ses constituants sur *Mucor ramannianus* Möll. *Compte Rendu des Séances de la Société Biologie* **205**, 687. (Cited by Schopfer, 1943).
Munnecke, D. M. and Hsieh, D. P. H. (1974). Microbial decontamination of parathion and *p*-nitrophenol in aqueous media. *Applied Microbiology* **28**, 212–217.
Nurmikko, V. (1956). Biochemical factors affecting symbiosis among bacteria. *Experientia* **12**, 245–249.
Oberman, H. and Libudzisz, Z. (1973). The growth of mixed population of *Lactobacillus casei* and nisin producing strain of *Streptococcus lactis* in batch and continuous cultures. *Acta Microbiologica Polonica* Series B**5**, 151–161.
Odum, E. P. (1953). "Fundamentals of Ecology". Saunders, Philadelphia.
Osman, A., Bull, A. T.,and Slater, J. H. (1976). Growth of mixed microbial populations on orcinol in continuous culture. Abstracts of Papers, 5th International Fermentation Symposium, p. 124. Berlin: Verlag Versuchs—und Lehranstalt für Spiritusfabrikation und Fermentationstechnologie.
Pirt, S. J. (1975). "Principles of Microbe and Cell Cultivation". Blackwell Scientific Publications, Oxford.
Rayner, A. D. M. and Todd, N. K. (1979). Population and community structure and dynamics of fungi in decaying wood. *Advances in Botanical Research* **7**, 333–420.
Roper, M. M. and Marshall, K. C. (1974). Modification of the interaction between *Escherichia coli* and bacteriophage in saline sediment. *Microbial Ecology* **1**, 1–13.
Salt, G. W. (1974). Predator and prey densities as controls of the rate of capture by the predator *Didinium nasutum*. *Ecology* **55**, 434–439.
Schopfer, W. H. (1943). "Plants and Vitamins". Chronica Botanica Company, Waltham, Mass.
Senior, E., Bull, A. T. and Slater, J. H. (1976). Enzyme evolution in a microbial community growing on the herbicide Dalapon. *Nature London* **263**, 476–479.
Slater, J. H. (1978). The role of microbial communities in the natural environment. *In* "The Oil Industry and Microbial Ecosystems", pp. 137–154. (Chater, K. W. A. and Somerville, H. J., eds.) Heyden & Sons, London.
Slater, J. H. (1981). Mixed cultures and microbial communities. *In* "Mixed Culture Fermentations", pp. 1–24. (Bushell, M. E. and Slater, J. H., eds.) Academic Press, London and New York.
Slater, J. H. and Bull, A. T. (1978). Interactions between microbial populations. *In* "Companion to Microbiology", pp. 181–206. (Bull, A. T. and Meadow, P. M., eds.) Longman, London.

Slater, J. H. and Godwin, D. (1980). Microbial adaptation and selection. *In* "Contemporary Microbial Ecology", pp. 137–160. (Ellwood, D. C., Hedger, J. N., Latham, M. J., Lynch, J. M., and Slater, J. H., eds.) Academic Press, London and New York.

Slater, J. H. and Somerville, H. J. (1979). Microbial aspects of waste treatment with particular attention to the degradation of organic compounds. *In* "Microbial Technology: Present State, Future Prospects", pp. 221–261. (Bull, A. T., Ellwood, D. C., and Ratledge, C., eds.) University Press, Cambridge.

Smith, D. C. (1978). Photosynthetic endosymbionts of invertebrates. *In* "Companion to Microbiology", pp. 387–414. (Bull, A. T. and Meadow, P. M., eds.) Longman, London.

Tilman, D. (1977). Resource competition between planktonic algae: an experimental and theoretical approach. *Ecology* **58**, 338–348.

Veldkamp, P. and Jannasch, H. (1972). Mixed culture studies with the chemostat. *Journal of Applied Chemistry and Biotechnology* **22**, 105–123.

Wilder, C. T., Cadman, T. W., and Hatch, R. T. (1980). Feedback control of a competitive mixed culture system. *Biotechnology and Bioengineering* **22**, 89–106.

Wilkinson, T. G., Topiwala, H. H., and Hamer, G. (1974). Interactions in a mixed bacterial population growing on methane in continuous culture. *Biotechnology and Bioengineering* **16**, 41–47.

Williams, F. M. (1980). On understanding predator–prey interactions. *In* "Contemporary Microbial Ecology", pp. 349–375. (Ellwood, D. C., Hedger, J. N., Latham, M. J., Lynch, J. M., and Slater, J. H., eds.) Academic Press, London and New York.

Williamson, M. (1972). "The Analysis of Biological Populations". Edward Arnold, London.

Yeoh, H. T., Bungay, H. R., and Krieg, N. R. (1968). A microbial interaction involving a combined mutualism and inhibition. *Canadian Journal of Microbiology* **14**, 491–492.

3

Methods for Enriching, Isolating, and Analysing Microbial Communities in Laboratory Systems

R. J. Parkes

1. INTRODUCTION

The ultimate aim of studying microbial communities is to understand how they function in the environment, whether the research is for pure or applied reasons. Therefore, it seems contradictory to advocate the study of these communities in the laboratory when such a study may not be relevant to the environment. However, study in the natural environment poses many problems which make field study difficult, and sometimes makes the interpretation of data and its subsequent application just as precarious as extrapolation from the laboratory.

The main problems in investigations *in situ* are caused by the very feature which separates the laboratory system from the natural environment, i.e. the complex and dynamic nature of the environment. This produces daunting methodological problems in field research, not least of which is the fundamental problem of sample collection that in itself may have a profound effect on the system being analysed. For example, Falk (1974) estimated the impact of the sampler in a large ecosystem and compared this effect with other "predators": the effect was significant and, furthermore, these sampling effects must be even more important on the microbial scale. Another problem is that there are many important factors controlling the environment that field study always involves selecting parameters for study and, although sound correlations may be obtained at the end of the study, it remains uncertain whether these are causal and what effect the unmeasured parameters have had on the system. The constantly changing conditions in the environment also make interpretation of field work difficult as the parameters influencing or controlling the reactions under study may vary throughout the period of the investigation. These problems become particularly important if the effects of pollution stress are being investigated, as it becomes difficult to separate natural variations from changes induced by the imposed pollution stress (see Fry, p. 103).

The degree of control that can be asserted in laboratory systems is the main asset: the number of variables to be included in the study can be selected and held constant throughout, or one variable changed with the remainder held constant, or a number of variables changed throughout the experiment in an ordered and controlled manner. This advantage makes the laboratory system extremely flexible and much easier to interpret. However, the problems arise when these results are extrapolated to the natural environment. The inapplicability of many laboratory results to the natural environment may be due to inappropriate conditions being applied in the laboratory systems (e.g., the use of unrealistic temperatures and concentrations of carbon substrates). Therefore, if laboratory systems are going to be of any use, more importance has to be given to the design of laboratory systems relevant to the particular ecosystem under study. It is obvious that the conflict of laboratory versus field is an artificial one, since both approaches have their particular advantages and disadvantages, and only through a combined use of both approaches will we increase our understanding of the microbe and its environment. The aim of this chapter is to indicate the types of laboratory systems available, the criteria implicit in their use and indicate, where possible, their potential relevance to specific environmental situations.

However, first I would like to underline the need to study mixed microbial communities, as opposed to the pure culture approach. Since the pioneering work of Koch, we have been dominated by the concept of pure culture work which has taken us far in our understanding of the effect of microbes in the environment. But now this approach may well be inhibiting further progress. It is now apparent that most microbial activity in the environment is a result of the action of mixtures of different microorganisms (Bungay and Bungay, 1968; Slater, 1978) and that pure culture activity is the exception rather than the rule. The types of mixed microbial interactions now recognized are many and varied (see Bull and Slater, p. 18), ranging from protocooperation (one organism only benefiting from a relationship) to symbiosis (complete inter-dependence) and we ignore the importance of these interactions at our peril. Indeed, the pure culture approach may to a certain extent be the reason for the poor extrapolations from laboratory to field. This problem has obviously been recognized for a long time and may have had much to do with the development of the chemical approach to microbial ecology—the overall chemical and biochemical effects of microorganisms being studied, rather than the organisms themselves. However, this approach is not immune to the limitations of pure culture techniques, as the chemical data have to be interpreted on the basis of the physiology of microorganisms, more often than not enriched in batch culture and studied in pure culture: as will become apparent throughout this chapter these organisms may not represent the organisms actually active in the environment. Hence, researchers concerned with studying microorganisms themselves and the effect of their metabolic activity in the environment should benefit from the use of more realistic laboratory systems for both the initial enrichment and the subsequent study of the microorganisms.

2. REVIEW OF ENRICHMENT TECHNIQUES

2.1. Terminology

Gradually systems used to enrich or study organisms in the laboratory have become known as microcosms, perhaps to indicate that they do have some relevance to the natural environment. However, this term is now used so widely that it describes anything from conical flasks (Bryfogle and McDiffet, 1979) to complex environment-simulating systems (Bond *et al.*, 1976) and its usage is confusing and ambiguous. Matsumura (1979) suggested that microcosms could be separated into four different types, but these divisions seem rather artificial and the term microcosm would still be used for each type. I would like to suggest the following definitions and these will be used throughout the chapter.

2.1.1 Microcosm

A laboratory system which attempts to simulate as far as possible the conditions prevailing in the whole or part of the environment under study is termed a "microcosm". For greater clarity a more descriptive phrase should be included with the word microcosm: e.g. microcosm, aquatic system with terrestrial:aquatic interface (Metcalf *et al.*, 1971); coniferous forest soil:litter microcosm (Bond *et al.*, 1976).

2.1.2 Model Systems

Any laboratory system not included in the above definition can be termed a model system, i.e. a laboratory system operating under conditions which are functionally similar to the whole or part of the environment to be studied and this term would also be followed by a more descriptive term: e.g. model system, chemostat. These model systems range from simple batch flasks to multiple stage and phase chemostats.

2.2 Methods

Most natural environments contain a large variety of microbes in numbers which reflect the habitat and the relative abilities of the individual organisms to compete for the available nutrients. Enrichment techniques are designed to change the environment in such a way that the organism or organisms of interest will successfully compete against all other organisms and hence become the dominant population(s). The process is highly selective and many species initially present will be lost due to their inability to compete under the prevailing enrichment conditions. There is a bewildering variety of enrichment systems available to microbiologists and the arguments for using a particular system to the exclusion of others is often lucidly and convincingly presented (e.g. chemostat techniques—Veldkamp (1977) and Veldkamp and Jannasch (1972)). However, it is obvious that one technique will not be applicable to all environmental situations and therefore more emphasis must be placed on the advantages and disadvantages of particular techniques in order to enable a technique to be selected which is most appropriate for a particular situation.

2.2.1 Model Closed Systems

In a closed system enrichment initially takes place in defined conditions, with no further input of growth substrates or removal of metabolic end products or cells. These systems include the most classical methods, such as plate culture, serial dilution techniques, shake flask systems, and soil slurries, and are often collectively termed batch enrichment systems. Control of the enrichment process is due to the design of growth conditions and a large number of selective media for the enrichment of nutritionally distinct organisms have been developed (these methods have been adequately described elsewhere—see Fry, p. 108; for a description of general methods, Norris and Ribbons (1969); general media design, Norris and Ribbons (1970b); media for autotrophs, anaerobes, and specialized bacteria, Norris and Ribbons (1970c); media for fungi, Booth (1971); and enrichment of mutants, Schlegel and Jannasch (1967)).

Problems associated with closed enrichment conditions

Although the initial enrichment conditions can often be defined, due to the metabolic activity of the enriched organisms, the environmental conditions in the system and hence selection conditions are constantly changing. This situation can lead to poor reproducibility in the enrichment procedure and even failure to enrich the desired organisms. However, if the microorganisms being enriched are nutritionally distinct (e.g. sulphate-reducing bacteria), the enrichment can be reproducible despite these problems. The constantly changing enrichment conditions also mean that enrichment conditions that actually apply during the selection process are often unknown. This must hinder the development of new selective media since this tends to be an empirical process in the absence of detailed knowledge of the enrichment conditions.

The initial concentrations of limiting nutrients need to be high if the enrichment is to be continued for any length of time, and time is an essential part of the selection process. It is common practice in batch cultures to use media containing 1 to 2% (w/v) substrate (Andrews, 1968) and this obviously causes problems in the enrichment of microorganisms capable of degrading toxic compounds. This problem may be overcome for toxic substrates by using a fed-batch system (Pirt, 1975) but this system would not alleviate inhibition caused by the production of a toxic metabolite from a non-toxic substrate.

When the substrate is in excess, as it must be in batch systems, competition, and hence selection of microbes, is based solely on their maximum specific growth rate (μ_{max}). Thus, organisms enriched in batch systems characteristically have a high growth efficiency at high substrate concentrations and low substrate specificity (Jannasch, 1967a). It is not surprising that these microorganisms have growth characteristics which are identical to the opportunistic or zymogenous organisms as defined by Winogradsky (1949). However, in the environment, which is often nutrient limited (Veldkamp, 1977), the important organisms may be those microorganisms which are able to utilize low substrate concentrations and have low maximum specific growth rates (Jannasch, 1967a). These autochthonous organisms are out-competed in the nutrient-rich batch systems due to their low μ_{max} values and therefore do not appear in this method of enrichment (Jannasch, 1967a).

The terms "zymogenous" and "autochthonous" are seemingly identical to r-strategists and k-strategists, respectively, which have recently been discussed by Carlile (1980), and Slater and Lovatt (1982). Carlile (1980) compared r-stragetists (opportunists) to prokaryotes and k-stragetists to eukaryotes, but, as indicated in the previous discussion, both types of stragetists can be found in the prokaryotes.

Another consequence of batch enrichment is the tendency to enrich for organisms requiring growth factors, i.e. auxotrophs. As a result of lysis of organisms in the initial inoculum and bacteria which dominate the early stages batch growth, the concentration of amino acids and vitamins increases, which leads eventually to the selective enrichment of bacteria which require one or more growth factors. Ayanaba and Alexander (1972) demonstrated this effect with the selection of soil bacteria with known nutritional requirements in batch culture. In a similar manner, batch requirements also tend to favour the enrichment of pure cultures, with any growth requirement which might lead to an interdependence on another organism being provided by lysis products.

From the above considerations it is apparent that batch systems enrich for microorganisms with characteristic growth parameters (zymogenous) and nutritional requirements (auxotrophs), but these only represent two of the many different types of microorganism which are present in the environment. It has been argued (Veldkamp and Jannasch, 1972; Veldkamp, 1977; and Jannasch, 1967a, b, 1974) that the types of organisms enriched in batch systems are unrepresentative of the microorganisms actually active in the environment where "substrate limitation is undoubtedly the rule" (Veldkamp, 1976). These views are probably reflected in the number of reports in the literature indicating poor extrapolation of the results of batch culture experiments to the natural environment and, indeed, failure to enrich for microorganisms with the desired physiological or metabolic properties. For example, MacRae and Alexander (1965) isolated a *Flavobacterium* species which could metabolize 2,4-dichlorophenoxybutyric acid and hence provide protection for alfalfa seedlings. The system only worked in sterile soil inoculated with the *Flavobacterium* species, and it was concluded that the *Flavobacterium* species was unable to function in the presence of the native soil population. Bartha and Pramer (1970) demonstrated the conversion of anilines to azo-compounds by soil suspensions but were unable to obtain, by conventional enrichment procedures, microorganisms capable of the transformation. Laanbroek *et al.* (1977) reported the isolation of a *Camplylobacter* species from an anaerobic digester by continuous culture techniques, whereas previously, using batch culture techniques, the *Camplylobacter* species had only been isolated from animal or human sources. Varon (1979) isolated a *Bdellovibrio*-resistant mutant of *Photobacterium leiognathi* using chemostat techniques. This type of mutant had not been shown in the previous 16 years and the author suggested that the conditions prevailing in batch cultures do not allow the evolution of such a mutant.

However, batch culture techniques despite the problems outlined above, do have other attributes besides their obvious simplicity.

(1) As outlined by Veldkamp (1976), certain environments are substrate limited and the zymogenous portion of the microbial community is likely to be dormant. But when nutrients are added to this environment the dormant, opportunistic microorganisms grow rapidly and then successful competition is determined by similar kinetic and nutritional characteristics as those governing classical, substrate excess, batch enrichments (Slater and Bull, 1978; Bull and Slater, p. 34) including the ability to cope with a continuously changing environment due to the metabolism of the added substrate. For example, consider the burial of leaves during the autumn in a carbon-limited sediment. The leaf is degraded sequentially, the most available carbon components of the leaf being utilized first, producing a whole succession of different microorganisms exploiting the continuously changing substrate. A similar situation occurs when algal blooms die off and sediment to the lake bottom. The incorporation of readily oxidizable carbon into lower layers of soil which are carbon limited, by the action of ploughing, would also promote the growth of the opportunistic, zymogenous organisms.

(2) The possible inapplicability of using batch systems for studying microbial growth on toxic compounds, due to problems of inhibition of the microorganisms, has been mentioned. But there are many microbial activities in the environment which may be controlled by the inhibitory products of community metabolism. Rumen metabolism, for example, is affected by the rate of removal of hydrogen gas by methanogenesis, since an accumulation of hydrogen gas in the rumen makes certain anaerobic reactions unfavourable (Hungate, 1966). A similar situation may also be present in carbon rich, anoxic marine sediments, although here sulphate reduction may be the mechanism of removal of potentially toxic metabolic end products (Parkes, 1978). Simi-

lar situations could easily develop on a smaller scale in microhabitats, with the localized build-up of inhibitory metabolic products, adverse pH or Eh gradients, and control microbial activity within the habitat. This effect may be quite widespread. Jørgenson (1977) gave some indication of the importance of microhabitats and their associated metabolism, by demonstrating that anaerobic activity can occur in a 100 μm thick detrital particle which was suspended in oxygenated sea water. The hydrogen sulphide produced by bacterial metabolism within the detrital particle assisted in maintaining the anaerobic conditions. In all these situations batch growth is a reasonable model for studying microbial growth.

Undoubtedly, batch culture studies have provided a wealth of data concerning the metabolism of different organisms, which indicate the potential mechanisms of microbial activity and the importance of microbial interaction in the environment. Another important aspect of batch experiments is the realization of the advantages and disadvantages of such methods and so their use may be even more rewarding in the future.

2.2.2 Model Open System

In open systems there are both inputs and outputs and they provide a totally different set of enrichment conditions to those prevailing in batch systems. The open nature of these systems reflects the dynamic interchange which is present in the natural environment, although in practice the types of inputs and outputs are usually limited to a few defined variables. The most common example of an open system in microbial enrichment is the continuous-flow culture system where the input is fresh growth medium and the output is spent medium and microbial biomass. When the microbial culture is homogeneous, the environmental conditions are constant and controllable and such a system is termed a chemostat.

Features of continuous-flow culture systems, particularly chemostats

(1) The conditions throughout the enrichment can be kept constant. Metabolic end-products, which if allowed to accumulate would change enrichment conditions, can be constantly removed to maintain constant conditions. A variety of parameters, e.g., pH, Eh and dissolved oxygen, can be automatically controlled thereby providing a stable enrichment environment. These conditions lead to the reproducible enrichment of desired microorganisms (Jannasch, 1967a; Senior, 1977) and enable the development of effective enrichment conditions for a de-

sired microorganism or activity as the prevailing conditions are known.

(2) The dilution rate of the system (flow rate of new media/culture volume), determines the growth rate of the enriched organism. Thus, low dilution rates coupled with low substrate concentrations in the medium input can be used to enrich for autochthonous organisms. The enrichment of this type of organism has been demonstrated by many workers (Jannasch, 1967a; b; Jannasch and Mateles, 1974; Mallory *et al.*, 1977; Veldkamp and Kuenen, 1973) and their presence in the environment has been found to be widespread. Kinetic analyses of the growth characteristics of autochthonous bacteria reflect the specialized nature of their scavenging role (low maximum specific growth rate and high substrate affinity). Although the enrichment of scavenging bacteria has been strongly associated with chemostat systems, these systems are very flexible and, unlike batch systems, are able to enrich for more than one type of ecologically important bacteria. For example, Gottschal *et al.* (1979) demonstrated that a versatile, mixotrophic thiobacillus (able to grow heterotrophically on acetate or glycollate, and autotrophically on thiosulphate) was able to maintain itself in the presence of both a specialized autotroph and heterotroph but by varying the culture conditions, each different nutritional type could be enriched.

(3) The constant removal of metabolites and cells in continuous-flow culture systems prevents the accumulation of potentially inhibitory products which, combined with the ability to use low concentrations of inhibitory substrates, makes continuous-flow culture systems ideal for the enrichment of organisms able to degrade toxic compounds. Conversely, high substrate concentrations and low dilution rates allow the inhibitory effects of metabolites to be investigated. The removal of microbial biomass is also an important feature of continuous-flow culture systems, as this simulates possible predation effects and ensures that the microbial cells are under constant growth conditions. The effects of grazing of microbial populations has been shown to increase the activity of the microbial population; for example, Barsdate *et al.* (1974) showed this effect for the cycling of phosphorus.

(4) Probably as a consequence of the ability to provide stable conditions during enrichment, coupled with the removal of microbial lysis products, continuous-flow culture systems are ideally suited for the enrichment of microbial communities which are ubiquitous in the environment (Slater, 1978).

(5) Continuous-flow culture systems enable the long term effects of potentially toxic compounds to be investigated. Low concentrations of toxic substances can be applied for long exposure periods, a feature

which is more relevant to the natural situation than the classical lethal dose batch experiments. Similar systems can also be used to study the degradation of these toxic compounds as the fate and the effect are interrelated. There are a number of this type of study using pesticides (Senior *et al.*, 1976; Daughton and Hsieh, 1977; Pritchard *et al.*, 1979), naturally-occurring recalcitrant compounds (Bott *et al.*, 1977, Pawlowsky and Howell, 1973), oils (Maccubbin and Kator, 1979) and metals (Tan, 1980). However, some of the most elegant and informative experiments have been conducted with herbicides and phytoplankton communities. For example, Fisher *et al.* (1974) demonstrated that small concentrations (0.1 $\mu g\ l^{-1}$) of polychlorinated biphenyls (PCB) had no effect either on the growth of pure or mixed batch cultures of a diatom and a green alga, or growing on their own in continuous culture. However, when the two organisms were competing in mixed continuous-flow culture, the presence of PCB greatly affected the competitive ability of the diatom and this effect also occurred in natural phytoplankton communities.

The long-term interaction of mixed microbial populations with recalcitrant compounds is a potentially permissive environment for genetic exchange and evolution, and the realization of this potential has been demonstrated (Senior *et al.*, 1976; Hartman *et al.*, 1979).

(6) In a chemostat a steady state develops when the microorganism's growth rate is equal to the dilution rate and biomass concentration, and environmental parameters remain constant. This state can be considered to be time independent since parameters such as culture conditions and biomass concentrations are constant. Time independence is of prime importance in chemostat studies as a system in this state lends itself to simple mathematical analysis and important kinetic parameters can be determined by measuring steady state kinetics at a variety of dilution rates and substrate concentrations (Pirt, 1975). Mixed microbial communities growing in a chemostat only approximate to a steady state and "stable state" is often used as a more appropriate term for mixed cultures. The mathematical implications of the analysis of mixed community stable states will be discussed later (see pp. 87–91).

Steady- or stable-state microbial growth produces an ideal system for the analysis of elemental biomass balances (e.g., carbon, nitrogen, oxygen, phosphorus, and other elements—Herbert, 1976).

Chemostat enrichments

As previously mentioned, a chemostat is a special type of continuous-flow culture system where the microorganisms are in a perfectly mixed suspension into which sterile medium is fed at a constant rate and the culture is removed at the same rate so that the culture volume remains constant (Pirt, 1975). The composition of the sterile medium is such that all components needed for growth except one, the growth-limiting substrate, are in excess. The growth rate of the culture is then determined by the rate of supply of the rate-limiting nutrient which can be any nutrient required by the organism for growth, e.g. carbon, nitrogen, phosphorus, magnesium, iron, or a trace element. For photoautotrophs light can be made the limiting factor and under these conditions the system is energy limited. Microorganisms growing on recalcitrant carbon compounds may also be energy limited rather than carbon limited.

Although the term chemostat is commonly used to describe a controlled single-stage continuous-flow controlled system (Fig. 1), there

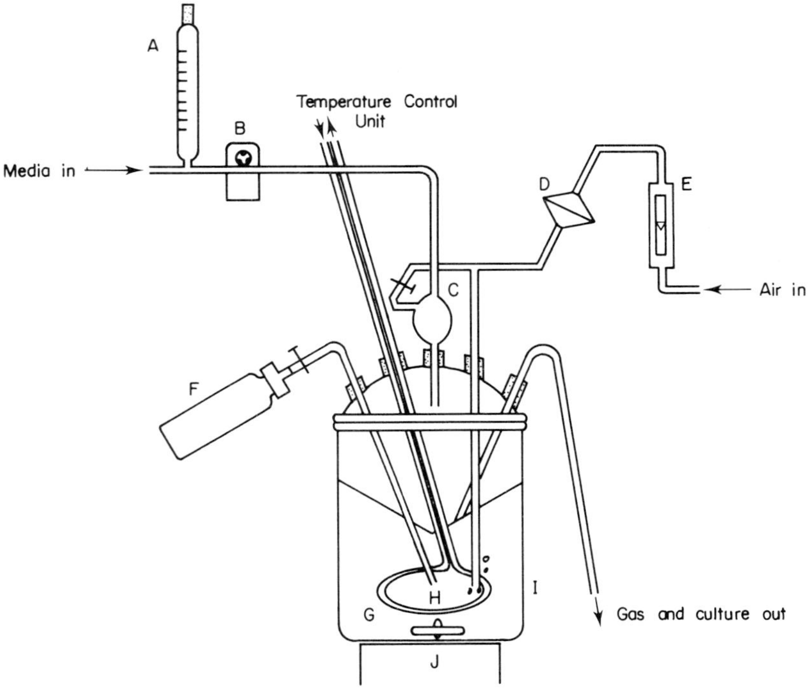

Fig. 1. Diagram of a simple chemostat. A = pipette (connected to medium reservoir) for measuring media flow rate; B = peristaltic pump; C = line-breaker to prevent "backgrowth" of culture; D = in-line air filter; E = rotameter to measure gas flow; F = sampling bottle; G = cooling/heating coil; H = magnetic stirring bar; I = quick-fit reaction vessel; J = magnetic stirrer.

are various modifications (turbidostat and multiple stage systems) and various degrees of culture control (pH, Eh, dissolved oxygen and others). Description of the practical aspects of chemostat design and control can be found in Norris and Ribbons (1970a) and Harder *et al.* (1974). Obviously the more control that is required, the more expensive the chemostat apparatus will be, but simple-flow controlled systems can be made inexpensively using Quickfit apparatus (Fig. 1) which are generally adequate for low biomass cultures. When high biomasses are involved culture control is often required although, as pointed out by Sinclair *et al.* (1971), these control systems are themselves dynamic and their interaction with the growing culture can produce spurious results. These authors point out that such devices can produce oscillations due to system variables in pure cultures and there is little doubt that they can do the same in mixed culture. Therefore, the simple, low biomass systems may have more attributes than were originally envisaged.

As enrichment conditions in a chemostat are constant, the selection of these conditions is crucial since the required organisms may be excluded by using inappropriate enrichment conditions. Microorganisms unable to maintain themselves in the chemostat under the prevailing conditions will be washed out of the system. There is an enormous amount of information concerning growth within chemostats of bacteria originally isolated and purified under batch conditions. But there is relatively little information about the use of chemostats for enrichment and the data which are available are almost exclusively concerned with the enrichment of microorganisms on simple, single carbon substrates. In these conditions microorganisms are enriched under competing conditions and normally, as predicted by Monod kinetics, a single organism dominates (Jannasch, 1967a; Jannasch and Mateles, 1974). The effects of enrichment conditions on the types of microorganisms obtained when complex or multiple substrates are used which tend to favour the development of interacting communities may be very different from those with competing conditions.

Possible mechanisms of enrichment of microbial communities

Due to the lack of information on the effect of chemostat conditions on the enrichment of interacting microbial communities, it may be useful to consider the possible mechanisms by which such communities may be enriched and, therefore, the subsequent section, which discusses the effect of enrichment conditions in general in a chemostat, may prove more valuable. There seem to be two, possibly contradictory, views on enrichment.

(1) Mixed microbial populations develop within a chemostat due to

a unique coming together of the biochemical potentials, present in the microorganisms in the inoculum, which are required to utilize the substrate provided. This could mean that the microbial community which is ultimately enriched in the chemostat, was not present in the environment as a community and is merely an artifact of the enrichment conditions. If this situation exists it is likely that the composition of the enriched community will be very susceptible to small changes in the enrichment conditions and, hence, reproducibility of the enrichment will be poor.

(2) A community is enriched which has some resemblance to a community which is active or potentially active in the environment utilizing similar substrates to those present during the enrichment. It is now realized that most microbial activity in the environment is the result of community activity, whether these are loosely or closely associated. The primary organisms involved in a particular activity have often been recognized for a long time, e.g. *Clostridium* species which utilize cellulose anaerobically, but the importance and effect of other members of the community in general on the process is still to be investigated (Weimer and Zeikus, 1977). It may be that some of the interrelationships are very close and represent a core community primarily responsible for a certain activity and super-imposed upon this core, may be numerous loosely associated microorganisms which aid the overall efficiency of the activity but are not essential for the process to occur. When a community is enriched in a chemostat it may be that the core community is selected together with certain of the loosely associated microorganisms, depending on the enrichment conditions. If this situation is true the core community should be capable of repeatable enrichment and minor changes in the enrichment conditions only affect the nature of the loosely associated organisms. The difficulties often encountered when trying to isolate pure cultures of organisms, particularly anaerobes, may indeed be a reflection of the existence of core communities.

The presence of a core community is obviously more likely if the substrate being utilized is a recalcitrant compound or, at least, difficult to degrade. It may be significant that the initial communities isolated in chemostat enrichments utilized complex compounds, e.g. dalapon (Senior *et al.*, 1976), orcinol (Osman *et al.*, 1976) and parathion (Daughton and Hsieh, 1977). There is little data concerning the repeatability of these enrichments but the seven-membered Dalapon community was repeatedly enriched from four different soils and with two different enrichment conditions (Senior, 1977). It may be that with the application of more environmentally relevant enrichment conditions (see below), microbial communities may be enriched which can utilize less recalcitrant and more ecologically significant substrates. If this is achieved a very powerful tool in microbial ecology will become available and give some indication as to which one of the views outlined above is nearer the truth.

The effect of enrichment conditions on the isolation of microorganisms

Media composition. The type of growth-limiting substrate, other nutrients, and physicochemical conditions used in the enrichment is determined by the activity or type of microorganism (or community) that is to be isolated. As with batch enrichment systems, nutritionally distinct bacteria are capable of repeatable enrichment (Jannasch, 1967). However, as chemostat systems have only been comparatively recently used for enrichment, there are few examples of this type of specific enrichment, although there are many examples of chemostat media for the growth of nutritionally distinct groups of bacteria, e.g., chemolithotrophs (Kuenen *et al.*, 1977; Gottschal *et al.*, 1979), sulphate-reducing and methanogenic bacteria (Cappenberg, 1975), phytoplankton (Fisher *et al.*, 1974) and photolithotrophs (van Gemerden and Jannasch, 1971). Media composition in the few examples of chemostat media for enrichment of specific microorganisms tend to be similar to the media have been found to be suitable for batch enrichment cultures, except that the concentration of growth rate limiting substrates are much lower than used in batch. These are likely to be conditions which are more representative of conditions in the natural environment. Therefore, adaptation of specific media for a particular nutritional group of microorganisms grown in batch culture seems a reasonable starting point for the development of selective chemostat media. However, even if an equivalent batch culture medium does exist it will be important to analyse the conditions prevailing in the environment where the desired organism or community is active, e.g. pH, Eh, DO_2, temperature and other parameters. This information is vital in the development of realistic enrichment conditions, e.g. the ratio of carbon : nitrogen should indicate whether the system should be nitrogen or carbon limited.

Jannasch (1967a) demonstrated for marine bacteria that the effect of different carbon concentrations has a profound effect on the types of microorganism isolated. At a fixed dilution rate ($D = 0.1\ h^{-1}$) with lactate concentrations of 0.1, 1.0 and 10.0 mg l^{-1} in the inflowing medium led to the selection of an *Achromobacter* sp., a *Micrococcus* sp. and a *Spirillum* sp., respectively. In the same paper a combination of different dilution rates and different substrate concentrations in the inflowing medium were used to enrich for bacteria. At low dilution rates and low substrate concentrations, organisms belonging to the genera *Spirillum*, *Achromobacter*, *Vibrio* or *Micrococcus* dominated, whilst high dilution rates and high substrate concentrations favoured the selection of organisms belonging to the genera *Pseudomonas* or *Aerobacter*. Analysis of the growth kinetics of these two groups showed that they were analogous to the autochthonous (low dilution rate, low substrate concentration) and zymogenous (high dilution rate, high substrate concentration) types of organism previously discussed (pp. 49–50). Jannasch found that these results were quite repeatable and competition experiments with pure cultures of the organisms were in agreement with initial isolation results. The autochthonous bacteria dominated at low dilution rates (i.e. low substrate concentrations) and the zymogenous bacteria dominated at high dilution rates (i.e. high substrate concentrations).

Similar results have been found by other workers. For example, Veldkamp and Kuenen (1973) isolated different freshwater organisms under a variety of carbon limiting conditions (lactate, glutamate and methanol; Harder *et al.* (1977)); Mateles and Chian (1969) used a mixture of glucose and butyrate; and Kuenen *et al.* (1977) with different levels of phosphate limitation. Kuenen *et al.* (1977) also investigated competition between two of the bacteria which were isolated, namely the *Spirillum* sp. at low substrate concentration and the *Pseudomonas* sp. at high substrate concentration, in a variety of limiting conditions (lactate, succinate, alanine, asparagine, magnesium, potassium, and ammonium as the limiting substrates). The *Spirillum* sp. out-competed the rod at low concentrations of any of these substrates. Therefore, under a variety of different conditions, at low limiting nutrient concentrations, spiral-shaped organisms were isolated with relatively high surface-to-volume ratios compared with the rod-shaped organisms which were often enriched at higher growth-limiting substrate concentrations. A high surface-to-volume ratio affects the maximum specific growth rate which gives these organisms a selective advantage at the lower substrate concentrations. An interesting point is that chemoorganotrophic bacteria isolated at different concentrations of a growth-limiting organic substrate often show similar growth requirements and it is difficult to find an organic substrate which can be attacked by one organism and not by the other. Veldkamp (1976) suggested that this factor indicated that substrate concentration, as well as substrate type, may play a vital role in species selection and hence species diversity. Competition experiments similar to those just described have been used to discriminate between autochthonous and zymogenous chemolithotrophic organisms under iron limitation (Kuenen *et al.*, 1977), although these bacteria were not enriched in a chemostat. Tilman (1977) showed that, unlike the competition experiments with bacteria previously described, competition between planktonic algae is affected by medium composition. *Asterionella* sp. was dominant under phosphate limitation and *Cyclotella* sp. was dominant under silicate limitation; and both species coexisted when each species was growth limited by a different resource.

The results of Boethling and Alexander (1979) are relevant to workers intending to use chemostat techniques to study degradation of xenobiotic compounds, although this work was conducted in batch culture. It was demonstrated that two herbicides (chlorobenzoate and chloroacetate) were degraded at faster rates at lower initial herbicide concentrations (47 pg ml^{-1}) although for 2,4-dichlorophenoxyacetate and 1-naphthyl *N*-methylcarbamate, the reverse was true, with little degradation taking place at low initial concentrations (2 to 3 ng ml^{-1} or less). These results indicated that the concentrations of xenobiotic compounds being used in laboratory studies ought to be similar to the field concentrations if realistic results are to be obtained. The results also indicate two main problems in studying xenobiotic compounds with chemostats: i.e. the possible toxicity of the xenobiotic compounds to the degrading communities (e.g. chlorobenzoate and chloroacetate) and the existence of threshhold concentrations below which microorganisms cannot metabolize a compound (e.g. 1-naphthyl *N*-methylcarbamate). Toxicity of xenobiotic compounds can be estimated by conventional plate systems containing different concentrations of the xenobiotic compound; however, the final xenobiotic compound concentration used is often a compromise between toxicity and the requirement for a reasonable biomass.

The existence of threshhold concentrations may reflect the need for a certain minimum amount of energy or reducing power required to initiate or sustain the metabolism of the xenobiotic compound. This can be supplied by higher concentrations of the compound itself (e.g. 2,4-dichlorophenoxyacetate) or a separate source of readily assimilable carbon (Huber *et al.*, 1975). More readily degradable analogues of the xenobiotic compound under study have been used either as carrier substrates or the sole carbon sources, in initial enrichments of bacterial communities. Subsequently, the community is gradually adapted to increased concentrations of the xenobiotic compound (Hovarth, 1972; Lovatt *et al.*, 1978; Hartman *et al.*, 1979). However, since suitable analogues are not always available, a non-related organic compound may have to be used as a carrier, the selection of which is important. This was shown by Dimock (1975) where a community developed which was able to utilize the carrier compound, but not the xenobiotic compound.

The effect of carbon or nitrogen limitation on the development of microbial communities can be quite marked. Wardel and Brown (1980) found that, under nitrogen limitation, communities enriched from river water were more diverse than those obtained under carbon limitation and produced extracellular material, which was not the case with carbon-limited cultures. The authors also noted that, with organisms enriched from an estuary, the concentration of sodium chloride used had a marked effect on population density and the dominant organisms present. Dunn *et al.* (1980) have also demonstrated that different concentrations of sodium chloride affected the dominant bacterial species obtained with chemostat enrichment from estuarine sediments. *Acinetobacter* sp. dominated when no sodium chloride was present, but with 0.2 M and 0.4 M sodium chloride enterobacteria (probably a *Klebsiella* sp.) dominated. The type of dominant bacteria were also affected by the type of organic substrate (glycerol or acetate) and whether the culture was aerobic or anaerobic.

The type of growth limitation used during enrichment can also be crucial in obtaining communities able to degrade xenobiotics. Thus, success in isolating Amitrole-degrading and Lontrel-degrading communities depended on these herbicides being used as growth-limiting sources of nitrogen rather than of carbon (Campacci *et al.*, 1977; Lovatt *et al.*, cited in Bull and Brown, 1979).

Megee *et al.* (1972) found that the competition between *Saccharomyces cerevisiae* and *Lactobacillus casei* for glucose was markedly affected by pH and the addition of riboflavin. These results showed clearly how the abiotic environment can influence competition, and therefore selection, during an enrichment process. M. Larkin (unpublished observations) also showed the importance of pH for the enrichment of a carbaryl (1-naphthyl *N*-methylcarbamate)-degrading community. At alkaline pH values the insecticide hydrolysed rapidly to 1-naphthol, methylamine and carbon dioxide and, if the enrichment was conducted at these pH values, only organisms able to degrade the hydrolysis products were obtained. However, if the enrichment was conducted at pH 5, where carbaryl is quite stable, a community able to degrade the insecticide itself was obtained.

There are fewer problems with basic media design for photoautotrophs as light is a universal energy source. However, light intensity is an important factor in determining the outcome of an enrichment (Bader *et al.*, 1976b) and the type of limitation is also important (Tilman, 1977).

An interesting problem arises when mixed microbial communities are enriched, since the type of growth-limiting conditions may not be the same for all the members of the community. For example, Wilkinson *et al.* (1974) suggested that, even though as a whole their methane-utilizing community was oxygen limited, the *Hyphomicrobium* sp. in the culture was effectively carbon limited, as it used nitrate in the medium as an electron acceptor. Hence, the occurrence of stable microbial communities may provide a unique system for the enrichment of micro-organisms under a variety of growth-limiting conditions, a feature which is probably common in natural environments.

Multiple substrates. Chemostats are usually operated with a single constituent of the culture medium present at a concentration which limits growth. This growth-limiting compound is more often than not a carbon substrate. However, in natural environments the growth of micro-organisms occurs in the presence of a diversity of mixed substrates and, therefore, it is remarkable that, while continuous-flow culture techniques have been encouraged as model systems (Veldkamp, 1976), few studies have focused on mixed substrates. Mateles *et al.* (1967), using mixed, pure cultures of bacteria grown in a carbon-limited chemostat, demonstrated that, at low dilution rates, glucose and fructose were simultaneously utilized but at higher dilution rates one substrate was preferred to the other, a situation which is analogous to diauxic growth. Later, Mateles and Chian (1969) showed a similar behaviour with mixed populations enriched from polluted river water, including a change in the dominant population when the dilution rates were changed. The natural population excreted acetate during growth, and triphasic growth was observed. Glucose was used first, followed by lactose and, finally, the excreted acetate. Similar results have been shown by Edwards (1969) and other workers (see Bull and Brown (1979) and Harder and Dijkhuizen (1976) who have reviewed the subject), which suggests that simultaneous use of multiple compounds in nutritionally poor natural environments (as simulated by low dilution rate chemostats) is a general phenomenon. It has also been observed that for most organisms, the substrate which supports the highest growth rate is preferentially utilized from a mixture.

Taylor and Williams (1975) analysed theoretical aspects of competition between bacterial species and concluded that competing species cannot coexist unless the number of growth-limiting substrates is equal to, or greater than, the number of different bacterial species. Thus, the use of a mixed substrate feed might allow the development of a heterogeneous, interacting community which is similar to the communities that occur in the natural environment. These authors did not consider the possibility of achieving a similar, but more natural, situation by the degradation of a complex substrate (e.g. cellulose or a xenobiotic compound) producing a variety of secondary metabolites/substrates, or the possibility of cell lysis or leakage compounds providing the required diversity of substrates to produce a heterogeneous community (Mateles and Chian, 1969).

However, even if mixed substrates are used, the remainder of the medium may not be appropriate for the selection of naturally occurring microbial communities. An obvious solution to this problem is to use natural water which has been filter-sterilized and which is modified by the addition of the desired substrates (Jannasch, 1967b; Veldkamp and Kuenen, 1973; Martin and Bianchi, 1980). There may be problems with this solution due to the presence of inhibitory compounds in the natural water supply and the concentration of nutrients in the supply varying with time. The findings of Jannasch (1967b) are also relevant in this context. He suggested that sea water ought to be considered a sub-optimal medium for microbial growth (due to the requirement of a minimum biomass concentration to maintain bacterial growth in sea water media, within the chemostat) and this may be true for other natural waters.

Temperature. The potential of the chemostat to select on the basis of differences in the response to temperature has been illustrated by Harder and Veldkamp (1971) with bacteria isolated from the North Sea. Competition was studied between a facultatively psychrophilic *Pseudomonas* sp. (optimum growth temperature = 30°C) and an obligately psychrophilic *Spirillum* sp. (optimum growth temperature = 14°C) with respect to the effect of two environmental parameters: growth-limiting substrate (lactate) concentration and temperature. At temperature extremes of 16°C and −2°C, one of the organisms grew faster at all the lactate concentrations studied: at the higher temperature the facultative psychrophile dominated, whilst the lower temperature favoured the selection of the obligate psychrophile. These and other experiments are discussed in full by Veldkamp and Jannasch (1972).

Other important factors. There are several problems arising from the traditional methods of using chemostats when the enrichment of natural communities is desired. The usual practice is to inoculate the chemostat vessel, containing its working volume of growth medium, and allow the organism to grow as a batch culture before establishing the required medium flow. This is a reasonable procedure with pure cultures since, at the end of the batch growth period, the biomass is sufficiently high to ensure that washout of the culture does not take place when the system is placed on flow. However, if the same procedure is used for the enrichment of mixed communities, initial selection is based on batch enrichment (with all its associated problems) and, even when the system is placed on flow, only those organisms which have been selected by the period of batch growth will grow, possibly producing erroneous results. However, even if the chemostat is placed directly on flow, the microorganisms in the inoculum may experience substrate-accelerated death (Postgate and Hunter, 1964), as the inoculum when taken from the environment was probably growing in a nutrient limiting situation. This problem can be solved simply for water sample inocula by using an inoculum volume which is equivalent to the working volume of the chemostat. Obviously this is not possible with soil or sediments and still maintain chemostat enrichment characteristics. Brown *et al.* (1978) suggested suspending the inoculum in a non-nutrient-containing solution, and then adding the growth medium at a slow rate. Another solution might be repeated reinoculation of the chemostat over a length of time, to ensure that those organisms which should predominate, under the conditions imposed by the chemostat, do so. Harrison and Wren (1976) described a chemostat that allowed multiple soil samples to be infused directly into the chemostat.

Inoculum size may provide problems when the chemostat is used to enrich algal communities, as demonstrated by Bader *et al.* (1976b). Algal cultures growing at high densities, and hence experiencing low light intensities on a per cell basis may suffer cellular damage due to light shock if the cultures are diluted more than a hundredfold when introduced into the chemostat and growth could subsequently be impaired. The intensity of light to which the algae were previously subjected will also influence the amount of dilution causing light shock. The authors suggested that this phenomena of light shock could explain the observed discrepancies between the mathematical models of algae being grazed by cilliates and the actual experimental results. The findings of Jannasch (1967b), suggesting that sea water is a sub-optimal medium for the growth of microorganisms, may indicate that, when natural water is used as chemostat media, the inoculation must produce a certain initial, critical biomass concentration if wash-out is not to occur.

This discussion of the variety of enrichment conditions obtainable within chemostat systems, may indicate that batch enrichment may have no significant advantages that cannot be produced in a chemostat;

for example, when operating the chemostat at high substrate concentrations and low dilution rates to enable the effects of inhibitory substrates, and possibly inhibitory metabolites, to be studied, hence simulating one of the features suggested as an advantage of batch systems. However, this is not the case: the chemostat cannot reproduce exactly the conditions prevailing during batch enrichment. Batch culture presents a spectrum of physiological states arranged in a temporal order, and the growth of organisms in such a system is influenced by the past environment, i.e. by the history of the culture. The main feature of a chemostat is its homogeneous, time independent state, where the growth rate is described by dilution rate alone, the past environment is always the same, and, therefore, the culture has no history (Powell *et al.*, 1967).

Homogeneity. The homogeneous state of a chemostat is essential if the normal features and properties of chemostat growth are to be directly applied and the application of simple kinetic analysis is to be meaningful. Therefore, great efforts are normally made to eliminate any heterogeneity from chemostat systems, such as, wall growth, imperfect mixing, pellet formation, and other factors. However, most natural environments are heterogeneous and it may be that when chemostats are used for enrichment, the presence of heterogeneity within the system is vital for the realistic enrichment of microbial communities. In the environment, a large proportion of bacteria may be associated with surfaces (e.g. clay and detrital particles, man-made structures, other micro or macroorganisms and surface films) which may provide a spatial permanence, which is vital for their survival. Attached bacterial populations are particularly significant in low nutrient waters, where the numbers of suspended organisms are generally quite low (Fletcher, 1979). Hence, when these microorganisms are introduced into a homogeneous system they may not be able to maintain themselves and so become washed out. Wardell and Brown (1980) demonstrated that glass slides placed in chemostats during enrichments showed a diverse bacterial population which often forms microcolonies and differs in terms of the predominant bacterial types, to those which are present in the liquid phase of culture. Atkinson and Fowler (1974) noted that thick films arise when mixed populations of microorganisms are involved, and so the lack of available surfaces within a conventional, homogeneous chemostat system may inhibit the development of mixed microbial communities.

Surfaces can also provide spatial arrangement which may play an important role in the development of microbial communities, through the effects of the metabolic activity of closely associated microorganisms over a prolonged period. This may provide the bacteria (or microcolony) with an opportunity to respond to the adjacent metabolic activity and develop a complimentary metabolism. The spatial organization of microorganisms is widespread in nature and may play a crucial role in controlling microbial activity. Often microorganisms are ordered along diffusion gradients, as typified by the classic Winogradsky column experiment (Winogradsky, 1949) where a variety of bacteria become ordered along a redox gradient. Examples of spatial order in the environment together with a discussion of its significance in terms of microbial activity can be found in the recent review by Wimpenny (1981).

Another form of heterogeneity experienced by microbial communities in the natural environment are due to environmental variations, such as temperature, salinity, pH, dissolved oxygen, and carbon concentrations. There may be regular (e.g. salinity and temperature changes in an estuary) or irregular (e.g. point source pollution loading). These environmental fluctuations obviously can have a strong selective effect in the environment and may play a vital role in controlling the activity of microbial communities. However, these fluctuations are usually ignored in current chemostat research.

The use of fluctuating enrichment conditions and heterogeneous phases within the chemostat may enable more diverse microbial populations to be maintained within the chemostat as well as enabling the use of more environmentally relevant enrichment conditions. This increased microbial diversity provides a more permissive environment for genetic exchange which could lead to the development of important, new microbial activities (Slater and Godwin, 1980). Increased heterogeneity within the chemostat, however, produces a more complex growth system to analyse (this will be discussed later, p. 88) and without the ideal chemostat characteristics. However, the general advantages of a continuous-flow culture system (previously outlined) should still be applicable.

Heterogeneity within continuous culture systems. The importance of surfaces and their associated films have been recognized by biochemical engineers for a long time, and many commercial processes are dominated by the kinetics of microbial biomass present in the form of films, e.g. trickling filter, rotating disc, "quick" vinegar process, bacterial leaching of ores (Atkinson and Fowler, 1974). Therefore, the different types of fermenter systems used by biochemical engineers may provide a rich source of ideas for microbiologists interested in heterogeneous enrichment. Simple methods of increasing heterogeneity within traditional fermenter systems will be discussed first and then more complex systems.

Surfaces. Jannasch and Pritchard (1972) described a series of experiments in which clay particles were continuously added with the medium input to a chemostat. The growth-limiting substrate was valeric acid and enrichment from natural sea water was found to be significantly influenced by the presence of the particles, although these were not available to microbial attack. A large, non-motile rod dominated the enrichment when clay (kaolinite) particles were present, but in the absence of these particles a small motile rod dominated the culture. These organisms were isolated as pure cultures and both inoculated into chemostats with and without particles. The results of these competition experiments confirmed the selective advantage of the large, motile rod in the presence of particles and, conversely, the small, motile rod in the absence of particles. The experiment also showed that the presence of particles strongly influenced the mode of valeric acid transformation by the large motile rod and decreased the efficiency of transformation by the small motile rod. Under batch conditions it was demonstrated that bacterial oxygen uptake was increased by the presence of particles at low carbon concentrations but not at higher carbon concentrations. This threshold effect was also dependent on the nature of and the concentration of the particulate matter and may have been due to the concentration of nutrients at the surface of particles, making them more readily available to microorganisms. Throughout the experiments no morphological mechanism for attachment was detected and, hence, it was suggested that attachment may have been due to the formation of a slime or some other excretory material. Similar results using chemostats have been described by Maigetter and Pfister (1975).

In earlier work, Jannasch (1960) only observed denitrification by *Pseudomonas stutzeri* in the presence of high dissolved oxygen concentrations when inert particulate matter was present. The particles generated heterogeneous conditions by the formation of bacterial clumps and increased uptake of oxygen within these aggregations led to the formation of anaerobic pockets. Jørgensen (1977) demonstrated a similar situation with anaerobic sulphate reduction taking place within detrital particles suspended in aerated sea water. In the natural environment there are many situations where aerobic and anaerobic conditions prevail within very close physical dimensions (e.g. at the sediment–water interface) and it may be that this marked heterogeneity plays an important part in the high activity often associated with such environments. The degradation of some xenobiotic compounds may even require fluctuating redox conditions. For example DDT was converted, under anaerobic conditions, to chlorinated diphenylethanes and dichlorobenzophenone by an *Alcaligenes* sp., by freshwater sediments and in sewage sludge communities (Pfaender and Alexander, 1972). Subsequent aeration of *Alcaligenes* sp. cultures led to ring cleavage and the appearance of *p*-chlorophenylacetic acid, which may be used by *Arthrobacter* spp. Hence, traditional homogeneous culture conditions (whether aerobic or anaerobic) may provide very misleading results for the degradation of some xenobiotic compounds. Williams (1977) has reviewed the importance of anaerobic microorganisms in the metabolism of pesticides.

The presence of surfaces may also be important in determining the outcome of microbial predator–prey relationships. Bader *et al.* (1976b) found in chemostat culture that a cilliate, *Colopoda steinii*, preying on a cyanobacterium, *Anacystis nidulans*, encysted when the numbers of the cyanobacteria declined, and the cysts attached to the walls of the vessel. Algal cells then began to stick to the attached cysts. This was, in turn, followed by an increase of both populations and the development of steady state conditions rather than the typical predator-prey oscillations, perhaps reflecting the continuous reinoculation of the liquid phase with both populations from the wall growth.

Discussion of the general significance of microbial films in fermenters can be found in reviews by Atkinson and Fowler (1974), and Atkinson and Daoud (1976).

Multiple stage systems. Continuous-flow culture systems can be linked together in series with each vessel (except the first) receiving a constant inoculum from the previous vessel. The conditions of each vessel can be controlled independently thereby providing a variety of conditions for microbial activity to take place. For example, one vessel could be anaerobic and another aerobic, thus simulating the conditions outlined above as possible prerequisites for DDT degradation. These systems also enable organisms with linked metabolism (e.g. sulphate-reducing bacteria and sulphide-oxidizing bacteria) but which require different growth conditions to be studied. Often the flexibility of such a system is only limited by the ingenuity of the experimenter and the number of vessels.

Multi-stage systems can also be operated with similar environmental conditions in each vessel. This mode of operation is useful with toxic substrates, as the toxic compound can be broken down in the first stage, releasing metabolites which can be utilized in the subsequent vessels by organisms that may be sensitive to the primary substrate and, hence, would not be enriched in a single-stage system.

Mateles and Chian (1969) used a two-stage system to study the metabolism of a mixture of lactose and glucose by bacteria enriched from river water. In the first stage, at dilution rates below 0.75 h^{-1}, all the glucose was utilized; hence the medium supplied to the second stage consisted of organisms and lactose. Increased lactose utilization was observed in stage two compared with stage one, probably due to the removal of glucose inhibition.

Daughton and Hsieh (1976, 1977) used a two-stage chemostat system to investigate the degradation of parathion by a microbial community. The core of the community consisted of *Pseudomonas stutzeri*, capable of hydrolysing parathion to *p*-nitrophenol (PNP) and diethylthiophosphate (DETP), and *Pseudomonas aeuruginosa* which utilized *p*-nitrophenol as a growth substrate. After more than 2 years' growth, the culture remained a multispecies association suggesting a symbiotic relationship between the two organisms. DETP however was not metabolized by the community and its increased concentration in the second stage exerted a toxic effect, resulting in a lower biomass in stage two compared to stage one. By comparison, operation of the chemostat on PNP resulted in similar biomass concentrations in both stages one and two.

As well as different metabolic activities being distributed throughout the system, the bacteria associated with each function may also be segregated within the system, thus providing a very useful analytical system as well as an enrichment technique (this is discussed later in the chapter).

Fluctuating environmental conditions. In the natural environment the irregular availability of growth-limiting substrates is the rule rather than the exception (Veldkamp, 1977) and, in the few studies of fluctuating growth-limiting substrate concentrations within chemostats, fluctuating conditions have been shown to have profound effects on microorganisms. Neijssel and Tempest (1976), for example, observed that the intermittent addition of a concentrated glucose solution to a glucose-limited culture of *Aerobacter aerogenes* manifested itself as an increased maintenance energy requirement. The loss of growth efficiency was greater at low growth rates than at high growth rates.

Chemical and physical parameters also fluctuate in the environment (e.g., temperature, salinity, pH and Eh) and enrichment under constant environmental conditions may lose many of the microbes actually active in the environment.

Elaborations on the basic continuous culture apparatus. Plug-flow culture is a means of simulating a batch culture in an open system and combines the advantages of both continuous-flow and batch culture. Ideally, the inoculum and the medium are mixed on entry into the system and the culture flows at a constant velocity through the fermenter without mixing. Thus, the batch culture phases, which are normally separated only in time, are separated spatially along the length of plug-flow culture.

On the laboratory scale it has proved difficult to realize true plug-flow culture, mainly due to laminar flow in tubular vessels causing both a gradient of culture across the tube and wall growth. A good approximation, however, can be obtained by using a series of chemostats (as discussed above).

Another form of series chemostats is the gradostat (Wimpenny, 1981) where the flow within the series is made bidirectional. Such a system was first devised by Cooper and Copeland (1973) in order to simulate estuarine conditions (Trinity Bay) in the laboratory. The system consisted of five plastic containers (nine-gallon) each gently stirred and connected to each other with glass tubing couplers. Fresh water was run into the first vessel and there was a single overflow in the last of the five vessels. There was input of salt water into the fifth vessel resulting in the formation of a salinity gradient in the system. Cooper and Copeland (1973) found that the organisms present in the model were qualitatively, but not quantitatively, the same as the populations present in Trinity Bay estuary.

The Wimpenny gradostat (Wimpenny, 1981) is based on a series of homogeneous chemostats with pumps used to transfer culture from one vessel to the next so that the desired diffusion profile can be programmed for the system by choosing appropriate flow rates between vessels. The system has been used to examine the growth of *Paracoccus denitrificans* in opposing gradients of nitrate and succinate, and the response of *Escherichia coli* to opposing gradients of oxygen plus nitrate versus glucose. The gradostat has also been used to study the response of a mixed culture of aerobic and anaerobic bacteria to gradients of oxygen versus glucose in a rich nutrient medium. The two physiological types are segregated in the expected directions. In the future the system may also be modified to introduce some of the features previously discussed, such as heterogeneous phases within the different fermenter vessels, e.g. particles; providing each vessel with a separate input (simulating point source pollution); and using fluctuating growth conditions. The number of different environments possible is only limited by the number of vessels.

Wimpenny (1981) commented that the separation of organisms from solutes is typical of a diffusion system where the organisms might be fixed in one position, e.g. Winogradsky's column, and suggested that this could easily be simulated in the gradostat by separating the vessels horizontally by a permeable filter, with the microorganisms in the lower section. A similar system was proposed by McClure (1973) as a method of treating xenobiotic compounds in waste water, although this was a single-stage device. The gradostat can also be used to examine single-ended systems, such as a film, sediment, or dental plaque, where nutrients and products enter or leave from the same vessel: this can be achieved by eliminating the source and sink at one end of the gradostat.

The importance of surfaces and particles within continuous-flow culture systems has already been mentioned (pp. 66–68). However, there are practical problems in maintaining particles within conventional continuous culture vessels and controlling the thickness of the microbial film (Atkinson and Davis, 1972). This has led to several different fermenter systems being designed specifically for use with particles and surfaces. One such system is the completely mixed microbial film fermenter (Atkinson and Davis, 1972; Atkinson and Knights, 1975). This system relied on the upward movement of medium in a column to support the particles. The culture medium, once through the column, entered a small reservoir (with facilities for aeration, temperature and pH control, and medium input and output) and was recycled to the base of the column. The frequent particle–particle contact in the column caused the biological film thickness to attain a dynamic steady state between growth and attrition of the microbial biomass. Particles used in the system were initially glass beads, but this was changed to stainless steel balls and, finally, to carefully engineered stainless steel filaments. Biomass, in excess of the equilibrium film thickness, was recirculated as a floc, so that the fermenter performance depended on the contributions of both film and flocs. In operation, the recirculation ratio is large and so the conditions throughout the fermenter were essentially uniform. Atkinson and his colleagues have worked largely on undefined mixed cultures, without the use of aseptic conditions, as their main consideration has been with the performance and growth characteristics of the system.

A fluidized bed system similar to Atkinson's has been used for treatment of effluents from coal conversion processes (specifically for the removal of phenol from such effluents), the enzymatic production of hydrogen, microbial denitrification, and general waste water treatment (Scott *et al.*, 1975; Hancher *et al.*, 1978; Genung *et al.*, 1979). These authors used tapered columns which allowed a wide range of flow rates to be used without loss of bed material, since the fluid velocity decreases with reactor height. These systems have been scaled down (R. J. Parkes, S. F. Minney, and A. T. Bull, unpublished experiments) for more convenient laboratory use (Fig. 2) and utilized to study the degradation of xenobiotic compounds by mixed microbial communities attached to clay particles. The method of Spycher and Young (1977) for obtaining minimally altered clay-sized soil and organic-mineral particles, may allow fluidized bed systems to be run with a variety of different clay fractions.

Floc and film formation is strongly influenced by both temperature and pH (Stanley and Rose, 1967; Atkinson and Fowler, 1974) and varying these parameters within a fluidized bed system may enable the effect of the degree of attachment on a particular activity to be investigated as well as providing a means of harvesting attached growth for analysis.

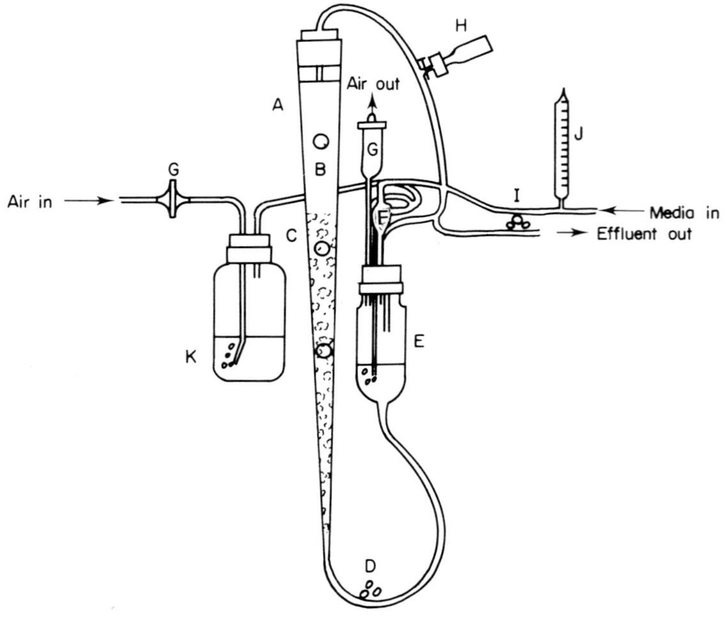

Fig. 2. Fluidized bed reactor system. A = tapered fluidized bed reactor; B = suba seal sampling ports; C = suspended particles; D = recycle pump; E = aeration and recycle vessel; F = line breaker; G = air filter; H = sampling vessel; I = input and output pumps; J = pipette; K = air saturator.

Effective growth models for these fluidized bed systems have now been developed (Atkinson and Fowler, 1974; Atkinson and Daoud, 1976; Scott and Hancher, 1976) and these systems are now open to exploitation under defined culture conditions and have great potential as investigative tools in microbial ecology.

Microbial film development is allowed in other types of continuous-flow culture apparatus that do not rely on the fluidized bed principle and, hence, the control of the film thickness by attrition of the suspended particles. Atkinson *et al.* (1968) described a fixed film reactor which used an inclined, roughened glass surface down which nutrient solutions flowed. Control of the film thickness was by mechanical scraping which made its use under aseptic conditions difficult. A laboratory disc fermenter, such as that described by Blain *et al.* (1979), in which microbial film growth occurred on rotating discs, is similar to the rotating disc fermenter used for biological treatment of industrial and domestic effluent. Control of biofilm thickness was by excess growth sloughing from discs and the system can be operated and sampled under aseptic conditions. Ashby and Bull (1977) described a column fermenter which was used with particles kept in suspension by an airlift principle. The system did not require a circulating pump and hence is probably cheaper to construct than systems using the fluidized bed principle. The amount of shearing within the system may also be less than in a fluidized bed system, enabling it to model environmental situations where only a limited degree of turbulence exists (e.g. particles suspended in the water column, rather than sediments being constantly fluidized by tidal action). The system could also be used without particles relying on microbial floc formation for heterogeneity as in commercial tower fermenters.

The main characteristic of soil is the downward flow of water, and the solutes dissolved in it, which results in the vertical stratification of microorganisms and their metabolites (Bazin *et al.*, 1976). These characteristics are reflected in soil perfusion systems where a tube is packed with soil or some other porous medium and down which nutrient solution is passed. Lees and Quastel (1946) originally adapted the perfusion techniques employed in experiments on isolated animal organs for use with soil. Their method consisted of recycling nutrient solution through a container of soil and sampling the solution at time intervals in order to detect any changes that may have taken place. The system as a whole must be regarded as closed but the soil itself can be thought of as being open to the external nutrient supply. A separate input and output can be incorporated into the recycling system (similar to the fluidized bed system described above) or the system can be used as a single passage flow-through column (Macura, 1961) giving the system as a whole open characteristics. The chief disadvantage of this method is that microbial biomass or cell numbers cannot be measured without disturbing the system, although this problem can be overcome to some extent by inserting ports down the length of the tube. With this modification changes in the medium as a function of depth down the column as well as time can be monitored. Hill and Arnold (1978) described a soil apparatus which was used for monitoring pesticide degradation and enabled the use of radiolabelled compounds and the trapping of volatile compounds and carbon dioxide. The system, however, did not allow the downward flow of water which, according to Bazin *et al.* (1976), is one of the main characteristics of soil.

A low cost percolator based on a milk bottle and efficiently operated in large numbers has been described by Sharp and Taylor (1969). Bazin *et al.* (1976) described the use of a simplified perfusion system using glass beads in addition to soil or instead of it. Such modifications, of course, make the system more artificial but provide the advantage of eliminating much of the complex behaviour caused by adsorption of ions and can be used to investigate relatively simple kinetic models. Glass bead percolating systems have also been used with defined mixed cultures in order to simplify the system even further (Bazin and Saunders, 1973). Glass beads (coated with hydroxylapatite) have also been used in the continuous-flow culture of mixed oral flora (Sudo, 1977).

Similar systems to those designed for soil studies have been used for sediment investigations. Pritchard *et al.* (1979) described a system (known as the model eco-core) in which an intact sediment core and its associated water column was collected from the field in a sterile glass tube, and was incubated in the laboratory and used to study the degradation of xenobiotic compounds. The system was aerated and the effluent gases passed through an alkali trap and polystyrene bead resin to trap radiolabelled carbon dioxide and volatile products, respectively. They also described a continuous-flow culture system where artificial sea water and a pesticide were added at a constant rate to the sediment water system. In this case the sediment was not collected intact but "structured" using settled sediment which consisted of a 3 to 4 cm layer of detritus over an equally thick layer of sand. In both of these systems, although temperature was controlled (24°C), no attempt was made to incubate the models under environmental conditions (e.g. appropriate temperature, light, salinity fluctuations, and other factors); I have therefore termed these "system models" although the authors used the term "microcosms". A similar continuous-flow culture sediment system has been recently described by Tan (1980) and has been used to study the effect of exposure to lead on sea water and sediment bacteria. Intact cores have also been used under continuous-flow conditions (R. J. Parkes, S. F. Minney and A. T. Bull, unpublished observations) to study the degradation of xenobiotic compounds, including a facility for sampling the water moving down the core and the sediment at different depths (via a series of ports on the side of the core holder).

2.2.3 Microcosms

In this review, microcosms have been defined as laboratory systems which attempt to simulate as far as possible the conditions prevailing in the environment, or the part of the environment, under study. Such systems can be either open or closed but, due to the complexity and heterogeneity of most of the systems, even when there is not an actual input and output, the interchange between different compartments of the system (e.g. the sediment–water interface is constantly turned over by any benthic organisms incorporated into the microcosm) must provide some open characteristics. These complex systems have the potential to assess the rates and routes of transformation of both natural and man-made compounds, the bioaccumulation of such compounds through a food chain and the long term effects of these compounds in the environment (e.g. toxicity, decrease in ecosystem stability, and diversity). Furthermore, the ability to use radiolabelled compounds within microcosms contributes greatly to their usefulness. As a thorough review of the large variety of microcosms used is outside the scope of this chapter, only examples of microcosms used to simulate different environments are discussed in order to indicate the wide applicability of the method.

Metcalf *et al.* (1971) developed a microcosm with a terrestrial–aquatic interface and a seven-element food chain. The basic unit was a glass aquarium (10 × 12 × 20 in) containing 15 kg washed, white quartz sand moulded into a sloping soil–air–water interface. Standard reference water (7 l) was added, as well as terrestrial plants, aquatic algae and plankton, snails, cladocera, mosquitoes and fish. The use of pure sand and water was more reproducible than soil and enabled the extraction and determination of radiolabelled compounds to be more efficient. However, it made the conditions within the microcosm quite different from those in the natural environment. This led Hill and Arnold (1978) to comment that, while microcosms may be valuable as indicators of potential pesticide transfer and accumulation along food chains, the technique provides little useful information on either the physical movement or the transformation of the chemical in the environment. However, Metcalf reported that the results obtained with DDT after one month in the food chains of the model system showed a remarkable similarity to the pattern observed after many years in nature. Similar results have now been shown for other insecticides, such as methoxychlor, fonofos and aldrin (Cole *et al.*, 1976). Other microcosms have incorporated soil (Isensee *et al.*, 1976); have been based on a percolating system where the effluent could be collected for analysis (Lichtenstein *et al.*, 1974); and Beall *et al.* (1976) described an enclosed "agro-ecosystem", in which a number of plants were grown in soil and the pesticide degradation studied under differing regimes of rainfall, wind velocity, and light.

Bond *et al.* (1976) described a sophisticated coniferous forest soil–litter microcosm system, in which oxygen, temperature, and humidity were kept constant, and oxygen generation, carbon dioxide, and heat evolution rates were monitored. An adjustment period of 7 to 10 days after microcosm preparation was necessary in order to approach relatively constant production rates. It was considered that this unstable period, which had high oxygen consumption and carbon dioxide evolution rates and high numbers of microbes and nematodes, was due to the growth of opportunist (zymogenous) organisms utilizing the altered nutrient supply and was not indicative of normal environmental situations. The zymogenous organisms declined after approximately 2 to 3 weeks. Bourquin (1977) described an artificial salt marsh environment similar to the model eco-cores previously described (p. 74) although the xenobiotic compound under study was applied at field rates (and 10× field rate) and the system was incubated under environmentally relevant lighting conditions. The results with malathion underlined the potential importance of cometabolism in xenobiotic compound degradation since only 10% of the malathion degraders were able to use the compound as a sole carbon source.

Jørgenson and Fenchel (1974) described a marine sediment microcosm which consisted of sediment (sand and eelgrass) in a plexiglass aquarium over which aerated sea water was constantly passed. At intervals the circulating sea water was connected to an external system and renewed with sea water pumped directly from the sea. The rates of sulphate reduction in the microcosm were similar during the first 2 to 3 months to those measured by other workers for coastal sediments. After this time, the rates of sulphate reduction decreased as the system slowly oxidized (after 7 months the rates of sulphate reduction were reduced by a third). Bott *et al.* (1977) described three different microcosms for studying the degradation of natural leaf litter and a test chemical, nitrilotriacetic acid (NTA), in freshwater streams. The microcosm used for studying leaf degradation was a plexiglass tube, closed at the ends with facilities for sampling, trapping carbon dioxide and apertures for thermometers. The microcosm for NTA degradation was a plexiglass tray which held microscope cover glasses. Stream water was passed once through each of the systems to establish naturally occurring periphytic microorganisms. At the start of the experiment either leaves or NTA were added and the systems converted to a recycling system to facilitate the use of labelled substrate and to simplify the methodological problems. Obviously these systems could be placed in the natural environment for initial seeding and, ultimately, the experiment itself could be conducted in the environment in order to obtain more realistic results (true experiments *in situ* are discussed by Fry pp. 127–128). The third microcosm was an artificial stream which permitted different current velocities to be used. The bottom substrate was sand, gravel and small rocks and silt where it had been deposited in pool areas. The sides of the system were of similar material. Natural aquatic communities were established in the streams over a 2-year period, including seasonal changes in organisms. By including different features in the replicate microcosms the importance of certain features, such as leaching, microbes, and invertebrates, to the degrading process were quantified.

Some important considerations in the design of microcosms

Composition of microcosms. The types and concentrations of organisms present in microcosms affect the processes occurring in the system, and this is closely associated with the choice of soil substratum used. In the Metcalf system the material chosen, for experimental expediency, was washed quartz sand. Giddings and Eddlemon (1977) studying arsenic transport within microcosms compared the use of sand and natural sediments and their results highlighted the danger of using artificial substrata. The lake sediment contributed to a significantly greater algal productivity than sand as measured in terms of biomass, pH, dissolved oxygen, dissolved nutrients, and chlorophyll *a*. This was due to the release of nutrients into the water column from the sediment. Furthermore, most of the arsenic was bound to the substrate in the sediment microcosms but remained in the water of the sand system. This resulted in a much greater arsenic accumulation in snails in the sediment system compared with the sand system (arsenic accumulation was directly proportional to substrate arsenic concentrations). It was concluded that substrata for microcosms ought to be selected from the particular environment to be studied or should be as similar as possible. Similar conclusions were reached by Kaplan (1977). Giddings and Eddlemon (1977) also commented that, due to the above considerations, the use of a standard test microcosm for fate and effect studies of pollutants will probably lead to results which will not be applicable to many environmental situations. It is likely that other components of a microcosm could have similar effects. For example, Bond *et al.* (1976) demonstrated that an increase in the water content of a terrestrial microcosm caused a direct increase in oxygen uptake, carbon dioxide evolution and bacterial density but a decrease in the number of fungi and actinomycetes.

Stabilization period. Most workers who use microcosms recognize the need for a stabilization period before the system provides reliable data. There seems, however, to be considerable disagreement with respect to the length of this stabilization period. Pritchard *et al.* (1979) suggested, with their batch eco-cores, that microbial activity decreased rapidly when the system was incubated in the laboratory (significant changes in 7 d). They suggested this was due to the rapid utilization of growth factors by the resident microbial community, after which these factors had to be supplied from within the system at much lower rates. This possibly indicated a change from a zymogenous to an autochthonous community, as postulated by Bond *et al.* (1976). Pritchard *et al.* (1979), therefore, argued that biodegradation studied during long periods of laboratory incubation, reflect the potential characteristics of the laboratory system and not the characteristics of the original natural environment. Thus the most informative results can only be generated with a fresh system. Conversely, Bond *et al.* (1976) commented that the stabilization period was probably due to zymogenous organisms and was not characteristic of the natural situation and they suggested that data from microcosms after the stabilization period, when autochthonous organisms dominated, would be more environmentally relevant. The optimum period for microcosm use obviously varies for different systems and has to be individually investigated, but if the different types of populations (zymogenous or autochthonous) are to be studied this should be made clear in order to avoid considerable confusion and discrepancy between results.

Microcosm size. Giddings and Eddlemon (1977) compared freshwater microcosms of two different sizes (7 and 70 l). Their results indicated that size affected the performance of the microcosms but this effect was much less than the effects of different substrata types (discussed above). The authors suggested that the observed size effects could be attributed to different ratios of substrata surface area to water volume. Due to the shorter time required for stabilization, the smaller likelihood of perturbations resulting from sampling activities and the greater probability of colonization by important groups, such as zooplankton in sufficient numbers, it was argued that the larger size microcosm ought to be used where possible.

Further discussion of the problems associated with the use of microcosms can be found in Matsumura (1979).

2.2.4 *Applicability of Results from Models and Microcosms to the Environment*

Throughout the preceding discussion of microcosms and model systems, their relevance, or otherwise, to the natural environment has been stated. It should therefore be apparent that there is considerable applicability of these systems to environmental problems, provided that the systems are functionally similar to the environment (models) or the dominant conditions are similar to those in the environment (microcosms) under study. Microcosms also exhibit basic properties which are common to all ecosystems (Gorden *et al.*, 1969) and have good reproducibility (Beyers, 1963; Abbot, 1966; Bond *et al.*, 1976; Isensee, 1976). Model systems have also been shown to be reproducible (Jannasch, 1967a; Senior, 1977). Both model systems and microcosms are also capable of predicting the effect of stresses on the environment, e.g. cadmium (Bond *et al.*, 1976), arsenic (Giddings and Eddlemon, 1977) and polychlorinated biphenyls (Fisher *et al.*, 1974). Falco *et al.* (1977) found that different model systems (flowing stream model, batch culture, and chemostat) produced comparable data when they were used to study malathion degradation. The chemostat, however, was found to describe more accurately malathion degradation at high concentrations (1 to 4 mg l^{-1}) than batch cultures because of the toxic effects of malathion at high concentrations in the latter (see p. 53). These results also underlined the need to extrapolate only from the laboratory to the field within the range of conditions that have been tested. It must, however, be recognized that, as useful as laboratory systems are, they are only tools for the study of the environment and not a substitute for it.

2.2.5 *Basic Factors Influencing Enrichment in Laboratory Systems*

There are many factors which influence enrichment in both model systems and microcosms, most of which have already been discussed in the relevant sections, and it is recommended that both sections on models and microcosms be referred to in order to obtain a comprehensive understanding of the effect of enrichment in laboratory systems. Only factors general to both systems not previously mentioned are discussed in this section.

The inoculum or substrata used should be relevant to the natural environment under study or, if the possible routes of transformation of a novel compound are being studied, the inoculum should be obtained from systems that exhibit a wide variety of different microbial activities, e.g. eutrophic sediments or water, activated sludge systems and others.

Treatment of the sample before use should be avoided if at all possible. For example, soil is often air dried before use to facilitate sifting to remove large particles, to be weighed accurately, or so it can be adjusted (by weight) to a known field moisture capacity. Bartha (1971) indicated the possible dangers of such drying when it was found that drying severely affected the outcome of propanil degradation. Jannasch (1967a) showed that the storage of sea water resulted in the loss of certain marine organisms, which he suggested may have been due to the effects of batch enrichment during storage. It seems reasonable to assume that similar storage effects take place with other environmental samples. Samples for inoculation into chemostats are often filtered prior to use in order to remove predatory protozoa (Veldkamp, 1977); this procedure may result in a significant change in the microbial community eventually obtained due to the removal of large or filamentous bacteria or the physical damage of appendaged bacteria, e.g. prosthecate bacteria. It may be more realistic to conduct the enrichment without filtering and when a stable community has been established, then to remove the protozoa by filtration. Any change in community composition due to filtration could then be recognized and lost members re-established. Indeed, the presence of protozoa may have a significant effect on the bacterial process being investigated and their removal may produce unrealistic experimental results (e.g. bacteria grazed by protozoa have a more rapid cycling of phosphorous—Barsdate *et al.*, 1974). Interestingly when xenobiotic compounds are utilized as substrates, protozoa often do not become established in the chemostat, presumably because they are more susceptible to the toxic effects of the xenobiotic compound than are the bacteria.

The possibility of pre-enrichment of the laboratory inoculum in the field is a procedure which has been little exploited, but which might result in the isolation of different microorganisms compared to those obtained by more conventional means and might remove some of the problems of obtaining and handling inocula. The substrate to be used in the enrichment can be placed in capillaries, or inside membrane filters if the substrate is insoluble, or in solid agar strips if soluble, and then introduced into the environment to be studied. These systems could be left in place in order to be colonized by the resident microbial community, i.e. in a similar manner to chemical "peepers" (semi-permeable membranes often placed in sediments in order to measure the concentration of small molecules without disturbing the sediment). Such systems could be placed in different depths in the soil or sediment in order to obtain some idea of the distribution of the isolated communities in nature. Obviously such systems would be batch enrichments, but a continuous system would not be difficult to design, perhaps along similar lines to the apparatus described by McClure (1973) or Caldwell and Hirsch (1973). Alternatively, the inoculum could be pumped directly and continuously into the chemostat (Jannasch, 1967a); see also pp. 91–92.

One important factor that is often neglected is the source of the inoculum. For example, soil types vary markedly in their ability to utilize xenobiotic compounds and in the route of transformation (Kaufman and Blake, 1973). Sediments are much more active in degrading organochlorines than the water column (Lee and Ryan, 1979) and the response of bacteria to herbicides may vary considerably with sediment depth (S. F. Minney, R. J. Parkes, and A. T. Bull, unpublished observations). The activity of the inoculum may also vary markedly with season. For example, Maccubbin and Kator (1979) found that petroleum-degrading activity varied with season in the surface waters of the middle Atlantic and similar variations have been found for carbon mineralization (Klein, 1977), 2,4-D degradation (Watson, 1977), inhibitory effects of herbicides and the isolation of herbicide-degrading communities (R. J. Parkes, S. F. Minney, and A. T. Bull, unpublished observations), phytoplankton diversity (Martin and Bianchi, 1980), and cellulose degradation in estuarine sediments (Parkes, 1978). The composition of the substrata for microcosms shows similar variation in activity but may also show differences in basic composition (e.g. the phosphorus content—Barsdate *et al.*, 1974), which will have a considerable inpact on the outcome of microcosm experiments. These considerations emphasize the need for initial field investigations both to identify the potential source of the desired microbial activity and to appreciate the possible seasonal variation of such activity. This is particularly true if modelling is to help clarify our understanding of processes in the natural environment rather than merely cause confusion. Initial field investigations are also essential for the development of realistic laboratory enrichment conditions, the importance of which has been emphasized throughout this chapter.

3. ANALYSIS OF MICROBIAL INTERACTIONS IN THE LABORATORY

3.1 Microbial and Chemical Analysis

When a microbial community has been enriched in the laboratory, the methods used to study their interactions are largely basic microbiological techniques, such as estimation of numbers, types, biomass, activity (oxygen uptake, carbon dioxide evolution, electron transport system, and enzyme activity, etc.), the disappearance of substrates and the appearance of metabolites. Hence, the methods used are similar to those applied to microbial communities in the natural environment. The main advantages of laboratory systems are that these communities can now be studied under controlled conditions and environmental variables can be changed to facilitate the investigation of the microbial interaction. These basic microbiological techniques have already been discussed by Fry (pp. 105–121) and will not be repeated here. Instead less standard techniques, including the use of enrichment systems themselves as tools for investigating, microbial interaction are discussed.

3.2 Batch Growth Analysis

Once a microbial community has been enriched, its component members can be isolated in pure culture by traditional agar plate procedures. Some members may prove difficult to purify by this procedure which may reflect close metabolic interdependence. The use of a medium containing different growth factors and possible substrates supplied by the other members of the community can often enable the purification of difficult or fastidious isolates, and also provide information which is useful in determining the mechanism of the community interdependence. This process of identification and purification is crucial to subsequent analysis and is often difficult since features such as colony shape and colour can often change when an organism is isolated from the community, or as a result of repeated sub-culture. This procedure therefore, has to be carried out with the utmost care. Microscopic examination of both the community and the isolates, together with observation (microscopic and plate growth) of a community reconstructed from the pure isolates, can help considerably in ensuring that all the community members have been isolated. Jannasch (1967b) emphasized the need to maintain isolates on media of equivalent substrate strength to that of the enrichment medium, since it was found that growth of the isolates on medium of higher substrate concentrations resulted in changes in growth parameters, presumably due to mutation.

If the community utilizes a complex organic compound, continued sub-culturing growth of the isolates on agar plates containing the complex compound can provide a good indication of primary-degrading organisms (Senior *et al.*, 1976). Growth of isolates on potential metabolic intermediates also produces useful information. However, data obtained in this manner can be misleading due to the carry-over of growth factors and substrates from previous plates (hence the need for repeated sub-culture). Furthermore, if co-operation is required in order to utilize a particular substrate, this capability may not be exhibited on the agar plate due to the spatial separation of colonies and the limited diffusion characteristics. Therefore, co-operation is better studied in liquid culture. Batch growth of the isolates and combinations of the isolates enable metabolites, which may not appear in the complete community (due to their rapid utilization by the different community members), to be quantitatively identified. The possible inhibitory effects of these transient metabolites on community members is also recognizable. Similar experiments could be conducted in a chemostat, but batch systems have the advantage that metabolites which may be produced in undetectably small quantities will gradually accumulate and eventually reach measurable concentrations.

Although the metabolites involved in community interaction can sometimes be identified (e.g., *p*-nitrophenol—Daughton and Hsieh, 1977), the biochemical basis of community interaction often remains unresolved (Senior *et al.*, 1976; Osman *et al.*, 1976). An indication of how subtle community interactions can be is given by Osman *et al.* (1976) who described a three-membered bacterial community growing on orcinol. Only one bacterium (*Pseudomonas stutzeri*) could metabolize orcinol. The two secondary species (*Brevibacterium linens* and *Curtobacterium sp.*) comprised less than 20% of the total community and were unable to cleave the aromatic ring. No metabolic intermediate between the primary and secondary organisms could be recognized and general cell lysis and leakage products were suggested as the possible cause of community interaction. However, kinetic analysis of growth revealed a strong basis for community interdependence. Batch growth of the community, the isolates and combinations of the three isolates at different substrate concentrations gave a measure of the maximum specific growth rate (μ_m) of the various combinations using the Monod relationship (Fig. 3).

$$\mu = \frac{\mu_m \cdot s}{(K_s + s)}$$

This relationship enables K_s to be calculated where μ = specific growth rate which is analogous to compound interest rate on an investment, e.g. $\mu = 0.1\ h^{-1}$ is equivalent to a compound interest rate of 10% per hour; K_s = saturation constant and is a measure of the affinity of the

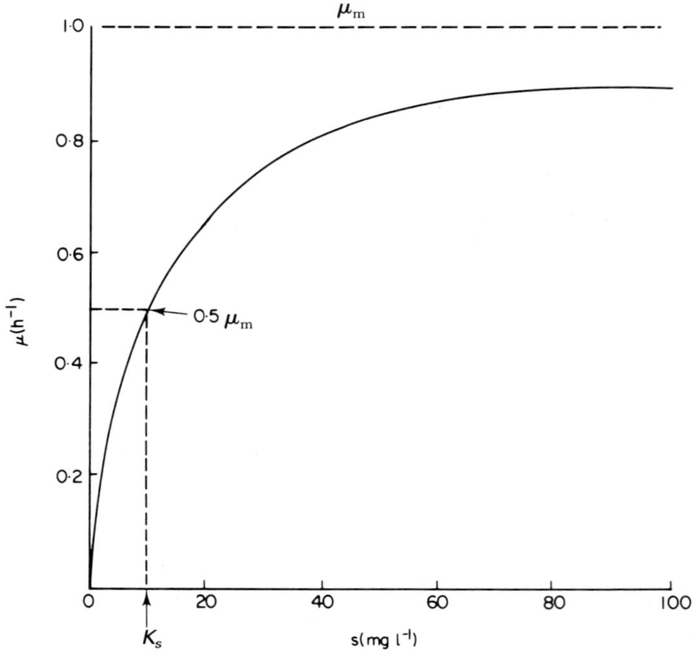

Fig. 3. Specific growth rate (μ) plotted as a function of substrate concentration (s) according to the Monod equation $\mu = \mu_m s/(K_s + s)$ where $\mu_m = 1.0\ h^{-1}$ and $K_s = 10\ mg\ l^{-1}$.

bacterium for its substrate. The lower the K_s the greater affinity the bacterium has for its substrate; and s = substrate concentration.

N.B. Specific growth rate from batch data is determined by a plot of log biomass against time, slope of straight line × 2.303 = μ.

From Table 1 it can be seen that if μ_m and K_s are the only factors governing selection, as suggested by Veldkamp (1976), then the *Pseudomonas* sp.:*Curtobacterium* sp. two-membered mixed culture should have been the most competitive combination. But the batch culture experiments, over a range of orcinol concentrations, revealed typical substrate-inhibited growth kinetics and the inhibition constants for the various bacterial combinations were calculated, assuming Haldane inhibition kinetics applied (Dixon and Webb, 1958):

$$\mu = \frac{\mu_m}{1 + \dfrac{K_s}{s} + \dfrac{s}{K_i}}$$

Table 1. Kinetic data obtained for a three-membered bacterial community growing on orcinol with an initial orcinol concentration $S_R = 1.0$ g l^{-1} (data of A. Osman, A. T. Bull and J. H. Slater, 1976).

	μ_m (h^{-1})	K_s (mg l^{-1})	K_i (mg l^{-1})
Three-membered community	0.28	72	1750
Pseudomonas stutzeri	0.29	100	690
Pseudomonas stutzeri + *Brevibacterium linens*	0.30	83	570
Pseudomonas stutzeri + *Curtobacterium*	0.35	59	790

where K_i = inhibition constant and when K_i is very much larger than K_s (the normal situation) K_i is numerically equal to the substrate concentration producing half maximal growth. This analysis showed that the community enriched had much greater tolerance to the inhibitory substrate than any other combination of isolates and hence gave a reason for its selective advantage in the presence of an inhibitory substrate, compared with the high affinity two-membered mixed culture.

A point of practical importance is that great care has to be taken if changes in culture absorbency are used as a measure of growth of a mixed community, both in batch culture and chemostats. The use of absorbence as a measure of growth is quite reasonable for pure cultures, but the biomass of a mixed population of microorganisms with different sizes may change quite markedly without being reflected in changes in absorbence.

3.3 Growth in Continuous Culture Systems as a Tool for Community Analysis

The constant environment prevailing in chemostat systems which is one of their main advantages for enrichment procedures, is a similar asset when such systems are used for community analysis. The chemostat enables a variety of factors to be altered, either singly or in combination, whilst keeping all the other factors constant. Hence the changes observed in the culture can be directly associated with experimental variables. Increases in the dilution rate of the system have the potential to separate loosely associated organisms from a community. For example, Senior *et al.* (1976) lost a loosely associated yeast from their dalapon community, by wash-out, the first time the dilution rate was raised above 0.2 h^{-1}; the remaining community proved to be very stable for further changes in the dilution rate. Similar effects could be achieved by varying both the concentration and the substrate provided to the community. This was done by Daughton and Hsieh (1976, 1977) and, although none of the community members washed out, sufficient changes in biomass occurred as a result of substrate changes to allow the community interaction to be identified. Wilkinson *et al.* (1974) added small amounts of methanol to a methane-utilizing community to confirm that the role of the *Hyphomicrobium* was to remove inhibitory levels of methanol produced by the community.

Cappenberg (1975) conducted a series of chemostat experiments which confirmed earlier extrapolations from field data (Cappenberg, 1974a, b; Cappenberg and Prins, 1974; see Fry p. 123) and emphasized the great potential of chemostat systems as experimental tools for studying mixed microbial interactions. Cappenberg suggested, on the basis of field data, that methanogenic bacteria in freshwater sediments utilized the acetate produced by sulphate-reducing bacteria for methane production. These two groups of bacteria were, however, spatially separated in the lake sediments, the sulphate-reducing bacteria being present near the surface (0 to 2 cm) and the methanogenic bacteria lower down (3 to 6 cm). It was suggested that this separation could be due to inhibition of methanogenic bacteria by the sulphide produced by sulphate-reducing bacteria. Cappenberg (1975) investigated this relationship by growing the organisms separately in a single-stage chemostat (carbon and energy limited) and then combining the effluents into a third vessel. In the third vessel there was an increase in both the concentration of methane and methanogenic bacteria, due to the presence of the acetate produced in the first stage by the sulphate-reducing bacteria. This confirmed the expected commensalism. However, when the substrate concentration in the reservoir was increased, high concentrations of sulphide were produced which inhibited the methanogenic bacteria in the mixed system and they were consequently washed out.

The use of membranes within continuous-flow culture systems to separate organisms from solutes, provides a useful system for identifying any possible biochemical interaction between community members. These can be placed horizontally within a culture vessel with the microorganism in the lower section. The upper layer is analysed for metabolites and exudates produced by the organisms, and pumped into a second-stage system, to investigate whether or not the effluent supports another member of the community. The variety of continuous-flow culture systems, previously discussed, provides an array of potential analytical system that have yet to be fully exploited. The experimenter can select a system most suitable for the environment or the community under investigation and can also select the degree of complexity required in the system. An initial approach might be to select a very simple system and then to add more complex features (e.g. particles and films), once the basic system is clearly understood. Thus the experimenter has available a potential battery of systems which ought to assist both in the community analysis and in the investigation of how the community adapts to varying environmental conditions.

Microbial growth within a chemostat is self-regulating (Veldkamp, 1977) and a "steady state" situation results when biomass concentration and environmental conditions remain constant and the growth rate of the organism becomes equivalent to the dilution rate of the system. This provides a system which is very amenable to mathematical analysis and Monod kinetics (previously described for batch growth) have been successfully applied to chemostat systems (Herbert *et al.*, 1956; Tempest, 1970). Once a steady state is reached, the culture has become time independent and the steady-state concentration of biomass ($\bar{x}$) and growth limiting substrate ($\bar{s}$) may be defined as:

$$\bar{x} = Y\,(S_R - \bar{s})$$

where Y is the growth-yield coefficient (weight of biomass formed/weight of substrate used) and

$$\bar{s} = K_s\left(\frac{D}{\mu_m - D}\right).$$

once μ_m, K_s and Y are known, the values of $\bar{x}$ and $\bar{s}$ can be predicted for any dilution rate. To obtain a steady state, the dilution rate should not exceed a certain critical value (D_c) which is determined by the concentration of growth-limiting substrate in the feed reservoir (S_R):

$$D_c = \mu_m\left(\frac{S_R}{K_s + S_R}\right).$$

The theoretical effect of increased dilution rate on steady-state bacterial population is shown in Fig. 4. However, this theoretical relationship is not always found in practice (Jannasch, 1976b). Jannasch (1969) found that bacteria growing on sterilized sea water at low dilution rates were unstable, thus making the determination of bacterial growth rates by conventional wash-out techniques difficult. He suggested that the growth rates in these conditions could be calculated from the difference between the imposed dilution rate and the actual wash-out rate of cells, and this method could be used to estimate

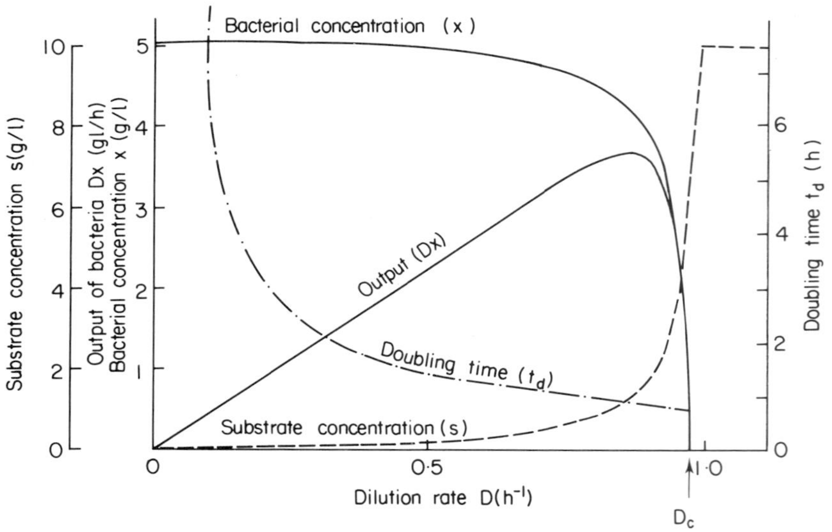

Fig. 4. Theoretical relations of the dilution rate (D) and the steady state values of bacterial concentration (x), substrate concentration (s), doubling time (t_d) and output (Dx). Data were collected with the following growth constants: $\mu_m = 1.0\ h^{-1}$, $Y = 0.5$ and $K_s = 0.2$ g l^{-1}, and a substrate concentration in the inflowing medium of $S_R = 10$ g l^{-1}. (From Herbert *et al.*, 1956.)

bacterial growth rates in natural waters. (For further details of chemostat theory see Pirt, 1975).

Simple Monod kinetics, however, are only applicable if there is only one growth-limiting nutrient and the only interaction between organisms is that of competition for this limiting nutrient. Under such circumstances successful or unsuccessful competition for a limiting substrate in a mixed microbial community is based on the particular growth parameters of the individual species present (Veldkamp and Jannasch, 1972) and one species should dominate. The importance of substrate affinity and maximum specific growth rate in determining the outcome of competition experiments (and so possibly survival in the environment) has been demonstrated by both Veldkamp and Jannasch (these have been previously discussed) but other factors are also important, such as actual specific growth rate (Meers, 1971), the presence of predators (Jost *et al.*, 1973), the availability of other limiting substrates (Meers and Tempest, 1968; Megee *et al.*, 1972), and probably many other unrecognized parameters which do not appear in the Monod expression. Therefore, it is not surprising that mixed cultures of microorganisms often stabilize under conditions which, according to simple Monod kinetics, a single population should dominate.

Several workers have modified the simple Monod relationship in order to include some of the factors mentioned above. Taylor and Williams (1975) commented that a stable mixed population of competing organisms cannot coexist unless the number of growth-limiting substrates is equal to or greater than the number of competing species, and they extend Monod growth kinetics to cover more than one limiting substrate. This consideration explains why the competition experiments of Jannasch (1967a) with a single rate-limiting nutrient regularly resulted in the dominance of a single organism. Often, however, no particular species dominated (especially at low dilution rates and low substrate concentrations) which suggested that interactions other than competition were involved. Multiple substrates can be provided by the degradation of a complex compound producing a variety of metabolites and it is significant that many of the microbial communities so far studied in the laboratory are communities degrading complex compounds, e.g. xenobiotic compounds (Senior *et al.*, 1976; Osman *et al.*, 1976; Daughton and Hsieh, 1977).

Meyer *et al.* (1975) considered the effect of interactions other than competition on a mixed population of organisms and predicted that the combination of two intrinsically unstable interactions (competition and pure mutualism) can give stable mixed populations in the chemostat. The importance of substrate inhibition in determining growth and hence selection is discussed by Andrews (1968) who suggested that the Monod relationship was only a special case of a more general relationship since even glucose can inhibit growth if present at high enough concentrations. He suggested a growth equation based on Haldane inhibition kinetics and simulation studies based on this growth equation showed that inhibition might lead to instability in continuous-flow culture systems. Similar results have been found by other workers (Yano and Koga, 1969; Edwards, 1970; Hill and Robinson, 1975; Yang and Humphrey, 1975). Meyer *et al.* (1975) included a term for potential product inhibition and showed that inhibition could stabilize an unstable, pure mutualistic relation and also commented that there must be many other situations of mutualism that would yield stable coexistence that have not yet been recognized. De Freitas and Fredrickson (1978) also emphasized the importance of inhibition in mixed microbial systems. Button (1978) described a theory of control of microbial growth kinetics which considered the importance of nutrient transport. Meyer *et al.* (1975) commented that it is not always possible to attain a mutualistic interaction in continuous-flow culture, even though the populations involved do produce exchangeable substrates in sufficient quantities, as only a limited range of initial cultural conditions enable the community to stabilize. This underlines the importance of realistic design of enrichment conditions. Further examples of behaviour of mixed microbial populations and analysis of their growth can be found in the review by Fredrickson (1977).

There are many mathematical relationships for analysing microbial growth and the Monod function may be just a special case of a more general relationship. This more general relationship seems not to have been developed yet as many of the more complex mathematical relationships only seem applicable to the system they were designed to describe (Wilkinson *et al.*, 1974). One of the main problems is that Monod kinetics are only applicable to systems with steady-state microbial populations (Bazin *et al.*, 1976) and hence transient growth (from one steady state to another) is not well described by Monod kinetics (Harrison and Topiwala, 1974) unless a number of extra constants are included (Wilder *et al.*, 1980). A feature of mixed microbial communities is often their inability to produce a true steady state (Contois and Yango, 1964; Bungay and Bungay, 1968) and a stable oscillating state often develops (Cassell *et al.*, 1966), the extent of these oscillations often becoming dampened during prolonged cultivation (Senior, 1977). The oscillations may be due to members of the community competing for a growth-limiting substrate, for which none of the competitors has a great selective advantage, hence small changes in culture conditions can cause considerable oscillations in biomass and its activity. Therefore, mixed communities may be constantly in transient growth phases and as such may not be amenable to simple mathematical analysis. However, transient states are probably very important in controlling many reactions which occur in the environment (Jannasch, 1974; Bazin *et al.*, 1976) and so the non-steady state of mixed communities in chemostats may be an important feature if such systems are to model successfully natural processes.

A further complication is that populations within the chemostat may not be constant. Loosely associated members of the community may be washed out, genetic mutation of an individual member may totally change the community structure, and wall growth may develop which is physiologically very different from the community in the supernatant. None of these factors can be successfully predicted by existing mathematical models.

The problem of wall growth is particularly important when the chemostat is used for kinetic analysis. The presence of wall growth, which is often associated with mixed microbial populations (Atkinson and Fowler, 1974), can have a considerable effect on washout kinetics. Sometimes wall growth only becomes apparent when the culture is close to washing out, as if it were a reaction by the community to maintain itself within the chemostat. Steady state conditions in open cultures at dilution rates higher than the maximum specific growth rate have frequently been reported, and one of the important factors causing this situation may be the presence of microbial films within the chemostat. Pawlowsky and Howell (1973), for example, found that wall growth increased by a factor of three the dilution rate at which 90% conversion of phenol could be obtained. These authors developed a mathematical model to incorporate wall growth effects and showed good agreement between predicted and experimental values. Bader *et al.* (1976b) also found significant effects of wall growth on the results of grazing of cyanobacteria by ciliates. For these reasons the majority of kinetic analysis of mixed microbial communities is often conducted under batch conditions (see previous Section 3.2). From theoretical considerations, Wilkinson and Hamer (1974) postulated that with a captive population on the walls of the chemostat, the stable operating range would be extended and the productivity of the system increased. This aspect of microbial film growth has been exploited fully by Atkinson and his colleagues with the development of systems designed specifically to encourage controlled film development. Some of these systems have already been discussed, but more detailed descriptions of film reactors, the kinetics of film growth and the significance of microbial films in general in fermenters can be found in Atkinson and Daoud (1976), and Atkinson and Fowler (1974).

3.4 Diffusion Gradient Systems

Mixed microbial communities, although integrated units, are made up of organisms with different optimum conditions for growth. Therefore, if a sufficiently discriminating range of different environmental conditions were to be provided, the community ought to become segregated. Also, measurement of the conditions which would allow each isolate to grow would be exceedingly useful in determining the basis of community interaction. Stabilized diffusion gradient systems can provide an infinite series of habitats within a chosen range and, hence, have great potential in the study of microbial communities. The potential of agar diffusion as an analytical tool in microbiology has been recognized for a long time: Beijerinck (1889) suggested its use for examining the nutrient requirements of yeasts, but it is still not widely used. One of the problems with early systems was that solute concentrations within any point in the gel varied with time. Caldwell and Hirsch (1973), however, devised a two-dimensional agar diffusion plate that generated steady-state gradients. The system was made of transparent plastic and the agar layer was 2.5 cm square and 0.3 cm deep. Two reservoirs fed nutrients to each of two sides of the agar square at right angles and acted as the sources of solutes. The upper surface of the agar was continuously washed with fresh aqueous medium, providing a sink for diffusing solutes, including unused substrates and waste products, and in addition giving the system open characteristics. The system took about 2 d to equilibrate, after which a thin layer of agar containing an inoculum was placed on the surface of the agar square and the structure incubated. These workers confirmed a mathematical model describing rates of diffusion in the system using labelled substrates (Caldwell *et al.*, 1973). They were able to separate a mixed population of genera of *Hyphomicrobium*, *Rhodomicrobium* and *Thiopedia* using this system with gradients of methylamine and sodium sulphide. The authors suggested that the system was capable of isolating stenobiotic organisms (only capable of growing in a narrow range of conditions) as well as separating members of a mixed population whose growth requirements differed only slightly. Wimpenny (1981) described a variety of diffusion gradient systems including stopped time dependent gradient plates (to overcome the problem of variation with time in diffusion systems) and gel stabilized systems, with both one dimensional and two dimensional solute flow. Using such systems Wimpenny (1981) has been able to separate bacteria along pH:sodium chloride gradients, showing banding of growth in pure cultures of *Bacillus cereus* believed to be equivalent to Leisegang rings, and to model the gradient dominated water ecosystem found below the oil layer in oil storage tanks. Such systems have not yet been used to segregate closely associated communities, such as those degrading xenobiotic compounds, but potentially these systems have great value for research into interactions between mixed microbial communities.

REFERENCES

Abbot, W. (1966). The replicability of microcosms. *Journal of the Water Pollution Control Federation* **38**, 258–270.

Andrews, J. F. (1968). A mathematical model for the continuous culture of microorganisms. *Biotechnology and Bioengineering* **10**, 707–723.

Ashby, R. E. and Bull, A. T. (1977). A column fermenter for the continuous cultivation of microorganisms attached to surfaces. *Laboratory Practice* **26**, 327–329.

Atkinson, B. and Davies, I. J. (1972). The completely mixed microbial film fermenter—a method of overcoming wash-out in continuous fermentation. *Transactions of the Institution of Chemical Engineers* **50**, 208–216.
Atkinson, B. and Daoud, I. S. (1976). Microbial flocs and floculation in fermentation process engineering. *Advances in Biochemical Engineering* **4**, 41–124.
Atkinson, B., Daoud, I. S. and Williams, D. A. (1968). A theory for the biological film reactor. *Transactions of the Institute of Chemical Engineers* **46**, T245–T250.
Atkinson, B. and Fowler, H. W. (1974). The significance of microbial film in fermenters. *Advances in Biochemical Engineering* **3**, 221–277.
Atkinson, B. and Knights, A. J. (1975). Bioengineering report—microbial film fermenters: their present and future applications. *Biotechnology and Bioengineering* **17**, 1245–1267.
Ayanaba, A. and Alexander, M. (1972). Changes in nutritional types in bacterial successions. *Canadian Journal of Microbiology* **18**, 1427–1430.
Bader, F. G., Tsuchiya, H. M., and Fredrickson, A. G. (1976a). Grazing of ciliates on blue-green algae: effects of ciliate encystment and related phenomena. *Biotechnology and Bioengineering* **18**, 311–332.
Bader, F. G., Tsuchiya, H. M., and Fredrickson, A. G. (1976b). Grazing of ciliates on blue-green algae: effects of light shock on the grazing relation and algal population. *Biotechnology and Bioengineering* **18**, 333–348.
Barsdate, R. J., Prentki, R. T., and Fenchel, T. (1974). Phosphorus cycle of model ecosystems: significance for decomposer chains and effect of bacterial grazers. *Oikos* **25**, 239–251.
Bartha, R. (1971). Altered propanil biodegradation in temporarily dried soil. *Journal of Agricultural Food Chemistry* **19**, 394–395.
Bartha, R. and Pramer, D. (1970). Metabolism of acylanide herbicides. *Advances in Applied Microbiology* **13**, 317–341.
Bazin, M. J. and Saunders, P. T. (1973). Dynamics of nitrification in a continuous flow system. *Soil Biology and Biochemistry* **5**, 531–543.
Bazin, M. J., Saunders, P. T., and Prosser, J. I. (1976). Models of microbial interactions in the soil. *CRC Critical Reviews in Microbiology* **4**, 463–498.
Beall, M. L., Nash, R. G., and Kearney, P. C. (1976). Agro ecosystem: a laboratory model ecosystem to simulate agricultural field conditions for monitoring pesticides. *Proceedings of the Environmental Protection Agency Conference on Modeling and Simulation*, 790–793.
Beijerinck, M. (1889). Auxanography, a method useful in microbiological research, involving diffusion in gelatin. *Archives Neerlandaises des Sciences Exactes et Naturelles, Haarlem* **23**, 367–372.
Beyers, R. J. (1963). The metabolism of twelve aquatic laboratory microecosystems. *Ecological Monographs* **33**, 281–306.
Blain, J. A., Anderson, J. G., Todd, J. R., and Divers, M. (1979). Cultivation of filamentous fungi in the disc fermenter. *Biotechnology Letters* **1**, 269–274.
Boethling, R. S. and Alexander, M. (1979). Effect of concentration of organic chemicals on their biodegradation by natural microbial communities. *Applied and Environmental Microbiology* **37**, 1211–1216.

Bond, H., Lighthart, B., Shimabuka, R., and Russel, L. (1976). Some effects of cadmium on coniferous forest soil and litter microcosms. *Soil Science* **121**, 278–287.

Booth, C. (1971). "Methods in Microbiology, Volume 4". Academic Press, New York and London.

Bott, T. L., Preslan, J., Finlay, J., and Brunker, R. (1977). The use of flowing-water microcosms and ecosystem streams to study microbial degradation of leaf litter and nitrilotriacetic acid (N.T.A.). *Developments in Industrial Microbiology* **18**, 171–184.

Bourquin, A. W. (1977). Effects of malathion on microorganisms of an artificial salt-marsh environment. *Journal of Environmental Quality* **6**, 373–378.

Brown, C. M., Ellwood, D. C., and Hunter, J. R. (1978). Enrichments in a chemostat. *In* "Techniques for the Study of Mixed Populations", pp. 213–222. (D. W. Lovelock and R. Davies, eds). Academic Press, New York and London.

Bryfogle, B. M. and McDiffett, W. F. (1979). Algal succession in laboratory microcosms. *The American Midland Naturalist* **101**, 344–354.

Bull, A. T. and Brown, C. M. (1979). Continuous culture applications to microbial biochemistry. *In* "International Review of Biochemistry, Microbial Biochemistry, Volume 21", pp. 177–226. (J. R. Quayle, ed.) University Park Press, Baltimore.

Bungay, H. R. and Bungay, M. L. (1968). Microbial interactions in continuous culture. *Advances in Applied Microbiology* **10**, 269–290.

Button, D. K. (1978). On the theory of control of microbial growth kinetics by limiting nutrient concentrations. *Deep Sea Research* **25**, 1163–1177.

Caldwell, D. E. and Hirsch, P. (1973). Growth of microorganisms in two dimensional steady state diffusion gradients. *Canadian Journal of Microbiology* **19**, 53–58.

Caldwell, D. E., Lai, S. H., and Tiedje, J. M. (1973). A two dimensional steady state diffusion gradient for ecological studies. *Bulletins from the Ecological Research Committee* (*Stockholm*) **17**, 151–158.

Campacci, E. F., New, P. B., and Tchan, Y. T. (1977). Isolation of amitrole degrading bacteria. *Nature, London* **266**, 164–165.

Cappenberg, Th. E. (1974a). Interrelations between sulphate-reducing and methane producing bacteria in bottom deposits of a fresh water lake. I. Field observations. *Antonie van Leeuwenhoek* **40**, 285–295.

Cappenberg, Th. E. (1974b). Interrelation between sulphate-reducing and methane producing bacteria in bottom deposits of a fresh water lake. II. Inhibition experiments. *Antonie van Leeuwenhoek* **40**, 297–306.

Cappenberg, Th. E. and Prins, R. A. (1974). Interrelations between sulphate-reducing and methane-producing bacteria in bottom deposits of a fresh-water lake. III. Experiments with ^{14}C-labelled substrates. *Antonie van Leeuwenhoek* **40**, 457–469.

Cappenberg, Th. E. (1975). A study of mixed continuous cultures of sulphate-reducing and methane producing bacteria. *Microbial Ecology* **2**, 60–72.

Carlile, M. J. (1980). From prokaryote to eukaryote: gains and losses. *In* "The Eukaryotic Microbial Cell", pp. 1–40. (G. W. Gooday, D. Lloyd, and A. P. J. Trinci, eds). Cambridge University Press, Cambridge.

Cassell, E. A., Sulzer, F. T., and Lamb, J. C. (1966). Population dynamics and selection in continuous mixed cultures. *Journal of the Water Pollution Control Federation* **38**, 1398–1409.

Cole, L. K., Metcalf, R. L., and Sanborn, J. R. (1976). Environmental fate of insecticides in terrestrial model ecosystems. *International Journal of Environmental Studies* **10**, 7–14.

Contois, D. E. and Yango, L. D. (1964). Studies of steady state, mixed microbial populations. *Abstracts of Papers of the American Chemical Society 148th Meeting*, Q17–18.

Cooper, D. C. and Copeland, B. J. (1973). Responses of continuous-series estuarine micro-ecosystems to point-source input variations. *Ecological Monographs* **43**, 213–236.

Daughton, C. G. and Hsieh, D. P. H. (1976). Interactions of ethyl parathion and an acclimated bacterial community. *In* "Abstracts of Papers, 5th International Fermentation Symposium", p. 123. (H. Dellweg, ed.). Berlin.

Daughton, C. G. and Hsieh, D. P. H. (1977). Parathion utilization by bacterial symbionts in a chemostat. *Applied and Environmental Microbiology* **34**, 175–184.

De Freitas, M. J. and Fredrickson, A. G. (1978). Inhibition as a factor in the maintenance of diversity of microbial ecosystems. *Journal of General Microbiology* **106**, 307–320.

Dimock, C. W. (1975). "The Effect of Soluble Organic Matter on the Utilization of Phenoxy Herbicides by *Pseudomonas* sp.". M.Sc. Thesis, University of Rhode Island, Kingston, U.S.A.

Dixon, M. and Webb, E. C. (1958). "Enzymes". Longman, London.

Dunn, G. M., Wardell, J. N., Herbert, R. A., and Brown, C. M. (1980). Enrichment, enumeration and characterisation of nitrate-reducing bacteria present in the sediments of the River Tay estuary. *Proceedings of the Royal Society of Edinburgh, Series B* **78**, s47–s56.

Edwards, V. H. (1969). Correlation of lags in the utilization of mixed sugars in continuous fermentation. *Biotechnology and Bioengineering* **11**, 99–102.

Edwards, V. H. (1970). The influence of high substrate concentrations on microbial kinetics. *Biotechnology and Bioengineering* **12**, 679–712.

Falco, J. W., Sampson, K. T., and Carsel, R. F. (1977). Physical modeling of pesticide degradation. *Developments in Industrial Microbiology* **18**, 193–202.

Falk, J. H. (1974). Estimating experimenter-induced bias in field studies: a cautionary tale. *Oikos* **25**, 374–378.

Fisher, N. S., Carpenter, E. J., Remsen, C. C., and Wurster, C. F. (1974). Effects of PCB on interspecific competition in natural and gnotobiotic phytoplankton communities in continuous and batch cultures. *Microbial Ecology* **1**, 39–50.

Fletcher, M. (1979). The attachment of bacteria to surfaces in aquatic environments. *In* "Adhesion of Microorganisms to Surfaces", pp. 87–108. (D. C. Ellwood, J. Melling, and P. Rutter, eds.) Academic Press, New York and London.

Fredrickson, A. G. (1977). Behaviour of mixed cultures of microorganisms. *Annual Review in Microbiology* **31**, 63–87.

Genung, R. K., Million, D. L., Hancher, C. W., and Pitt, W. W. Jnr. (1979). Pilot plant demonstration of an anaerobic fixed bioreactor for waste water treatment. *In* "Biotechnology in Energy Production and Conservation", pp. 329–344. (C. D. Scott, ed.), John Wiley & Sons, New York.

Giddings, J. M. and Eddlemon, G. K. (1977). The effects of microcosm size and substrate type on aquatic microcosm behaviour and arsenic transport. *Archives of Environmental Contamination and Toxicology* **6**, 491–505.

Gordon, R. W., Beyers, R. J., Odum, E. P., and Eagon, R. G. (1969). Studies of a simple laboratory microecosystem: bacterial activities in a heterotrophic succession. *Ecology* **50**, 86–100.

Gottschal, J. C., DeVries, S., and Kuenen, J. G. (1979). Competition between the facultatively chemolithotrophic *Thiobacillus* A2, an obligately chemolithotrophic *Thiobacillus* and a heterotrophic spirillum for inorganic and organic substrates. *Archives of Microbiology* **121**, 241–249.

Hancher, C. W., Taylor, P. A., and Napier, J. M. (1978). Operation of a fluidised bed reactor for denitrification. *In* "Biotechnology in Energy Production and Conservation", pp. 361–378. (C. D. Scott, ed.) John Wiley & Sons, New York.

Harder, W. and Dijkhuizen, L. (1976). Mixed substrate utilization. *In* "Continuous Culture 6. Applications and New Fields", pp. 297–311. (A. C. R. Dean, D. C. Ellwood, C. G. T. Evans, and J. Melling, eds.) Ellis Horwood, Chichester, England.

Harder, W., Visser, K., and Kuenen, J. G. (1974). Laboratory fermenter with an improved magnetic drive. *Laboratory Practice* **23**, 644–645.

Harder, W., Kuenen, J. G., and Matin, A. (1977). A review: microbial selection in continuous culture. *Journal of Applied Bacteriology* **43**, 1–24.

Harder, W. and Veldkamp, H. (1971). Competition of marine psychrophilic bacteria at low temperatures. *Antonie van Leeuwenhoek* **37**, 51–63.

Harrison, D. E. F. and Topiwala, H. H. (1974). Transient and oscillatory states of continuous culture. *Advances in Biochemical Engineering* **3**, 167–219.

Harrison, D. E. F. and Wren, S. J. (1976). Mixed microbial cultures as a basis for future fermentation processes. *Process Biochemistry* **11**, 30–32.

Hartmann, J., Reineke, W., and Knackmuss, H. J. (1979). Metabolism of 3-chloro, 4-chloro, and 3,5-dichlorobenzoate by a Pseudomonad. *Applied and Environmental Microbiology* **37**, 421–428.

Herbert, D. (1976). Stoicheiometric aspects of microbial growth. *In* "Continuous Culture 6. Applications and New Fields", pp. 1–30. (A. C. R. Dean, D. C. Ellwood, C. G. T. Evans, and J. Melling, eds.) Ellis Horwood, Chichester, England.

Herbert, D., Elsworth, R., and Telling, R. C. (1956). The continuous culture of bacteria: a theoretical and experimental study. *Journal of General Microbiology* **14**, 601–622.

Hill, I. R. and Arnold, D. J. (1978). Transformations of pesticides in the environment—the experimental approach. *In* "Pesticide Microbiology", pp. 203–245. (I. R. Hill and S. J. L. Wright, eds.) Academic Press, New York and London.

Hill, G. A. and Robinson, C. W. (1975). Substrate inhibition kinetics: phenol degradation by *Pseudomonas putida*. *Biotechnology and Bioengineering* **17**, 1599–1615.

Horvath, R. S. (1972). Cometabolism of the herbicide 2,3,6-trichlorobenzoate by natural microbial populations. *Bulletin of Environmental Contamination and Toxicology* **7**, 273–276.

Huber, T. J., Street, J. R., Bull, A. T., Cook, K. A., and Cain, R. B. (1975). Aromatic metabolism in the fungi: growth of *Rhodotorula mucilaginosa* in p-hydroxybenzoate-limited chemostats and the effect of growth rate on the synthesis of enzymes of the 3-oxoadipate pathway. *Archives of Microbiology* **102**, 139–144.

Hungate, R. E. (1966). "The Rumen and its Microbes". Academic Press, New York and London.

Isensee, A. R. (1976). Variability of aquatic model ecosystem derived data. *International Journal of Environmental Studies* **10**, 35–41.

Isensee, A. R., Holden, E. R., Woolson, E. A., and Jones, G. E. (1976). Soil persistence and aquatic bioaccumulation potential of hexochlorobenzene (HCB). *Journal of Agricultural Food Chemistry* **24**, 1210–1217.

Jannasch, H. W. (1960). Versuche über denitrification und die verfügbarkeit des sauerstoffes in wasser und schlamm. *Archiv für Mikrobiologie* **56**, 355–369.

Jannasch, H. W. (1967a). Enrichments of aquatic bacteria in continuous culture. *Archiv für Mikrobiologie* **59**, 165–173.

Jannasch, H. W. (1967b). Growth of marine bacteria at limiting concentrations of organic carbon in seawater. *Limnology and Oceanography* **12**, 264–271.

Jannasch, H. W. (1969). Estimations of bacterial growth rates in natural waters. *Journal of Bacteriology* **99**, 156–160.

Jannasch, H. W. (1974). Steady state and the chemostat in ecology. *Limnology and Oceanography* **19**, 716–720.

Jannasch, H. W. and Mateles, R. I. (1974). Experimental bacterial ecology studied in continuous culture. *Advances in Microbial physiology* **11**, 165–212.

Jannasch, H. W. and Pritchard, P. H. (1972). The role of inert particulate matter in the activity of aquatic microorganisms. *Memorie dell'Istituto italiano di idrobiologia Dott. Marco de Marchi* **29** suppl., 289–308.

Jørgenson, B. B. and Fenchel, T. (1974). The sulfur cycle of a marine sediment model system. *Marine Biology* **24**, 189–201.

Jørgenson, B. B. (1977). Bacterial sulfate-reduction within microniches of oxidized marine sediments. *Marine Biology* **41**, 7–17.

Jost, J. L., Drake, J. F., Fredrickson, A. G., and Tsuchiya, H. M. (1973). Interactions of *Tetrahymena pyriformis, Escherichia coli, Azotobacter* and glucose in a minimal medium. *Journal of Bacteriology* **113**, 834–840.

Kaplan, A. M. (1977). Microbial degradation of materials in laboratory and natural environments. *Developments in Industrial Microbiology* **18**, 203–211.

Kaufman, D. D. and Blake, J. (1973). Microbial degradation of several acetamide, acylanilide, carbamate, toluidine and urea pesticides. *Soil Biology and Biochemistry* **5**, 297–308.

Klein, D. A. (1977). Seasonal carbon flow and decomposer parameter relationships in a semi-arid grassland soil. *Ecology* **58**, 184–190.

Kuenen, J. G., Boonstra, J., Schröder, H. G. J., and Veldkamp, H. (1977). Competition for inorganic substrates among chemoorganotrophic and chemolitho trophic bacteria. *Microbial Ecology* **3**, 119–130.

Laanbroek, H. J., Kingma, W., and Veldkamp, H. (1977). Isolation of an aspartate-fermenting, free-living *Campylobacter* species, *FEMS Letters* **1**, 99–102.

Lee, R. F. and Ryan, C. (1979). Microbial degradation of organochlorine compounds in estuarine water and sediments. *In* "Microbial Degradation of Pollutants in Marine Environments", pp. 443–450. (A. W. Bourquin and P. H. Pritchard, eds.) United States Environmental Protection Agency, Gulf Breeze, Florida.

Lees, H. and Quastel, J. H. (1946). Biochemistry of nitrification in soil 1. Kinetics of, and the effects of poisons on soil nitrification, as studied by a soil perfusion technique. *Biochemical Journal* **40**, 803–814.

Lichtenstein, E. P., Fuhremann, T. W., and Schulz, K. R. (1974). Translocation and metabolism of [^{14}C]-phorate as affected by percolating water in a model soil–plant ecosystem. *Journal of Agricultural Food Chemistry* **22**, 991–996.

Lovatt, D., Slater, J. H., and Bull, A. T. (1978). The growth of a stable mixed culture on picolinic acid in continuous-flow culture. *Society for General Microbiology Quarterly* **6**, 27.

MacRae, I. C. and Alexander, M. (1965). Microbial degradation of selected herbicides in soil. *Journal of Agricultural Food Chemistry* **13**, 72–76.

Maccubbin, A. E. and Kator, H. (1979). Distribution, abundance and petroleum-degrading potential of marine bacteria from Middle Atlantic continental shelf waters. *In* "Microbial Degradation of Pollutants in Marine Environments", pp. 380–395. (A. W. Bourquin and P. H. Pritchard, eds.) U.S. Environmental Protection Agency, Gulf Breeze, Florida.

Macura, J. (1961). Continuous-flow method in soil microbiology. I. Apparatus. *Folia Microbiologia* **6**, 328–334.

Maigetter, R. Z. and Pfister, R. M. (1975). A mixed bacterial population in a continuous culture with and without kaolinite. *Canadian Journal of Microbiology* **21**, 173–180.

Mallory, L. M., Austin, B., and Colwell, R. R. (1977). Numerical taxonomy and ecology of oligotrophic bacteria isolated from an estuarine environment. *Canadian Journal of Microbiology* **23**, 733–750.

Martin, Y. P. and Bianchi, M. A. (1980). Structure, diversity and catabolic potentialities of aerobic heterotrophic bacterial populations associated with continuous culture of natural marine phytoplankton. *Microbial Ecology* **5**, 265–279.

Mateles, R. I. and Chian, S. K. (1969). Kinetics of substrate uptake in pure and mixed culture. *Environmental Science and Technology* **3**, 569–574.

Mateles, R. I., Chian, S. K., and Silver, R. (1967). Continuous culture on mixed substrates. *In* "Microbial Physiology and Continuous Culture", pp. 232–239. (E. O. Powell, C. G. T. Evans, R. E. Strange, and D. W. Tempest, eds.) HMSO, London.

Matsumura, F. (1979). Report of task group IV: Microcosms. *In* "Microbial Degradation of Pollutants in the Marine Environment", pp. 520–524. (A. W. Bourquin and P. H. Pritchard, eds.) U.S. Environmental Protection Agency, Gulf Breeze, Florida.

McClure, G. W. (1972). Degradation of phenylcarbamates in soil by mixed suspensions of IPC-adapted microorganisms. *Journal of Environmental Quality* **1**, 177–180.

McClure, G. W. (1973). A membrane biological filter device for reducing water borne biodegradable pollutants. *Water Research* **7**, 1683–1690.

Meers, J. L. (1971). Effect of dilution rate on the outcome of chemostat mixed culture experiments. *Journal of General Microbiology* **67**, 359–361.

Meers, J. L. and Tempest, D. W. (1968). The influence of extracellular products on the behaviour of mixed microbial populations in magnesium limited chemostat cultures. *Journal of General Microbiology* **52**, 309–317.

Megee, R. D., Drake, J. F., Fredrickson, A. G., and Tsuchiya, H. M. (1972). Studies of intermicrobial symbiosis: *Saccharomyces cerevisiae* and *Lactobacillus casei*. *Canadian Jounal of Microbiology* **18**, 1733–1742.

Metcalf, R. L., Songha, G. K., and Kapor, I. P. (1971). Model ecosystem for the evaluation of pesticide biodegradability and ecological magnification. *Environmental Science and Technology* **5**, 709–713.

Meyer, J. S., Tsuchiya, H. M., and Fredrickson, A. G. (1975). Dynamics of mixed populations having complementary metabolism. *Biotechnology and Bioengineering* **17**, 1065–1081.

Neijssel, O. M. and Tempest, D. W. (1976). The role of energy-spilling reactions in the growth of *Klebsiella aerogenes* NCTC 418 in aerobic chemostat culture. *Archives of Microbiology* **110**, 305–311.

Norris, J. R. and Ribbons, D. W. (1969). "Methods in Microbiology, Volume 1". Academic Press, New York and London.

Norris, J. R. and Ribbons, D. W. (1970a). "Methods in Microbiology, Volume 2". Academic Press, New York and London.

Norris, J. R. and Ribbons, D. W. (1970b). "Methods in Microbiology Volume 3A". Academic Press, New York and London.

Norris, J. R. and Ribbons, D. W. (1970c). "Methods in Microbiology Volume 3B". Academic Press, New York and London.
Osman, A., Bull, A. T., and Slater, J. H. (1976). Growth of mixed microbial populations on orcinol in continuous culture. *In* "Abstracts of Papers, 5th International Fermentation Symposium", p. 124. (H. Dellweg, ed.) Berlin.
Parkes, R. J. (1978). "The seasonal variation of bacteria within the sediments of a polluted estuary". Ph.D. Thesis, University of Aberdeen.
Pawlowsky, U. and Howell, J. A. (1973). Mixed culture biooxidation of phenol. II. Steady state experiments in continuous culture. *Biotechnology and Bioengineering* **15**, 897–903.
Pfaender, F. K. and Alexander, M. (1972). Extensive microbial degradation of DDT *in vitro* and DDT metabolism by natural communities. *Agricultural and Food Chemistry* **20**, 842–846.
Pirt, S. J. (1975). "Principles of Microbe and Cell Cultivation". Blackwell, Oxford.
Postgate, J. R. and Hunter, J. R. (1964). Accelerated death of *Aerobacter aerogenes* starved in the presence of growth-limiting substrates. *Journal of General Microbiology* **34**, 459–473.
Powell, E. O., Evans, C. G. T., Strange, R. E., and Tempest, D. W. (1967). Editorial Comment. *In* "Microbial Physiology and Continuous Culture", pp. 209–210. (E. O. Powell, C. G. T. Evans, R. E. Strange, and D. W. Tempest, eds.) HMSO, London.
Pritchard, P. H., Bourquin, A. W., Frederickson, H. L., and Maziarz, T. (1979). System design factors affecting environmental fate studies in microcosms. *In* "Microbial Degradation of Pollutants in the Marine Environment", pp. 251–272. (A. W. Bourquin and P. H. Pritchard, eds.) United States Environmental Protection Agency, Gulf Breeze, Florida.
Schlegel, H. G. and Jannasch, H. W. (1967). Enrichment cultures. *Annual Review of Microbiology* **21**, 49–70.
Scott, C. D. and Hancher, C. W. (1976). Use of a tapered fluidised bed as a continuous bioreactor. *Biotechnology and Bioengineering* **18**, 1393–1403.
Scott, C. D., Hancher, C. W., Holladay, D. W., and Dinsmore, G. B. (1975). A tapered fluidised bed reactor for treatment of aqueous effluents from coal conversion processes. *In* "Proceedings of the Symposium on Environmental Aspects of Fuel Conversion Technology II", pp. 233–240. U.S. Environmental Protection Agency.
Senior, E. (1977). "Characterisation of a microbial association growing on the herbicide dalapon". Ph.D. Thesis, University of Kent.
Senior, E., Bull, A. T., and Slater, J. H. (1976). Enzyme evolution in a microbial community growing on the herbicide dalapon. *Nature, London* **263**, 476–479.
Sharp, R. F. and Taylor, B. P. (1969). A new soil percolator for the elective culture of soil organisms. *Soil Biology and Biochemistry* **1**, 191–194.
Sinclair, C. G., King, W. R., Ryder, D. N., and Topiwala, H. H. (1971). Some difficulties in fitting dynamic models to experimental transient data in continuous culture. *Biotechnology and Bioengineering* **13**, 451–452.

Slater, J. H. (1978). The role of microbial communities in the natural environment. *In* "The Oil Industry and Microbial Ecosystems", pp. 137–154. (K. W. A. Chater and H. J. Somerville, eds.) Heyden & Son, London.

Slater, J. H. and Lovatt, D. (1982). Biodegradation and the significance of microbial communities. *In* "Biochemistry of Microbial Degradation", in press. (D. T. Gibson, ed.) Marcel Dekker, New York.

Slater, J. H. and Bull, A. T. (1978). Interactions between microbial populations. *In* "Companion to Microbiology", pp. 181–206. (A. T. Bull and P. M. Meadow, eds.) Longman, London.

Slater, J. H. and Godwin, D. (1980). Microbial adaptation and selection. *In* "Contemporary Microbial Ecology", pp. 137–160. (D. C. Ellwood, J. N. Hedges, M. J. Latham, J. M. Lynch, and J. H. Slater, eds.) Academic Press, New York and London.

Spycher, G. and Young, J. L. (1977). Density fractionation of water dispersible soil organic-mineral particles. *Communications in Soil Science and Plant Analysis* **8**, 37–48.

Stanley, S. O. and Rose, A. H. (1967). On the clumping of *Corynebacterium xerosis* as affected by temperature. *Journal of General Microbiology* **48**, 9–23.

Sudo, S. Z. (1977). Continuous culture of mixed oral flora on hydroxyapatite coated glass beads. *Applied and Environmental Microbiology* **33**, 450–458.

Tan, T. L. (1980). Effect of long term lead exposure on seawater and sediment bacteria from heterogeneous continuous flow cultures. *Microbial Ecology* **5**, 295–311.

Taylor, P. A. and Williams, P. J. LeB. (1975). Theoretical studies on the coexistence of competing species under continuous flow conditions. *Canadian Journal of Microbiology* **21**, 90–98.

Tempest, D. W. (1970). The continuous culture of microorganisms: 1. Theory of the chemostat. *In* "Methods in Microbiology, Volume 2", pp. 259–276. (J. R. Norris and D. W. Ribbons, eds.) Academic Press, New York and London.

Tilman, D. (1977). Resource competition between planktonic algae: An experimental and theoretical approach. *Ecology* **58**, 338–348.

van Gemerden, H. and Jannasch, H. W. (1971). Continuous culture of *Thiorhodaceae*. *Archiv für Mikrobiologie* **79**, 345–353.

Varon, M. (1979). Selection of predation resistant bacteria in continuous culture. *Nature, London* **277**, 386–388.

Veldkamp, H. (1976). "Continuous Culture in Microbial Physiology and Ecology". Meadowfield Press, Durham.

Veldkamp, H. (1977). Ecological studies with the chemostat. *Advances in Microbial Ecology* **1**, 59–94 (M. Alexander, ed.).

Veldkamp, H. and Jannasch, H. W. (1972). Mixed culture studies with the chemostat. *Journal of Applied Chemistry and Biotechnology* **22**, 105–123.

Veldkamp, H. and Kuenen, J. G. (1973). The chemostat as a model system for ecological studies. *Bulletins from the Ecological Research Committee (Stockholm)* **17**, 347–355.

Wardell, J. N. and Brown, C. M. (1980). Growth of bacteria at glass surfaces in continuous flow enrichments. *Society for General Microbiology Quarterly* **7**, 111.

Watson, J. R. (1977). Seasonal variation in the biodegradation of 2,4-D in river water. *Water Research* **11**, 153–157.

Weimer, P. J. and Zeikus, J. G. (1977). Fermentation of cellulose and cellobiose by *Clostridium thermocellum* in the absence and presence of *Methanobacterium thermoautotrophicum*. *Applied and Environmental Microbiology* **33**, 289–297.

Wilder, C. T., Cadman, T. W., and Hatch, R. T. (1980). Feedback control of a competitive mixed culture system. *Biotechnology and Bioengineering* **22**, 89–106.

Wilkinson, T. G. and Hamer, G. (1974). Wall growth in mixed bacterial cultures growing on methane. *Biotechnology and Bioengineering* **16**, 251–260.

Wilkinson, T. G., Topiwala, H. H., and Hamer, G. (1974). Interactions in a mixed bacterial population growing on methane in continuous culture. *Biotechnology and Bioengineering* **16**, 41–59.

Williams, P. P. (1977). Metabolism of synthetic organic pesticides by anaerobic microorganisms. *Residue Reviews* **66**, 63–135.

Wimpenny, J. W. T. (1981). Spatial order in microbial ecosystems. *Biological Reviews* (in press).

Winogradsky, S. (1949). "Microbiologie du Sol. Oeuvres complètes". Masson, Paris.

Yang, R. D. and Humphrey, A. E. (1975). Dynamic and steady state studies of phenol biodegradation in pure and mixed cultures. *Biotechnology and Bioengineering* **17**, 1211–1235.

Yano, T. and Koga, S. (1969). Dynamic behaviour of the chemostat subject to substrate inhibition. *Biotechnology and Bioengineering* **11**, 139–153.

4

The Analysis of Microbial Interactions and Communities *in situ*

J. C. Fry

1. INTRODUCTION

Microorganisms interact with each other whenever they come into close proximity and, consequently, the study of microbial interactions is of vital importance to those interested in ecology. The methods used to analyse interactions are important since the interpretation of the way in which microorganisms interact can depend upon the precise technique used to study the interaction. Similarly, microbe–microbe interactions cannot be quantified unless the whole community of microorganisms can be adequately described. Hence this chapter deals, firstly, with methods of describing communities in terms of the species present, their numbers, biomass, activities and growth rates; and, secondly, discusses the ways in which microbial interactions are studied. Apart from these techniques an investigator of communities and interactions will also need to use techniques to determine the chemical and physical nature of the environment, as the way microorganisms react with their environment determines how they interact with each other.

The chapter will not consider laboratory methods (see Parkes, pp. 45–102) but is confined to the analysis of interactions and communities *in situ*. By this is meant the study of microorganisms as they are in their natural environments. The natural environments considered in this chapter are mainly freshwater and marine, but soil, waste waters, the rumen, and some other habitats will also be considered. Studies *in situ* sometimes demand undertaking measurements actually in the environment, such as the estimation of chlorophyll *a in situ* described in Section 2.3.2, but more commonly it means taking a sample and making a measurement which can be directly related to the natural habitat. This means that samples must be taken speedily and dealt with in the laboratory, or close to the sampling site, as quickly as possible and certainly within 24 h. Such speed is necessary as it is well known, for example, with enclosed aquatic samples that microorganisms quickly multiply and adapt to their new environment: this is the so called "bottle effect" (Jones, 1977a).

The details of sampling will not be discussed further as they have been described fully elsewhere. Such descriptions exist for soil (Parkinson *et al.*, 1971; Williams and Gray, 1973) and the aquatic habitats of water and sediment (Collins *et al.*, 1973; Collins, 1977; Sieburth, 1979). Once samples are taken they often need to be pre-treated to make them more amenable to study, including techniques such as breaking up aggregations in waste waters (Gayford and Richards, 1970; Banks and Walker, 1977), removal of bacteria from plant surfaces (Fry and Humphrey, 1978a) and concentration of water samples (Jones, 1979a; Sieburth, 1979). Other pretreatment techniques are found in the general references given above for general sampling. Pretreatment should be used with great care as it can drastically change the nature of the sample and so render the results useless for interpretation in terms of what occurs *in situ*.

Before studies *in situ* can be undertaken and a sampling programme planned (Montgomery and Hart, 1974), a working knowledge of the habitat is required. Once again excellent general descriptions of most habitats exist in the literature, such as those for fresh waters (Wetzel, 1975), the human skin (Noble and Pitcher, 1978), the rumen (Hungate, 1966), and soil (Hattori and Hattori, 1976).

2. METHODS FOR DESCRIBING COMMUNITIES

2.1 Isolation and Identification

Isolation and identification methods for microorganisms are familiar to most microbiologists and are well documented, so they need only brief mention here. Although the isolation of microorganisms from natural habitats can be accomplished by most microbiologists, many naturally occurring organisms are difficult to culture after primary isolation and consequently are hard to study or identify: oligotrophic bacteria provide a good example of this phenomenon (Kuznetsov *et al.*, 1979). Many general texts give details of isolation procedures from natural habitats (Skerman, 1967; Aaronson, 1970). There are also available many reviews of more specific methods, such as those for soil microorganisms (Parkinson *et al.*, 1971), autotrophic bacteria (Collins, 1969), and algae (Stein, 1973).

Unlike isolation, identification takes more skill and practice. Eukaryotic microorganisms are easier to identify than the prokaryotes, and many good keys are available for some groups, such as the algae (George, 1976). Simpler keys, designed for specific purposes, are often most useful, such as those for freshwater algae (Prescott, 1954; Bellinger, 1974; Belcher and Swale, 1976), common diatoms (Weber, 1971), aquatic hyphomycetes (Ingold, 1975), ciliated protozoa in activated sludge (Curds, 1969), and the amoebae of soil and water (Page, 1976). Bacteria are particularly difficult to identify and, even with the help of the computer-based techniques of numerical taxonomy, many natural isolates are unidentifiable (Kaneko *et al.*, 1979) and have to be given imprecise names, such as "yellow chromogens", "coryneform bacteria" or "Gram-negative rods" (Austin *et al.*, 1978). Bergey's Manual (Buchanan and Gibbons, 1974) remains the main text used for identifying bacteria, but many identification schemes for smaller bacterial groups have been developed, such as heterotrophic bacteria (Bonde, 1977) and aquatic, Gram-negative heterotrophic bacteria (Gibson *et al.*, 1977). Examination of species diversity within communities is difficult at present because of the problems of identifying bacteria; consequently, the use of numerical profiles might enable quicker progress to be made in this type of work (Griffiths and Lovitt, 1980). Numerical profiles require only the results of standard taxonomic tests and give organisms numbers but not names; so, diversity can be examined but not species structure; of course, dominant types could be identified.

2.2 Enumeration

Counting microorganisms is probably the commonest way in which microbial communities are described by microbial ecologists. The two main types of enumeration procedure used are the total and the viable counting methods. Eukaryotic microorganisms are almost exclusively enumerated by total counting methods and this is particularly true for algae and protozoa which are difficult to grow reliably from individual cells making viable counting procedures difficult. Bacteria are counted by both methods. However, with most natural communities of bacteria, the conventional viable count is only a small proportion of the total count. This proportion is about 1 to 15% in soil (Paul and Van Veen, 1978), less than 10% in rumen (Hobson and Summers, 1978), 0.9 to 22.2% in fresh water (Sorokin, 1972a), 5 to 10% for freshwater epiphytic bacteria (Ramsay and Fry, 1976) and only 0.001 to 0.1% in some sea water samples (Kogure *et al.*, 1979). It has been suggested that for soil this discrepancy is due to the total counts overestimating the true viable count (Parkinson *et al.*, 1971); similar suggestions have been made for aquatic habitats where large numbers of bacteria are thought by some to be dormant (Stevenson, 1978). Other studies have shown that 10 to 90% of the freshwater epiphytic bacteria (Ramsay and Fry, 1976) and 2 to 56% of the coastal water bacteria (Meyer-Reil, 1978) are able to take up radioactively labelled glucose, whilst 5 to 36% of the bacteria from coastal waters are able to respire (Zimmerman *et al.*, 1978). These results can be interpreted as meaning that many more bacteria are alive and active in natural systems, particularly water, than has been previously thought. This view is substantiated by the recent developments of microscopic cultural methods which enable much higher, and perhaps more realistic, viable counts to be obtained than has previously been possible. These techniques are substantially better than the older techniques involving the growth of microcolonies on membrane filters (Straskrabova, 1973) and have given higher viable to total count ratios of about 1:10 in seawater, by counting the long cells formed after nalidixic acid inhibition of cell division (Kogure *et al.*, 1979), and 1:5 to 1:2 in freshwater samples, by counting micro-colonies by a technique similar to that of Postgate *et al.* (1961; J. C. Fry and T. Zia, unpublished observations). An ideal enumeration procedure ought to count all the viable organisms but such a method has yet to be achieved.

It has been suggested that 1×10^6 bacteria ml^{-1} have to be present before they can contribute substantially to an ecosystem (Brock, 1966). Also in the rumen, where total counts are about 1 to 5×10^{10} per g rumen contents, it has been suggested (Wolin, 1979) that only bacteria which occur above 1 × 10^8 per g rumen contents should be considered as part of the dominant community. These suggestions show that microbiologists should be careful when interpreting changes which occur in small numbers of bacteria in natural ecosystems. However, it should not be forgotten that small numbers of bacteria can quickly grow to larger, significant populations given the right conditions.

2.2.1 Total Counts

Total or direct counts of microorganisms attempt to count all microorganisms of a particular group in a sample. They ought to be undertaken in all ecological studies since they can give useful baseline information against which all other results can be assessed.

Total counts of most eukaryotic microorganisms and the larger prokaryotes are commonly performed on suspensions observed in counting chambers (Jones, 1979a). Detailed methods for the algae have often been described (Edmondson, 1969; Margalef, 1969; Wetzel and Westlake, 1969) and sometimes involve concentrating the algae in sedimentation chambers, preserving the samples with preservatives like Lugol's iodine and viewing by bright field or phase contrast microscopy with inverted or conventional microscopes. Both protozoa and fungi can be counted by similar methods using, for water samples, specialized counting chambers (Goulder, 1972, 1975), algal counting chambers (Willoughby, 1978) and microslide tubes (Poole and Fry, 1980). Stained agar films (Jones and Mollison, 1948; Jenkinson *et al.*, 1976) and other techniques (Parkinson *et al.*, 1971) are often used in soil. For the larger prokaryotes, such as the bacterium *Leptothrix ochrarea*, concentration by membrane filtration and enumeration on the filter is often necessary for water samples (Jones, 1975, 1979a). More recently total numbers of small aquatic eiphytic bacteria have been enumerated successfully by direct staining on the plant (Hossell and Baker, 1979a,b). With most of these techniques it is possible to identify eukaryotic microorganisms whilst counting and so record the numbers of dominant species or groups of organisms. This is rarely possible for bacteria unless they are morphologically distinct.

Recently for bacteria in soil and aquatic habitats, the most commonly used and best total counting procedures have been those using fluorescent stains or optical brighteners. These are much better than the older methods which involved, for example, conventional stains for soil bacteria (Parkinson *et al.*, 1971), erythrosine stained aquatic bacteria on membrane filters (Sorokin and Overbeck, 1972) and phase contrast counts of bacterioplankton (Salonen, 1977). Two of the most popular methods for soil bacteria are those using fluorescein isothiocyanate (Babiuk and Paul, 1970; Greaves *et al.*, 1978) and a europium chelate-fluorescent brightener combination (Anderson and Slinger, 1975a,b). In aquatic systems the best techniques are those which stain the bacteria for 3 to 5 min with 5 to 10 mg acridine based fluorochrome l^{-1}, such as acridine orange or euchrysine 2GNX, before filtering onto suitable black membrane filters, either commercially purchased or specially dyed (Jones and Simon, 1975a), and counting green fluorescing bacteria by epifluorescence microscopy. These techniques have been described for water and discussed in detail elsewhere (Daley and Hobbie, 1975; Hobbie *et al.*, 1977) with full practical details sometimes given (Jones, 1979a). However, it is possible to stain the bacteria after filtration (Ramsay, 1978) and this allows storage of the filters for up to 6 d and is thought to give higher counts. Techniques similar to these have been used for epiphytic bacteria (Fry and Humphrey, 1978a) and sediments (Jones, 1979a) and, similarly, epifluorescence microscopy has been used to count unstained algae directly on stones (Jones, 1974, 1978a). By using fluorescent antibodies it is now possible to count some species of bacteria in natural habitats (Schmidt, 1973a), such as soil (Greaves *et al.*, 1976) and recently even the difficult methanogenic bacteria have been enumerated in sediment with fluorescent antisera (Ward and Frea, 1979a).

2.2.2 Viable Counts

Enumeration techniques for viable organisms usually rely on a cultural index of viability, such as the ability to grow in a liquid medium or to form a microcolony on agar or a membrane filter. Consequently they cannot, and almost certainly never will, count all organisms in a sample. I believe, as do many microbiologists, that viable counts should only be used with care in natural habitats in limited circumstances, and this view is supported by the excellent short summary of a panel discussion given by Schmidt (1973b). These circumstances are:

(1) When the results are supported by other data, such as direct counts, activity or biomass estimates.
(2) When specific physiological or other groups of organisms are being counted.
(3) With proper controls, when an experimental approach is adopted, such as an examination of a treatment or perturbation.

There are many methods available for viable counting, the most commonly used are described for soil and water by Parkinson *et al.*

(1971), Hopton *et al.* (1972), Melchiorri-Sahtolini (1972) and Jones (1979a) and, more generally, in a wide variety of microbiological texts. The plate count technique is the most regularly used for aerobic organisms but it can be modified for anaerobes by incubation with reducing agents in anaerobic jars, although the roll tube technique is often preferred for these organisms (Hungate, 1969) and has been extensively used in rumen microbiology. There are often large errors associated with plate counts of natural samples, but these can be reduced with increased replication (Kaper *et al.*, 1978) or by using a surface spotting technique (Bousfield *et al.*, 1973) instead of the normal spread plate. Membrane filtration techniques enable organisms to be counted when only small numbers are present in water samples but offer few other advantages. Most probable number (MPN) techniques, which can be performed in liquid or on solid media (Harris and Sommers, 1968), allow lower concentrations of organisms to be enumerated and also allow growth in liquid medium to be used as a viability index. MPN methods in liquid media are tedious but can be made easier and quicker to perform by automating them with a method like that developed by Darbyshire (1973) for soil protozoans.

Many selective media have been designed for the enumeration of specific groups of microorganisms from natural habitats and some of the organisms which can be counted with these media and methods are given in Table 1. However, despite their attractiveness in purporting to estimate numbers of sometimes very well-defined groups of organisms, recovery and growth of all the desired organisms rarely occurs (Pratt and Reynolds, 1973; Hoadley and Cheng, 1974); consequently, results obtained with them should be interpreted with caution. Media and methods are often used which attempt to count the maximum number of organisms obtainable from a particular environment and such counts are often called total viable or heterotroph counts. However, as has been argued earlier, they can never be in any sense true total counts. Casein–peptone–starch agar has considerable advantages over other media for enumerating heterotrophic bacteria in freshwater samples (Jones, 1970; Staples and Fry, 1973b; Fry and Humphrey, 1978a). Similarly for sewage samples casitone–glycerol–yeast extract agar has advantages (Pike *et al.*, 1972). However, in other habitats, one new and better medium constantly replaces another, and this has occurred, for example, in marine bacteriology (Vaatanen, 1977). Consequently, it would appear that media and methods need to be chosen carefully and be fully tested before most ecological studies are undertaken.

Comparatively recently some newer methods for viable counting have been suggested; they are more rapidly performed than conven-

Table 1. Methods for the enumeration of some specific groups of viable microorganisms *in situ*; PC = plate count, MF = membrane filtration, RT = roll tube, MPN = most probable number, PFU = plaque forming unit.

Group of microorganism	Method	Habitat	Reference
Actinomycetes and fungi	PC	Soil	Greaves *et al.* (1978)
Aeromonas hydrophila	MF	Water	Rippey and Cabelli (1979)
Anaerobic cellulolytic bacteria	RT	Sediment	Parkes *et al.* (1979)
Aquatic bacteria, many types	PC	Water	Collins *et al.* (1973)
Aromatic degraders	MPN	Soil and sewage	DiGeronimo *et al.* (1978)
Beggiatoa sp.	MPN	Water and sediment	Strohl and Larkin (1978)
Bdellovibrio bacteriovorus	PFU	Water, sediment and sewage	Staples and Fry (1973a)
Candida albicans	MF	Water	Buck and Bubucis (1978)
Coliforms	MPN	Water	Olson (1978)
Clostridium perfringens	MF	Water	Bisson and Cabelli (1979)
Denitrifying bacteria	MPN	Soil	Focht (1978)
Enteroviruses	PFU	Sewage	Hurst *et al.* (1978)
Escherihia coli	MF	Water	Halls and Ayres (1974)
Exoenzyme producers	PC	Water	Jones (1971) Fry and Humphrey (1978a)
Faecal streptococci	MF, MPN	Water	Dutka and Kwan (1978)
Fluorescent pseudomonads	PC	Soil	Simon *et al.* (1973)
Gingival bacteria, several species	PC	Mouth	van Palenstein Helderman (1975)
Gram-negative bacteria	PC	Water, sediment and sewage	Fry and Staples (1974)
Halophilic bacteria	PC	Water	Gibbons (1969)
Indicator bacteria	MF, MPN	Water	Anon (1969)
Methanogenic bacteria	MPN	Sewage	Iannotti *et al.* (1978)
Micrococcus luteus	PC	Skin	Murphy (1975)

Petroleum-degrading bacteria	MPN	Water	Maccubbin and Kator (1979)
Pseudomas aeruginosa	MF	Water	Brodsky and Ciebin (1978)
Rumen bacteria, many types	PC, RT	Rumen	Hobson (1969)
Sulphate reducing bacteria	RT	Sediment	Parkes *et al.* (1979)
Thermoactinomyces	MF	Water	Al-Diwany *et al.* (1978)
Thiobacillus	MF	Soil	Mouraret and Baldensperger (1977)

tional methods but none has yet been used extensively. Apart from the microscopic methods mentioned previously (p. 106), these newer methods include dip slides for counting bacterial indicators of water pollution (Mara, 1972; Cotton *et al.*, 1975), and a β-galactosidase assay for faecal coliforms (Warren *et al.*, 1978).

2.3 Estimates of Biomass

Enumeration is often considered an inadequate description of microorganisms in communities when microbes of different sizes are being investigated (Brock, 1966). To overcome this problem the biomass of a microbial community is often estimated. Estimates of total biomass or the biomass of specific groups of organisms can be made. Jones (1979a) gives a particularly good discussion and description of the major techniques used in fresh waters and many of these methods are thought to be applicable to other habitats, such as the sea (Sieburth, 1979) and soil (Parkinson *et al.*, 1971; Greaves *et al.*, 1976; Paul and van Veen 1978; Dommergues *et al.*, 1978).

2.3.1 Direct Methods

These methods all involve mutiplying a total count of the relevant microorganisms by a function dependent on their size. Tables of average sizes are available for some planktonic algae (Findenegg, 1969; van Heusden, 1972; Wetzel, 1975) and protozoa (Michiels, 1974; Poole and Fry, 1980) but not for bacteria, although average cell volumes of bacteria for some fresh waters have been suggested (Straskrabova and Sorokin, 1972). These tables should be used with great caution and it is usually better to measure the size of organisms in a suitable sample. Sizes of microorganisms are usually estimated as volumes from measurements after observation with light (Casida, 1971) or transmission electron microscopes (Watson *et al.*, 1977). The dimensions of individual bacteria can be measured directly or with a split image eyepiece (Powell and Errington, 1963) and a microprojector can also be used to make measuring easier (Straskrabova and Sorokin, 1972). The volume is calculated for regular shapes using basic formulae (Findenegg, 1969; Parkinson *et al.*, 1971; Rodina, 1972; Jones, 1979a) or, for irregularly shaped organisms, by the water displacement of plasticine scale models.

The Coulter counter can also be used for fresh water (Mulligan and Kingsbury, 1968; Evans and McGill, 1969) or marine (Sheldon and Parsons, 1967; Sieburth, 1979) algae and perhaps protozoa (Curds *et al.*, 1978). However this method is only applicable *in situ* for large organisms in water samples containing little debris. Once the average size has been determined the biomass can be expressed directly as biovolume (Jenkinson *et al.*, 1976) or converted to biomass by multiplying by the specific gravity to obtain the wet weight and by other factors to obtain the dry weight or the cell carbon content (Jones, 1979a). Specific gravities of 1.0 (Jones, 1979a), 1.1 (van Veen and Paul, 1979) and 1.5 (Parkinson *et al.*, 1971) have been suggested although van Veen and Paul (1979) gave warnings about the use of a constant. Although these techniques are time consuming and tedious to use, they allow the biomass of specific microbial groups or morphologically distinct species to be estimated in almost any habitat and are often used to validate other methods (Jenkinson *et al.*, 1976).

2.3.2 Indirect Methods

These methods usually involve the estimation of a chemical component of the microbial cell which is present in direct proportion to the amount of cellular material. Many methods have been devised and consequently only the most commonly used or promising are discussed here. They are often claimed to have the advantage over direct methods as they sometimes measure the biomass of viable microorganisms present whereas direct methods cannot distinguish between living and dead microorganisms.

The most commonly used indirect method of biomass estimation is the adenosine triphosphate (ATP) method. This involves extracting the cells to remove the ATP and destroy the cellular ATPase and measuring the amount of ATP present by the emission of light from its reaction with luciferin, mediated by the fire-fly enzyme luciferase. The details of the method have been described many times (Holm-Hansen and Booth, 1966; Holm-Hansen, 1973; Holm-Hansen and Karl, 1978; Jones, 1979a) so need not be described here. The light emitted from the reactions can be measured with a liquid scintilation counter (Stanley and Williams, 1969; Rudd and Hamilton, 1973) or more popularly, and with greater accuracy, by a sensitive photometer (Chappelle and Levin, 1968; Sharpe, 1973; Jones and Simon, 1977). Either the peak height of the light emission or the amount of light emitted over a short length of time is measured and is a function of the amount of ATP present. This technique allows the total microbial biomass from water samples to be measured relatively easily, but it cannot distinguish between bacterial and algal biomass (Jassby, 1975a). The organisms are usually concentrated onto membrane filters but this can lead to underestimates of the biomass (Sutcliffe *et al.*, 1976). This problem can be overcome by measuring the ATP directly in water, without concentration, using an assay of improved sensitivity (Jones and Simon, 1977; Karl, 1979a; Jones, 1979b). Other problems have also arisen due to, for example, extraction of resistant algal cells (Frischknecht and Schneider, 1979), the accumulation of detritus on filters (Perry *et al.*, 1979) and interference due to non-adenine triphosphates (Karl, 1978). Microbial biomass has been measured by the ATP method in dental plaque (Robrish *et al.*, 1978), epiphytic communities (Clark *et al.*, 1978) and soils (Ausmus, 1973). The latter two habitats present special technical difficulties due to extraction problems, inhibition of the enzyme assay and adsorption onto charged sites. These problems have to some extent been overcome by dilution (Jones and Simon, 1977; Karl, 1979b), removal of inhibitors by cation exchange resins (Lee *et al.*, 1971; Cunningham and Wetzel, 1978) and with special extraction procedures using special buffers (Bulleid, 1978) and paraquat to prevent ATP adsorption onto clays in soils (Jenkinson and Oades, 1979; Oades and Jenkinson, 1979). The amount of ATP can be converted to a microbial biomass concentration, expressed as cellular carbon, by multiplication with an appropriate correction factor: factors of 250 (Jones, 1979a) and 286 (Holm-Hansen, 1973) for water and 120 (Oades and Jenkinson, 1979) for soil have been suggested.

Chlorophylls can be used as a biomass indicator for phototrophic microorganisms. To measure chlorophylls in samples taken from the natural habitat, extraction in acetone or methanol is necessary and organisms normally have to be concentrated by filtration. The amount of chlorophyll is measured spectrophotometrically or by fluorescence. These methods have been fully described for planktonic algae (Talling,

1969; Marker, 1972; Youngman, 1978) and periphyton (Wetzel and Westlake, 1969). Bacteriochlorophylls have also been measured by similar techniques (Takahashi and Ichimura, 1968). Biomasses of phototrophic populations *in situ* without the removal of samples can also be obtained by spectrophotometry for both algae (Truper and Yentsch, 1967) and bacteria (Bauld and Brock, 1973) and for algae with a more sensitive fluorometric method (Kiefer, 1973; Tunzi *et al.*, 1974; Heaney, 1978). It is also possible to distinguish between the populations of different types of phototrophic organisms using a fluorescence method *in situ* which measures accessory pigments (Caldwell, 1977).

A promising, specific method for the estimation of bacterial biomass is the lipopolysaccharide assay in which an aqueous extract of amoebocytes from the horseshoe crab, *Limulus polyphemus*, reacts with an endotoxin from Gram-negative bacteria. The amount of complex formed is estimated either by turbidity (Watson *et al.*, 1977; Coates, 1977) or colorimetrically (Nakamura *et al.*, 1977; Maeda and Taga, 1979) and has been applied to sea water and waste water samples (Jorgensen *et al.*, 1979) and is thought to show promise for soil bacterial populations (Dommergues *et al.*, 1978).

Other indirect methods recommended for biomass estimation include those which involve: deoxyribonucleic acid analysis for total biomass in water (Jones, 1979a); chitin analysis for fungal biomass in terrestrial habitats (Swift, 1973) and water (Willoughby, 1978); muramic acid analysis for bacteria in water (King and White, 1977); sediments (Moriarty, 1977) and soil (Millar and Casida, 1970); lipid analysis for total biomass in sediments (White *et al.*, 1979); and the flush of respiration after fumigation (Jenkinson and Powlson, 1976) or glucose addition (Anderson and Domsch, 1978) for total biomass in soil.

2.4 Measuring Activities

Many methods have been used for measuring activities of microorganisms in natural habitats and consequently several different types of activity have been measured: some of the most important methods are discussed in this section. Some techniques are true measurements *in situ* where samples are isolated, treated in some way and then reincubated in the habitat; e.g., most of the carbon dioxide fixation methods are of this type. Others, like those for measuring carbon uptake and mineralization, use laboratory incubation of samples and these are only measurements *in situ* if the samples are treated within a few hours and short incubation times are used so that the natural population has little time to adapt to the laboratory conditions. Another group of methods measures whole community activities but these are not discussed in this section since, as with the measurements of photosynthesis and respiration obtained from diurnal oxygen curves (Edwards *et al.*, 1978), the activities of microorganisms, animals, and plants cannot easily be separated. Microbiologists often wish to partition the activities of different microorganisms in ecological studies but this is very difficult to do either by size fractionation or using inhibitors, and techniques currently available have been discussed in detail elsewhere (Jones, 1977a).

2.4.1 Carbon Dioxide Fixation

The measurement of algal photosynthesis which results in carbon dioxide fixation has been described and used extensively by limnologists, and so a detailed description is unnecessary. Two methods are generally used for measurements *in situ* in both of which water samples are incubated in light and dark bottles. The first method, known as the oxygen technique (Soeder and Talling, 1969), measures the change in oxygen concentration during incubation and is most suitable for relatively productive waters. The second method, the [^{14}C]-carbon technique (Goldman *et al.*, 1969) measures the uptake of [^{14}C]-bicarbonate from solution. It is a very sensitive method and most suitable for unproductive waters. However, it presents more problems than the oxygen method (Talling and Fogg, 1969; Steedmann-Nielsen, 1977) and these should be understood fully before the method is used. Methods very similar to the [^{14}C]-carbon method for algal photosynthesis have been used to estimate bacterial photosynthesis and chemosynthesis, again using light and dark incubations (Sorokin, 1969; Romanenko *et al.*, 1972; Sorokin, 1972b). These methods have been used to assess bacterial production (Anderson and Dokulil, 1977) and, although they are apparently simple, they are also likely to have severe problems associated with them (Hobbie and Rublee, 1975). Carbon dioxide fixation estimates in soil are rare (Hubbard, 1973; Smith *et al.*, 1973) because radiotracers are needed to obtain sensitivity and the processing necessary for measuring the incorporated radioactivity is more difficult in soil than water.

2.4.2 Respiration and Electron Transport Activity

It is possible to obtain an overall estimate of microbial activity *in situ* by measuring respiration, either by oxygen uptake or by carbon dioxide production. The methods, however, do not provide unequivocal results as it is often difficult to separate microbial respiration from that of other organisms. Consequently these techniques often claim to measure total community respiration, but despite this problem they are often used by microbiologists. Chemical oxygen uptake in the sample nearly always occurs and has to be allowed for in these estimations by techniques involving formalin treated controls, which are discussed by Jones (1977a) and described in several of the references quoted later in the section.

Respiration has been extensively estimated as an overall measure of microbial activity in soils (Stotzky, 1972). Common techniques involve oxygen uptake measured in Gilson and Warburg respirometers and carbon dioxide production measured by collecting evolved carbon dioxide from the samples incubated in respirometer flasks (Parkinson *et al.*, 1971; Greaves *et al.*, 1978). In the latter case large soil samples are needed and the evolved carbon dioxide is measured either titrimetrically or by more sophisticated methods, such as infra-red gas analysis. Sealed containers placed over soil in the field can also be used to estimate respiration by carbon dioxide evolution (Parkinson *et al.*, 1971).

Measuring respiration in unmodified aquatic samples is more difficult as they are often less active than soil samples. Sediment respiration is estimated by two methods which both use oxygen uptake as the index of respiration. In the first, samples of sediment and its overlying water are removed and incubated for a short period, not longer than 24 h, and oxygen uptake is followed in sub-samples of overlying water by a sensitive oxygen titration method, such as the ampherometric back-titration proposed by Talling (1973). This method (Collins, 1977) has been successfully used in many studies such as that in which hypolimnetic oxygen depletion in a lake was explained in terms of sediment respiration (Jones, 1976). The second method involves placing a chamber over the sediment in which the overlying water is stirred, to prevent localized oxygen depletion, and either samples are removed for titrimetric oxygen estimation or electrodes are used to indicate the oxygen concentration. The oxygen uptake rate is estimated from the change in oxygen concentration. Many systems are described in the literature (Stein and Denison, 1967; Pamatmat, 1971; Edberg and Hofsten, 1973) but one novel design is the submersible operated grab respirometer (Smith *et al.*, 1978) which allows oxygen uptake and nutrient exchange to be measured and also has the capacity to remove a sediment sample to estimate the relative contribution of the various sediment organisms. Short incubations are needed for water samples which are relatively inactive (Straskrabova, 1979) since this prevents "bottle effects" which contribute to overestimates of respiration (Jones, 1977a). To obtain measureable oxygen changes in a short period, sensitive methods of oxygen determination are needed, although the ampherometric titration has been used (Vaatanen, 1979). It is probably better to use the photometric titration which is ideal for this purpose because it has a constant coefficient of variation of 0.1% (Bryan *et al.*, 1976).

Another overall estimate of microbial activity is electron transport system (ETS) activity which has been used for benthic and planktonic algal and bacterial communities. This activity is estimated from the reduction of 2-(*p*-iodophenyl) 3-(*p*-nitrophenyl) 5-phenyl tetrazolium chloride (INT) by enzymes from disrupted microorganisms. The method has been used primarily for marine samples and a recent comparison of methods has shown that the addition of triton X-100 greatly improved the assay (Christensen and Packard, 1979). It has also been shown (Jones and Simon, 1979) that with freshwater samples sonication increased enzyme yields as does appropriate dilution of sediment. Other electron acceptors have been used, such as resazurin for sediments (Liu and Strachan, 1979) and triphenyl tetrazolium chloride for activated sludge (Klapwijk *et al.*, 1974) but the assay based on INT and Triton X-100 seems the best currently available. Although ETS activity is a useful supplement to the currently available techniques for measuring overall activity and is easier to perform than the respiration methods, it cannot replace them because when ETS activity results are converted to oxygen uptake unrealistically high values are obtained (Jones and Simon, 1979).

2.4.3 Uptake and Mineralization of Carbon Compounds

These techniques have been more carefully studied and more commonly applied to aquatic systems than to soils or other habitats. Consequently, all the methods in this section apply to habitats in water, however there seems no logical reason why they should not be applied in other habitats like soil.

The best accepted technique is the kinetic method proposed by Wright and Hobbie (1965) which has been fully described in many of the succeeding references. The method involves incubation of natural samples for short periods with [^{14}C]-labelled organic substrates at a variety of concentrations. Acid is added after incubation to stop microbial activity and controls before incubation. Microorganisms are removed by filtration or [^{14}C]-carbon dioxide collected with β-phenylethylamine to correct for respiration (Hobbie and Crawford, 1969) or estimate for mineralization. Activity estimates are obtained by an appropriate Michaelis-Menten kinetic plot of the results and the most useful parameters estimated are the maximum potential rate

(V_{max}) and the natural turnover time (T_t). The method has been used in water and sediments (Harrison *et al.*, 1971; Litchfield *et al.*, 1979) and with epiphytic bacteria (Fry and Ramsay, 1977; Fry and Humphrey, 1978a). However, the method is not without its problems which have been extensively discussed (Williams, 1973; Wright, 1973; Hobbie and Rublee, 1977) and in particular the method does not work well in nutrient poor waters, like the open ocean (Vaccaro and Jannasch, 1967). The method measures bacterial activity even in the presence of algae, since algae have higher saturation coefficients than bacteria, a fact which was established soon after the method's proposal (Wright and Hobbie, 1966) and has since been confirmed by conventional techniques (Bennett and Hobbie, 1972) and autoradiography (Hoppe, 1978).

The main problem with the kinetic method is that it can only estimate V_{max} and T_t: the actual rate of uptake (v_n) cannot be estimated unless the natural substrate concentration (S_n) is known. Although v_n is ecologically more useful than V_{max} it is technically difficult to determine since S_n is normally very low and difficult to measure (Hicks and Carey, 1968; Jones and Simon, 1975b). Occasionally v_n has been estimated when S_n was estimated chemically (Crawford *et al.*, 1973; Cavari *et al.*, 1978; Cavari and Hadas, 1979) and by a dilution bioassay (Hobbie and Wright, 1965; Rai, 1979).

Assays of activity using single concentrations of substrates at tracer levels have also been undertaken with [^{14}C]-isotopes (Williams and Askew, 1968; Gocke, 1977) and high specific activity [^{3}H]-substrates (Azam and Holm-Hansen, 1973). These methods offer little advantage over the kinetic method, except perhaps ease of use. However, one method using [^{3}H]-substrates does enable v_n to be estimated although several substrate concentrations are used again (Dietz *et al.*, 1977). It is often useful to combine total counts of bacteria obtained by epifluorescence with activity estimates to calculate a value of activity per bacterium, and three ways of doing this have been suggested using kinetic and single concentration activity estimates (Wright, 1978; Goulder, 1979).

One of the problems with the use of specific carbon compounds in estimating the rates of uptake and mineralization, is that natural habitats contain an unknown mixture of dissolved carbon compounds and, consequently, only relative rates can be obtained. The uptake of [^{35}S]-sulphate by bacteria and algae in the light has been suggested as a measure of their total carbon uptake, as there is a constant C : S ratio of 500 : 1 in these organisms and sulphate is the main source of sulphur for planktonic microorganisms. The original method involved correcting the uptake for the phytoplankton by simultaneous measurement of [^{14}C]-carbon dioxide assimilation to obtain bacterial uptake (Monheimer, 1974, 1975). However, a modification of the method by dark incubation (Jassby, 1975b) estimated bacterial carbon uptake more easily and, consequently, bacterial production could be calculated from this (Jordan and Peterson, 1978). However, problems have recently been found with both methods due to the variability of algal sulphate uptake in the light and dark (Monheimer, 1978). Campbell and Baker (1978a,b) have apparently overcome these problems with a modification of the technique involving size fractionation to partially separate algae and bacteria, and the sulphate uptake is then corrected with the ratio of [^{3}H]-glucose uptake between the size fractions to estimate the bacterial uptake of the whole sample. Suggestions have also been made for the correction of certain purely methodological problems (Jordan *et al.*, 1978; Campbell and Baker, 1978a).

2.4.4 Other Activities

Apart from those methods for measuring microbial activities in natural habitats already discussed, there are many others described in the literature. Most are based on techniques developed for pure cultures and many microbial ecologists have developed these methods for use in natural environments. Space does not permit a complete discussion of all the different types of methods available and so Table 2 lists some which have been used in aquatic and terrestrial studies. These are given only as examples of some of the kinds of techniques that might be used in estimates of microbial activity *in situ*.

2.5 Growth Rate Measurement

Few estimates of microbial growth rates in nature exist. Some workers have attempted to estimate natural growth rates by following changes in numbers of bacteria in soil over long periods (Parinkina, 1973; Bohlool and Schmidt, 1973). However, these estimates are probably unrealistic as account is not taken of death, predation or other losses. Brock (1971) reviewed a much better series of methods which take these and other factors into consideration. These methods include those involving counts of dividing cells, counts of dividing microcolonies or filaments, thymidine-labelling techniques and methods involving loss of algal cells from darkened matts in hot springs. Despite the intrinsic appeal of many of these methods, they have rarely been used. Thymidine-labelling techniques have been used to study cell prolifera-

Table 2. Some methods for measuring microbial activities in the natural environment.

Activity	Habitat	Method	Reference
Cellulolysis	Soil	Glucose production	Benefield (1971)
Cellulolysis	Soil	Loss of tensile strength in cloth	Walton and Allsopp (1977)
Denitrification	Soil	Review of methods	Focht (1978)
Denitrification	Sediment	Loss of $^{15}NO_3$	Chatarpaul and Robinson (1979)
Heterotrophic	Soil	Microcalorimetry	Mortensen *et al.* (1973)
Heterotrophic	Water	Microcalorimetry	Tiefenbrunner (1979)
Methanogenesis	Sediment	Gas evolution	Ward and Frea (1979b)
Nitrate reduction	Sediment	Nitrate reductase assay	Jones (1979c)
Nitrification and nitrate reduction	Sediment	^{15}N dilution	Koike and Hatton (1978)
Nitrogen fixation	Water	$^{15}N_2$ fixation and acetylene reduction	Burris (1972)
Nitrogen fixation	Various	Review of various isotopic and acetylene reduction	Sprent (1979)
Phosphatase	Water	Nitrophenol release from *p*-nitrophenol phosphate	Jones (1972)
Phosphatase	Soil	Phenol release from disodium phenyl phosphate	Greaves *et al.* (1978)
Proteolysis	Water	Dye loss from azure dyed hide powder	Little *et al.* (1979)
Sulphate reduction	Sediment	S^{35} sulphide production from $S^{35}O_4$	Jørgensen (1979)
Urease	Soil	Ammonia evolution from urea	Pettit *et al.* (1976)

tion in soil (Thomas *et al.*, 1974) and lake sediments (Tobin and Anthony, 1978) but organism growth rates were not always calculated. Counts of dividing cells have recently been successfully used in aquatic environments to calculate growth rates *in situ* from the frequency of dividing cells (Hagstrom *et al.*, 1979). Also bacterial counts *in situ* on plant tissue have been recently used to estimate the growth rate of epiphytic bacteria (Hossell and Baker, 1979c), although the method used has much of the appeal of some of the Brock's methods (Bott and Brock, 1970), and the growth rates estimated appeared very low.

The most popular methods used are those which depend on counts obtained from communities of microorganisms isolated from their natural habitat over short lengths of time. With these methods it is hoped that, by incubating the natural sample for a short time to avoid disturbing the microflora, it should be possible to estimate the real growth rate *in situ*. These methods include batch culture techniques, such as those used for rumen bacteria (El-Shazly and Hungate, 1965), and continuous-flow culture methods for bacteria in water (Jannasch, 1969; Hendricks, 1972). The population changes used in these methods have been estimated with techniques other than counting, such as ATP estimation for marine microplankton (Sheldon and Sutcliffe, 1978) and various parameters indicative of soil bacteria (Nannipieri *et al.*, 1978).

It appears obvious from the paucity of microbial growth rate estimate available that more attempts to estimate natural microbial growth rates are badly needed.

3. METHODS FOR STUDYING INTERACTIONS WITHIN NATURAL COMMUNITIES

Compared with other subjects in microbiology, interactions between organisms have been rarely studied and, even when interactions have been investigated, laboratory work has provided the most information. More recently, however, modern microbiologists have started to follow the lead of the earlier pioneers like Winogradsky (1947) and study interactions between microbes actually in natural environments. There are exceptions: for example, interactions in the rumen have been comprehensively examined (Wolin, 1979) and the methods for their study comparatively well documented (Coleman, 1978; Hobson and Summers, 1978; Latham and Wolin, 1978), both *in situ* and in the laboratory. Consequently, this section concentrates on other habitats, particularly water and sediments, but also soil, and attempts to give some examples of the heterogeneous techniques available for studying these interactions in a logical framework.

3.1 Simultaneous Estimates of Number or Biomass

In this group of techniques, counts and biomass estimates of microorganisms within a natural habitat are recorded over a length of time and conclusions drawn about the interactions between the organisms from the way in which the estimates vary with time. This is one of the most straightforward techniques and, consequently, has proved to be most popular.

Silvey and Wyatt (1971) followed this approach by counting the numbers of algae, actinomycetes, and other bacteria in several reservoirs, using mainly plate counts for the bacteria. They concluded that associations often existed between bacteria and algae, and they also claimed that the actinomycetes helped reduce algal populations and Gram-positive bacteria subsequently helped decompose the actinomycetes or their products. Total counts of bacteria on erythrosin stained membrane filters and chlorophyll *a* estimates to indicate algal biomass were used by Menon *et al.* (1972) for a similar purpose. They showed simultaneous peaks of algae and bacteria and suggested a nutritional dependency of the bacteria on the algae due to either photosynthetic secretion of dissolved organic carbon or algal degradation. Similarly, in Abbot's Pond occasional relationships between total counts of algae and viable counts of aerobic heterotrophic bacteria were demonstrated (Hickman and Penn, 1977). Such studies are, however, not generally convincing as several other investigations have failed to show similar interactions. One such study (Jones, 1971), using chlorophyll *a*, plate counts of heterotrophic and exoenzyme-producing bacteria, showed that although some algal peaks were accompanied by bacterial increases many were not, suggesting that the bacteria were not always dependent on algae.

In an elegant study of brackish water and marine bacteria, Hoppe (1977) showed that bacteria enumerated by colony counts were often especially associated with algal blooms and more polluted waters. He also showed that actively metabolizing bacteria determined by autoradiography, which he demonstrated represented the small truly aquatic bacteria, were not controlled by nutrient supply from algal blooms and organic enrichment, but were influenced mainly by temperature. Hence, the saprophytic colony formers could not compete favourably with the actively metabolizing bacteria in the offshore waters. Plate counts were also used to study the interactions between two species of marine luminous bacteria (Ruby and Nealson, 1978) over a 23-month period. Enumeration was performed after 80 to 100 colonies were identified to check the balance between the two species. The counts showed that numbers of each species oscillated seasonally in a complementary way. However, laboratory growth rate studies showed that the population variations were due to their different response to temperature and not due to competition or other interactions.

Anaerobic membrane filter counts of heterotrophic and methane-oxidizing bacteria together with double layer plate counts of sulphate-reducing bacteria in Lake Vechten water were used by Cappenberg (1972) to study their interactions. He suggested, from the patterns of counts with time and depth, that the heterotrophic bacteria reduced oxygen in the hypolimnion, the methane oxidizers scavenged the remaining oxygen which allowed the sulphate reducers to grow just above the sediment. Other similar interactions were studied in the sediment using a roll-tube technique to enumerate methane-producing bacteria (Cappenberg, 1974a). This work suggested that the methane producers were confined to the lower sediment layers by the inhibiting action of hydrogen sulphide produced by sulphate reducers in the upper layers.

Methods for studying the interactions of nitrifying bacteria in soil have been described (Belser, 1979) and once piece of work is of particular interest since it indicated the care needed in interpreting results of enrichments in terms of interactions (Belser and Schmidt, 1978a). In this work nitrifiers were enumerated by a MPN technique and showed that, whereas *Nitrosomonas* sp. was normally considered dominant (Belser, 1979) and was found only in the lower dilution MPN tubes, *Nitrosospira* sp. was present in all the higher dilution tubes. This suggested that *Nitrosomonas* sp. outcompeted other genera in laboratory culture but that *Nitrosospira* sp. was most numerous in the soil. Further work on these interactions *in situ* could be carried out with fluorescent antibody techniques (Belser and Schmidt, 1978b; Belser, 1979). Another study in which MPN enumeration was used suggested that prosthecate bacteria competed better with heterotrophs, and hence attained higher relative populations, in oligotrophic rather than eutrophic waters (Staley *et al.*, 1980). However, the numbers of prosthecate bacteria may have been underestimated as the methods used relied on the growth of both types of bacteria in the same MPN tubes. Thus, only prosthecate bacteria able to compete satisfactorily against heterotrophs in culture were counted.

The next two studies to be discussed also involve direct observation of microbial associations and, hence, are examples of mixtures of the techniques in this and the next section. Interactions of cyanobacteria with bacteria and cyanophages are well known (Stewart and Daft, 1977) and enumeration of these algae and algal-lysing myxobacteria by plaque count on *Anabaena* sp. with double layer plates, have shown the two species can coexist in lakes for most of the time, with only occasional agal population crashes caused by the myxobacteria. Laboratory studies showed that they coexisted because both required alkaline conditions and high temperatures and direct observations indicated that the algae provide an ideal substrate for the myxobacteria to glide upon. Thus this interaction helps the myxobacteria as many hundreds were seen on some cyanophycean colonies (Daft *et al.*, 1975). In another investigation McFeters *et al.* (1978) used membrane filter counts to observe unusually high numbers of coliforms in a pristine stream, thought to be due to algal matts on stones, which grew at the same time, providing nutrients for the coliforms. Additional field work showed the algal matts harboured 8.4×10^6 coliforms m^{-2} and this, together with laboratory work, confirmed the suspected symbiotic relationship.

These examples show some of the strengths and weaknesses of this type of investigation for studying microbial interactions. Two of the major problems are that a change of sampling period influences the interpretation of the results and the irregularity of the apparent interactions which is sometimes observed. These two and other factors can greatly effect the confidence which can be placed in the interactions suggested by these methods.

3.2 Studying Close Microbial Associations

These methods involve qualitative and quantitative observation of two or more microorganisms in direct contact or association. Normally in these cases it is clear that an interaction is occurring but the nature and significance of the interaction needs to be studied. In one study (Caldwell and Caldwell, 1978), 2.6×10^{11} bacteria ml^{-1} were observed in the mucilage around *Anabaena flosaquae* and disruption of the association, by homogenization at 50 000 rpm with glass beads, revealed a single bacterial isolate (observed over two consecutive years) which could not grow in liquid medium without the cyanobacterium. Thus, a firm association between two species was demonstrated from the natural samples.

Another close physical relationship and probable nutritional interaction was reported by Geesey *et al.* (1978) who used epifluorescence total counts of bacteria, electron microscopy and biomass estimates with chlorophyll *a* and ATP. With the electron microscope, they noted that a two-membered community, composed of epilithic algae and sessile bacteria, was held in close physical relationship by a network of fibres. In support of this evidence, other methods showed that the concentration of sessile bacteria coincided with fluctuations of algal biomass.

Another close association which consisted of bacteria, algae, protozoa, and fungi was the sewage fungus complex (Curtis, 1969) and in this case, although counts and biomass estimates of the species in the complex at a large number of sites have adequately described the community (Curtis and Curds, 1971), the nature of the interactions which occur has not been explained.

Predatory interactions between protozoa and bacteria in natural habitats are well known (Curds, 1977), and Fenchel (1975) attempted to quantify this interaction in the sediment of an arctic tundra pond. He measured digestion rates of bacteria in various protozoa, counted the numbers of bacteria in the vacuoles of protozoa collected from the sediment, and counted the total numbers of protozoa which occurred naturally. This demonstrated that protozoal grazing was reponsible for turnover of 5 to 10% of the bacterial population per day. However, it has needed laboratory work in microcosms (Fenchel, 1977), batch and continuous-flow culture (Curds and Bazin, 1977) to obtain a fuller understanding of this type of interaction.

There is an obvious parasitic interaction with some diatoms, desmids, and colonial green algae acting as hosts for fungi and protozoa. The study of this interaction has been reviewed recently by Canter (1979) who described how early work centred on observation and description of the interacting species, whilst more recent work, based on total counts of host and parasite *in situ* during epidemics, has shown that the parasites can reduce natural populations by 99% in two weeks. Once again it has needed laboratory work, using mainly *Rhizophydium planktonicum* which parasitizes *Asterionella* sp. to obtain more detailed information about the interaction.

3.3 Biochemical Approaches

These approaches to the study of interactions *in situ* involve the use of biochemical techniques to examine the relative activity of microorganisms within a natural community. It is often necessary to partition the community by using inhibitors or other methods to separate the activities of microbial groups and this has been recently discussed in detail (Jones, 1977a) but some other examples are given here.

Methods of size fractionation of microorganisms have been investigated using glass fibre or membrane filters (Salonen, 1974; Coveney, 1978) and these techniques have been used to study algal–bacterial interactions. The principle of the approach is to examine the activity of filtered fractions and compare them with the overall activity of the initial sample. Complete separation is not necessary because if one fraction contains only bacteria, the rates of the algal and bacterial activity can be calculated by proportion to biomass, total counts or another activity (Campbell and Baker, 1978a,b). This technique together with measurements of glucose uptake, photosynthesis, and dark carbon dioxide fixation, has been used by Overbeck (1979) to show that bacterial glucose uptake was slow at depths in a lake where algal photosynthesis was highest and was faster below the depth limit for algal photosynthesis, whereas bacterial carbon dioxide fixation occurred in both regions. These interactions, however, have not been explained. Similar techniques were also used by Williams (1970) to show that bacteria competed better than algae for dissolved organic material in seawater. However, it has taken laboratory studies (Jolley *et al.*, 1976; Jolley and Jones, 1977) to describe more precisely a similar interaction between a *Flavobacterium* sp. and *Navicula pelliculosa*. Kinetic methods measuring the uptake rate of several organic compounds on the same samples have also been used to study interactions of heterotrophic bacteria (Hobbie *et al.*, 1968; Robinson *et al.*, 1973). In these studies the uptake rates of different organic compounds have been used to determine the relative contribution of different types of activity, and hence bacterial types, to the overall population activity.

Biochemical interactions between nitrifying bacteria, assumed to be *Nitrosomonas* sp. and *Nitrobacter* sp., in waste waters have been studied by measuring levels of NH_3-nitrogen, NO_2^--nitrogen and $(NO_2^- + NO_3^-)$-nitrogen after 5 h incubation (Hall and Murphy, 1980). This work has resulted in a definition of the rates of nitrifying activities in the presence of heterotrophic bacteria which may lead to the design of better combined carbon and nitrogen removing wastewater treatment plants.

One example of the use of inhibitors comes from the work of Cappenberg (1974b). The interactions between methane-producing and sulphate-reducing bacteria were studied in sediments by adding inhibitors at 1 cm intervals down undisturbed sediment cores. This work showed that, after 10 d incubation, β-fluorolactate which inhibited the sulphate-reducing bacteria, allowed methanogenesis to occur in the upper sediment layers and stopped hydrogen sulphide production. This confirmed the earlier suggestions that hydrogen sulphide produced by sulphate-reducing bacteria controlled where the populations of methanogens were able to grow. Also in this work Cappenberg (1974b) suggested, from the accumulation of organic acids in the inhibitor treated sediment cores, that nutritional interactions occurred between the two bacteria. An elegant series of laboratory experiments provided further confirmation of these suggestions (Cappenberg, 1975) (see Parkes, p. 86).

3.4 Experimentation *in situ*

There are many experimental techniques which have been used to study the interactions actually in natural habitats and the limits to the methodology are set only by the ingenuity of the microbiologist planning the study. The transfer of carbon between organisms in a community has been followed in bottles incubated in the natural environment. Saunders (1972a,b; 1976; Saunders and Storch, 1971) used this approach effectively to demonstrate the transfer of organic materials produced by photosynthesis from planktonic algae to bacteria. Some of his experiments involved incubation *in situ* of 3.8 l bottles of lake water with [^{14}C]-sodium bicarbonate to demonstrate the secretion of photosynthetically fixed [^{14}C]-carbon dioxide as [^{14}C]-labelled organic material and its subsequent transfer to bacteria. The transfer to bacteria was also demonstrated by simulation experiments in which [^{14}C]-glucose was continually added to the bottles to represent the algal secretion of organic material. These experiments also showed the importance of light in generating both the photosynthetic release and consequently the bacterial uptake of organic materials in a regular diurnal cycle.

The predatory interaction between *Bdellovibrio bacteriovorus* and Gram-negative bacteria in natural habitats has been studied both directly and indirectly (Varon and Shilo, 1980). Flasks and dialysis sacs, which allow limited nutrient exchange, have been incubated in a river to study this interaction (Fry and Staples, 1974). In these experiments the numbers of *Bdellovibrio* sp. and its potential prey were followed during the incubations *in situ* and showed that population decreases of *Escherichia coli* were not due to *Bdellovibrio* sp. but to competition with the natural water flora. These experiments also indicated that between 1×10^6 and 1×10^7 prey cells ml^{-1} were needed for *Bdellovibrio* sp. growth and this suggestion was later confirmed in laboratory experiments (Varon and Zeigler, 1978). Dialysis sacs have been used in many other studies including one in which the importance of natural assemblages of organisms in dissolved and particulate organic carbon decomposition was examined (Paerl, 1978).

Other experiments *in situ* have involved litter bags. One investigation (Mason, 1976) incubated *Phragmites* sp. leaves in polyster bags in Alderfen Broad. The respiration of fungi and bacteria was studied after incubation with the aid of the anti-fungal agents, nystatin and actidione, and penicillin and streptomycin as anti-bacterials. The results showed that fungi and bacteria were of similar importance in leaf decomposition after 35 d.

Long-term experiments in aquatic habitats using small containers are useless because of the effects of phenomena such as microbial growth on the surface of the container (Jones, 1977a). To overcome these problems large experimental tubes which are 45 m diameter and which enclose a complete column of water have been designed (Lack and Lund, 1974). These have been used successfully over many years to study, for example, the effect of agricultural fertilizers on algal and bacterial populations (Lund, 1975; Jones, 1977b); hypolimnetic deoxygenation by communities of sediment microorganisms (Jones, 1976) and the interactions of various morphologically distinct bacteria (Jones, 1978b). Such apparatus should, in principle, be useful in the future for studying a wide range of microbial interactions. These tubes are expensive items of equipment and the microbial populations within them are not identical to those in the lake as a whole (Jones, 1973; Lund, 1975), and consequently care should be taken in their use (Lund, 1978) and the interpretation of results obtained from them.

3.5 Mathematical Analysis of Communities

Studies on natural communities often produce large amounts of data from which drawing conclusions can be both difficult and uncertain. Mathematical and statistical analysis of such data can help support the subjective judgements of the investigator with objective statements.

Many techniques are available to microbiologists for the analysis of data from communities and its interpretation in terms of interactions. These range from simple to complex, but few are beyond the grasp of microbial ecologists, although the help of a mathematician or statistician is sometimes required. The simpler techniques include the almost ubiquitously applicable analysis of variance (Sokal and Rohlf, 1969) which has often been applied to ecological data (Ashby and Rhodes-Roberts, 1976) and should almost always be used to decide if sets of mean values are different from each other, as interactions can only be established if it is known if populations change or not.

Correlation coefficients are also often used to test for associations between variables (Sokal and Rohlf, 1969) and to interpret data in terms of interactions between groups of organisms (Ramsay and Fry, 1976). However, large data matrices are better analysed by other methods since large tables of correlation coefficients are almost impossible to interpret without mathematical help. The multivariate techniques of principle component and factor analysis can help with this interpretation. These techniques have been used for analysing populations of microorganisms, such as soil bacteria (Rosswall and Kvillner, 1978) and phytoplankton (Allen and Koonce, 1973). They can help to distinguish patterns of association between organisms isolated from natural habitats, and consequently generate hypotheses about interactions which can be tested in specially designed experiments. These methods have also been found useful in this laboratory for comparing the microbiological methods used for analysing communities, determining relationships between the results from different techniques, and deciding which methods give similar information and which consequently might be redundant in future studies.

The various methods of cluster analysis used in numerical taxonomy (Sokal and Sneath, 1963) are also useful for detecting patterns of association between organisms. Hence, they have been used to show, for example, the dominance of *Xanthamonas campestris* and *Pseudomonas fluorescens* on leaf surfaces (Austin *et al.*, 1978) and the close association between bacteria and protozoa, but not algae and fungi, in the sewage fungus complex (Curtis and Curds, 1971). Once highlighted the reasons for such interactions can then be investigated.

Mathematical models can also be used to describe interactions and there are a wide number of different types of models which can be used and have been simply described (Jeffers, 1978). The use of models for examining microbial interactions in soils has been discussed by Bazin *et al.* (1976), showing that most models are formulated and tested with laboratory work and microcosms or other habitat-simulating experiments. Only rarely have models been tested with data from the natural environment. Curds (1971) formulated a kinetic model which described the interactions between protozoa and bacteria in activated sludge plants and, although the model seemed to fit the limited data available, many of the predictions made with this model have yet to be confirmed (Curds, 1973a,b). Despite this the model has received fairly general acceptance as a good description of protozoa–bacteria interactions in these waste treatment systems.

Multiple regression and time series analysis are two methods of empirical model-building which could perhaps be useful in understanding microbial interactions *in situ*. Although multiple regression has been used to examine natural communities of organisms (Jones, 1978b), it has mainly been used in natural habitats as a method of explaining microbial changes in terms of environmental variables (Brasfeild, 1972; Fry and Humphrey, 1978b). There is no reason, however, why variation in microbial populations should not be explained in terms of other microbial parameters by this technique and, hence, help understand microbial interactions. Although time series analysis has rarely been used in microbial ecology, it has been used to examine predator–prey relationships in mammal populations (Williamson, 1975) and consequently should be useful in understanding microbial interactions. Of particular promise is the cross-spectral analysis (Platt and Denman, 1975) which allows multiple time series to be examined to test whether one series regularly leads or follows another, so attempting to determine causal relationships. If the series were for microbial variables interaction between microorganisms could be examined with this technique.

4. CONCLUSIONS

Natural microbial communities are relatively easy to study and describe and the most successful studies often use a wide variety of methods. Techniques must be chosen to suit the aims of the investigation and the necessity of measuring environmental variables at the same time in order to help the interpretation of results. Although some of the methods *in situ* are inadequate, constant effort from microbiologists should help to provide a large battery of reliable and well-understood techniques from which to choose. The maintenance of this effort is essential if standards of research in microbial ecology are to continue improving.

The study of microbial interactions *in situ* is in its infancy and in some habitats, such as the skin (Noble and Pitcher, 1978) very little work *in situ* has been undertaken although a larger body of laboratory work exists. Despite this, studying interactions between naturally occurring microorganisms is really little different from studying communities because these communities are made up of numerous interacting microorganisms. Consequently it is essential to understand the interactions if the community structure is to be appreciated. It is common with investigations in natural habitats to study, firstly, the interactions between microorganisms and their physico-chemical environment and, secondly, the interactions between organisms. Most studies *in situ* of microbial interactions use standard techniques to examine members of communities and their activities and then interpret the results in terms of the association being studied. It would appear that our knowledge of these interactions will only improve substantially if microbiologists design specific experiments *in situ* to study interactions. Adequate methods for partitioning microorganisms by size and inhibitors are also urgently required. Field studies very rarely prove the association between microorganisms and it will have been evident from the studies discussed earlier (Section 3) that laboratory experiments are normally required to provide definitive evidence in support of the hypotheses generated from the natural environment (see Parkes p. 81). Although many microbiologists begin studying natural interactions in the laboratory, I believe it is essential to start an investigation with a field study. If this is done only those interactions relevant to a natural habitat will be studied and effort will not be spent studying interesting interactions which are irrelevant to those occurring *in situ*.

The nearest methodology *in situ* has come to defining natural interactions is probably in the studies of interactions between the phytosynthetic, iron-oxidizing, sulphate-reducing and other bacteria which occur at the thermocline and oxyline of eutrophic lakes and similar waters. These studies (Kuznetsov, 1977; Jones, 1978b; Parkin and Brock, 1980) demonstrate the value of using many techniques as counts, isolations, identifications, activity, and biomass measurements have all been used.

Mathematical techniques must not be forgotten as they are invaluable in unravelling the complexities which exist within the large bodies of data which are often produced from research in natural habitats.

REFERENCES

Aaronson, S. (1970). "Experimental Microbial Ecology". Academic Press, New York and London.

Al-Diwany, L. J., Unsworth, B. A., and Cross, T. (1978). A comparison of membrane filters for counting *Thermoactinomyces* endospores in spore suspensions and river water. *Journal of Applied Bacteriology* **45**, 249–258.

Allen, T. F. H. and Koonce, J. F. (1973). Multivariate approaches to algal stratagems and tactics in systems analysis of phytoplankton. *Ecology* **54**, 1234–1246.

Anderson, J. P. E. and Domsch, K. H. (1978). A physiological method for the quantitative measurement of microbial biomass in soils. *Soil Biology and Biochemistry* **10**, 215–221.

Anderson J. R. and Slinger, M. S. (1975a). Europium chelate and fluorescent brightner staining of soil propagules and their photomicrographic counting. I. Methods. *Soil Biology and Biochemistry* **7**, 205–209.

Anderson, J. R. and Slinger, M. S. (1975b). Europium chelate and fluorescent brightner staining of soil propagules and their photomicrographic counting. II. Efficiency. *Soil Biology and Biochemistry* **7**, 211–215.

Anderson, R. S. and Dokulil, M. (1977). Assessments of primary and bacterial production in three large mountain lakes in Alberta, Western Canada. *International Revue der Gesamten Hydrobiologie* **62**, 97–108.

Anon (1969). "The Bacterial Examination of Water Supplies". Reports in Public Health and Medical Subjects No. 71. HMSO, London.

Ashby, R. E. and Rhodes-Roberts, M. E. (1976). The use of analysis of variance to examine the variations between samples of marine bacterial populations. *Journal of Applied Bacteriology* **41**,439–451.

Ausmus, B. S. (1973). The use of the ATP assay in terrestrial decomposition studies. *Bulletins from the Ecological Research Committee (Stockholm)* **17**, 223–234.

Austin, B., Goodfellow, M. and Dickinson, C. H. (1978). Numerical taxonomy of phylloplane bacteria isolated from *Lolium perenne*. *Journal of General Microbiology* **104**, 139–155.

Azam, F. and Holm-Hansen, O. (1973). Use of tritiated substrates in the study of heterotrophy in seawater. *Marine Biology* **23**, 191–196.

Babuik, L. A. and Paul, E. A. (1970). The use of fluorescein isothiocyanate in the determination of the bacterial biomass of grassland soil. *Canadian Journal of Microbiology* **16**, 57–62.

Banks, C. J. and Walker, I. (1977). Sonication of activated sludge flocs and the recovery of their bacteria on solid media. *Journal of General Microbiology* **98**, 363–368.

Bauld, J. and Brock, T. D. (1973). Ecological studies of *Chloroflexis*, a gliding photosynthetic bacterium. *Archiv für Mikrobiologie* **92**, 267–284.

Bazin, M. J., Saunders, P. T., and Prosser, J. I. (1976). Models of microbial interactions in the soil. *CRC Critical Reviews in Microbiology* **4**, 463–498.

Belcher, H. and Swale, E. (1976). "A Beginner's Guide to Freshwater Algae". HMSO, London.

Bellinger, E. G. (1974). A key to the identification of the more common algae found in British freshwaters. *Water Treatment and Examination* **23**, 76–131.

Belser, L. W. (1979). Population ecology of nitrifying bacteria. *Annual Reviews of Microbiology* **33**, 309–333.

Belser, L. W. and Schmidt, E. L. (1978a). Diversity in the ammonia-oxidising nitrifier population of a soil. *Applied and Environmental Microbiology* **36**, 584–588.

Belser, L. W. and Schmidt, E. L. (1978b). Serological diversity within a terrestrial ammonia oxidising population. *Applied and Environmental Microbiology* **36**, 589–593.

Benefield, C. B. (1971). A rapid method for measuring cellulase activity in soils. *Soil Biology and Biochemistry* **3**, 325–329.

Bennett, M. E. and Hobbie, J. E. (1972). The uptake of glucose by *Chlamydomonas* sp. *Journal of Phycology* **8**, 392–398.

Bisson, J. W. and Cabelli, V. J. (1979). Membrane filter enumeration method for *Clostridium perfringens*. *Applied and Environmental Microbiology* **37**, 55–66.

Bohlool, B. B. and Schmidt, E. L. (1973). A fluorescent antibody technique for determination of growth rates of bacteria in soil. *Bulletins from the Ecological Research Committee (Stockholm)* **17**, 336–338.

Bonde, G. J. (1977). Bacterial indication of water pollution. *Advances in Aquatic Microbiology* **1**, 273–364.

Bott, T. L. and Brock, T. D. (1970). Growth and metabolism of periphytic bacteria: methodology. *Limnology and Oceanography* **15**, 333–342.

Bousfield, I. J., Smith, G. L., and Trueman, R. W. (1973). The use of semi-automatic pipettes in the viable counting of bacteria. *Journal of Applied Bacteriology* **36**, 297–299.

Brasfeild, H. (1972). Environmental factors correlated with size of bacterial populations in a polluted stream. *Applied Microbiology* **24**, 349–352.

Brock, T. D. (1966). "Principles of Microbial Ecology". Prentice-Hall, New Jersey.

Brock, T. D. (1971). Microbial growth rates in nature. *Bacteriological Reviews* **35**, 39–58.

Brodsky, M. H. and Ciebin, B. W. (1978). Improved medium for recovery and enumeration of *Pseudomonas aeruginosa* from water using membrane filters. *Applied and Environmental Microbiology* **36**, 36–42.

Bryan, J. R., Riley, J. P., and Williams, P. J. LeB. (1976). A Winkler procedure for making precise measurements of oxygen concentration for productivity and related studies. *Journal of Experimental Marine Biology and Ecology* **21**, 191–197.

Buchanan, R. E. and Gibbons, N. E. (1974). "Bergey's Manual of Determinative Bacteriology". Williams and Wilkins, Baltimore.

Buck, J. D. and Bubucis, P. M. (1978). Membrane filter procedure for enumeration of *Candida albicans* in natural waters. *Applied and Environmental Microbiology* **35**, 237–242.

Bulleid, N. C. (1978). An improved method for the extraction of adenosine triphosphate from marine sediment and seawater. *Limnology and Oceanography* **23**, 174–178.

Burris, R. H. (1972). Measurement of biological N_2 fixation with $^{15}N_2$ and acetylene. *In* "Techniques for the Assessment of Microbial Production and Decomposition in Freshwaters", pp. 3–14. (Y. I. Sorokin and H. Kadota, eds.) Blackwell Scientific Publications, Oxford.

Caldwell, D. E. (1977). Accessory pigment fluorescence for quantitation of photosynthetic microbial populations. *Canadian Journal of Microbiology* **23**, 1594–1597.

Caldwell, D. E. and Caldwell, S. J. (1978). A *Zoogloea* sp. associated with blooms of *Anabaena flos-aquae*. *Canadian Journal of Microbiology* **24**, 922–931.

Canter, H. M. (1979). Fungal and protozoan parasites and their importance in the ecology of phytoplankton. *Annual Reports of the Freshwater Biological Association* **47**, 43–50.

Campbell, P. G. C. and Baker, J. H. (1978a). Estimation of bacterial production in freshwaters by the simultaneous measurement of [^{35}S]-sulphate and D-[^{3}H]-glucose uptake in the dark. *Canadian Journal of Microbiology* **24**, 939–946.

Campbell, P. G. C. and Baker, J. H. (1978b). Measurement of sulphate uptake in the dark by suspended microorganisms: application to running waters. *Verhandlung der Internationalen Vereinigung fur theoretische und angewandte Limnologie* **20**, 1423–1428.

Cappenberg, T. E. (1972). Ecological observations on heterotrophic, methane-oxidising and sulphate-reducing bacteria in a pond. *Hydrobiologia* **40**, 471–485.

Cappenberg, T. E. (1974a). Interrelations between sulphate reducing and methane producing bacteria in bottom deposits of a freshwaer lake. I. Field observations. *Antonie van Leeuwenhoek* **40**, 285–295.

Cappenberg, T. E. (1974b). Interrelations between sulphate-reducing and methane-producing bacteria in bottom deposits of a freshwater lake. II. Inhibition experiments. *Antonie van Leeuwenhoek* **40**, 297–306.

Cappenberg, T. E. (1975). Relationships between sulphate-reducing and methane-producing bacteria. *Plant and soil* **43**, 125–139.

Casida, L. E. (1971). Microorganisms in unamended soil as observed by various forms of microscopy and staining. *Applied Microbiology* **21**, 1040–1045.

Cavari, B. Z. and Hadas, O. (1979). Heterotrophic activity, glucose uptake and primary productivity in Lake Kinneret. *Freshwater Biology* **9**, 329–338.

Cavari, B. Z., Phelps, G., and Hadas, O. (1978). Glucose concentrations and heterotrophic activity in Lake Kinneret. *Verhandlung der Internationalen Vereinigung fur theoretische und angewandte Limnologie* **20**, 2249–2254.

Chappelle, E. W. and Levin, G. V. (1968). Use of the firefly bioluminescent reaction for rapid detection and counting of bacteria. *Biochemical Medicine* **2**, 41–52.

Chatarpaul, L. and Robinson, J. B. (1979). Nitrogen transformations in stream sediments: ^{15}N studies. *In* "Methodology for Biomass Determinations and Microbial Activities in Sediments", pp. 119–127. (C. P. Litchfield and P. L Seyfried, eds.) American Society for Testing and Materials, Philadelphia.

Christensen, J. P. and Packard, T. T. (1979). Respiratory electron transport activities in phytoplankton and bacteria: Comparison of methods. *Limnology and Oceanography* **24**, 576–583.

Clark, J. R., Mesenger, D. I., Dickson, K. L., and Cairns, J. (1978). Extraction of ATP from Aufwachs communities. *Limnology and Oceanography* **23**, 1055–1059.

Coates, D. A. (1977). Enhancement of the sensitivity of the *Limulus* assay for the detection of Gram-negative bacteria. *Journal of Applied Bacteriology* **42**, 445–449.

Coleman, G. S. (1978). Methods for the study of the metabolism of rumen ciliate protozoa and their closely associated bacteria. *In* "Techniques for the Study of Mixed Populations", pp. 143–163. (D. W. Lovelock and R. Davies, eds.) Academic Press, London and New York.

Collins, V. G. (1969). Isolation, cultivation and maintenance of autotrophs. *In* "Methods in Microbiology, Vol. 3B", pp. 1–52. (J. R. Norris and D. W. Ribbons, eds.) Academic Press, London and New York.

Collins, V. G. (1977). Methods in sediment microbiology. *Advances in Aquatic Microbiology* **1**, 219–272.

Collins, V. G., Jones, J. G., Hendrie, M. S., Shewan, J. M., Wynn-Williams, D. D., and Rhodes, M. E. (1973). Sampling and estimation of bacterial populations in the aquatic environment. *In* "Sampling-Microbiological

Monitoring of Environments", pp. 77–110. (R. G. Board and D. W. Lovelock, eds.) Academic Press, London and New York.
Cotton, R. A., Sladek, K. J., and Sohn, B. I. (1975). Evaluation of a single step bacterial pollution monitor. *Journal of the American Water Works Association* **67**, 449–451.
Coveney, M. F. (1978). Separation of algae and bacteria in lake water by size fractionation. *Verhandlung der Internationalen Vereinigung fur theoretische und angewandte limnologie* **20**, 1264–1269.
Crawford, C. C., Hobbie, J. E., and Webb, K. L. (1973). Utilization of dissolved organic compounds by microorganisms in an estuary. *In* "Esturine Microbial Ecology", pp. 169–180. (L. H. Stevenson and R. R. Colwell, eds.) University of South Carolina Press, Columbia.
Cunningham, H. W. and Wetzel, R. G. (1978). Fulvic acid interferences on ATP determinations in sediments. *Limnology and Oceanography* **23**, 166–173.
Curds, C. R. (1969). "An illustrated key to the British Freshwater Ciliated Protozoa Commonly Found in Activated Sludge". Water Pollution Research Technical Paper No. 12. HMSO, London.
Curds, C. R. (1971). Computer simulations of microbial population dynamics in the activated-sludge process. *Water Research* **5**, 1049–1066.
Curds, C. R. (1973a). A theoretical study of factors influencing the microbial population dynamics of the activated sludge process. I. The effects of diurnal variations of sewage and carnivorous ciliated protozoa. *Water Research* **7**, 1269–1284.
Curds, C. R. (1973b). A theoretical study of factors influencing the microbial population dynamics of the activated-sludge process. II. A computer simulation study to compare two methods of plant operation. *Water Research* **7**, 1439–1452.
Curds, C. R. (1977). Microbial interactions involving protozoa. *In* "Aquatic Microbiology", pp. 69–105. (F. A. Skinner and J. M. Shewan, eds.) Academic Press, London and New York.
Curds, C. R. and Bazin, M. J. (1977). Protozoan predation in batch and continuous culture. *Advances in Aquatic Microbiology* **1**, 115–176.
Curds, C. R., Roberts, D. M., and Wu, C. (1978). The use of continuous cultures and electronic sizing devices to study the growth of two species of ciliated protozoa. *In* "Techniques for the Study of Mixed Populations", pp. 165–177. (D. W. Lovelock and R. Davies, eds.) Academic Press, London.
Curtis, E. J. C. (1969). Sewage fungus: its nature and effects. *Water Research* **3**, 289–311.
Curtis, E. J. C. and Curds, C. R. (1971). Sewage fungus in rivers in the United Kingdom: The slime community and its constituent organisms. *Water Research* **5**, 1147–1159.
Daft, M. J., McCord, S. B., and Stewart, W. D. P. (1975). Ecological studies on algal lysing bacteria in freshwaters. *Freshwater Biology* **5**, 577–596.
Daley, R. J. and Hobbie, J. E. (1975). Direct counts of aquatic bacteria by a modified epifluorescence technique. *Limnology and Oceanography* **20**, 875–882.

Darbyshire, J. F. (1973). The estimation of soil protozoan populations. *In* "Sampling Microbiological Monitoring of Environments", pp. 175–188. (R. G. Board and D. W. Lovelock, eds.) Academic Press, London and New York.

Dietz, A. S., Albright, L. J., and Tuominen, T. (1977). Alternative model and approach for determining microbial heterotrophic activities in aquatic systems. *Applied and Environment Microbiology* **33**, 817–823.

DiGeronimo, M. J., Nikaido, M., and Alexander, M. (1978). Most-probable-number technique for the enumeration of aromatic degraders in natural environments. *Microbial Ecology* **4**, 263–266.

Dommergues, Y. R., Belser, L. W., and Schmidt, E. L. (1978). Limiting factors for microbial growth and activity in soil. *Advances in Microbial Ecology* **2**, 49–104.

Dutka, B. J. and Kwan, K. K. (1978). Comparison of eight media-procedures for recovering faecal streptococci from water under winter conditions. *Journal of Applied Bacteriology* **45**, 333–340.

Edberg, N. and Hofsten, B. (1973). Oxygen uptake of bottom sediments studied *in situ* and in the laboratory. *Water Research* **7**, 1285–1294.

Edmondson, W. T. (1969). A simplified method for counting phytoplankton. *In* "A Manual on Methods for Measuring Primary Production in Aquatic Environments", pp. 14–15. (R. A. Vollenweider, ed.) Blackwell Scientific Publications, Oxford.

Edwards, R. W., Duffield, A. N., and Marshall, E. J. (1978). Estimates of community metabolism of drainage channels from oxygen distributions. *Proceedings of the European Weed Research Society Symposium on Aquatic Weeds* **5**, 295–302.

El-Shazly, K. and Hungate, R. E. (1965). Fermentation capacity as a measure of net growth of rumen microorganisms. *Applied Microbiology* **13**, 574–582.

Evans, J. H. and McGill, S. M. (1969). An investigation of the Coulter counter in "Biomass" determinations of natural freshwater phytoplankton populations. *Hydrobiologia* **35**, 401–419.

Fenchel, T. (1975). The quantitative importance of the benthic microflora of an arctic tundra pond. *Hydrobiologia* **46**, 445–464.

Fenchel, T. (1977). The significance of bacteriovorous protozoa in the microbial community of detrital particles. In "Aquatic Microbial Communities", pp. 529–544. (J. Cairns, ed.) Garland, New York.

Findenegg, I. (1969). Expressions of populations. *In* "A Manual on Methods for Measuring Primary Production in Aquatic Environments", pp. 16–17. (R. A. Vollenweider, ed.) Blackwell Scientific Publications, Oxford.

Focht, D. D. (1978). Methods for analysis of denitrification in soils. *In* "Nitrogen in the Environment, Vol. 2", pp. 433–490. (D. R. Nielsen and J. G. MacDonald, eds.) Academic Press, New York and London.

Frischknecht, K. and Schneider, K. (1979). An improved bioluminescence ATP analysis for extraction-resistant algal cells. *Archiv für Hydrobiologie* **87**, 19–36.

Fry, J. C. and Humphrey, N. C. B. (1978a). Techniques for the study of bacteria epiphytic on aquatic macrophytes. *In* "Techniques for the Study of Mixed Populations", pp. 1–29. (D. W. Lovelock and R. Davies, eds.) Academic Press, London and New York.

Fry, J. C. and Humphrey, N. C. B. (1978b). The effect of paraquat induced death of aquatic plants on heterotrophic activity of freshwater bacteria. *In* "Proceedings British Crop Protection Conference Weeds, Vol. 2", pp. 595–601.

Fry, J. C. and Ramsay, A. G. (1977). Changes in the activity of epiphytic bacteria of *Elodea canadensis* and *Chara vulgaris* following treatment with the herbicide, paraquat. *Limnology and Oceanography* **22**, 556–561.

Fry, J. C. and Staples, D. G. (1974). The occurrence and role of *Bdellovibrio bacteriovorus* in a polluted river. *Water Research* **8**, 1029–1035.

Gayford, C. G. and Richards, J. P. (1970). Isolation and enumeration of aerobic heterotrophic bacteria in activated sludge. *Journal of Applied Bacteriology* **33**, 342–350.

Geesey, G. G., Mutch, R., Costerton, J. W., and Green, R. B. (1978). Sessile bacteria: an important component of the microbial population in small mountain streams. *Limnology and Oceanography* **23**, 1214–1223.

George, E. A. (1976). A guide to algal keys (excluding seaweeds). *British Phycological Journal* **11**, 49–55.

Gibbons, N. E. (1969). Isolation, growth and requirements of halophilic bacteria. *In* "Methods in Microbiology, Vol. 3B", pp. 169–183. (J. R. Norris and D. W. Ribbons, eds.) Academic Press, London and New York.

Gibson, D. M., Hendrie, M. S., Houston, N. C., and Hobbs, G. (1977). The identification of some Gram-negative heterotrophic aquatic bacteria. *In* "Aquatic Microbiology", pp. 135–159. (F. A. Skinner and J. M. Shewan, eds.) Academic Press, London and New York.

Gocke, K. (1977). Comparison of methods for determining the turnover times of dissolved organic compounds. *Marine Biology* **42**, 131–141.

Goldman, C. R., Steemann Nielsen, E., Vollenweider, R. A., and Wetzel, R. G. (1969). The ^{14}C light and dark bottle technique. *In* "A Manual on Methods for Measuring Primary Production in Aquatic Environments", pp. 70–73. (R. A. Vollenweider, ed.) Blackwell Scientific Publications, Oxford.

Goulder, R. (1972). The vertical distribution of some ciliated protozoa in the plankton of a eutrophic pond during summer stratification. *Freshwater Biology* **2**, 163–176.

Goulder, R. (1975). The effects of photosynthetically raised pH and light on some ciliated protozoa in a eutrophic pond. *Freshwater Biology* **5**, 313–322.

Goulder, R. (1979). V_{max} per bacterium and turnover rate per bacterium for glucose mineralisation in natural waters. *Current Microbiology* **2**, 365–368.

Greaves, M. P., Davies, H. A., Marsh, J. A. P., and Wingfield, G. I. (1976). Herbicides and soil microorganisms. *CRC Critical Reviews in Microbiology* **5**, 1–38.

Greaves, M. P., Croper, S. L., Davies, H. A., Marsh, J. A. P., and Wingfield,

G. I. (1978). "Methods of Analysis for Determining the Effects of Herbicides on Soil Microorganisms". Technical report Agricultural Research Council Weed Research Organisation No. 45. Agricultural Research Council Weed Research Organisation, Oxford.

Griffiths, A. J. and Lovitt, R. (1980). Use of numerical profiles for studying bacterial diversity. *Microbial Ecology* **6**, 35–43.

Hagstrom, A., Larsson, U., Horstedt, P., and Normark, S. (1979). Frequency of dividing cells, a new approach to the determination of bacterial growth rates in aquatic environments. *Applied and Environmental Microbiology* **37**, 805–812.

Hall, E. R. and Murphy, K. L. (1980). Estimation of nitrifying biomass and kinetics in wastewater. *Water Research* **14**, 297–304.

Halls, S. and Ayres, P. A. (1974). A membrane filtration technique for the enumeration of *Escherichia coli* in sea water. *Journal of Applied Bacteriology* **37**, 105–109.

Harris, R. F. and Sommers, L. E. (1968). Plate-dilution frequency technique for assay of microbial ecology. *Applied Microbiology* **16**, 330–334.

Harrison, M. J., Wright, R. T., and Morita, R. Y. (1971). Method for measuring mineralization in lake sediments. *Applied Microbiology* **21**, 698–702.

Hattori, T. and Hattori, R. (1976). The physical environment in soil microbiology: an attempt to extend principles of microbiology to soil microorganisms. *CRC Critical Reviews in Microbiology* **4**, 423–461.

Heaney, S. I. (1978). Some observations on the use of the *in vivo* fluorescence technique to determine chlorophyll-*a* in natural populations and cultures of freshwater phytoplankton. *Freshwater Biology* **8**, 115–126.

Hendricks, C. W. (1972). Enteric bacterial growth rates in river water. *Applied Microbiology* **24**, 168–174.

Hickman, M. and Penn, I. D. (1977). The relationship between planktonic algae and bacteria in a small lake. *Hydrobiologia* **52**, 213–219.

Hicks, S. E. and Carey, F. G. (1968). Glucose determination in natural waters. *Limnology and Oceanography* **13**, 361–363.

Hoadley, A. W. and Cheng, C. M. (1974). The recovery of indicator bacteria on selective media. *Journal of Applied Bacteriology* **37**, 45–57.

Hobbie, J. E., Crawford, C. C., and Webb, K. L. (1968). Amino acid flux in an estuary. *Science* **159**, 1463–1464.

Hobbie, J. E. and Crawford, C. C. (1969). Respiration corrections for bacterial uptake of dissolved organic compounds in natural waters. *Limnology and Oceanography* **14**, 528–532.

Hobbie, J. E., Daley, R. J., and Jasper, S. (1977). Use of nucleopore filters for counting bacteria by fluorescence microscopy. *Applied and Environmental Microbiology* **33**, 1225–1228.

Hobbie, J. E. and Rublee, P. (1975). Bacterial production in an arctic pond. *Verhandlung der Internationalen Vereinigung fur theoretische und angewandte Limnologie* **19**, 466–471.

Hobbie, J. E. and Rublee, P. (1977). Radioisotope studies of heterotrophic bacteria in aquatic ecosystems. *In* "Aquatic Microbial Communities", pp. 441–476. (J. Cairns, ed.) Garland, New York.

Hobbie, J. E. and Wright, R. T. (1965). Bioassay with bacterial uptake kinetics: glucose in freshwater. *Limnology and Oceanography* **10**, 471–474.

Hobson, P. N. (1969). Rumen bacteria. *In* "Methods in Microbiology", vol. 3B, pp. 122–149. (J. R. Norris and D. W. Ribbons, eds.) Academic Press, London and New York.

Hobson, P. N. and Summers, R. (1978). Anaerobic bacteria in mixed cultures; ecology of the rumen and sewage digesters. *In* "Techniques for the Study of Mixed Populations", pp. 125–141. (D. W. Lovelock and R. Davies, eds.) Academic Press, London and New York.

Holm-Hansen, O. (1973). Determination of total microbial biomass by measurement of adenosine triphosphate. *In* "Estuarine Microbial Ecology", pp. 73–89. (L. H. Stevenson and R. R. Colwell, eds.) University of South Carolina Press, Columbia.

Holm-Hansen, O. and Booth, C. R. (1966). The measurement of adenosine triphosphate in the ocean and its ecological significance. *Limnology and Oceanography* **11**, 510–519.

Holm-Hansen, O. and Karl, D. M. (1978). Biomass and adenylate energy charge determination in microbial cell extracts and environmental samples. *In* "Methods in Enzymology, Vol. 57", pp. 73–85. (M. A. DeLuca, ed.) Academic Press, London and New York.

Hoppe, H. G. (1977). Analysis of actively metabolizing bacterial populations with the autoradiographic method. *In* "Microbial Ecology of a Brackish Water Environment", pp. 179–197. (G. Rheinheimer, ed.) Springer-Verlag, Berlin.

Hoppe, H. G. (1978). Relations between active bacteria and heterotrophic potential in the sea. *Netherlands Journal of Sea Research* **12**, 78–98.

Hopton, J. W., Melchiorri-Santolini, U., and Sorokin, Y. I. (1972). Enumeration of viable cells of microorganisms by plate count technique. *In* "Techniques for the Assessment of Microbial Production and Decomposition in Fresh Waters", pp. 59–64. (Y. I. Sorokin and H. Kadota, eds.) Blackwell Scientific Publications, Oxford.

Hossell, J. C. and Baker, J. H. (1979a). A note on the enumeration of epiphytic bacteria by microscopic methods with particular reference to two freshwater plants. *Journal of Applied Bacteriology* **46**, 87–92.

Hossell, J. C. and Baker, J. H. (1979b). Epiphytic bacteria of the freshwater plant *Ranunculus penicillatus*: enumeration, distribution and identification. *Archiv für Hydrobiologie* **86**, 322–337.

Hossell, J. C. and Baker, J. H. (1979c). Estimation of the growth rates of epiphytic bacteria and *Lemna minor* in a river. *Freshwater Biology* **9**, 319–327.

Hubbard, J. S. (1973). Radio respirometric methods in measurement of metabolic activities in soil. *Bulletins from the Ecological Research Committee (Stockholm)* **17**, 199–206.

Hungate, R. E. (1966). "The Rumen and its Microbes". Academic Press, London and New York.

Hungate, R. E. (1969). A roll tube method for cultivation of strict anaerobes. *In* "Methods in Microbiology, Vol. 3B", pp. 117–132. (J. R. Norris and D. W. Ribbons, eds.) Academic Press, London and New York.

Hurst, C. J., Farrah, S. R., Gerba, C. P., and Melrick, J. L. (1978). Development of quantitative methods for the detection of enteroviruses in sewage sludges during activation and following land disposal. *Applied and Environmental Microbiology* **36**, 81–89.

Iannotti, E. L., Fischer, J. R., and Sievers, D. M. (1978). Medium for the enumeration of bacteria from swine waste digester. *Applied and Environmental Microbiology* **36**, 555–566.

Ingold, C. T. (1975). "An Illustrated Guide to Aquatic and Water-borne Hyphomycetes (fungi imperfecti) with Notes on their Biology". Freshwater Biological Association Scientific Publication No. 30. Freshwater Biological Association, Windermere, U.K.

Jannasch, H. W. (1969). Estimations of bacterial growth rates in natural waters. *Journal of Bacteriology* **99**, 156–160.

Jassby, A. D. (1975a). An evaluation of ATP estimations of bacterial biomass in the presence of phytoplankton. *Limnology and Oceanography* **20**, 646–648.

Jassby, A. D. (1975b). Dark sulphate uptake and bacterial productivity in a subalpine lake. *Ecology* **56**, 627–636.

Jeffers, J. N. R. (1978). "An Introduction to Systems Analysis: with Ecological Applications". Edward Arnold, London.

Jenkinson, D. S. and Oades, J. M. (1979). A method for measuring adenosine triphosphate in soil. *Soil Biology and Biochemistry* **11**, 193–199.

Jenkinson, D. S., Powlson, D. S., and Wedderburn, R. W. M. (1976). The effects of biocidal treatments on metabolism in soil. III. The relationship between soil biovolume, measured by optical microscopy, and the flush of decomposition caused by fumigation. *Soil Biology and Biochemistry* **8**, 189–202.

Jenkinson, D. S. and Powlson, D. S. (1976). The effects of biocidal treatments on metabolism in soil. V. A method for measuring soil biomass. *Soil Biology and Biochemistry* **8**, 209–213.

Jolley, E. T. and Jones, A. K. (1977). The interaction between *Navicula muralis* Grunow and an associated species of *Flavobacterium*. *British Phycological Journal* **12**, 315–328.

Jolley, E. T., Jones, A. K., and Hellebust, J. A. (1976). A description of glucose uptake in *Navicula pelliculosa* (Breb) Hilse including a brief comparison with an associated *Flavobacterium* sp. *Archives of Microbiology* **109**, 127–133.

Jones, J. G. (1970). Studies on freshwater bacteria: effect of medium composition and method on estimates of bacterial population. *Journal of Applied Bacteriology* **33**, 679–686.

Jones, J. G. (1971). Studies on freshwater bacteria: factors which influence the population and its activity. *Journal of Ecology* **59**, 593–613.

Jones, J. G. (1972). Studies on freshwater microorganisms: phosphatase activity in lakes of differing degrees of eutrophication. *Journal of Ecology* **60**, 777–791.

Jones, J. G. (1973). Studies on freshwater bacteria: the effect of enclosure in large experimental tubes. *Journal of Applied Bacteriology* **36**, 445–456.
Jones, J. G. (1974). A method for observation and enumeration of epilithic algae directly on the surface of stones. *Oecologia* **16**, 1–8.
Jones, J. G. (1975). Some observations on the occurrence of the iron bacterium *Leptothrix ochracea* in fresh water, including reference to large experimental enclosures. *Journal of Applied Bacteriology* **39**, 63–72.
Jones, J. G. (1976). The microbiology and decomposition of seston in open water and experimental enclosures in a productive lake. *Journal of Ecology* **64**, 241–278.
Jones, J. G. (1977a). The study of aquatic microbial communities. *In* "Aquatic microbiology", pp. 1–30. (F. A. Skinner and J. M. Shewan, eds.) Academic Press, London and New York.
Jones, J. G. (1977b). The effect of environmental factors on estimated viable and total populations of planktonic bacteria in lakes and experimental enclosures. *Freshwater Biology* **7**, 67–91.
Jones, J. G. (1978a). Spatial variation in epilithic algae in a stony stream (Wilfin Beck) with particular reference to *Cocconeis placentula*. *Freshwater Biology* **8**, 539–546.
Jones, J. G. (1978b). The distribution of some freshwater planktonic bacteria in two stratified eutrophic lakes. *Freshwater Biology* **8**, 127–140.
Jones, J. G. (1979a). "A Guide to Methods for Estimating Microbial Numbers and Biomass in Freshwater". Freshwater Biological Association Scientific Publication No. 39. Freshwater Biological Association, Windermere, U.K.
Jones, J. G. (1979b). Reply to the comments by D. M. Karl. *Freshwater Biology* **9**, 285–287.
Jones, J. G. (1979c). Microbial nitrate reduction in freshwater sediments. *Journal of General Microbiology* **115**, 27–35.
Jones, J. G. and Simon, B. M. (1975a). An investigation of errors in direct counts of aquatic bacteria by epifluorescence microscopy, with reference to a new method for dyeing membrane filters. *Journal of Applied Bacteriology* **39**, 317–329.
Jones, J. G. and Simon, B. M. (1975b). Some observations on the fluorometric determination of glucose in freshwater. *Limnology and Oceanography* **20**, 882–887.
Jones, J. G. and Simon, B. M. (1977). Increased sensitivity in the measurement of ATP in freshwater samples with a comment on the adverse effect of membrane filtration. *Freshwater Biology* **7**, 253–260.
Jones, J. G. and Simon, B. M. (1979). The measurement of electron transport system activity in freshwater benthic and planktonic samples. *Journal of Applied Bacteriology* **46**, 305–315.
Jones, P. C. T. and Mollison, J. E. (1948). A technique for the quantitative estimation of soil microorganisms. *Journal of General Microbiology* **2**, 54–69.
Jordan, M. J., Daley, R. J., and Lee, K. (1978). Improved filtration procedures for freshwater [^{35}S]SO_4 uptake studies. *Limnology and Oceanography* **23**, 154–157.

Jordan, M. J. and Peterson, B. J. (1978). Sulphate uptake as a measure of bacterial production. *Limnology and Oceanography* **23**, 146–150.
Jørgensen, B. B. (1979). A comparison of methods for the quantification of bacterial sulphate reduction in coastal marine sediments. *Geomicrobiology Journal* **1**, 11–28.
Jørgensen, J. H., Lee, J. C., Alexander, G. A., and Wolf, H. W. (1979). Comparison of *Limulus* assay, standard plate count and total coliform count for microbiological assessment of renovated wastewater. *Applied and Environmental Microbiology* **37**, 928–931.
Kaneko, T., Krichevsky, M. I., and Atlas, R. M. (1979). Numerical taxonomy of bacteria from the Beaufort Sea. *Journal of General Microbiology* **110**, 111–125.
Kaper, J. B., Mills, A. L., and Colwell, R. R. (1978). Evaluation of the accuracy and precision of enumerating aerobic heterotrophs in water samples by the spread plate method. *Applied and Environmental Microbiology* **35**, 756–761.
Karl, D. M. (1978). Occurrence and ecological significance of GTP in the ocean and in microbial cells. *Applied and Environmental Microbiology* **36**, 349–355.
Karl, D. M. (1979a). Comments on a paper by J. G. Jones and B. M. Simon: Increased sensitivity in the measurement of ATP in freshwater samples with a comment on the adverse effect of membrane filtration. *Freshwater Biology* **9**, 281–284.
Karl, D. M. (1979b). Adenosine triphosphate and guanosine triphosphate determinations in inter tidal sediments. *In* "Methodology for Biomass Determinations and Activities in Sediments", pp. 5–20. (C. D. Litchfield and P. L. Seyfried, eds.) American Society for Testing and Materials, Philadelphia.
Klapwijk, A., Drent, J., and Steenvoorden, J. H. A. M. (1974). A modified procedure for the TTC-dehydrogenase test in activated sludge. *Water Research* **8**, 121–125.
Kiefer, D. A. (1973). The *in vivo* measurement of chlorophyll by fluorometry. *In* "Estuarine Microbial Ecology", pp. 421–430. (L. H. Stevenson and R. R. Colwell, eds.) University of South Carolina Press, Columbia.
King, J. D. and White, D. C. (1977). Muramic acid as a measure of microbial biomass in estuarine and marine samples. *Applied and Environmental Microbiology* **33**, 777–783.
Kogure, K., Simidu, U., and Taga, N. (1979). A tentative direct microscopic method for counting living marine bacteria. *Canadian Journal of Microbiology* **25**, 415–420.
Koike, I. and Hattori, A. (1978). Simultaneous determinations of nitrification and nitrate reduction in coastal sediments by a ^{15}N dilution technique. *Applied and Environmental Microbiology* **35**, 853–857.
Kuznetsov, S. I. (1977). Trends in the development of ecological microbiology. *Advances in Aquatic Microbiology* **1**, 1–48.
Kuznetsov, S. I., Dubinina, G. A., and Lapteva, N. A. (1979). Biology of oligotrophic bacteria. *Annual Review of Microbiology* **33**, 377–387.

Lack, T. J. and Lund, J. W. G. (1974). Observations and experiments on the phytoplankton of Blelham Tarn, English Lake District. I. The experimental tubes. *Freshwater Biology* **4**, 399–415.

Latham, M. J. and Wolin, M. J. (1978). Use of a serum bottle technique to study interactions between strict anaerobes in mixed culture. *In* "Techniques for the Study of Mixed Populations", pp. 113–124. (D. W. Lovelock and R. Davies, eds.) Academic Press, London and New York.

Lee, C. C., Harris, R. F., Williams, J. D. H., Armstrong, D. E., and Syers, J. K. (1971). Adenosine triphosphate in lake sediments: I. Determination. *Soil Science Society of America, Proceedings* **35**, 82–86.

Litchfield, C. D., Devanas, M. A., Zindulis, J., Carty, C. E., Nakas, J. P., and Martin, E. L. (1979). Application of the ^{14}C organic mineralisation technique to marine sediments. *In* "Methodology for Biomass Determinations and Microbial Activities in Sediments", pp. 128–147. (C. D. Litchfield and P. L. Seyfried, eds.) American Society for Testing and Materials, Philadelphia.

Little, J. E., Sjogren, R. E., and Carson, G. R. (1979). Measurement of proteolysis in natural waters. *Applied and Environmental Microbiology* **37**, 900–908.

Liu, D. and Strachan, W. M. J. (1979). Characterization of microbial activity in sediment by resazurin reduction. *Ergebnisse der Limnologie* **12**, 24–31.

Lund, J. W. G. (1975). The uses of large experimental tubes in lakes. *In* "The Effects of Storage on Water Quality", pp. 291–311. Water Research Centre, Medmenham.

Lund, J. W. G. (1978). Experiments with lake phytoplankton in large enclosures. *Annual Reports of the Freshwater Biological Association* **46**, 31–39.

Maccubbin, A. E. and Kator, H. (1979). Distribution, abundance and petroleum degrading potential of marine bacteria from middle Atlantic continental shelf water. *In* "Microbial Degradation of Pollutants in Marine Environments", pp. 380–395. (A. W. Bourquin and P. H. Pritchard, eds.) US Environmental Protection Agency, Gulf Breeze.

Maeda, M. and Taga, N. (1979). Chromogenic assay method of lipopoly saccharide (LPS) for evaluating bacterial standing crop in seawater. *Journal of Applied Bacteriology* **47**, 175–182.

Mara, D. D. (1972). The use of agar dip-slides for estimates of bacterial numbers in polluted waters. *Water Research* **6**, 1605–1607.

Margalef, R. (1969). Counting. *In* "A Manual on Methods for Measuring Primary Production in Aquatic Environments", pp. 7–14. (R. A. Vollenweider, ed.) Blackwell Scientific Publications, Oxford.

Marker, A. F. (1972). The use of acetone and methanol in the estimation of chlorophyll in the presence of phaeophytin. *Freshwater Biology* **2**, 361–385.

Mason, C. F. (1976). Relative importance of fungi and bacteria in the decomposition of Phragmites leaves. *Hydrobiologia* **51**, 65-69.

McFeters, G. A., Stuart, S. A., and Olson, S. B. (1978). Growth of heterotrophic bacteria and extracellular products in oligotrophic waters. *Applied and Environmental Microbiology* **35**, 383–391.

Melchiorri-Santolini, U. (1972). Enumeration of microbial concentration in dilution series (MPN). *In* "Techniques for the Assessment of Microbial Production and Decomposition in Fresh Waters", pp. 64–70. (Y. I. Sorokin and H. Kadota, eds.) Blackwell Scientific Publications, Oxford.

Menon, A. S., Glooschenko, W. A., and Burns, N. M. (1972). Bacteria-phytoplankton relationships in Lake Erie. *In* "Proceedings Fifteenth Conference on Great Lakes Research", pp. 94–101. International Association of Great Lakes Research, Ann Arbor.

Meyer-Reil, L. A. (1978). Autoradiography and epifluorescence microscopy combined for the determination of number and spectrum of actively metabolising bacteria in natural waters. *Applied and Environmental Microbiology* **36**, 506–512.

Michiels, M. (1974). Biomass determination of some freshwater ciliates. *Biologisch jaarboek* **42**, 132–136.

Millar, W. N. and Casida, L. E. (1970). Evidence for muramic acid in soil. *Canadian Journal of Microbiology* **18**, 299–304.

Monheimer, R. H. (1974). Sulphate uptake as a measure of planktonic microbial production in freshwater ecosystems. *Canadian Journal of Microbiology* **20**, 825–831.

Monheimer, R. H. (1975). Planktonic microbial heterotrophy: its significance to community biomass production. *Verhandlung der Internationalen Vereinigung fur theoretische und angewandte Limnologie* **19**, 2658–2663.

Monheimer, R. H. (1978). Difficulties in interpretation of microbial heterotrophy from sulphate uptake data: laboratory studies. *Limnology and Oceanography* **23**, 150–153.

Montgomery, H. A. C. and Hart, I. C. (1974). The design of sampling programs for rivers and effluents. *Water Pollution Control* **73**, 77–101.

Moriarty, D. J. W. (1977). Improved method using muramic acid to estimate biomass of bacteria in sediments. *Oecologia* **26**, 317–323.

Mortensen, U., Noren, B., and Wadso, I. (1973). Microcalorimetry in the study of the activity of microorganisms. *Bulletins from the Ecological Research Committee (Stockholm)* **17**, 189–197.

Mouraret, M. and Baldensperger, J. (1977). Use of membrane filters for the enumeration of autotrophic thiobacilli. *Microbial Ecology* **3**, 345–359.

Mulligan, H. F. and Kingsbury, J. M. (1968). Application of an electronic particle counter in analysing natural populations of phytoplankton. *Limnology and Oceanography* **13**, 499–506.

Murphy, C. T. (1975). Nutrient materials and the growth of bacteria on human skin. *Transactions of St. John's Hospital Dermatological Society* **61**, 51–57.

Nakamura, S., Morita, T., Iwanaga, S., Niwa, M., and Takahashi, K. (1977). A sensitive substrate for clotting enzyme in horseshoe crab hemocytes. *Journal of Biochemistry* **81**, 1567–1569.

Nannipieri, P., Johnson, R. L., and Paul, E. A. (1978). Criteria for measurement of microbial growth and activity in soil. *Soil Biology and Biochemistry* **10**, 223–229.

Noble, W. C. and Pitcher, D. G. (1978). Microbial ecology of human skin. *Advances in Microbial Ecology* **2**, 245–289.

Oades, J. M. and Jenkinson, D. S. (1979). Adenosine triphosphate content of the soil microbial biomass. *Soil Biology and Biochemistry* **11**, 201–204.
Olson, B. H. (1978). Enhanced accuracy of coliform testing in seawater by a modification of the most probable number method. *Applied and Environmental Microbiology* **36**, 438–444.
Overbeck, J. (1979). Studies on heterotrophic functions and glucose metabolism of microplankton in Plußsee. *Ergenbnisse der Limnologie* **13**, 56–76.
Paerl, H. W. (1978). Microbial organic carbon recovery in aquatic ecosystems. *Limnology and Oceanography* **23**, 927–935.
Page, F. C. (1976). "An Illustrated Key to Freshwater and Soil Amoebae with Notes on Cultivation and Ecology". Freshwater Biological Association Scientific Publication No. 34. Freshwater Biological Association, Windermere, UK.
Pamatmat, M. M. (1971). Oxygen consumption by the seabed. IV. Shipboard Laboratory experiments. *Limnology and Oceanography* **16**, 536–550.
Parinkina, O. M. (1973). Determination of bacterial growth rates in tundra soils. *Bulletins from the Ecological Research Committee* (*Stockholm*) **17**, 303–309.
Parkes, R. J., Bryder, M. J., Madden, R. H., and Poole, N. J. (1979). Techniques for investigating the role of anaerobic bacteria in estuarine sediments. *In* "Methodology for Biomass Determinations and Microbial Activities in Sediments", pp. 107–118. (C. D. Litchfield and P. L. Seyfried, eds.) American Society for Testing and Materials, Philadelphia.
Parkin, T. B. and Brock, T. D. (1980). The effects of light quality on the growth of phototrophic bacteria in lakes. *Archives of Microbiology* **125**, 19–27.
Parkinson, D., Gray, T. R. G., and Williams, S. T. (1971). "Methods for Studying the Ecology of Soil Microorganisms", IBP Handbook No. 19. Blackwell Scientific Publications, Oxford.
Paul, E. A. and van Veen, J. A. (1978). The use of tracers to determine the dynamic nature of organic matter, Vol. 3. *In* "Proceedings 11th International Congress of Soil Science", pp. 61–102. International Society of Soil Science, Edmonton.
Pettit, N. M., Smith, A. R. J., Freedman, R. B., and Burns, R. G. (1976). Soil urease: activity, stability and kinetic properties. *Soil Biology and Biochemistry* **8**, 479–484.
Perry, W. B., Boswell, J. T., and Stanford, J. A. (1979). Critical problems with extraction of ATP for bioluminescence assay of plankton biomass. *Hydrobiologia* **65**, 155–163.
Pike, E. B., Carrington, E. G., and Ashburner, P. A. (1972). An evaluation of procedures for enumerating bacteria in activated sludge. *Journal of Applied Bacteriology* **35**, 309–321.
Platt, T. and Denman, K. L. (1975). Spectral analysis in ecology. *Annual Review of Ecology and Systematics* **6**, 189–210.
Pratt, D. B. and Reynolds, J. W. (1973). The use of selective and differential media in the analyis of marine and estuarine bacterial populations. *In*

"Estuarine Microbial Ecology", pp. 37–44. (L. H. Stevenson and R. R. Colwell, eds.) University of South Carolina Press, Columbia.
Prescott, G. W. (1954). "How to know the Freshwater Algae". Brown, Dubuque.
Poole, J. E. P. and Fry, J. C. (1980). A study of the protozoan and metazoan populations of three oxidation ditches. *Water Pollution Control* **79**, 19–27.
Postgate, J. R., Crumpton, J. E., and Hunter, J. R. (1961). The measurement of bacterial viabilities by slide culture. *Journal of General Microbiology* **24**, 15–24.
Powell, E. O. and Errington, F. P. (1963). The size of bacteria, as measured with the Dyson image-splitting eyepiece. *Journal of the Royal Microscopical Society* **82**, 39–49.
Rai, H. (1979). Glucose in freshwater of central Amozan lakes: natural substrate concentrations determined by dilution bioassay. *Internationale Revue der Gesamten Hydrobiologie* **64**, 141–146.
Ramsay, A. J. (1978). Direct counts of bacteria by a modified acridine orange method in relation to their heterotrophic activity. *New Zealand Journal of Marine and Freshwater Research* **12**, 265–269.
Ramsay, A. J. and Fry, J. C. (1976). Response of epiphytic bacteria to the treatment of two aquatic macrophytes with the herbicide paraquat. *Water Research* **10**, 453–459.
Rippey, S. R. and Cabelli, V. J. (1979). Membrane filter procedure for enumeration of *Aeromonas hydrophila* in freshwaters. *Applied and Environmental Microbiology* **38**, 108–113.
Robinson, G. G. C., Hendzel, L. L., and Gillespie, D. C. (1973). A relationship between heterotrophic utilisation of organic acids and bacterial populations in West Blue Lake, Manitoba. *Limnology and Oceanography* **18**, 264–269.
Robrish, S. A., Kemp, C. W., and Bowen, W. H. (1978). Use of extractable adenosine triphosphate to estimate the viable cell mass in dental plaque samples obtained from monkeys. *Applied and Environmental Microbiology* **35**, 743–749.
Rodina, A. G. (1972). "Methods ın Aquatic Microbiology". University Park Press, Baltimore.
Romanenko, V. I., Overbeck, J., and Sorokin, Y. I. (1972). Estimation of production of heterotrophic bacteria using ^{14}C. *In* "Techniques for the Assessment of Microbial Production and Decomposition in Fresh Waters", pp. 82–85. (Y. I. Sorokin and H. Kadota, eds.) Blackwell Scientific Publications, Oxford.
Rosswall, T. and Kvillner, E. (1978). Principal-components and factor analysis for the description of microbial populations. *Advances in Microbial Ecology* **2**, 1–48.
Ruby, E. G. and Nealson, K. H. (1978). Seasonal changes in the species composition of luminous bacteria in nearshore water. *Limnology and Oceanography* **23**, 530–533.
Rudd, J. W. M. and Hamilton, R. D. (1973). Measurement of adenosine triphosphate (ATP) in two precambian shield lakes of north western Ontario. *Journal of the Fisheries Research Board of Canada* **30**, 1537–1546.

Salonen, K. (1974). Effectiveness of cellulose ester and perforated polycarbonate membrane filters in separating bacteria and phytoplankton. *Annales Botanici Fennici* **11**, 133–135.

Salonen, K. (1977). The estimation of bacterioplankton numbers and biomass by phase contrast microscopy. *Annales Botanici Fennici* **14**, 25–28.

Saunders, G. W. (1972a). The transformation of artificial detritus in lake water. *Memorie dele'Istituto Italiano di Idrobiologia Dott Marco de Marchi* **29** Suppl., 261–288.

Saunders, G. W. (1972b). The kinetics of extracellular release of soluble organic matter by plankton. *Verhandlung der Internationalen Vereinigung fur theoretische und angewandte Limnologie* **18**, 140–146.

Saunders, G. W. (1976). Decomposition in freshwater. *In* "The Role of Terrestrial and Aquatic Organisms in Decomposition Processes", pp. 341–373. (J. M. Anderson and A. Macfadyen, eds.) Blackwell Scientific Publications, Oxford.

Saunders, G. W. and Storch, T. A. (1971). Coupled oscillatory control mechanism in a planktonic system. *Nature, London* **230**, 58–60.

Schmidt, E. L. (1973a). Fluorescent antibody techniques for the study of microbial ecology. *Bulletins from the Ecological Research Committee* (*Stockholm*) **17**, 67–76.

Schmidt, E. L. (1973b). The traditional plate count technique among modern methods—chairman's summary. *Bulletins from the Ecological Research Committee* (*Stockholm*) **17**, 453–454.

Sharpe, A. N. (1973). Automation and instrumentation developments for the biological laboratory. *In* "Sampling—Microbiological Monitoring of Environments", pp. 197–232. (R. G. Board and D. W. Lovelock, eds.) Academic Press, London and New York.

Sheldon, R. W. and Parsons, T. R. (1967). "A Practical Manual on the Use of the Coulter Counter in Marine Research". Coulter Electronic Sales Company Canada, Toronto.

Sheldon, R. W. and Sutcliffe, W. H. (1978). Generation times of 3 h for Sargasso sea microplankton determined by ATP analysis. *Limnology and Oceanography* **23**, 1051–1055.

Sieburth, J. M. (1979). "Sea Microbes". Oxford University Press, New York.

Silvey, J. K. G. and Wyatt, J. T. (1971). The inter-relationship between freshwater bacteria, algae and actinomycetes in Southwestern reservoirs. *In* "The Structure and Function of Freshwater Microbial Communities", pp. 249–275. (J. Cairns, ed.) Virginia Polytechnic Institute and State University, Blacksburg.

Simon, A., Rovira, A. D., and Sands, D. C. (1973). An improved selective medium for isolating fluorescent pseudomonads. *Journal of Applied Bacteriology* **36**, 141–145.

Skerman, V. B. D. (1967). "A Guide to the Genera of Bacteria: with Methods and Digests of Generic Characteristics". Williams and Wilkins, Baltimore.

Smith, D. W., Fliermans, C. B., and Brock, T. D. (1973). An isotopic technique for measuring the autotrophic activity of soil microorganisms *in*

situ. *Bulletins from the Ecological Research Committee* (*Stockholm*) **17**, 243–246.
Smith, K. L., White, G. A., Laver, M. B., and Haugsness, J. A. (1978). Nutrient exchange and oxygen consumption by deep sea benthic communities: preliminary *in situ* measurements. *Limnology and Oceanography* **23**, 997–1005.
Soeder, C. J. and Talling, J. F. (1969). The enclosure of phytoplankton communities. *In* "A Manual on Methods of Measuring Primary Production in Aquatic Environments", pp. 62–70. (R. A. Vollenweider, ed.) Blackwell Scientific Publications, Oxford.
Sokal, R. R. and Rohlf, F. J. (1969). "Biometry". W. H. Freeman, San Francisco.
Sokal, R. R. and Sneath, P. H. A. (1963). "Principles of Numerical Taxonomy". W. H. Freeman, San Francisco.
Sorokin, Y. I. (1969). Bacterial production, general methods. *In* "A Manual on Methods of Measuring Primary Production in Aquatic Environments", pp. 128–146. (R. A. Vollenweider, ed.) Blackwell Scientific Publications, Oxford.
Sorokin, Y. I. (1972a). Application of different methods of enumeration of microorganisms. *In* "Techniques for the Assessment of Microbial Production and Decomposition in Freshwaters", pp. 70–71. (Y. I. Sorokin and H. Kadota, eds.) Blackwell Scientific Publications, Oxford.
Sorokin, Y. I. (1972b). Production of autotrophic microorganisms. *In* "Techniques for the Assessment of Microbial Production and Decomposition in Freshwaters", pp. 86–88. (Y. I. Sorokin and H. Kadota, eds.) Blackwell Scientific Publications, Oxford.
Sorokin, Y. I. and Overbeck, J. (1972). Direct microscopic counting of microorganisms. *In* "Techniques for the Assessment of Microbial Production and Decomposition in Freshwaters", pp. 44–47. (Y. I. Sorokin and H. Kadota, eds.) Blackwell Scientific Publications, Oxford.
Sprent, J. I. (1979). "The Biology of Nitrogen-fixing Organisms". McGraw-Hill, Maidenhead.
Staley, J. T., Marshall, K. C., and Skerman, V. B. D. (1980). Budding and prosthecate bacteria from freshwater habitats of various trophic states. *Microbial Ecology* **5**, 245–251.
Stanley, P. E. and Williams, S. G. (1969). Use of liquid scintilation counting for determining adenosine triphosphate by the luciferase enzyme. *Analytical Biochemistry* **29**, 381–392.
Staples, D. G. and Fry, J. C. (1973a). Factors which influence the enumeration of *Bdellovibrio bacteriovorus* in sewage and river water. *Journal of Applied Bacteriology* **36**, 1–11.
Staples, D. G. and Fry, J. C. (1973b). A medium for counting aquatic heterotrophic bacteria in polluted and unpolluted waters. *Journal of Applied Bacteriology* **36**, 179–181.
Steemann Nielsen, E. (1977). The carbon-14 technique for measuring organic production by plankton algae. A report on the present knowledge. *Folia Limnologica Scandinavica* **17**, 45–48.
Stein, J. R. (1973). "Handbook of Phycological Methods, Culture Methods and Growth Measurements". Cambridge University Press, Cambridge.

Stein, J. E. and Denison, J. G. (1967). *In situ* benthal oxygen demand of cellulosic fibers. *Advances in Water Pollution Research* **3**, 181–197.

Stevenson, L. H. (1978). A case for bacterial dormancy in aquatic systems. *Microbial Ecology* **4**, 127–133.

Stewart, W. D. P. and Daft, M. J. (1977). Microbial pathogens of cyanophycean blooms. *Advances in Aquatic Microbiology* **1**, 177–218.

Stotzky, G. (1972). Activity, ecology and population dynamics of microorganisms in soil. *CRC Critical Reviews in Microbiology* **2**, 59–137.

Straskrabova, V. (1973). Methods for counting water bacteria—comparison and significance. *Acta hydrochimica et hydrobiologica* **1**, 433–454.

Straskrabova, V. (1979). Oxygen methods for measuring the activity of water bacteria. *Ergebnisse der Limnologie* **12**, 3–10.

Straskrabova, V. and Sorokin, Y. I. (1972). Determination of cell size of microorganisms for the calculation of biomass. *In* "Techniques for the Assessment of Microbial Production and Decomposition in Freshwaters", pp. 48–50. (Y. I. Sorokin and H. Kadota, eds.) Blackwell Scientific Publications, Oxford.

Strohl, W. R. and Larkin, J. M. (1978). Enumeration, isolation and characterisation of *Beggiatoa* from freshwater sediments. *Applied and Environmental Microbiology* **36**, 755–770.

Sutcliffe, W. H., Orr, E. A., and Holm-Hansen, O. (1976). Difficulties with ATP measurements in inshore waters. *Limnology and Oceanography* **21**, 145–149.

Swift, M. J. (1973). The estimation of mycelial biomass by determination of the hexosamine content of wood tissue decayed by fungi. *Soil Biology and Biochemistry* **5**, 321–322.

Takahashi, M. and Ichimura, S. (1968). Vertical distribution of organic matter production of photosynthetic sulphur bacteria in Japanese lakes. *Limnology and Oceanography* **13**, 644–655.

Talling, J. F. (1969). General outline of spectrophotometric methods. *In* "A Manual on Methods for Measuring Primary Production in Aquatic Environments", pp. 22–25. (R. A. Vollenweider, ed.) Blackwell Scientific Publications, Oxford.

Talling, J. F. (1973). The application of some electrochemical methods to the measurement of photosynthesis and respiration in freshwaters. *Freshwater Biology* **3**, 335–362.

Talling, J. F. and Fogg, G. E. (1969). Possible limitations and artificial modifications. *In* "A Manual on Methods for Measuring Primary Production in Aquatic Environments", pp. 73–78. (R. A. Vollenweider, ed.) Blackwell Scientific Publications, Oxford.

Thomas, D. R., Richardson, J. A., and Dicker, R. J. (1974). The incorporation of tritiated thymidine into DNA as a measure of the activity of soil microorganisms. *Soil Biology and Biochemistry* **6**, 293–296.

Tiefenbrunner, F. (1979). Estimation of microbial activities in polluted surface waters by microcalorimetry. *Ergebnisse der Limnologie* **12**, 136–145.

Tobin, R. S. and Anthony, D. H. J. (1978). Tritiated thymidine incorporation

as a measure of microbial activity in lake sediments. *Limnology and Oceanography* **23**, 161–165.
Truper, H. G. and Yentsch, C. S. (1967). Use of glass fiber filters for the rapid preparation of *in vivo* absorption spectra of photosynthetic bacteria. *Journal of Bacteriology* **94**, 1255–1256.
Tunzi, M. G., Chu, M. Y., and Bain, R. C. (1974). *In vivo* fluorescence, extracted fluorescence and chlorophyll concentrations in algal mass measurements. *Water Research* **8**, 623–636.
Vaatanen, P. (1977). Effects of composition of substrate and inoculation technique on plate counts of bacteria in the northern Baltic sea. *Journal of Applied Bacteriology* **42**, 437–443.
Vaatanen, P. (1979). Microbial activity in brackish water determined as oxygen consumption. *Ergebnisse der Limnologie* **12**, 32–37.
Vaccaro, R. F. and Jannasch, H. W. (1967). Variation in uptake kinetics for glucose by natural populations in seawater. *Limnology and Oceanography* **12**, 540–542.
van Heusden, G. P. H. (1972). Estimation of biomass of plankton. *Hydrobiologia* **39**, 165–208.
van Palenstein Helderman, W. H. (1975). Total viable count and differential count of *Vibrio* (*Campylobacter*) *sputorum*, *Fusobacterium nucleatum*, *Selenommas sputigena*, *Bacteriodes ochraveus* and *Veillonella* in the inflamed and non-inflamed human gingival crevice. *Journal of Periodontal Research* **10**, 294–305.
van Veen, J. A. and Paul, E. A. (1979). Conversion of biovolume measurements of soil organisms grown under various moisture tensions, to biomass and their nutrient content. *Applied and Environmental Microbiology* **37**, 686–692.
Varon, M. and Shilo, M. (1980). Ecology of aquatic bdellovibrios. *Advances in Aquatic Microbiology* **2**, 1–48.
Varon, M. and Zeigler, B. P. (1978). Bacterial predetor-prey interaction at low prey density. *Journal of Applied and Environmental Microbiology* **36**, 11–17.
Walton, D. W. H. and Allsopp, D. (1977). A new test cloth for soil burial trials and other studies on cellulose decomposition. *International Biodeterioration Bulletin* **13**, 112–115.
Ward, T. E. and Frea, J. I. (1979a). Determining the sediment distribution of methanogenic bacteria by direct fluorescent antibody methodology. *In* "Methodology for Biomass Determinations and Microbial Activities in Sediments", pp. 75–86. (C. D. Litchfield and P. L. Seyfried, eds.) American Society for Testing and Materials, Philadelphia.
Ward, T. E. and Frea, J. I. (1979b). Estimation of microbial activities in lake sediments by measurement of sediment gas evolution. *In* "Methodology for Biomass Determinations and Microbial Activities in Sediments", pp. 156–166. (C. D. Litchfield and P. L. Seyfried, eds.) American Society for Testing and Materials, Philadelphia.
Warren, L. S., Benoit, R. E., and Jessee, J. A. (1978). Rapid enumeration of fecal coliforms in water by a colorimetric β-galactosidase assay. *Applied and Environmental Microbiology* **35**, 136–141.

Watson, S. W., Novitsky, T. J., Quinby, H. L., and Valois, F. W. (1977). Determination of bacterial number and biomass in the marine environment. *Applied and Environmental Microbiology* **33**, 940–946.

Weber, C. I. (1971). "A Guide to the Common Diatoms at Water Pollution Surveillance System Stations". US Environmental Protection Agency, Cincinnati.

Wetzel, R. G. (1975). "Limnology". W. B. Saunders, Philadelphia.

Wetzel, R. G. and Westlake, D. F. (1969). Periphyton. *In* "A Manual on Methods for Measuring Primary Production in Aquatic Environments", pp. 33–40. (R. A. Vollenweider, ed.) Blackwell Scientific Publications, Oxford.

White, D. C., Bobbie, R. J., King, J. D., Nickels, J., and Amoe, P. (1979). Lipid analysis of sediments for microbial biomass and community structure. *In* "Methodology for Biomass Determinations and Activities in Sediments", pp. 87–103. (C. D. Litchfield and P. L. Seyfried, eds.) American Society for Testing and Materials, Philadelphia.

Williams, P. J. LeB. (1970). Heterotrophic utilisation of dissolved organic compounds in the sea. I. Size distribution of population and relationship between respiration and incorporation of growth substances. *Journal of the Marine Biological Association of the United Kingdom* **50**, 859–870.

Williams, P. J. LeB. (1973). The validity of the application of simple kinetic analysis to heterogenous microbial populations. *Limnology and Oceanography* **18**, 159–165.

Williams, P. J. Le B. and Askew, C. (1968). A method of measuring the mineralisation by micro-organisms of organic compounds in sea-water. *Deep-sea Research* **15**, 365–375.

Williams, S. T. and Gray, T. R. G. (1973). General principles and problems of soil sampling. *In* "Sampling—Microbiological Monitoring of Environments", pp. 111–121. (R. G. Board and D. W. Lovelock, eds.) Academic Press, London and New York.

Williamson, M. (1975). The biological interpretation of time series analysis. *Bulletin of the Institute of Mathematics and its Applications* **11**, 67–69.

Willoughby, L. G. (1978). Methods for studying microorganisms on decaying leaves and wood in freshwater. *In* "Techniques for the Study of Mixed Populations", pp. 31–50. (D. W. Lovelock and R. Davies, eds.) Academic Press, London and New York.

Winogradsky, S. N. (1947). Principles de la microbiologie ecologique. *Antonie van Leeuwenhoek* **12**, 5–15.

Wolin, M. J. (1979). The rumen fermentation: a model for microbial interactions in anaerobic ecosystems. *Advances in Microbial Ecology* **3**, 49–77.

Wright, R. T. (1973). Some difficulties in using ^{14}C-organic solutes to measure heterotrophic bacterial activity. *In* "Estuarine Microbial Ecology", pp. 199–217. (L. H. Stevenson and R. R. Colwell, eds.) University of South Carolina Press, Columbia.

Wright, R. T. (1978). Measurement and significance of specific activity in the heterotrophic bacteria in natural waters. *Applied and Environmental Microbiology* **36**, 297–305.

Wright, R. T. and Hobbie, J. E. (1965). The uptake of organic solutes in lake water. *Limnology and Oceanography* **10**, 22–28.

Wright, R. T. and Hobbie, J. E. (1966). Use of glucose and acetate by bacteria and algae in aquatic ecosystems. *Ecology* **47**, 447–464.

Youngman, R. E. (1978). The measurement of chlorophyll. Water Research Centre Technical Report TR 82. Water Research Centre, Medmenham.

Zimmermann, R., Iturriaga, R., and Becker-Birck, J. (1978). Simultaneous determination of the total number of aquatic bacteria and the number thereof involved in respiration. *Applied and Environmental Microbiology* **36**, 926–935.

5

Competition among Chemolithotrophs and Methylotrophs and their Interactions with Heterotrophic Bacteria

J. G. Kuenen and J. C. Gottschal

1. INTRODUCTION

Chemolithotrophic bacteria are able to derive metabolically useful energy from the oxidation of reduced, or incompletely oxidized inorganic compounds, such as sulphide, sulphur, ammonia, nitrite, hydrogen, carbon monoxide, and ferrous iron. The electron acceptor for these oxidation processes is usually molecular oxygen, although in some cases nitrate can be used instead (Table 1).

This chapter deals mainly with the chemolitho(auto)trophic bacteria which are able to use carbon dioxide as their major carbon source for the synthesis of cell material. Furthermore, this discussion is limited to (facultatively) aerobic organisms only. This excludes strictly anaerobic, chemolitho(auto)trophic hydrogen bacteria, such as

Table 1. Examples of oxidation reactions carried out by chemolithotrophic bacteria

Sulphur bacteria
$S^{2-} + 2O_2 \rightarrow SO_4^{2-}$
$2S^0 + 3O_2 + 2H_2O \rightarrow 2SO_4^{2-} + 4H^+$
$S_2O_3^{2-} + 2O_2 + H_2O \rightarrow 2SO_4^{2-} + 2H^+$
$4S_2O_3^{2-} + O_2 + 2H_2O \rightarrow 2S_4O_6^{2-} + 4OH^-$
$5S_2O_3^{2-} + 8NO_3^- + H_2O \rightarrow 10SO_4^{2-} + 2H^+ + 4N_2$

Iron-oxidizing bacteria
$4Fe^{2+} + O_2 + 4H^+ \rightarrow 4Fe^{3+} + 2H_2O$

Hydrogen bacteria
$2H_2 + O_2 \rightarrow 2H_2O$
$5H_2 + 2NO_3^- + 2H^+ \rightarrow N_2 + 6H_2O$

Carboxydobacteria
$2CO + O_2 \rightarrow 2CO_2$

Nitrifying bacteria
$2NH_4^+ + 3O_2 \rightarrow 2NO_2^- + 2H^+ + H_2O$
$2NO_2^- + O_2 \rightarrow 2NO_3^-$

Methanobacterium thermoautotrophicum (Zeikus and Wolfe, 1972). For a detailed survey of the well-known representatives of the various chemolithotrophs the reader is referred to several recent reviews (Kelly, 1971; Kuenen, 1975; Schlegel, 1975; Smith and Hoare, 1977).

Nature harbours a great diversity of aerobic chemolithotrophs, constituting an almost complete spectrum of different physiological types, ranging from obligate chemolithotrophs, through facultative chemolithotrophs to chemolithotrophic heterotrophs. The obligate chemolitho(auto)trophs are normally restricted to reduced inorganic compounds as their energy source, while they use carbon dioxide as their major carbon source under all growth conditions. The facultative chemolitho(auto)trophs not only can grow as true chemolithoautotrophs but also as heterotrophs and, given the proper growth conditions, they can also grow mixotrophically which implies the simultaneous operation of chemolitho(auto)trophic and heterotrophic metabolism. The chemolithotrophic heterotrophs have been defined by Rittenberg (1969) as organisms which are able to derive metabolically useful energy from the oxidation of reduced inorganic compounds, but are unable to grow autotrophically.

The habitats of all chemolithotrophs must provide both inorganic energy sources and the appropriate electron acceptor (oxygen, nitrate). Since many of the reduced inorganic energy sources originate from the anaerobic breakdown of organic matter (fermentation, dissimilatory sulphate reduction, ammonification), these organisms are found most commonly at the interfaces of anaerobic and aerobic environments, for example, the top sediment of ditches, tidal mudflats and the interface of stratified lakes (Kuenen, 1975).

One of the ecologically most important questions is how to explain the coexistence of the enormous spectrum of physiologically related organisms in one and the same habitat. In the last decade, the study of microbial interactions, and particularly microbial competition for growth-limiting substrates, has contributed significantly to answering this question. Until recently, studies on microbial interactions among chemolithotrophs and heterotrophs have been possible only on a descriptive and rather qualitative scale. With the introduction of the flow-controlled continuous cultivation systems, such as the chemostat, a much more quantitative analysis of microbial interactions has become possible. The main focus of these studies has been on competition for growth-limiting substrates. In the course of these studies it has been shown that many basic features of selection and survival of species may be explained in terms of competition for one growth-limiting substrate. But, at the same time, anomalies in the expected behaviour of microbial populations have been observed in the chemostat and have drawn attention to other microbial interactions than competition, such as mutualism and commensalism (Bungay and Bungay, 1968; Frederickson, 1977). It is the aim of this chapter to summarize what is known of competition and other microbial interactions among chemolithotrophs and heterotrophic bacteria.

2. CHEMOLITHOTROPHIC SULPHUR BACTERIA

The different physiological types of colourless sulphur bacteria which play a role in the recycling of reduced inorganic sulphur compounds in nature are depicted in Fig. 1. The spectrum consists of highly specialized, obligately chemolithotrophic species, such as *Thiobacillus neapolitanus*, *Thiomicrospira pelophila* and the acidophilic *Thiobacillus thiooxidans*; the versatile facultatively chemolithotrophs, such as *Thiobacillus intermedius*, *Thiobacillus* sp. strain A2, *Sulfolobus* and some *Beggiatoa* spp.; and the chemolithoheterotrophic organisms, such as *Thiobacillus perometabolis* and similar organisms (Tuttle and Jannasch,

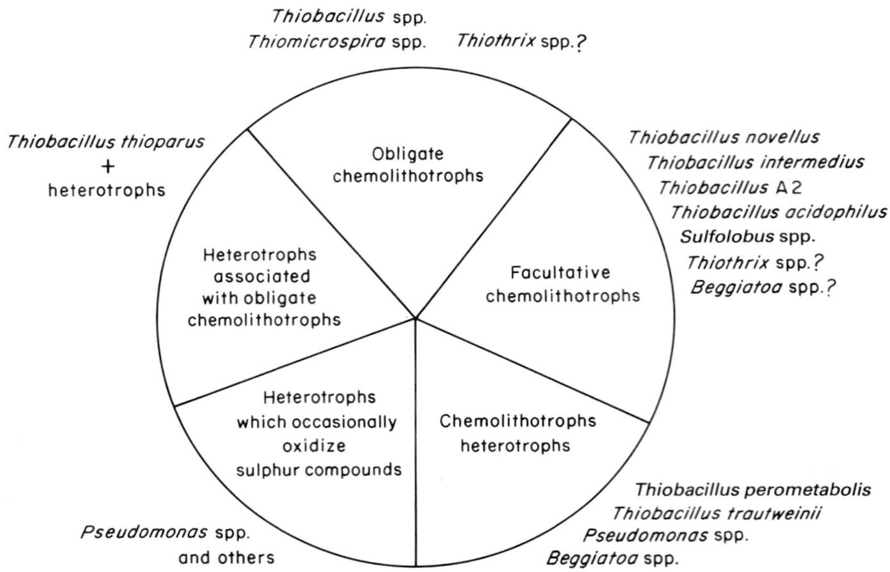

Fig. 1. Spectrum of organisms involved in the aerobic metabolism of inorganic reduced sulphur compounds (after Kuenen, 1975).

1972; Tuttle *et al.*, 1974) and *Beggiatoa* spp. In addition many heterotrophic bacteria are known to be able to oxidize sulphur compounds but are unable to gain energy from this oxidation process. These may be considered as "occasional" sulphur oxidizers. It is to be expected that, apart from the primary utilizers of sulphur compounds, many other bacteria have a significant effect on the recycling of sulphur compounds through a variety of interactions with the chemolithotrophic, colourless sulphur bacteria. For example, it has often been reported that the growth of chemolithotrophs was stimulated by the presence of heterotrophic contaminants living in close association with these organisms (Kuenen, 1975).

In order to gain insight into the ecology of these different organisms studies have been made in the field (Kuenen, 1975) and in the laboratory. It is particularly from these studies that some principles of the interactions occurring among chemolithotrophic sulphur oxidizers and between these chemolithotrophs and heterotrophic bacteria, have become apparent. The following sections discuss in detail some of the recent findings.

3. COMPETITION FOR REDUCED SULPHUR COMPOUNDS

The principle of competition for growth-limiting substrates has been discussed by several authors (Veldkamp and Jannasch, 1972; Harder *et at.*, 1977; Slater and Bull, 1978). A point crucial to the understanding of this principle is that, at least in theory, the outcome of the competition is entirely dependent on the growth characteristics described by the specific growth rate (μ) and the substrate concentration s relationship, determined by the saturation constant (K_s) and maximum specific growth rate (μ_{max}) of the competing species. Thus, the outcome of competition between two organisms for a growth-limiting substrate may give valuable information on the relative position of the two μ–s curves of the two competing species. Following the pioneering work of Jannasch, a number of studies (Veldkamp and Jannasch, 1972; Harder *et al.*, 1977) have shown that many environments harbour a variety of organisms with very similar physiological properties, but possessing crossing μ–s curves. These organisms appear to have either a combination of a high substrate affinity (low K_s) and a relatively low maximum specific growth rate (μ_{max}) or a combination of a low substrate affinity and a high μ_{max}. Depending on the fluctuations of the growth-limiting substrate concentration in a particular environment, the two types of organisms may have an ecological advantage alternately, and this may explain why these organisms can coexist. This principle has now been shown to exist not only for organic substrates but also for inorganic substrates and applies to competition for inorganic substrates among chemolithotrophs.

Competition for thiosulphate as an energy source or for iron as an iron source, by two obligate chemolithotrophic sulphur bacteria has been studied by Kuenen (1972) and by Kuenen *et al.* (1977). A comparative study was made of a marine strain of *Thiobacillus thioparus* and a physiologically very similar, spiral-shaped organism, *Thiomicrospira pelophila*, which had been isolated from the same marine sediment. The question was raised as to how these organisms could coexist in nature. It was shown that *Tms. pelophila* had a remarkably high sulphide tolerance, compared to *T. thioparus*, and was particularly abundant in sulphide rich areas. On the other hand *T. thioparus* had a higher μ_{max} than *Tms. pelophila*. The pH optima of the two organisms were also somewhat different. Competition experiments under thiosulphate-limitation in the chemostat revealed that *T. thioparus* rapidly outcompeted *Tms. pelophila* at pH values below 6.5, whereas *Tms. pelophila* became dominant above pH 7.5. Competition for iron was also studied since it was ex-

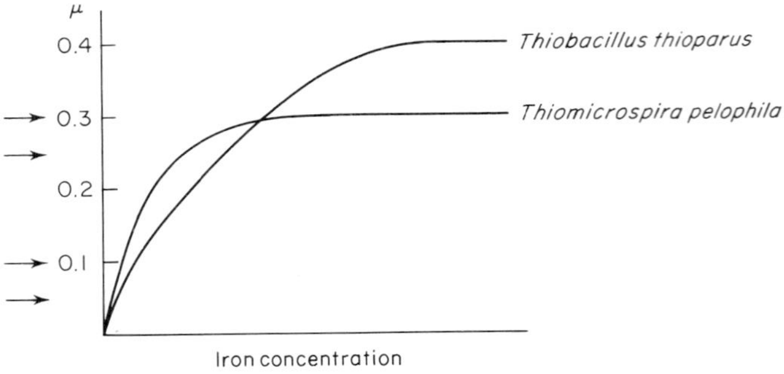

Fig. 2. Specific growth rate (μ) of *Thiomicrospira pelophila* and *Thiobacillus thioparus* as a function of the growth-limiting iron concentration. The results were obtained from competition experiments carried out in the chemostat. Bacteria were grown in a thiosulphate-minerals medium at 25°C. Arrows indicate dilution rates at which competition experiments were carried out. μ_{max} for *Thiomicrospira pelophila* = 0.35 h^{-1}; μ_{max} for *Thiobacillus thioparus* = 0.45 h^{-1} (after Kuenen *et al.*, 1977).

pected that in their marine habitat iron might be a growth limiting factor. This might have been true particularly at high sulphide concentrations, since ferrous sulphide is very insoluble. Figure 2 shows that at low growth-limiting iron concentrations, *Tms. pelophila* became dominant, whereas at high iron concentrations *T. thioparus* came to be the dominant organism. Thus, the high sulphide tolerance of *Tms. pelophila* might, at least in part, be explained by its ability to grow relatively fast at low iron concentrations. Taken together, the observed differences between the two organisms may very well account for the coexistence of the two organisms in the same sediment where pH, sulphide concentration and the available iron may fluctuate as a result of the diurnal cycle, tidal movement and variations in the rate of sulphide production.

Another example of competition between chemolithotrophic sulphur oxidizers was studied by Timmer-ten Hoor (1975; 1977). Anaerobic chemostat enrichments were carried out to select for marine organisms of the *Thiobacillus denitrificans* type, using a nitrate-limited, thiosulphate, mineral salts medium. Somewhat unexpectedly, a spirillum-shaped organism became dominant. This organism, which was named *Thiomicrospira denitrificans* was a facultatively anaerobic microaerophile, unable to grow in air-saturated medium (Timmer-ten Hoor, 1975). In this respect the new organism was very different from *T. denitrificans* which grew well under aerobic conditions.

When small quantities of oxygen were introduced into the culture a *T. denitrificans*-type organism became dominant, but this organism was not further identified. It was assumed that the apparent success of *Tms. denitrificans* under strictly anaerobic conditions was due to this organism's higher affinity for nitrate, compared to *T. denitrificans*. However, when a competition experiment was carried out in continuous-flow culture under identical conditions between *Tms. denitrificans* and a laboratory strain of *T. denitrificans*, the *Thiomicrospira* sp. was washed out (A. Timmer-ten Hoor, unpublished observations). The explanation for the outcome of this experiment is not known but it might be related to the extreme sensitivity of *Tms. denitrificans* to oxygen. In the anaerobic chemostat cultures traces of oxygen inevitably were introduced through stoppers, glass joints, and tubing, thereby favouring *T. denitrificans*. In the original enrichment culture set up to isolate *Tms. denitrificans*, any oxygen present might have been scavenged effectively by the small satellite population of heterotrophic organisms growing on excretion or lysis products of the autotroph. In the mixed culture of *T. denitrificans* and *Tms. denitrificans*, the presence of trace amounts of oxygen might have led to a competitive advantage for *T. denitrificans*, not only because *Tms. denitrificans* is very sensitive to oxygen, but perhaps also because the physiology of *T. denitrificans* seems particularly suited for mixotrophic growth on both oxygen and nitrate as terminal electron acceptors. As pointed out by Timmer-ten Hoor (1977) *Tms. denitrificans* would have its ecological niche at a lower range of redox potentials than *T. denitrificans*. Although further work is undoubtedly necessary to explain these observations, the experiments do show that competition for the terminal electron acceptor may also be an important factor in the selection of chemolithotrophic sulphur oxidizers.

4. INTERACTIONS BETWEEN CHEMOLITHOTROPHIC SULPHUR OXIDIZERS AND HETEROTROPHS

In both the examples mentioned above, a purely competitive interaction between two obligately chemolithotrophic sulphur oxidizers led to the exclusion of one of the two species. However, recent results from competition experiments in which two *Thiobacillus* species competed for a single growth-limiting substrate (thiosulphate), have shown that stable coexistence nevertheless was possible if one of the two organisms was a facultatively chemolithotrophic *Thiobacillus* sp. (Gottschal *et al.*, 1979). This result was not an exception to the generally accepted rule that one growth-limiting substrate permits only one species to become established in a continuous-flow culture (Powell, 1958; Taylor and Williams, 1975; Yoon *et al.*, 1977), but instead points to the presence of other interactions in addition to pure competition. In this particular case it has been shown that the obligate chemolithotrophic *Thiobacillus neapolitanus* excreted substantial amounts of organic compounds into the culture (up to 15% of the carbon dioxide fixed). Glycollate constituted the major part of the excreted organic material, presumably produced as a result of the oxygenase activity of the carbon dioxide assimilating enzyme, ribulose-1,5-bisphosphate carboxylase (RuBPCase) (Cohen *et al.*, 1979). Glycollate could serve as a carbon and energy source for the facultative chemolithotrophic *Thiobacillus* sp. strain A2 which was the second organism present in the mixed culture. In this mixed culture glycollate was not detected, and thus it seems likely that growth on this excretion product was the explanation for the coexistence of the two species. As will be pointed out later, it is very likely that *Thiobacillus* sp. strain A2 consumed not only the glycollate but also some of the available thiosulphate.

A further detailed analysis of the mixed culture revealed an interesting feature concerning the amount of cell material formed per mole thiosulphate consumed (i.e. the growth yield). It appeared that the mixed culture yielded more organic carbon than was predicted from growth yield data of pure cultures of *Thiobacillus neapolitanus* grown on thiosulphate media and a pure culture of *Thiobacillus* sp. strain A2 grown on the carbon excreted by *T. neapolitanus*. A comparable increase in cell yield was found in mixed cultures of *T. neapolitanus* with unidentified glycollate-metabolizing heterotrophs which were unable to oxidize thiosulphate. These mixed cultures had developed after the addition of freshwater samples to a pure culture of *T. neapolitanus* (Table 2; J. G. Kuenen, unpublished observations). Thus, mixed cultures of the obligate chemolithotroph with either a facultative chemolithotroph or heterotrophs, utilized the available thiosulphate 5 to 11% more efficiently for the production of cell material than did pure cultures of *T. neapolitanus*. This increase may seem marginal, but it should be realized that during nutrient-limited growth even the slightest advantage in growth economy may be of importance in the selection (and survival) of species (Harder *et al.*, 1977). The explanation for the more efficient use of thiosulphate is unknown. It also is not known whether the growth of both components of the two-membered culture was stimulated or only one of them. It is tempting to speculate on the possibility that an inhibitory effect of the excreted products on the growth of *T. neapolitanus* (Pan and Umbreit, 1972b; Rittenberg, 1969) was relieved by the growth of the secondary population. For instance, amino acids and keto acids have been shown to be toxic at relatively low levels (1 mM) (Borichewski and Umbreit, 1966; Kelly, 1971; Lu *et al.*, 1971).

Table 2. Carbon and glycollate analysis of thiosulphate-limited chemostat cultures of *Thiobacillus neapolitanus* grown in the presence or absence of heterotrophic bacteria. (Minerals medium with 38.5 mM thiosulphate, D = 0.07 h^{-1}.) The heterotrophic bacteria in the mixed culture removed 11 mg organic C l^{-1}. Assuming that about 50% of this carbon had been dissimilated, an increase of 5.5 mg C l^{-1} in the total cell mass was predicted, compared to the pure culture of *T. neapolitanus*.

		Thiobacillus neapolitanus alone	Mixed culture	
			experimental	calculated
Growth yield	mg C l^{-1}	79.0	91.4	84.5
Organic carbon content supernatant	mg C l^{-1}	15.3	4.3	—
Glycollate in supernatant	mg C l^{-1}	7.8	1.6	—

5. INTERACTIONS BETWEEN CHEMOLITHOTROPHIC SULPHUR OXIDIZERS AND DENITRIFIERS

Enrichments in anaerobic thiosulphate-nitrate medium for the facultatively anaerobic, denitrifying *Thiobacillus denitrificans* both in batch culture (Taylor *et al.*, 1971; Karavaiko *et al.*, 1973; Vainshtein, 1975, 1976) and in the chemostat (Timmer-ten Hoor, 1977) often yielded mixed populations of the aerobic *Thiobacillus thioparus* (*Thiobacillus intermedius*) and heterotrophs. A possible explanation is that these "aerobic" chemolithotrophs, although being unable to produce dinitrogen from nitrate can reduce nitrate to nitrite. Nitrite is always present at high concentrations in such cultures and some of the growth-inhibitory nitrite produced would be converted further to dinitrogen gas by denitrifying heterotrophs growing on the excretion products of the autotroph or on lysis products of this organism. On theoretical grounds it can be shown that the amount of cell material produced by *T. thioparus* during growth on thiosulphate-nitrate medium is not sufficient to allow substantial reduction of the nitrite to dinitrogen gas by heterotrophs (Timmer-ten Hoor, 1977). In the natural environment, however, such an association of autotrophs and heterotrophic denitrifiers might be more successful, since additional organic substrates might be simultaneously present.

6. THE EFFECT OF ORGANIC COMPOUNDS ON THE COMPETITION FOR INORGANIC SULPHUR COMPOUNDS

It has been proposed that the ecological advantage of facultatively chemolithotrophic thiobacilli would be their ability to utilize both inorganic and organic compounds (Rittenberg, 1969; Smith and Hoare, 1977; Matin, 1978). However, a superficial inspection of their growth characteristics suggests that this would be unlikely, since the maximum specific growth rate of facultative chemolithotrophs is usually lower than that of specialized, obligate chemolithotrophs or heterotrophs. An example to illustrate this point is shown in Table 3. In addition, it is known that facultatively chemolithotrophic thiobacilli almost never become dominant in enrichment cultures containing both inorganic and organic substrates. Yet the mere existence of these organisms in nature, even in habitats with specialized organisms, must mean that their type of metabolism has survival value.

This question was investigated by Gottschal *et al.* (1979) who chose the three model organisms given in Table 3 for a series of competition experiments. It was shown that chemostat grown cultures of *Thiobacillus* sp. strain A2 were able to utilize thiosulphate and acetate simultaneously during dual growth limitation by these two compounds (Gottschal and Kuenen, 1980b). In contrast, when grown in batch culture in the presence of excess acetate and thiosulphate, *Thiobacillus* sp. strain A2 showed diauxic growth. In the first phase acetate was the preferred substrate during which a strong repression of the chemolithotrophic potential was observed (Gottschal and Kuenen, 1980b). The hypothesis was that during dual substrate limited growth *Thiobacillus* sp. strain A2 might be able to compete successfully with the specialist *T. neapolitanus* and the heterotrophic *Spirillum* sp. strain G7. Figure 3 shows the outcome of the first series of competition experiments between *Thiobacillus* A2 and *T. neapolitanus*. As expected, during thiosulphate limitation, *T. neapolitanus* outcompeted *Thiobacillus* A2. As discussed in the previous paragraph, *Thiobacillus* A2 was not completely eliminated from the culture since it could use the glycollate excreted by *T. neapolitanus*. When an organic compound was added to the inflowing medium, in this case acetate or glycollate, the number of *Thiobacillus* A2 increased and at the same time the number of *T. neapolitanus* decreased. At a medium reservoir concentration of approximately 8 mM acetate or glycollate only 10% of the culture consisted of *T. neapolitanus*. At approximately 12 mM of the organic compounds, *T. neapolitanus* was completely eliminated from the culture.

Table 3. Maximum specific growth rates (h^{-1}) of the specialized, obligately chemolitho(auto)trophic *Thiobacillus neapolitanus*, the versatile, facultatively chemolithotrophic *Thiobacillus* sp. strain A2 and the specialized heterotrophic *Spirillum* sp. strain G7, during growth in defined media supplemented with thiosulphate, acetate or mixtures of both substrates. *T. neapolitanus* and *Spirillum* sp. strain G7 cannot grow on acetate or thiosulphate respectively.

	Maximum specific growth rate (h^{-1}) on		
Organism	Thiosulphate	Acetate	Thiosulphate + acetate
Thiobacillus neapolitanus	0.35	—	0.35
Thiobacillus sp. strain A2	0.10	0.22	0.22
Spirillum sp. strain G7	—	0.43	0.43

The outcome of this set of experiments can be explained by the ability of *Thiobacillus* A2 to grow mixotrophically, i.e. to metabolize the organic compound and the thiosulphate simultaneously. Since *T. neapolitanus* can assimilate only limited amounts of the acetate (Kuenen and Veldkamp, 1973), *Thiobacillus* sp. strain A2 rapidly increased in numbers as the acetate concentration increased. As the number of *Thiobacillus* sp. strain A2 increased, its total capacity to oxidize thiosulphate also increased, resulting in an increased contribution of *Thiobacillus* A2 population to the total turnover of thiosulphate.

Independently Smith and Kelly (1979) carried out a similar experiment with the same organisms using a mixture of 50 mM thiosulphate and 2.3 mM glucose and the outcome was essentially the same. In addition, it was shown that the pH value during the competition was crucial since *T. neapolitanus* had a lower pH-optimum than *Thiobacillus* A2. Thus the outcome of competition could also be manipulated by increasing or lowering the culture pH.

In order to investigate the general principle of these experiments Gottschal *et al.* (1979) also performed analogous competition experiments with *Thiobacillus* sp. strain A2 and the specialist heterotroph, *Spirillum* sp. strain G7 (Table 3) again using dual growth limitation by acetate and thiosulphate. As expected, the specialist heterotroph outcompeted *Thiobacillus* A2 in acetate (10 mM) defined medium, whilst *Thiobacillus* A2 completely eliminated *Spirillum* sp. G7 from the culture when more than 10 mM thiosulphate was present in the supply. In nature, bacteria of the *Thiobacillus* A2 type, of course, would have to com-

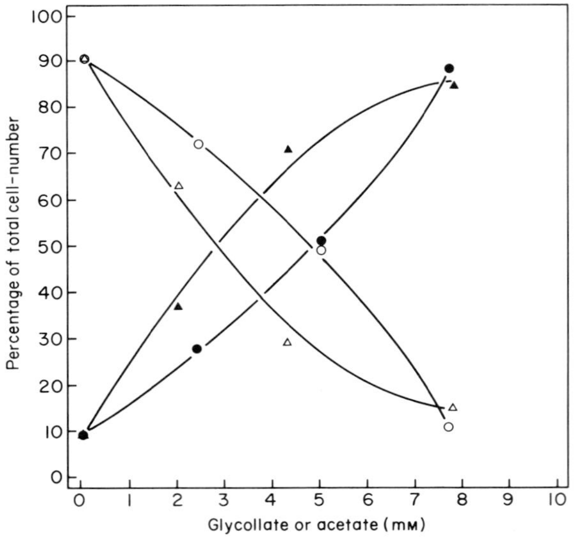

Fig. 3. Effect of different concentrations of organic substrates on the outcome of the competition between the versatile, facultatively chemolithotrophic *Thiobacillus* sp. strain A2 and the specialized obligately chemolithotrophic *T. neapolitanus* for growth-limiting thiosulphate. The chemostat was operated at a dilution rate of 0.07 h^{-1}. The inflowing medium contained 40 mM thiosulphate together with either acetate or glycollate at concentrations ranging from 0 to 7 mM. Relative cell numbers were determined after a steady state had been established. The percentage of *Thiobacillus* sp. strain A2 cells in cultures with thiosulphate plus acetate (▲) or glycollate (●) in the feed. The percentage of *T. neapolitanus* cells in cultures with thiosulphate plus acetate (△) or glycollate (○) in the feed. (After Gottschal *et al.*, 1979).

pete with both types of specialists and, therefore, further competition experiments were carried out using a three-membered culture. Figure 4 shows that in the mixed culture *Thiobacillus* A2 again became the dominant organism over a large range of acetate : thiosulphate ratios in the inflowing medium. In contrast to the predictions of mathematical models, which indicated that a maximum of two organisms would be able to coexist on the two growth-limiting substrates provided, coexistence of the three organisms was observed (Fig. 4). Given the fact that at more extreme acetate : thiosulphate ratios, no adverse effects of *T. neapolitanus* or the *Spirillum* sp. on *Thiobacillus* A2 has been apparent, it seems unlikely that the drop in numbers of *Thiobacillus* A2 can be explained by negative interactions between the organisms. Other possibilities could include cross feeding (Meyer *et al.*, 1975) but this seems unlikely since none of the organisms required growth factors, or the

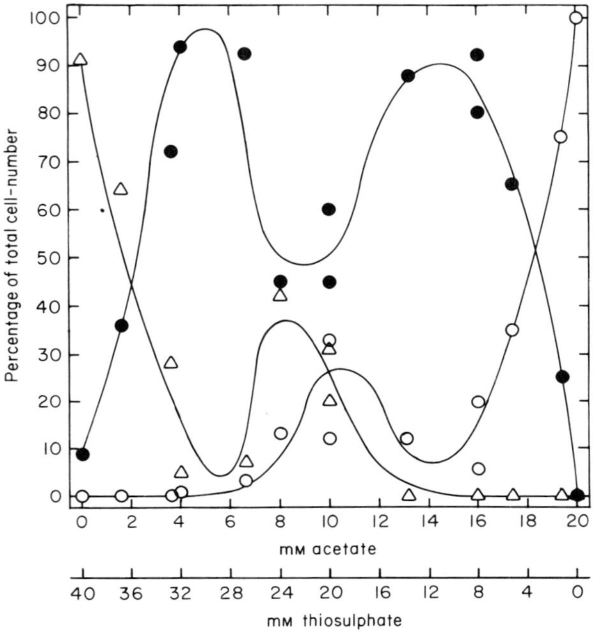

Fig. 4. Competition between the versatile *Thiobacillus* sp. strain A2, the specialist heterotroph *Spirillum* sp. strain G7 for thiosulphate and acetate as growth-limiting substrates in the chemostat at a dilution rate of 0.075 h^{-1}. Concentrations of the substrates in the inflowing medium ranged from 0 to 20 mM for acetate and 40 to 0 mM for thiosulphate. After steady states with different ratios of acetate and thiosulphate had been established relative cell numbers of *Thiobacillus* sp. strain A2 (●), *Thiobacillus neapolitanus* (△), and *Spirillum* sp. strain G7 (○) were determined by microscopically counting microcolonies grown on appropriate agar. (After Gottschal *et al.*, 1979).

excretion of auto-inhibitory compounds (de Freitas and Frederickson, 1978). The latter possibility can be ruled out since no such effects were observed in pure cultures. A more likely explanation has been suggested by J. C. Gottschal and T. F. Thingstad (unpublished observations) who indicated, with the aid of computer simulations, that the steady states observed in the experiments in Fig. 4 may have been pseudo-steady states; in the laboratory experiments steady states were considered to have been established when the ratios of organisms remained constant for at least 1 or 2 volume changes of the chemostat. From the computer simulation it appeared that under certain conditions steady states might not have been reached even after 40 volume changes. This was particularly true if the growth parameters of

Thiobacillus sp. strain A2, such as the saturation constant (K_s) for the substrates, changed with changing concentrations of acetate and thiosulphate in the medium. That this may indeed have occurred is indicated by the typical response of the thiosulphate and acetate respiration capacity of *Thiobacillus* A2, to changing ratios of the two substrates (Gottschal and Kuenen, 1980b).

The outcome of these competition experiments clearly showed the potential ecological advantage of a mixotrophic type of metabolism as found in this facultative chemolithotroph, *Thiobacillus* sp. strain A2. In order to show the general validity of this finding Gottschal and Kuenen (1980a), established chemostat enrichments using dual limitation by acetate and thiosulphate. Employing different ratios of thiosulphate and acetate in the inflowing medium, it proved possible to enrich selectively various facultatively chemolithotrophic thiobacilli from freshwater samples. Table 4 shows that in four cases facultative chemolitho(auto)trophic organisms dominated. In one case, at a mixture of 10 mM acetate and 15 mM thiosulphate in the medium reservoir, a chemolithotrophic heterotroph became dominant. This was not surprising since, at this particular ratio, enough acetate was available to satisfy the organic carbon requirement of the organism. In fact, one might predict that such a chemolithotrophic heterotroph would have an advantage over a facultative chemolithoautotroph, because the latter organism would carry unnecessary genetic information. This has been shown in several cases to be a disadvantage in the competition for growth-limiting substrates (Zamenhof and Eichhorn, 1967; Godwin and Slater, 1979; Mason and Slater, 1979). Table 4 shows that, again, the actual outcome of the selection in the chemostat did not yield pure cultures of one or two organisms as predicted from the theory. Apparently non-competitive interactions also occurred. One example appeared to be due to a vitamin requirement of the dominant organism which needed *p*-amino-benzoic acid for growth. In later experiments similar phenomena were often observed.

In the study on the ecological niche of facultatively chemolithotrophic sulphur oxidizers, the interaction between the facultative chemolithotroph, *Thiobacillus* sp. strain A2 and the obligately chemolithotrophic *Thiobacillus neapolitanus* was also studied under conditions of alternating supply of thiosulphate and acetate. It was argued that during alternating or fluctuating supply of organic and inorganic substrates the metabolically versatile *Thiobacillus* A2 should have an advantage over the specialist organism, since only the versatile organism would be able to grow continuously. *Thiobacillus* A2 was shown to be able to grow well in pure culture in the chemostat during a regime of 4 h

Table 4. Results of enrichment cultures from freshwater samples (I–V) after 15 to 20 volume changes in the chemostat under dual substrate limitation (thiosulphate and acetate) at a dilution rate of 0.05 h^{-1}. The total cell number in the cultures was 2 to 3 × 10^9 organisms as determined by plate counts. Autotrophic growth was tested in batch cultures in basal medium supplemented with thiosulphate (20 mM) and yeast extract (0.01% w/v), or in thiosulphate-limited chemostat cultures (III and V). Heterotrophic growth was tested in batch cultures in basal medium supplemented with acetate (10 mM) and yeast extract (0.01% w/v). The thiosulphate-oxidizing potential was determined polarographically after growth in batch cultures in the presence of both acetate (5 mM) and thiosulphate (10 mM). ND, not determined. (After Gottschal and Kuenen, 1980a).

				Dominant population				Secondary population			
Enrichment	Sample	Substrate concentrations in the inflowing medium (mM)	Total cell density (mg dry wt)	Percent of total cell number	Autotrophic growth	Heterotrophic growth	Thiosulphate oxidation	Percent of total cell number	Autotrophic growth	Heterotrophic growth	Thiosulphate oxidation
I	Pekeler	thiosulphate (30)						8	−	+	+
	Hoofddiep	+	90	82	+	+	+				
		acetate (5)						10	−	+	+
II	Pekeler	thiosulphate (10)									
	Hoofddiep	+	146	75	+	+	+	25	−	+	+
		acetate (15)									
III	Small ditch	thiosulphate (30)						1	+	−	+
		+	94	85	+	+	+	7	+	+	+
		acetate (5)						7	+	+	+
IV	Small ditch	thiosulphate (20)						5	−	+	ND
		+	135	50	+	+	+				
		acetate (10)						45	−	+	−
V	Small ditch	thiosulphate (10)									
		+	139	86	−	+	+	14	−	+	−
		acetate (15)									

acetate (10 mM) and 4 h thiosulphate (40 mM) (Gottschal, Nanninga, and Kuenen, 1981). During this regime, *Thiobacillus* A2 repressed its thiosulphate oxidizing potential during the acetate period and it was derepressed in the absence of acetate. After the thiosulphate period the organism possessed overcapacity to oxidize acetate, but after the acetate period the capacity to oxidize thiosulphate was just high enough to oxidize completely the fresh supply of thiosulphate. In competition with *T. neapolitanus* under these conditions, *Thiobacillus* A2 was unable to profit from its metabolic versatility. Figure 5 shows that during an alternate supply of thiosulphate and acetate *T. neapolitanus* maintained itself in the culture at a level of approximately 50% of the total population. This indicated that *T. neapolitanus* metabolized virtually all thiosulphate present in the feed. Apparently, *Thiobacillus* A2 had lost (i.e. repressed) its capacity to metabolize thiosulphate and thus was forced to grow as a heterotroph under these conditions. As pointed out by the authors the explanation might be sought in the high overcapacity for thiosulphate oxidation which *T. neapolitanus* maintained during the starvation (=acetate) period, whilst *Thiobacillus* A2 repressed its much lower thiosulphate oxidizing capacity even further

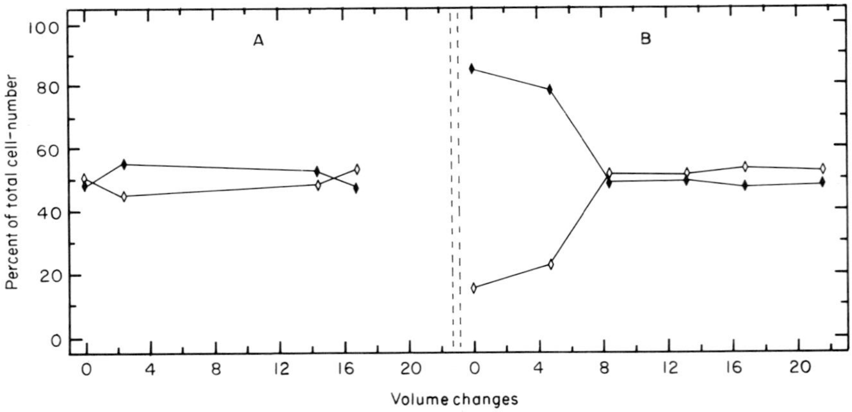

Fig. 5. Competition in continuous culture between the versatile *Thiobacillus* sp. strain A2 and the specialist *Thiobacillus neapolitanus* for thiosulphate and acetate as growth-limiting substrates supplied alternately to the chemostat in 4-hour cycles. The inflowing medium contained 10 mM acetate or 40 mM thiosulphate. *Thiobacillus* sp. strain A2 (◆) and *T. neapolitanus* (◇) as the percentage of the total cell number present in the culture. The dilution rate was 0.05 h^{-1}. Fig. 4 (a and b) represents identical experiments which were started with different initial ratios of *Thiobacillus* A2 and *T. neapolitanus*. (Gottschal, Nanninga and J. G. Kuenen, 1981).

during the acetate period. Because of its higher overcapacity for thiosulphate consumption at the beginning of each thiosulphate period, *T. neapolitanus* was able to lower quickly the concentration of thiosulphate to the point where insufficient induction of the capacity for thiosulphate oxidation in *Thiobacillus* A2 took place. In the context of this chapter it should be emphasized that this example clearly demonstrated that the presence of a second organism can have a dramatic influence on the physiological condition of an organism. In their mathematical analysis of competition of specialists and versatile organisms for two substrates, J. C. Gottschal and T. F. Thingstad (unpublished observations) came to the same conclusion.

During competition with alternating substrate supply and in the presence of the heterotrophic *Spirillum* sp. strain G7, *Thiobacillus* sp. strain A2 was more successful. During a 4 h alternating supply of acetate and thiosulphate, *Thiobacillus* A2 established a 70% dominant population in a mixed culture with *Spirillum* G7. The success of *Thiobacillus* A2 was probably due to the fact that during growth on thiosulphate minerals medium, its rate of acetate respiration was never repressed lower than to 30% of its full potential. This was in contrast to the thiosulphate oxidation capacity which was completely suppressed during growth on acetate alone.

In the competition with both *Thiobacillus neapolitanus* and the *Spirillum* sp. strain G7 in a three-membered culture operated under the same regime, *Thiobacillus* sp. strain A2 was rapidly eliminated by the specialists (Fig. 6). This must be due to the fact that *Thiobacillus* A2 was forced by *T. neapolitanus* to grow heterotrophically on acetate and could no longer successfully compete for acetate with the specialized heterotroph. When an alternating supply of varying mixtures of thiosulphate and acetate was presented to the three-membered culture *Thiobacillus* A2 was able to maintain itself in the culture for a prolonged time (Fig. 6).

In spite of the apparent inability of the "model" organism, *Thiobacillus* sp. strain A2, to take advantage of its metabolic versatility under alternately autotrophic and heterotrophic conditions, enrichment cultures performed under the same regime showed that facultatively chemolithotrophic thiobacilli could be enriched for with this technique (J. C. Gottschal, H. J. Nanninga, and J. G. Kuenen, unpublished observations). The general conclusion from this work is that nature harbours a large and diverse variety of facultative chemolithotrophs and chemolithoheterotrophs, each having its ecological niche under slightly different environmental conditions. With this knowledge in mind, field work should be done to further assess and quantify the role of the different metabolic types in the turnover of sulphate compounds in nature.

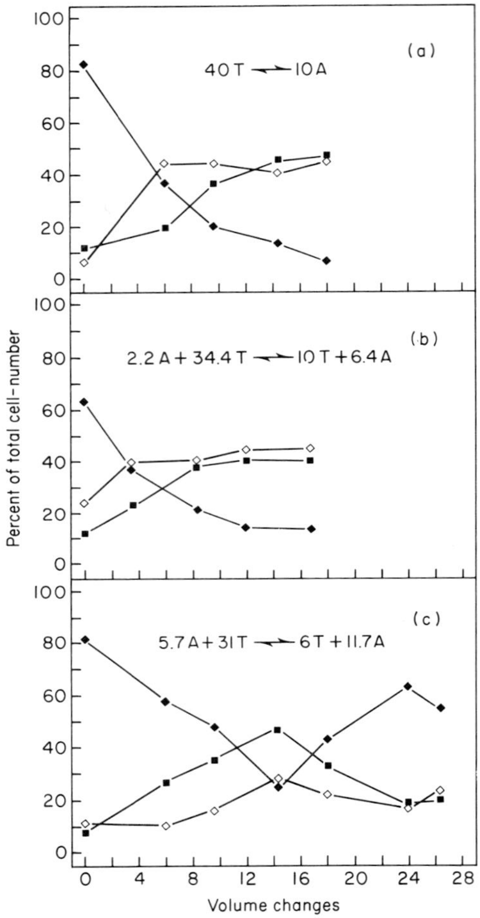

Fig. 6. Competition in continuous-flow culture between *Thiobacillus* sp. strain A2, *Thiobacillus neapolitanus*, and *Spirillum* sp. strain G7 for thiosulphate and acetate as the growth-limiting substrates. The dilution rate was 0.05 h^{-1} with intermittent feeding of two media containing either thiosulphate, acetate, or different combinations of thiosulphate and acetate. These media were supplied alternately to the culture in 4-hour cycles. *Thiobacillus* A2 (◆), *Spirillum* sp. strain G7 (■), and *T. neapolitanus* (◇) as the percentage of the total cell number present in the culture. (a) One medium contained 10 mM acetate, the other 40 mM thiosulphate; (b) one medium contained 2.2 mM acetate + 34.4 mM thiosulphate (I), the other 10 mM thiosulphate + 6.4 mM acetate (II) (compare with Fig. 4); (c) one medium contained 5.7 mM acetate + 31.0 mM thiosulphate (III), the other 6 mM thiosulphate + 11.7 mM acetate (IV) (compare with Fig. 4). (Gottschal *et al.*, 1981).

In a broad perspective one would expect that "specialists" chemolithotrophs would have an ecological advantage when the rate of turnover of the sulphur compounds is high relative to that of organic compounds, whereas the "versatile" organisms might have an advantage during periods when the average turnover of inorganic and organic substrates is of the same order of magnitude.

7. IRON-OXIDIZING BACTERIA

The last decade has seen an increasing interest in the interactions of acidophilic iron-oxidizing bacteria and other related bacteria. This is particularly due to the economic importance of these organisms in the leaching of metals from sulphide ores (Kelly *et al.*, 1979). The significance of microbial interactions in these organisms is discussed elsewhere in this volume (see Norris and Kelly, pp. 443–474). Therefore only a few examples are discussed here.

The breakdown of pyrite (FeS_2) by pure and mixed cultures of iron- and sulphur-oxidizing acidophilic bacteria was studied in the laboratory (Norris and Kelly, 1978). *Thiobacillus ferrooxidans* was able to grow well on pyrite in pure culture. In contrast, *Leptospirillum ferrooxidans*, another iron-oxidizing bacterium, grew well on ferrous iron but not on pyrite. *Thiobacillus thiooxidans*, a acidophilic, obligate chemolithotrophic sulphur oxidizer, grew well on sulphur compounds but was also unable to grow on pyrite. However, a mixed culture of *T. thiooxidans* and *L. ferrooxidans* was able to break down pyrite at a faster rate than *T. ferrooxidans*, indicating that the "mixotrophic" growth on pyrite by a single versatile organism was not necessarily as efficient as that of a mixture of two specialist organisms.

Brief mention should be made here of the very close associations which have been shown to exist between *T. ferrooxidans* and facultatively chemolithotrophic thiobacilli. From reputedly pure cultures of *T. ferrooxidans*, *T. acidophilus* (Guay and Silver, 1975) or *T. organoparus* (Markosyan, 1973) have been isolated. The two contaminants were unable to grow on ferrous iron as an energy source, but grew well autotrophically on reduced inorganic sulphur compounds and mixotrophically on organic and inorganic substrates. This finding throws new light on the reported ability of cultures of *T. ferrooxidans* to grow heterotrophically (Tuovinen *et al.*, 1978). The explanation for the tight association of these organisms with *T. ferrooxidans* remains unknown. It is known that *T. ferrooxidans* excretes organic acids during autotrophic growth. Particularly at low pH these organic acids (such as pyruvate) may become growth inhibitory unless they are removed by an accompanying organism (Schnaitman and Lundgren, 1965; Shivajo Rao and Berger, 1970).

8. INTERACTIONS AMONG HYDROGEN AND CARBON MONOXIDE UTILIZING FACULTATIVE CHEMOLITHOAUTOTROPHS

The group of organisms called hydrogen bacteria and carbon monoxide oxidizers ("carboxydobacteria") are all facultatively chemolithoautotrophic bacteria. According to Zavarzin and Nozhevnikova (1977) this group of organisms comprises a spectrum of bacteria some of which are able to utilize both hydrogen and carbon monoxide, while the majority of the known hydrogen bacteria are unable to utilize carbon monoxide. One strain was isolated which was able to utilize carbon monoxide but not hydrogen. An explanation for the fact that all known hydrogen and carbon monoxide oxidizers are facultative, and not obligate, chemolithotrophs is not available at present.

To our knowledge controlled competition experiments between autotrophic hydrogen or carbon monoxide utilizers have never been performed. It must be expected that similar principles will operate as have been discussed with the sulphur oxidizers, implying that these organisms would have a definite ecological advantage over other heterotrophs under conditions of mixed substrate supply. In the competition with other non-autotrophic hydrogen utilizers, mixed substrate utilization might be, again, of crucial importance. As long as enough carbon was available as a carbon source, the possession of the autotrophic potential might be of no advantage and, in fact, even be a disadvantage. As pointed out before, the possession of superfluous information would require unnecessary energy expenditure and thus might become a selective disadvantage.

One example of interactions between carboxydobacteria has been described by Noshevnikova and Zavarzin (1973). Growth yields of a mixture of two carboxydobacteria, strain Z 1155 and strain Z 1156, grown in batch culture, were close to the yields of the original enrichment culture, whereas pure cultures of the two organisms did not grow at all in the medium used for the enrichment (Table 5). Strain Z 1155 was shown to require thiamin and strain Z 1156 was found to be vitamin B_{12} dependent; but even when provided with these vitamins, the yields of the pure cultures were much lower than that of the original enrichment. An explanation for this phenomenon is not available at

Table 5. Growth yields (mg dry weight ml^{-1}) of carboxydobacteria incubated for 20 d under an atmosphere of carbon monoxide and oxygen. (After Zavarzin and Nozhevnikova, 1977.) nt, not tested.

Culture	Growth conditions: Mineral salts medium	Mineral salts medium with B vitamins
Enrichment culture	1.0 to 1.5	nt
Strain Z-1155	0	0.34
Strain Z-1156	0	0.64
Combined culture of Z-1155 plus 1156	0.94	nt

present. To our knowledge, interactions between hydrogen or carbon monoxide-utilizing bacteria and heterotrophs have never been studied.

Excretion of organic compounds by the hydrogen bacteria has been described only recently (Codd *et al.*, 1976; Vollbrecht *et al.*, 1978) but the total amount excreted was only significant for mutants unable to synthesize poly-β-hydroxybutyric acid (Vollbrecht *et al.*, 1978). Given the fact that facultative chemolithotrophs can be expected to be able to utilize their own excretion products, associations of these organisms and heterotrophs may be less common than those of obligate chemolithotrophs and heterotrophs.

9. INTERACTIONS BETWEEN CHEMOLITHOTROPHIC NITRIFIERS AND HETEROTROPHS

The chemolithotrophic nitrifying bacteria are able to oxidize ammonium to nitrite, or nitrite to nitrate, and to use the energy liberated in this process for their growth with carbon dioxide as the main carbon source. Although the family of the *Nitrobacteriaceae* consists of at least five genera of ammonium-oxidizing bacteria and three genera of nitrite-oxidizing bacteria (Watson, 1974; Harms *et al.*, 1976; Focht and Verstraete, 1977), most studies have been concerned with the physiology and the ecology of the genera *Nitrosomonas* (ammonium-oxidizer) and *Nitrobacter* (nitrite-oxidizer). The opinion that species of these two genera are usually predominant in nature (Focht and Verstraete, 1977) has recently been criticized by Belzer and Schmidt (1978a,b) who discovered that the number of the bacteria belonging to the genus *Nitrosospira* is usually underestimated because of the lower growth rate exhibited by these organisms.

The enrichment and subsequent isolation in pure culture of the chemolithotrophic nitrifying bacteria is a laborious procedure. This is due to the fact that the slowly growing nitrifiers are always accompanied by rapidly propagating heterotrophic organisms, even in simple inorganic media (Kingma Boltjes, 1934; Stapp, 1940; Gundersen, 1955; Clark and Schmidt, 1966; Golovacheva, 1975). A further problem in studying nitrifying bacteria is the susceptibility of pure cultures to contamination with heterotrophic bacteria (Clark and Schmidt, 1966; Pan and Umbreit, 1972a) which is a difficulty also encountered in the cultivation of other chemolithotrophic microorganisms. This can probably be explained by the fact that obligately chemolithotrophic bacteria are known to excrete organic compounds into their growth medium (Cohen *et al.*, 1979; Pan and Umbreit, 1972a).

In the following paragraph some examples of mixed cultures composed of nitrifying bacteria and heterotrophs are discussed which show that commensal and sometimes mutualistic (Bungay and Bungay, 1968) interactions appear to exist between heterotrophs and nitrifying bacteria.

Gundersen (1955) tried to isolate *Nitrosomonas* sp. in pure culture from soil by means of enrichment cultures followed by serial dilutions in inorganic, ammonium-containing medium. This isolation procedure appeared unsuccessful in that in all the dilutions in which nitrite could be detected, heterotrophic bacteria were also present and all nitrite-negative cultures appeared to be sterile. From the nitrite-positive cultures three distinct heterotrophic bacteria could be isolated on nutrient agar. Two of these organisms were tentatively identified as *Pseudomonas* species. The third bacterium was probably *Hyphomicrobium vulgare*, an organism also recognized by Kingma Boltjes (1934) in enrichment cultures of *Nitrobacter* sp. These organisms neither grew on, or oxidized, ammonium in an inorganic medium. Whether oxidation of ammonium took place in nutrient broth, in which growth of these bacteria was possible, was not tested. The two *Pseudomonas* species reduced nitrite to ammonium and one of them was able to reduce nitrate to nitrite and ammonium. The *Hyphomicrobium* species removed nitrite from the culture medium but ammonium was not detected in this case. These results do not produce direct evidence of interactions between the heterotrophs and the nitrifier. But it is not unreasonable to speculate that the removal of nitrite, nitrate, and organic excretion products may have a beneficial effect on the growth of *Nitrosomonas* sp. Both these inorganic and organic compounds have been demonstrated to inhibit the growth of nitrifying bacteria (Pan and Umbreit, 1972b; Bock, 1978). Moreover, the heterotrophic satellite population obviously profits from excreted organic carbon when this limits growth in a given environment.

Clark and Smith (1966) reported results which indicated a considerable stimulation of the growth of *Nitrosomonas europaea* in the presence of a heterotroph. The heterotroph was isolated from a contaminated culture of *N. europaea*. Addition of a washed cell suspension of the heterotroph to a freshly inoculated *N. europaea* culture resulted in a rapid development of this culture. Maximum levels of nitrite were reached before the pure control culture emerged from the lag phase, i.e. about 20 d after inoculation in this particular case. The lag phase in the pure cultures of *N. europaea* could be shortened by using larger inocula. Yet the addition of the heterotrophic organisms to the cultures always caused a further reduction in the length of the lag phase. The accumulation of nitrite in the mixed culture was the result of growth of *N. europaea* alone since in further experiments the heterotroph was shown to be unable to produce nitrite from either ammonium or nitrate. Furthermore, the heterotroph was shown to require a few amino acids for its growth and in that respect was dependent on *N. europaea*. The excretion products of *N. europaea* were not determined.

During further work on the nature of the stimulation of *N. europaea* by the heterotroph, it was found that pyruvate could reproduce all the essential features of this stimulation. This suggested that pyruvate or a comparable organic compound was produced by the heterotroph either during growth or during cell lysis which was shown to occur in early stages of growth of the mixed cultures.

Comparable mutualistic interactions have been found (Pan and Umbreit, 1972a) to exist between *Nitrobacter agilis* and some heterotrophic species. Most significant in this respect was the mixed culture of *N. agilis* and *Candida albicans* in inorganic nitrite-containing medium. About eight times more *Nitrobacter* sp. (4.0×10^7 organisms ml^{-1}) were found in this culture after an incubation time of 22 d compared with a pure culture of *Nitrobacter* sp. (5.0×10^6 organisms ml^{-1}) which was apparently caused by a shorter lag phase and an extended exponential phase of the mixed culture. *Candida albicans* had been inoculated quite heavily (1.7×10^7 organisms ml^{-1}) and did not grow at all during the incubation period. Instead, the viable cell number of the yeast decreased in a logarithmic fashion both in the absence and in the presence of *Nitrobacter* sp. Similar experiments indicated that other heterotrophs

(*Escherichia coli* strain B/r, *Hydrogenomonas eutropha* and *Sacharomyces cerevisiae*) exhibited a lower death rate in the presence of *Nitrobacter* sp. than in its absence. Although no effort was put into determining the nature of the interactions, it was proposed that species-specific mutualistic interactions did occur. Although the authors did not speculate on the reasons for the observed extended exponential phase of *Nitrobacter* sp. a rather trivial explanation might be that *Nitrobacter* sp. grew mixotrophically or even heterotrophically on lysed *Candida* sp. cells. At present it is not yet generally accepted that *N. agilis* is able to grow heterotrophically (Matin, 1978) but it has been proved unequivocally for at least one strain of *N. agilis* used by Smith and Hoare (1968).

Steinmüller and Bock (1976) found a slight stimulation in the rate of nitrite oxidation by a strain of *Nitrobacter agilis* in inorganic medium supplemented with yeast extract and peptone compared with a simple inorganic nitrite-containing medium. This stimulation by organic matter was considerably enhanced when heterotrophic bacteria had been allowed to grow in the yeast extract peptone broth and their culture filtrate subsequently added to the inorganic medium. The stimulation varied from 50 to 200% as compared to the autotrophic control cultures, depending on the heterotrophic species which had grown in the nutrient broth. The lag phase was not affected in these experiments. Out of 14 different species *Pseudomonas fluorescens* and *Proteus mirabilis* showed the highest stimulation (200%).

The enhancement of the rate of nitrite oxidation clearly did not only depend on the heterotrophic species used but also on the *Nitrobacter* species. *N. winogradski* was little stimulated and a new isolate, *Nitrobacter* sp. strain K4, was not stimulated at all. Although the mechanism of the observed stimulation remains unknown, it was suggested that the extent of the stimulation depended on differences in the capacity to grow mixotrophically.

A quite different type of interactions between heterotrophs and chemolithotrophic nitrifiers deserves attention, namely competition for the available reduced nitrogen compounds. It has been known for more than 50 years that many different heterotrophic bacteria are able to oxidize reduced nitrogen compounds, including ammonium and nitrite (for a survey on this particular topic, the reader is referred to Focht and Verstraete, 1977). Unfortunately, no competition experiments between chemolithotrophic and heterotrophic nitrifiers for limiting reduced nitrogen sources have been described. However, experiments with pure cultures of heterotrophic nitrifiers have shown that their rate of nitrification is 1.0×10^3 to 1.0×10^4 times lower than that of the chemolithotrophic nitrifiers (Focht and Verstraete, 1977). This seems to suggest that these organisms do not play a very important role in nature in the oxidation of reduced nitrogen compounds unless they occur in far greater numbers than do the chemolithotrophic nitrifiers. In this respect it might be relevant to note that recent reports suggest that still unknown populations were responsible for the major part of the conversion of nitrogen compounds into nitrate in soil (Ghiorse and Alexander, 1977; Belser, 1977) since the chemolithotrophic nitrifier population appeared to be too small to account for the observed activity. These observations, however, must be considered with great care as the basic parameters for these calculations were derived from pure culture experiments and from field enumeration techniques (mostly most probable number counts) and probably overlook large numbers of certain nitrifying species (Fliermans *et al.*, 1974; Belser and Schmidt, 1978a,b). Moreover, as will be clear from the differences between mixed cultures and pure cultures, growth kinetics determined in laboratory experiments are bound to deviate considerably from those found in "natures mixed cultures".

Another possibly important interaction which certainly deserves attention is the competition between ammonia oxidizers and methane oxidizers for ammonia. Many methane oxidizers are known to be able to co-oxidize ammonium (Whittenbury and Kelly, 1977), but the reverse has not been observed. Rates of oxidation of ammonia may be up to 50% of the methane oxidation rate (R. Hanson, personal communication) and so it must be expected that in environments containing ammonia and high levels of methane, a considerable amount of the available ammonia may be oxidized by methane oxidizers rather than by ammonia oxidizers.

10. METHYLOTROPHS IN MIXED CULTURE

The close analogy which exists between the type of metabolism of chemolithotrophs and that of methylotrophs is well known (Whittenbury and Kelly, 1977). Among the methane- and methanol-oxidizing bacteria it is possible to isolate highly specialized organisms which are virtually restricted to methane and/or methanol and related compounds as the sole carbon and energy sources, and on the other hand relatively versatile methylotrophs able to grow on a large variety of organic compounds. The methylotrophs with a restricted metabolism lack an operative tricarboxylic acid cycle and their physiological behaviour is very similar in this respect to that of obligate chemolithotrophs. The most significant difference between most methylotrophs and chemolithotrophs lies in the ability of the first group to assimilate the reduced carbon as "active" formaldehyde. However, some methylotrophs, such as *Pseudomonas denitrificans*, are unable to assimilate a reduced one carbon compound (Cox and Quayle, 1975). During growth on methanol *P. denitrificans* which can also grow autotrophically on molecular hydrogen, assimilates carbon as carbon dioxide via the reductive pentose phosphate cycle. One methane oxidizer, *Methylococcus capsulatus* strain Bath, has been reported to possess not only the ability to assimilate reduced one carbon compounds via the hexulose monophosphate pathway and the serine pathway, but also the capacity to reduce carbon dioxide through the Calvin-Benson cycle (Taylor, 1977).

In the light of these facts it seems appropriate to discuss briefly in this chapter some of the literature which has appeared on interactions among methylotrophs and heterotrophs. For an extensive recent review the reader should consult Harrison (1978).

Similar to the obligate chemolithotrophs, the specialized methylotrophs are relatively sensitive to high concentrations of organic compounds. Notably single amino acids sometimes may be growth inhibitory at concentrations of less than 1 mM (Kelly, 1971; Lu *et al.*, 1971; Eccleston and Kelly, 1972). It must be expected that the presence of heterotrophic bacteria able to metabolize these growth inhibitory compounds would be able to relieve the growth inhibition and thus exert a "growth stimulating" effect. That this seems to be the case indeed has been demonstrated by Malashenko *et al.* (1974). This work was prompted by the observation that methane-oxidizing bacteria often contained satellite populations of heterotrophic bacteria which were very persistent during isolation and purification procedures. Freshly isolated pure strains of methane oxidizers were found to be more sensitive to organic compounds than laboratory strains which had been maintained in pure culture for prolonged periods. These observations indicated that in the natural environment associations between methylotrophs and heterotrophs may indeed occur.

Interest in the interactions between methylotrophs and heterotrophs has been raised particularly because of the explorations made into the production of bacterial protein (single cell protein—SCP) from natural gas (CH_4) or methanol (see Linton and Drozd, pp. 357–406). It appeared that growth yields (unit of biomass produced per unit of methane or methanol oxidized) of mixed cultures often were higher than those of pure cultures (Harrison *et al.*, 1976) and that in some cases mixed cultures grown in the chemostat were more stable than pure cultures. In a few cases this has been traced back to commensal relationships in the culture and, in one case, to a mutualistic interaction.

One of the first studies in this area was reported by Wilkinson *et al.* (1974). It was discovered that in a mixed culture of a methane oxidizer and other bacteria a clear mutualistic relationship existed between the primary methane utilizer and a methanol utilizing *Hyphomicrobium* sp. During growth on methane the methane oxidizer produced traces of methanol which inhibited its growth. As a result a pure culture of this organism did not grow at all in liquid medium. In mixed cultures the *Hyphomicrobium* sp. utilized the methanol and thus relieved the growth inhibition of the methane oxidizer. In this mixed culture the two other organisms were present at levels of approximately 1% which were believed to grow on complex products of growth or lysis of the primary utilizer.

Linton and Buckee (1977) studied interactions in mixed culture of a *Methylococcus* sp. growing on methane and four heterotrophic bacteria unable to use methane. In ammonia-limited chemostat cultures of the pure *Methylococcus* sp., very high concentrations of organic compounds were detected. At a low dilution rate ($0.04\ h^{-1}$) approximately 50% of the organic carbon was found in the supernatant, whereas at a dilution rate of $0.2\ h^{-1}$ 10 to 15% of the organic carbon was present in the culture supernatant. Clear evidence was presented that this organic carbon originated from lysis of the primary utilizer and that little or no intermediary products of methane oxidation, such as methanol or formaldehyde, were formed by the *Methylococcus* sp. The heterotrophs were able to grow on the lysis product and between them possessed the necessary extracellular enzymes to degrade the major organic polymers originating from the *Methylococcus* sp., such as proteins, nucleic acids, and lipids. The presence of the heterotrophs, of course, had an effect on the total growth yield but had little or no influence on the saturation constant for methane or methanol of the *Methylococcus* sp. This indicated that only a commensal relationship existed between the primary and secondary utilizers. Advantages from a practical point of view of the mixed culture as compared to the pure culture are, in this case, somewhat higher yields, less foaming of the culture because protein was removed, and presumably also less susceptibility to contamination by other heterotrophic bacteria, such as pathogens (Harrison, 1978; Rokem *et al.*, 1980).

As pointed out by Harrison (1978) another advantage of mixed cultures in the production of SCP from methane may be their higher stability. Instability may be caused by the sensitivity of the methane utilizers to organic compounds (amino acids) which may accumulate to toxic levels due to cell lysis, particularly in dense cultures and/or due to recycling of the process water. Heterotrophs could stabilize the culture by consuming the organic compounds. A further advantage of mixed cultures could be that vitamins need not be added and that other substrates in the feed, for example ethane in natural gas, will be consumed also and thus lead to high yields.

Competition among methylotrophs has been experimentally investigated by Rokem *et al.* (1980). They studied competition for methanol between *Pseudomonas* sp. strain C and *Pseudomonas* sp. strain 1 during growth in the chemostat. *Pseudomonas* sp. strain C assimilated methanol via the hexulose monophosphate pathway for formaldehyde fixation and thus had a high growth yield on methanol, whereas *Pseudomonas* sp. strain 1 assimilated methanol via the serine pathway. The latter pathway is less efficient in terms of energy expenditure and leads to significantly lower growth yields. In spite of this lower efficiency *Pseudomonas* sp. strain 1 outcompeted *Pseudomonas* sp strain C in mixed culture during methanol-limited growth at a dilution rate of 0.1 h^{-1}. But at a dilution rate of 0.3 h^{-1}, *Pseudomonas* sp. strain C became dominant. As *Pseudomonas* sp. strain C had a lower affinity constant for methanol than *Pseudomonas* sp. strain 1 these results once more illustrated the importance of the K_s-value and/or μ_{max} for a single growth-limiting substrate and that the yield term was not decisive in the competition in these simple model systems (Table 6; Rokem *et al.*, 1980). Rokem *et al.* (1980) also studied mixed cultures of *Pseudomonas* sp. strain C and heterotrophic bacteria. Heterotrophic satellite populations could be established in methanol-limited chemostat grown cultures of *Pseudomonas* sp. strain C at levels of no more than 1% of the total population. Apparently little organic material was present in the culture supernatant of *Pseudomonas* sp. strain C. Yields remained unchanged or were slightly lower in these cultures, indicating that no mutualistic interactions took place. It is interesting to note that the beneficial effect of heterotrophic satellite populations in cultures of methanol utilizers become particularly conspicuous when the yield of the primary utilizer is relatively low: Harrison *et al.* (1976) reported a growth yield of 0.30 of biomass (g methanol)$^{-1}$ for pure cultures of their methanol utilizer, strain EN, improving to 0.46 of biomass (g methanol)$^{-1}$ in mixed culture. *Pseudomonas* sp. strain C had almost the same yield in pure culture (Table 6). The yield of strain EN was greatly improved by optimizing the growth medium and growth conditions (Wren, 1978; Roken *et al.*, 1980). Therefore it may be that in some cases the beneficial effects of hetero-

Table 6. Growth yields for different pure and mixed cultures of the methanol-utilizing *Pseudomonas* sp. strain C and *Pseudomonas* sp. strain I and heterotrophs. The data given were values obtained after a steady state had been established. In the mixed culture, at d = 0.1 h^{-1}, *Pseudomonas* sp. strain C was outcompeted by *Pseudomonas* sp. strain I. (After Rokem *et al.*, 1980).

Composition of the culture	Dilution rate (h^{-1})	Growth yield (g bacterial dry weight (g methanol)$^{-1}$)
1. *Pseudomonas* C	0.1	0.43
2. *Pseudomonas* I	0.14	0.38
3. *Pseudomonas* C + *Pseudomonas* I	0.1	0.35
4. *Pseudomonas* C + *Moraxella* sp.	0.1	0.43
5. *Pseudomonas* C + *Bacillus* sp.	0.1	0.35
6. *Pseudomonas* C + *Escherichia coli*	0.1	0.35

trophic "contaminants" must be attributed to peculiarities of the medium employed (see Linton and Drozd, pp. 357–406).

A most interesting but unexplained observation was made by Cremiaux *et al.* (1977) who obtained stable, mixed chemostat cultures of four methylotrophs growing on methanol as the only carbon and energy source. The three methylotrophs in the minority (comprising 20% of the total population) required yeast extract for growth in pure culture. As pointed out in the section on the sulphur oxidizers (see Section 3, p. 157), such a stable coexistence is possible only if interactions other than competition for the growth-limiting substrate also occur in the mixed culture. That this was indeed the case is evident from the fact that the satellite organisms required yeast extract for growth in pure culture. The most simple explanation for the stability of the culture would be excretion of a variety of products by the primary utilizer. In fact this might be an interrelation analogous to that found for the coexistence of the specialist *Thiobacillus neapolitanus* and the versatile *Thiobacillus* sp. strain A2 during thiosulphate-limited growth in the chemostat. As pointed out previously, *T. neapolitanus* excreted glycollate which could be utilized by *Thiobacillus* sp. strain A2. Under these conditions *Thiobacillus* sp. strain A2 must be expected to utilize some of the available thiosulphate also. An analogous situation may exist in the case of the four methylotrophs, since this explanation would account for the presence of facultative methylotrophs rather than heterotrophs.

A general point in these studies which greatly influences the possible importance of interactions with heterotrophic secondary populations is the nature of the growth limitation. In the case of methanol limitation the culture is limited by the carbon and energy course, and, therefore, as long as lysis does not occur, high concentrations of extracellular organic compounds will be unlikely. On the other hand, if, for example, nitrogen is growth limiting significant quantities of organic compound, such as keto acids, may be excreted. In pure culture this might lead to growth inhibition, whilst in mixed culture it may lead to high percentages of heterotrophs.

Very recently facultative methylotrophs have been selectively enriched for with the procedure originally designed for the enrichment of facultative chemolithotrophs (Gottschal and Kuenen, 1980a). The selective enrichment was performed in a chemostat under dual limitation by methanol and an organic compound (W. Harder, personal communication). The success of this procedure for the enrichment of versatile (facultative) methylotrophs underlines the general importance of mixed substrate utilization for versatile organisms in the competition for growth-limiting substrate with specialist organisms.

REFERENCES

Belser, L. W. (1977). Nitrate reduction to nitrite, a possible source of nitrite for growth of nitrite-oxidizing bacteria. *Applied and Environmental Microbiology* **34**, 403–410.

Belser, L. W. and Schmidt, E. L. (1978a). Diversity in the ammonia-oxidizing nitrifier population of a soil. *Applied and Environmental Microbiology* **36**, 584–588.

Belser, L. W. and Schmidt, E. L. (1978b). Serological diversity within a terrestrial ammonia-oxidizing population. *Applied and Environmental Microbiology* **36**, 589–593.

Bock, E. (1978). Lithoautotrophic and chemoorganotrophic growth of nitrifying bacteria. *Microbiology* 1978, 310–314 (American Society of Microbiology).

Borichewski, R. M. and Umbreit, W. W. (1966). Growth of *Thiobacillus thiooxidans* on glucose. *Archives of Biochemistry and Biophysics* **116**, 97–102.

Bungay, H. R. and Bungay, M. L. (1968). Microbial interactions in continuous culture. *Advances in Applied Microbiology* **19**, 269–290.

Clark, C. and Schmidt, E. L. (1966). Effect of mixed culture on *Nitrosomonas europaea* simulated by uptake and utilization of pyruvate. *Journal of Bacteriology* **91**, 367–373.

Codd, G. A., Bowien, B., and Schlegel, H. G. (1976). Glycollate production and excretion by *Alcaligenes eutrophus*. *Archives of Microbiology* **110**, 167–171.

Cohen, Y., de Jonge, I., and Kuenen, J. G. (1979). Excretion of glycolate by *Thiobacillus neapolitanus* grown in continuous culture. *Archives of Microbiology* **122**, 189–194.

Cox, R. B. and Quayle, J. R. (1975). The autotrophic growth of *Micrococcus denitrificans* on methanol. *Biochemical Journal* **150**, 569–571.

Crémieux, A., Chevalier, J., Combet, M., Dumenil, G., Parlouar, D., and Ballerini, D. (1977). Mixed culture of bacteria utilizing methanol for growth. I. Isolation and identification. *European Journal of Applied Microbiology* **4**, 1–9.

Eccleston, M. and Kelly, D. P. (1972). Assimilation and toxicity of exogenous amino acids in the methane-oxidizing bacterium *Methylococcus capsulatus*. *Journal of General Microbiology* **71**, 541–554.

Fliermans, C. B., Bohlool, B. N., and Schmidt, E. L. (1974). Autoecological study of the chemoautotroph *Nitrobacter* by immunofluorescence. *Applied Microbiology* **27**, 124–129.

Focht, D. D. and Verstraete, W. (1977). Biochemical ecology of nitrification and denitrification. *Advances in Microbial Ecology* **1**, 135–214.

Frederickson, A. G. (1977). Behaviour of mixed cultures of microorganisms. *Annual Review of Microbiology* **31**, 63–87.

de Freitas, M. J. and Frederickson, A. G. (1978). Inhibition as a factor in maintenance of diversity of microbial systems. *Journal of General Microbiology* **106**, 307–321.

Ghiorse, W. C. and Alexander, M. (1977). Nitrifying populations and the destruction of nitrogen dioxide in soil. *Microbial Ecology* **4**, 233–240.

Godwin, D. and Slater, J. H. (1979). The influence of the growth environment on the stability of a drug resistance plasmid in *Escherichia coli* K12. *Journal of General Microbiology* **111**, 201–210.

Golovacheva, R. S. (1975). Thermophilic nitrifying bacteria from hot springs. *Microbiologiya* (*transl.*) **45**, 377–379.

Gottschal, J. C. and Kuenen, J. G. (1980a). Selective enrichment of facultatively chemolithotrophic thiobacilli and related organisms in continuous culture. *FEMS Microbiology Letters* **7**, 241–247.

Gottschal, J. C. and Kuenen, J. G. (1980b). Mixotrophic growth of *Thiobacillus* A2 in acetate and thiosulfate as growth-limiting substrates in the chemostat. *Archives of Microbiology* **126**, 33–42.

Gottschal, J. C., de Vries, S., and Kuenen, J. G. (1979). Competition between the facultatively chemolithotrophic *Thiobacillus* A2, an obligately chemolithotrophic *Thiobacillus* and a heterotrophic spirillum for inorganic and organic substrates. *Archives of Microbiology* **121**, 241–249.

Gottschal, J. C., Nanninga, H. J., and Kuenen, J. G. (1981). Growth of *Thiobacillus* A2 under alternating growth conditions in the chemostat. *Journal of General Microbiology* **126**, 85–96.

Guay, R. and Silver, M. (1975). *Thiobacillus acidophilus* sp. nov.; isolation and some physiological characteristics. *Canadian Journal of Microbiology* **21**, 281–288.

Gundersen, K. (1955). Observations on mixed cultures of *Nitrosomonas* and heterotrophic soil bacteria. *Plant and Soil* **7**, 26–34.

Harder, W., Kuenen, J. G. and Matin, A. (1977). Microbial selection in continuous culture. *Journal of Applied Bacteriology* **43**, 1–24.

Harms H., Koops, H. P., and Wehrmann, H. (1976). An ammonia-oxidizing bacterium *Nitrosovibrio tenuis*, nov. gen. nov. sp. *Archives of Microbiology* **108**, 105–112.

Harrison, D. E. F. (1978). Mixed cultures in industrial fermentation processes. *Advances in Applied Microbiology* **24**, 129–164.

Harrison, D. E. F., Wilkinson, T. G., Wren, S. J., and Harwood, J. H. (1976). Mixed bacterial cultures as a basis for continuous production of single cell protein from C_1 compounds. *In* "Continuous Culture", Vol. 6, pp. 122–134. (A. C. R. Dean, D. C. Ellwood, C. G. T. Evans, and J. Melling, eds.) Ellis Horwood, Chichester.

Karavaiko, G. I., Shchetinina, E. V., Pivovarova, T. A., and Mubarakova, K. Y. (1973). Denitrifying bacteria isolated from deposits of sulfide ores. *Mikrobiologiya* (*transl.*) **42,** 109–114.

Kelly, D. P. (1971). Autotrophy: concepts of lithotrophic bacteria and their organic metabolism. *Annual Review of Microbiology* **25**, 177–210.

Kelly, D. P., Norris, P. R., and Brierley, C. L. (1979). Microbiological methods for the extraction and recovery of metals. *In* "Microbial Technology, Current Status, Future Prospects", pp. 263–307. (A. T. Bull, D. C. Ellwood, and C. Ratledge, eds.) Cambridge University Press, Cambridge.

Kingma Boltjes, T. Y. (1934). Onderzoekingen over nitrificerende bacteriën. Thesis, University of Delft, The Netherlands.

Kuenen, J. G. (1972). Kleurloze zwavelbacteriën uit het Groninger Wad. Thesis, University of Groningen, The Netherlands.

Kuenen, J. G. (1975). Colourless sulfur bacteria and their role in the sulfur cycle. *Plant and Soil* **43**, 49–76.

Kuenen, J. G. and Veldkamp, H. (1973). Effects of organic compounds on growth of chemostat cultures of *Thiomicrospira pelophila*, *Thiobacillus thioparus* and *Thiobacillus neapolitanus*. *Archiv für Mikrobiologie* **94**, 173–190.

Kuenen, J. G., Boonstra, J., Schröder, H. G. J., and Veldkamp, H. (1977). Competition for inorganic substrates among chemoorganotrophic and chemolithotrophic bacteria. *Microbial Ecology* **3**, 119–130.

Linton, J. P. and Buckee, J. C. (1977). Interactions in a methane utilizing mixed bacterial culture in a chemostat. *Journal of General Microbiology* **101**, 219–225.

Lu, M. C., Matin, A., and Rittenberg, S. C. (1971). Inhibition of growth of obligately chemolithotrophic thiobacilli by amino acids. *Archiv für Mikrobiologie* **79**, 354–366.

Malashenko, Yu. R., Romanovskaya, V. A., Bogachenko, V. N., and

Kryshtab, T. P. (1974). Influence of organic substances on the assimilation of methane by obligate methylotrophs. *Mikrobiologiya* (*transl.*) **43**, 290–294.
Markosyan, G. E. (1973). A new mixotrophic sulfur bacterium developing in acid media *Thiobacillus organoparus* sp. n. *Doklady Akademii Nauk USSR* (transl.) **211**, 318–320.
Mason, T. G. and Slater, J. H. (1979). Competition between an *Escherichia coli* tyrosine auxotroph and a prototrophic revertant in glucose- and tyrosine-limited chemostat. *Antonie van Leeuwenhoek* **45**, 253–263.
Matin, A. (1978). Organic nutrition of chemolithotrophic bacteria. *Annual Review of Microbiology* **32**, 433–468.
Meyer, J. S., Tsuchiya, H. M., and Frederickson, A. G. (1975). Dynamics of mixed populations having complementary metabolism. *Biotechnology and Bioengineering* **17**, 1065–1081.
Norris, P. R. and Kelly, D. P. (1978). Dissolution of pyrite (FeS_2) by pure and mixed cultures of some acidophilic bacteria. *FEMS Microbiology Letters* **4**, 143–146.
Nozhevnikova, A. N. and Zavarzin, G. A. (1973). Symbiontic oxidation of carbon oxide by bacteria. *Mikrobiologiya* (*transl.*) **42**, 134–135.
Pan, P. and Umbreit, W. W. (1972a). Growth of mixed cultures of autotrophic and heterotrophic organisms. *Canadian Journal of Microbiology* **18**, 153–156.
Pan, P. and Umbreit, W. W. (1972b). Growth of obligate autotrophic bacteria on glucose in a continuous flow-through apparatus. *Journal of Bacteriology* **109**, 1149–1155.
Powell, E. O. (1958). Criteria for the growth of contaminants and mutants in continuous culture. *Journal of General Microbiology* **18**, 259–268.
Rittenberg, S. C. (1969). The roles of exogenous organic matter in the physiology of chemolithotrophic bacteria. *Advances in Microbial Physiology* **3**, 159–196.
Rokem, J. S., Goldberg, I., and Mateles, R. I. (1980). Growth of mixed cultures of bacteria on methanol. *Journal of General Microbiology* **116**, 225–232.
Schlegel, H. G. (1975). Mechanisms of chemo-autotrophy. *In* "Marine Ecology", Vol. 2, pp. 9–60. (O. Kinne, ed.) John Wiley, London.
Schnaitman, C. and Lundgren, D. G. (1965). Organic compounds in the spent medium of *Ferrobacillus ferrooxidans*. *Canadian Journal of Microbiology* **11**, 23–27.
Shivaji Rao, G. S. and Berger, L. R. (1970). Basis of pyruvate inhibition in *Thiobacillus ferrooxidans*. *Journal of Bacteriology* **102**, 462–466.
Slater, J. H. and Bull, A. T. (1978). Interactions between microbial populations. *In* "Companion to Microbiology", pp. 181–206. (A. T. Bull and P. M. Meadow, eds.) Longman, London.
Smith, A. J. and Hoare, D. S. (1968). Acetate assimilation by *Nitrobacter agilis* in relation to its obligate autotrophy. *Journal of Bacteriology* **95**, 844–855.
Smith, A. J. and Hoare, D. S. (1977). Specialist phototrophs, lithotrophs and methylotrophs: a unity among a diversity of procaryotes. *Bacteriological Reviews* **41**, 419–448.

Smith, A. L. and Kelly, D. P. (1979). Competition in the chemostat between an obligately and a facultatively chemolithotrophic Thiobacillus. *Journal of General Microbiology* **115**, 377–384.

Stapp, C. (1940). Ueber Begleitorganismen der Nitrifikationsbakterien. *Zentralblatt für Bakteriologie und Parasitenkunde, Abteilung II* **102**, 193–214.

Steinmüller, W. and Bock, E. (1976). Growth of *Nitrobacter* in the presence of organic matter. I. Mixotrophic growth. *Archives of Microbiology* **108**, 299–304.

Taylor, B. F., Hoare, D. S., and Hoare, S. L. (1971). *Thiobacillus denitrificans* as an obligate chemolithotroph. Isolation and growth studies. *Archiv für Mikrobiologie* **78**, 193–204.

Taylor, P. A. and Williams, P. J. LeB. (1974). Theoretical studies on the coexistence of competing species under continuous flow conditions. *Canadian Journal of Microbiology* **21**, 90–98.

Taylor, S. (1977). Evidence for the presence of ribulose-1,5-bisphosphate carboxylase and phosphoribulokinase in *Methylococcus capsulatus* (Bath). *FEMS Microbiology Letters* **2**, 305–307.

Timmer-ten Hoor, A. (1975). A new type of thiosulphate oxidizing nitrate reducing microorganism: *Thiomicrospira denitrificans* sp. nov. *Netherlands Journal of Sea Research* **9**, 344–350.

Timmer-ten Hoor, A. (1977). Denitrificerende kleurloze zwavelbacteriën. Thesis, University of Groningen, The Netherlands.

Tuovinen, O. H., Kelly, D. P., Dow, C. S., and Eccleston, M. (1978). Metabolic transitions in cultures of acidophilic thiobacilli. *In* "Metallurgical Applications of Bacterial Leacking and Related Microbiological Phenomena", pp. 61–81. (L. E. Murr, A. E. Torma, and J. A. Brierley, eds.) Academic Press, New York and London.

Tuttle, J. H. and Jannasch, H. W. (1972). Occurrence and types of thiobacillus-like bacteria in the sea. *Limnology and Oceanography* **17**, 532–543.

Tuttle, J. H., Holmes, P. E., and Jannasch, H. W. (1974). Growth rate stimulation of marine pseudomonads by thiosulfate. *Archiv für Mikrobiologie* **99**, 1–14.

Vainshtein, M. B. (1975). Characteristics of thionic bacteria in the lakes of the Marian ASSR. *Mikrobiologiya (transl.)* **44**, 125–128.

Vainshtein, M. B. (1976). Systematic position of *Thiobacillus trautweinii*. *Mikrobiologiya (transl.)* **45**, 125–129.

Veldkamp, H. and Jannasch, H. W. (1972). Mixed culture studies with the chemostat. *Journal of Applied Chemistry and Biotechnology* **22**, 105–123.

Vollbrecht, D., El Newawy, M. A., and Schlegel, H. G. (1978). Excretion of metabolites by hydrogen bacteria. I. Autotrophic and heterotrophic fermentations. *European Journal of Applied Microbiology and Biotechnology* **6**, 145–155.

Watson, S. D. (1974). Bergey's Manual of Determinative Bacteriology, pp. 450–456. (R. E. Buchanan and N. E. Gibbons, eds.)

Whittenbury, R. and Kelly, D. P. (1977). Autotrophy: a conceptual phoenix. *In* "Microbial Energetics", pp. 121–149. (B. A. Haddock and W. A. Hamilton, eds.) Cambridge University Press, Cambridge.
Wilkinson, T. G., Topiwala, H. H., and Hamer, G. (1974). Interactions in a mixed bacterial population growing on methane in continuous culture. *Biotechnology and Bioengineering* **16**, 41–49.
Yoon, H., Klinzing, G. and Blanch, H. W. (1977). Competition for mixed substrate by microbial populations. *Biotechnology and Bioengineering* **19**, 1193–1211.
Zamenhof, S. and Eichhorn, H. H. (1967). Study of microbial evolution through loss of biosynthetic functions: establishment of "defective" mutants. *Nature, London* **216**, 456–458.
Zavarzin, G. A. and Nozhevnikova, A. N. (1977). Aerobic carboxydobacteria. *Microbial Ecology* **3**, 305–326.
Zeikus, J. G. and Wolfe, R. S. (1972). *Methanobacterium thermoautotrophicum* sp. n., an anaerobic, autotrophic, extreme thermophile. *Journal of Bacteriology* **109**, 707–713.
Wren, S. J. (1978). PhD Thesis, University of London.

6

The Interaction of Algae and Bacteria

A. K. Jones

1. INTRODUCTION

One of the earliest records of interactions between algae and bacteria in laboratory experiments is that of Englemann (1884) who employed bacteria to study photosynthesis in a species of *Spirogyra*. The discussion by Zobell (1946) of the effect of phytoplankton on marine bacteria revealed a controversial situation that is still unresolved. Then, as now, the argument existed as to whether bacteria lived on organic matter excreted by living algae or on their dead and dying remains. The possible effects of antibiotic substances produced by one organism on another in the marine environment had also started to attract some attention.

The present review considers these and related topics both in relation to freshwater and marine microscopic algae and seaweeds, and their associated bacteria. In this context I have decided to treat the cyanobacteria as blue-green algae, as they are predominantly aerobic organisms producing organic material via photosynthesis, and are most conveniently considered with the eukaryotic algae. The generic and specific identities of the bacteria and algae quoted here are those given by the authors in the individual papers, and some might possibly not conform to current taxonomic practice. I have, however, deliberately excluded any treatment of the hot-spring associations, as these constitute a specialized area of study.

2. METHODS OF ISOLATION

The central problem in attempting to understand the behaviour of natural communities of microorganisms is that of translating data based on laboratory cultures to natural systems and also of obtaining valid physiological data in the field (Caldwell, 1977). Caldwell has discussed the problems inherent to this area of study in the context of the lake environment, but his discussion is relevant to the whole sphere of microbial ecology.

Most of the methods so far employed to isolate bacteria and algae for laboratory studies have involved enrichment and plating techniques. Unfortunately, such procedures have previously failed to yield viable counts of bacteria within two orders of magnitude of the direct counts. Caldwell (1977) has suggested, therefore, that some generalizations concerning microbial activities in nature have been made without the organisms responsible for them having been studied in pure culture. It is probable that these criticisms apply to situations involving algae and bacteria, and that the artificial way in which the organisms have been separated from their environment casts doubt on the direct applicability of laboratory studies to nature. Nevertheless, axenic cultures are needed to ascertain the type of interaction to be looked for in natural communities and the avenues of approach to be adopted.

Techniques currently employed to obtain bacterial cultures from the smaller algae are shown in Table 1. Collins *et al.* (1973) have described methods of estimating populations of aquatic bacteria and give details of media which may be used to culture them. Isolation of bacteria from macroscopic algae was achieved after homogenization to separate them (Chan and McManus, 1967). Laycock (1974) homogenized twenty 16 cm^2 pieces from the frond of *Laminaria longicruris* in 100 ml of sterile water to do this. Factors which reduce counts and could therefore affect the bacterial flora isolated are temperatures in excess of 30°C and the possibility of mechanical injury in homogenizers used at speeds over 6,000 rev min^{-1}. Therefore, both Chan and McManus (1967) and Laycock (1974) employed homogenization times and speeds in an apparatus which ensured that these factors were not deleterious.

Fry and Humphrey (1978) reported that stomaching was a better procedure for the removal of bacterial epiphytes than homogenizing.

Table 1. Methods employed in the isolation of bacteria in recent algal-bacterial investigations involving smaller algae

Methods of Isolation	Notes	Reference
Plating using Marine Agar 2216 (Difco).	Three bacterial isolates obtained from cultures of *Thallasiosira pseudonana*.	Baker and Herson (1978)
Enrichment using Isol medium, 5.0 g peptone (Difco), 1.0 g yeast extract (Difco), 0.001 g $K_2 HPO_4$, filtered or synthetic sea water, 1 litre, 5 ml sea water incubated with 50 ml of Isol at 27°C for 16 to 18 hours before use. Then the technique of Adler (1969) was employed, with algal culture filtrates in 1 μl capillaries to chemotactically attract bacteria. Three bacterial types were subsequently isolated by plating out from the capillaries onto solidified Isol.	Authors suggested that the bacteria isolated were ones specifically responding to algal species. Original habitat—marine.	Bell and Mitchell (1972)
Water sample inoculated into axenic culture of *Skeletonema costatum*, followed by plating on solidified Isol medium, grown at room temperature.	A pseudomonad isolate was obtained which was later compared with a spirillum obtained by inoculation of 5 ml of water into 50 ml of Isol medium. Original habitat—marine.	Bell *et al*. (1974)
Plating using Plate Count Agar.	*Arthrobacter* and other lacustrine bacteria isolated. Original habitat—fresh-water reservoir in Massachusetts.	Berger *et al*. (1979)
Plating using Plate Count Agar (Difco) or skim milk agar plates (Skim milk 2.5 g, yeast extract 0.25 g, sodium acetate 0.1 g, 500 ml distilled water, 7.5 g agar). Plates and liquid media maintained at 10° and 22°C.	Bacteria were isolated from cultures of *Anabaena flos-aquae* and water blooms. Original habitat—fresh water.	Carmichael and Gorham (1977)

Table 1—*Continued*

Methods of Isolation	Notes	Reference
Plating using Chu 10 medium (Chu, 1942) modified to contain 10.0 g yeast extract and solidified with 17.5 g agar, plates incubated at 25°C.	Two groups of Gram -ve rods, *Pseudomonas* III and IV, from cultures of *Anabaena cylindrica*. Original habitat—fresh water.	Chrost and Brzeska (1978)
Bacteria were obtained by drawing washed *Oscillatoria* filaments through semi-soft nutrient agar in a sterile petri dish. Further isolation was carried out by plating using nutrient agar, the plates being incubated at 37°C.	Temperature seems high for isolation of aquatic bacteria but strains isolated were fairly typical (see later). See also Allen and Garrett (1977) for observations on temperature of incubation and effect on bacteria isolated. Original habitat—fresh water.	Delucca and McCracken (1977)
Plating, using CPS medium (Collins and Willoughby, 1962) or 5% ASW f/2 tryptone agar (see ref.) or diluted artificial seawater containing 0.1% (w/v) tryptone. Plates incubated aerobically at 20°C for 5 d.	*Flavobacterium* sp. isolated from *Navicula muralis* cultures. On the basis of its resistance to conventional separation techniques was deemed to be closely associated with the diatom. Original habitat—brackish water drainage ditch.	Jolley *et al.* (1976)

Stomaching involves the use of a simple instrument to obtain easily, and without aerial contamination, a homogeneous macerated suspension of plant or animal material in a presterilized robust plastic bag. The bag and contents are pounded vigorously by flat paddles until the material is finely comminuted or "stomached". Higher viable counts were obtained with this technique which was quick and easy to use aseptically, did not require the alga to be cut up before treatment and did not cause the heat generation associated with homogenization.

Pathogens of cyanobacteria were isolated (Stewart and Daft, 1977) by preparing cyanobacterial lawns and looking for plaques caused by a non-filterable agent. The choice of cyanophycean host was not absolutely critical as such bacteria have a wide host range but Stewart and Daft (1977) recommended that lawns of unicellular, non-heterocystous and heterocystous filamentous forms be employed. These workers usually included *Anabaena flos-aquae* and *Nostoc muscorum* which were susceptible to a wide range of bacterial isolates, produced uniform lawns and grew rapidly. If a sample from a developed plaque was shown to be non-filterable and able to grow independently of the host, it was assumed not to be a virus. After characterization using routine bacteriological techniques and if an axenic culture of the agent on addition to the cyanobacterium caused lysis, then it was accepted as being an "algal" pathogen.

The cultivation of marine organisms has been reviewed in detail by Kinne (1976), with chapters by Gundersen (1976) on bacteria; Ukeles (1976) on unicellular plants (algae) and Bonotto (1976) on multicellular plants (seaweeds). Stein (1973) and Droop (1969) provided detailed reviews of the methods for cultivating algae and the use of antibiotics for eliminating bacteria is specifically discussed by Spencer (1952), Berland and Maestrini (1969), Droop (1967), Bednarz (1972) and Jones *et al.* (1973). To obtain axenic cultures is often difficult. Bonneau (1977) brushed sections of *Ulva lactuca* with a sterile brush and dragged them through agar prior to incubation for four days on agar media containing various antibiotics, sulfa drugs, and fungicides. Transfer to similar fresh, inhibitory agar involved manipulation to ensure contact between the inhibitors and the previously exposed side. Swarmers were released and an isolated germling selected but a *Nocardia* sp. still persisted. Only after incubating eighteen pieces of *Ulva* sp. in sterile 0.005% (v/v) Jodopax in seawater for a few minutes were three viable axenic cultures obtained (Fries, 1963).

The only continuous culture study employing natural populations rather than mixtures of originally axenic cultures of algae and bacteria known to this reviewer is that of Martin and Bianchi (1980). The continuous culture apparatus was a large volume tank exposed to external climatic conditions into which natural oligotrophic seawater was pumped along with nutrients. The residence time was 48 h with a 1 $m^3 h^{-1}$ flow rate. Counts of bacteria were made via spread plates employing Oppenheimer and Zobell (1952) 2216E medium incubated at 25°C. This method was chosen rather than direct counts to allow identification of the bacterial isolates, but is also open to the criticisms of plating made earlier.

3. MICROSCOPIC ALGAE AND BACTERIA

3.1 The Concept of the "Phycosphere" and the Bacterial Flora of Microscopic Algae

The term phycosphere was coined by Bell and Mitchell (1972) to denote a zone surrounding phytoplankton, created by their production of extracellular products, which would be an aquatic equivalent of the rhizosphere of land plants. They demonstrated a chemotactic attraction of bacteria to filtrates of old algal cultures. However, the concentrations of compounds commonly found in algal extracellular material needed to elicit such chemotactic responses, were very high compared to concentrations found in seawater, which could indicate why filtrates from younger algal cultures failed to cause a response. Later, Bell *et al.* (1974) found that a *Pseudomonas* species would grow well in both batch and continuous culture with *Skeletonema costatum*, whereas a *Spirillum* sp., which could only take up 2.5% of the carbon from algal extracellular material as would the *Pseudomonas*, grew poorly with actively growing algal cells which also seemed to inhibit it. They, therefore, extended the phycosphere concept to include both inhibitory and stimulatory effects.

The specificity of such inhibition and/or stimulation is important as it may help to resolve apparently contradictory reports in the literature. For example, Droop and Elson (1966) found that bacterial growth associated with *Skeletonema* sp. populations in the River Clyde only increased after the diatom growth peak. This coincided with the observations of Sieburth (1968), using phase contrast microscopy and SEM studies (Sieburth, 1979), that bacteria associated with marine phytoplankton were free rather than attached to living algae and that bacterial attachment increased as the algae lost viability. Whereas Taga and Matsuda (1974) found more bacteria apparently attached to phytoplankton than were free living in seawater. The biochemical properties of isolates from the two niches were also different, suggesting a difference in the bacterial floras. Kogure *et al.* (1979) found that *Skeletonema costatum* generally inhibited the growth of a number of *Pseudomonas* spp. and *Vibrio* spp. isolates, whereas the growth of *Flavobacterium* spp. was generally stimulated. The inhibition was greatest during the exponential phase growth of the alga, thereby confirming the results of Simidu *et al.* (1977) (Table 2) where members of the Flavobacterium-Cytophaga group and *Acinetobacter sp.* were the most important part of the bacterial flora associated with the phytoplankton; this partly coincided with the observations of Sieburth (1968).

In fresh waters, Mikhaylenko and Kulikova (1973) have estimated that 97% of the bacteria in blue-green phytoplankton were epiphytic with only 3% free living, whereas Caldwell (1977) suggested that freshwater diatoms were rarely associated with bacteria and also included in this category desmids, *Scenedesmus* species and *Aphanizomenon* species.

The most important genera of bacteria isolated from cultures of marine phytoplankton by Berland *et al.* (1969) were *Pseudomonas*, *Flavobacterium*, and *Achromobacter*. From a monospecific bloom of *Bacillaria paxillifer*, Dhevendaran *et al.* (1978) found the genera *Vibrio*, *Bacillus*, *Micrococcus*, and *Alcaligenes* with 11% of the population unidentified, whereas from a bloom of *Coscinodiscus* sp., 80% of the bacteria were *Vibrio* spp. Delucca and McCracken (1977) obtained species of

Table 2. Bacterial flora of seawater and phytoplankton samples collected in Sagami Bay and Saruga Bay, Japan. The numbers show the percentage of the total population for each sampling area: a = non-aerogenic, non-luminescent strains (after Simidu *et al.*, 1977).

	Sea water		Phytoplankton	
Bacterial Genera	Sagami Bay (0–50 m)	Suruga Bay (0–400 m)	Sagami Bay	Suruga Bay
Vibrionaceae[a]	32.0	36.6	7.7	7.9
Pseudomonas	40.1	22.0	15.4	21.5
Caulobacter	0.0	0.0	15.4	0.0
Acinetobacter	22.3	14.6	38.4	4.6
Flavobacterium	5.6	7.3	23.1	66.0
Bacillus	0.0	0.0	0.0	0.0
Not identified	0.0	19.5	0.0	0.0
No. of strains tested	21	41	13	28

Pseudomonas, *Xanthomonas*, and *Flavobacterium* from a freshwater *Oscillatoria* sp. It seems that the bacteria normally correlated with algae are Gram-negative (Caldwell and Caldwell, 1978a; Godlewska-Lipowa and Jablonska, 1972; Simidu *et al.*, 1971; Zagallo, 1953). Chróst (1972, 1975a,b) has suggested that this relates to the production by the algae of inhibitors specific for Gram-positive organisms.

An interesting association has recently been reported by Tison *et al.* (1980) where *Legionella pneumophila*, the purported aetiological agent of Legionnaires disease, was isolated with the cyanobacterium *Fischerella* sp. from a matt community of cyanobacteria in a thermal effluent. Bershova *et al.* (1968) reported that the predominant "satellite" bacteria associated with cyanobacteria were from the genera *Pseudomonas* and *Chromobacterium*. Caldwell and Caldwell (1978b) isolated a Gram-negative bacterium (*Zoogloea* sp.) which only grew in liquid culture when the *Anabaena flos-aquae*, from which it was originally isolated, was present.

Thus far the concept of the phycosphere has been outlined and the nature of the bacteria associated with microscopic algae noted. Ukeles and Bishop (1975) suggested that there may also be a "bactosphere" or zone of enrichment surrounding some bacteria in which the growth of other microorganisms may be enhanced. This indicates the idea, or at least the possibility, of a two way process of algae affecting bacteria and *vice versa*. This is now explored in more detail.

3.1.1 Positive Interactions

The cycling of photosynthate from phytoplankton to bacteria and its ecological significance.

The relationship that exists between aquatic bacteria and organic material produced by phytoplankton is of general ecological importance. Much of the organic matter in the sea may exist as stable, unidentified or very old humic fractions (Wagner, 1969). What is of more interest is the much smaller fraction including sugars, organic acids, amino acids, and other known organic substances that could serve as a food source for bacteria. Wangersky (1978) has recently reviewed the production of organic matter in the sea.

The extracellular products of algae and their liberation has been discussed by Fogg (1962, 1966, 1971) and Hellebust (1974) in their reviews. These substances include carbohydrates, lipids, peptides, organic phosphates, volatile substances, sex factors, growth inhibitors and stimulators, enzymes, phenolic substances, vitamins, and toxins.

However, controversy surrounds the source of this material in sea water with Duursma (1963) arguing that dead and decaying phytoplankton, and not healthy cells, were the major contributors. It is the production of photosynthetically derived organic carbon (PDOC) by photosynthesising algae which is more central to the concept of the phycosphere if the term is used analogously with the rhizosphere. Generally, algal cells in lag and maximum population phases release more PDOC than exponentially growing cells (Guillard and Wangersky, 1958; Nalewajko *et al.*, 1963; Marker, 1965; Jolley and Jones, 1977) and probably much of the maximum population phase material is the result of autolysis.

However, Fogg (1966) stated that between 5 and 35% of the carbon fixed by marine phytoplankton was released and others have reported values up to 50% depending on species and growth conditions (Hellebust, 1974). Sharp (1977) concluded that much of the evidence relating to PDOC production was suspect due to inadequate assessment of control blanks in [^{14}C]-carbon field experiments and also because of cultural shocks caused by transferring algae from nutrient poor to nutrient rich conditions, subjecting them to centrifugation, resuspension, and different light conditions. This, he suggested, leads to the release of anomalously large amounts of PDOC whereas healthy cells would only release zero plus or minus 5% of the photoassimilated carbon. This is not very different from the 6 to 11% released by *Navicula muralis* in exponential growth phase (Jolley and Jones, 1977) and the 5% released by *Thalassiosira fluviatilis* (Hellebust ,1967) and lower than the 32% and 20%, respectively, released by maximum population phase cultures of these algae. Fogg (1977) pointed out that a low percentage release was only to be expected in exponential phase laboratory batch cultures; as outlined in a kinetic model (Fogg, 1966), free diffusion of a low molecular weight substance between the interior of the cells and the culture medium would be limited by equilibrium considerations when at high cell densities. Anderson and Zeutchel (1970), Thomas (1971) and Berman and Holm-Hansen (1974) appeared to demonstrate that sparse phytoplankton populations excreted a higher percentage of their fixed carbon than did denser populations. However, a reverse trend was exhibited in the total amount of material liberated and it is probable that the lower particulate fixation values recorded for the nutrient poor locations caused the elevation of the percentage release data. Fogg (1977) agreed with Sharp (1977) that insufficient care may have been paid to experimental error in a number of studies. However, this did not indicate that healthy algae never excrete organic material and Mague *et al.* (1980), paying particular attention to experimental technique, concluded that release of PDOC was a normal function of healthy algal cells and was closely related to photosynthetic rate. However, they also suggested that PDOC was a minor component of primary productivity in the coastal waters studied, but they measured net rather than gross release of PDOC.

In field studies, percentage extracellular release is usually determined according to the following generalized procedure. To the water sample is added specially prepared $NaH^{14}CO_3$ containing no organic impurities, samples are incubated for a few hours and filtered through 0.45 μm pore size membrane filters using minimal suction pressure. The filtrate is acidified and bubbled with a gas to drive off unassimilated [^{14}C]-carbon dioxide. The residual activity in the filtrate is that due to excretion and that retained by the filters attributed to that fixed carbon which is incorporated into algal biomass.

If aquatic bacteria in such samples can rapidly utilize the PDOC from algae, then:

(1) excretion, as measured above, could be the net excretion, representing the balance remaining after production by algae and uptake by bacteria. Nalewajko *et al.* (1976) found that in *Chlorella* sp. cultures with added bacteria, extracellular release showed a plateau with time as did lake water samples, whereas in axenic cultures release remained linear and identical to the initial rate observed in the mixed culture system. This indicated an initial lag in the metabolism of photosynthate by the added bacteria. Thus short incubations of one hour would give values near to the gross release and incubations of several hours would estimate net release by algae. In marine studies, Anderson and Zeutschel (1970) recognized the potential for underestimating PDOC release but could not detect any significant difference between control samples and ones containing penicillin and streptomycin, included to inhibit the bacteria. Lancelot (1979) concluded, on the basis of time course data, that four hour incubations could underestimate the gross excretion due to a lag phase in phytoplankton release and/or to heterotrophic uptake during the incubation period. Smith and Higgins (1978) managed to reduce bacterial contamination of a culture of *Isochrysis galbana* by adding either rifampicin or penicillin G three days before the experiments. Thus, instead of a plateau in production of PDOC after 100 min incubation, excretion continued to increase for at least 300 min after the antibiotic treatment. Chróst (1978) found that gentamycin rapidly and efficiently inhibited lake water bacteria and that its addition to lake water caused an increase of PDOC excretion from 5% to between 9.9 and 17.3% of photosynthetic carbon fixation, indicating considerable bacterial utilization of PDOC. Although Hellebust (1974), Jones (1977b), Jones *et al.* (1973) and A. K. Jones (unpublished observations) suggest that antibiotics may be used with varying success in preventing

bacterial activity in short-term experiments, the data above would seem to substantiate the findings of Nalewajko *et al.* (1976). Mague *et al.* (1980) also showed that with natural populations a plateau of PDOC release was obtained after 8 to 9 h at 68% of surface light intensity, compared with a linear response over the same time with axenic cultures of *Skeletonema costatum*. However, they concluded on the basis of only one selective filtration experiment that there was no heterotrophic uptake of PDOC.

(2) [14C]-carbon dioxide method of Steemann-Nielsen (1952) estimates the net fixation of inorganic carbon by phytoplankton plus the incorporation of PDOC into heterotrophs. Wright (1974) also suggested the possibility that a small amount of [^{14}C]-carbon dioxide may be released by heterotrophs and refixed by phytoplankton. This, besides other factors, complicates our assessment of exactly what the Steemann-Nielsen method measures and whether it estimates something in between net and gross photosynthesis. The fact that some PDOC is incorporated into heterotrophs may not significantly alter a gross estimate of primary production. However, as there is evidence that certain zooplankton can select potential food particles of different size, the complete analysis of the trophic relationships in a water body requires the measurement of PDOC formation and heterotroph production as well as algal carbon dioxide fixation. Derenbach and Williams (1974) using both the Steemann-Nielsen (1952) and Parsons and Strickland (1962) procedures in parallel have estimated that bacterial production in Southampton Waters varied between 1 and 30% of the primary production and was, therefore, an important factor in energy flow in that ecosystem.

Glycollic acid has attracted much attention as an algal extracellular product, probably because it is an initial product in photosynthesis and can be immediately lost from photosynthesising cells. Hellebust (1974) listed the algae, from seven different classes, shown to release glycollic acid. Its presence has been demonstrated in freshwater (Fogg and Nalewajko, 1964; Fogg and Watt, 1965; Watt, 1966; Nalewajko and Marin, 1969) and in seawater was shown to fluctuate between 0 to 80 $\mu g\ l^{-1}$ in the Gulf of Maine (Shah and Wright, 1974) and 0 to 60 $\mu g\ l^{-1}$ in the Menai Straits (Al-Hasan *et al.*, 1975).

Wright (1970) showed that [^{14}C]-glycollate was taken up by natural populations of microorganisms with (Kt + Sn) values between 0.05 to 2.0 $mg\ l^{-1}$ which compared well with a Kt of 0.13 $mg\ l^{-1}$ shown by a bacterium isolated using glycollate as sole carbon and energy source (where Kt = concentration of substrate giving ½ maximal uptake rate of substrate by a natural plankton population; Sn = natural substrate concentration). The pattern of glycollate uptake in the water seemed to be correlated with algal photosynthesis and the production and uptake of glycollate were of the same order of magnitude. Nalewajko and Lean (1972) found that glycollate was taken up by bacteria present in cultures of freshwater algae preventing its accumulation during algal growth. Tanaka *et al.* (1974) showed that the distribution peak of glycollate-utilizing bacteria, as distinct from the total heterotrophic population, corresponded closely to the depth at which maximum excretion of [^{14}C]-labelled PDOC occurred in Lake Biwa. The major component of the PDOC was glycollic acid and it was, therefore, attractive to conclude that the population maximum for glycollate-utilizing bacteria was caused by this. In Upper Klamath Lake, 69% of the glycollate taken up by heterotrophs was mineralized to carbon dioxide (Wright, 1975) and cultures of bacteria isolated from lake water exhibited uptake and mineralization patterns similar to the natural plankton. Under natural conditions it would seem that glycollic acid can be rapidly mineralized and the carbon dioxide made available for reincorporation by algae.

Berland *et al.* (1970) found that only one *Pseudomonas* sp. of 25 bacterial strains isolated from marine algal cultures grew with glycollate as sole carbon source. None of the 145 facultatively anaerobic Gram-negative rods tested by Baumann *et al.* (1971) would do so and only 22 of 218 aerobic strains had this capability (Baumann *et al.*, 1972), although the possibility of the presence of other substrates in the medium casts some doubt on the results (Wright and Shah, 1975). As 63% of the 141 colonies of marine bacteria tested by Wright and Shah (1977) could take up and respire glycollate but could not grow on it as the sole carbon source, it was suggested that glycollate might be used as an energy source for the active transport of other substrates. They also concluded that the use of glycollate by microbes is not tied to its production but to the presence of other substrates, leading to erratic fluctuations in concentration in natural waters. Coughlan and Al-Hasan (1977) concurred with this since the only factor that correlated with glycollate concentration was the rate of PDOC release and then only in their first field study. It was also noted that there were very long turnover times for glycollate (2636 h). Wright and Shah (1977) found a statistically significant correlation between V_{max} for glycollate uptake and glycollic acid concentrations ($r = 0.64$, $P = 0.005$) on a seasonal basis, suggesting a cause and effect trophic relationship. However, this hypothesis was rejected because of the inability of many marine bacteria to grow on glycollic acid and the lack of compatibility between the long turn over time for glycollic acid and the known growth rates of bacteria.

Wright and Shah (1977) finally concluded that, despite the considerable evidence in the literature for excretion of glycollate by algae, in Ipswich Bay the glycollate flux was responsible for only about 0.5% of phytoplankton production. Moreover, they contended that glycollic acid flux was quantitatively similar to that of other substrates and that glycollic acid was not a dominant factor in the flow of energy from algae to bacteria but contributed only as much energy to bacterial production as other substrates.

The previous discussion indicates that a close coupling between bacterial and algal production might exist in the plankton. Saunders and Storch (1971) suggested that bacterial utilization of PDOC would lag behind its release by phytoplankton and have demonstrated, by the addition of [^{14}C]-glucose to simulate algal release, a peak of bacterial activity some three hours after the peak of simulated algal excretion. However, Sorokin (1969, 1972) and Sorokin and Paveljeva (1972) have estimated that in the lakes of Central Russia heterotrophic metabolism much exceeded autotrophic primary production in midsummer between the spring and autumn phytoplankton blooms. Sorokin (1977) showed that a similar situation occurred in the North Japan Sea and suggested that heterotrophs at this stage used energy derived from organic matter accumulated during the previous spring phytoplankton bloom. Goldman *et al.* (1968) suggested that no direct relationship existed between algal and bacterial populations on a seasonal basis, whereas Silvey and Roach (1964) were able to demonstrate one. Jones (1972) established a positive correlation between bacterial numbers and chlorophyll *a* estimates. Very interestingly, there was variation with the species involved. The degree of attachment by bacteria was shown to diminish along the series, colonial green algae and cyanobacteria to filamentous green algae and filamentous cyanobacteria and diatoms to dinoflagellates, the trend being shown using both viable and direct count methods. Jones (1977a) also showed that components relating to algal productivity could account for 10 to 20% of the horizontal variability of viable bacteria from six lake sites. Overbeck (1967, 1968) demonstrated a correlation between peaks of algae and bacteria in depth profiles, whereas Guseva (1967), Tilzer (1972), Makisimova (1973), and Straskrabova (1975) found that peaks of bacterial numbers succeeded those of algae.

Coveney *et al.* (1977) showed that during the *Stephanodiscus* sp. bloom in spring and the *Aphanizomenon* sp. bloom in late summer, after an initial increase in bacterial numbers, there was a decrease, particularly in the case of *Aphanizomenon* sp., then a sharp increase in bacterial numbers as the algal numbers fell. They also detected a low net release of

PDOC (0 to 8% of total, integral phytoplankton productivity), which they believed to reflect high bacterial utilization rather than low release rates by the algae. It was therefore suggested that during the algal blooms the bacterial flora changed from one living on products excreted from live algae to one utilizing products of algal autolysis and decomposition.

It would seem reasonable to conclude that, as shown by Tanaka *et al.* (1974), the distribution of the total heterotrophic bacterial population can differ from those using specific algal extracellular products. This may account for the variable and apparently contradictory reports in the literature where attempts were made to correlate total heterotrophic activity or bacterial numbers with primary productivity or with algae. It was not surprising if only a part of the bacterial population readily responded to specific algal extracellular products that, based on glucose and amino acid utilization by heterotrophs and from estimates of the net PDOC excretion only (some may have been used by specific heterotrophs), Williams and Yentsch (1976) concluded that PDOC was insufficient to supply the heterotrophic population and that some other source or route of production for the dissolved organic carbon (DOC) was of major importance. Wiebe and Smith (1977) confirmed that PDOC only formed a small proportion (0.1%) of the DOC; this was explained on the basis of a tight coupling between production of PDOC and its utilization, so preventing any accumulation in the environment. Bacteria attached to particles of between 20 to 63 μm were implicated as responsible for the greatest uptake of PDOC. The hypothesis that organisms of 20 to 63 μm could be involved was dismissed on the grounds of lack of supporting evidence. This contention may not be entirely justified. They also concluded, in contrast to the findings of Wright (1975) with glycollate, that the PDOC was not used primarily for energy metabolism but was assimilated into heterotroph cell material before being respired to carbon dioxide. Williams and Yentsch (1976) also noted that 42% of the glucose and only 21% of the amino acid mixture taken up was respired by the population tested. This may again suggest that different algae produce varying proportions of several substrates and that different bacteria might take up only some of these and that the further metabolism of various substrates is also variable. Such tight coupling of utilization to production of PDOC was not observed by Iturriaga and Hoppe (1977); but they did find that between 8 to 17.5% of the PDOC was taken up by bacteria per hour. However, Smith and Higgins (1978), employing a technique where all the inorganic carbon supplied was radioactive to produce highly radioactive PDOC, found that in water samples and non-axenic algal cultures, the rate of production of PDOC was equalled by the rate of its incorporation into particulate organic carbon. Organisms producing PDOC were of a different size to those utilizing it, they interpreted the results as representing the functioning of an ecological unit consisting of at least two interacting species (alga and heterotroph) linked via a feedback mechanism to give a homeostatic system for the balanced production and uptake of PDOC. This was consistent with findings of Wiebe and Smith (1977) and Larson and Hagström (1979) who also found that PDOC represented about 0.1% of total DOC in their sea water samples and that labelled organic carbon in both bacterial and phytoplankton fractions increased linearly with time. The former represented a very rapid utilization of the PDOC released by the algae and it was calculated that the annual bacterial production resulted from the release of 45% of the corrected annual primary production as PDOC. Their fractionation experiments indicated that this was achieved without a firm physical association between the algae and bacteria. Hoppe and Horstmann (1981), using a new [^{3}H]–[^{14}C]-double labelling technique found that 20 to 30% of the bacteria in growths of *Nodularia spumigena* were attached to the cyanobacterial filaments and that 76% of the algal PDOC was immediately incorporated into the bacteria.

Vitamin exchange

The vitamin requirements of algae have been reviewed by Carlucci (1974), Provasoli (1958, 1963, 1971), Droop (1962), Provasoli and Carlucci (1974). Burkholder (1963) examined the role of vitamins in the nutritional relationships of microorganisms and Berland *et al.* (1978) reviewed the importance of vitamins in the sea and their utilization by algae and bacteria. Vitamin production by bacteria and some algae and their utilization by algae has been used by Provasoli (1971) to further substantiate the concept due to Lucas (1947) of ectocrine substances.

Algae appear to be unique amongst microorganisms as only three vitamins have been shown to be required by those so far cultured axenically (Provasoli and Carlucci, 1974). Vitamins have been detected in all aquatic environments and it was thought that they originated primarily from the activities of bacteria in the sediments (Burkholder and Burkholder, 1956) and in the sea (Provasoli, 1963). Burkholder (1963) demonstrated the production of various B vitamins by algae and Carlucci and Bowes (1970a) showed that *Skeletonema costatum*, *Stephanopyxis turris* and *Gonyaulax polyedra* produced thiamine and biotin whilst being auxotrophic for B_{12} and noted (Carlucci and Bowes, 1970b) that vitamins produced by one alga could be used by another.

Therefore the dependence of algae on bacterially produced vitamins or *vice versa* is not easily demonstrated in the environment. Undoubtedly vitamins are important factors in the interactions between bacteria and algae; Jolley and Jones (1977) showed that *Navicula muralis* produced biotin and three amino acids for which an associated *Flavobacterium* sp. was auxotrophic. However, as these nutrients could be part of the phycosphere of a number of algae it was not taken to imply a great deal with respect to the specificity of the association. Berland *et al.* (1978) concluded that the concentration of vitamins in seawater often exceeded the requirements of many organisms and, therefore, that they were important in terms of species composition of a population, and Hellebust (1974) suggested that vitamin release and production of vitamin-binding factors (Droop, 1968) were important in species succession. Stewart *et al.* (1977) showed that in the series of lochs they studied, a possible correlation existed between the concentration of vitamin B_{12} in the lochs and the abundance of vitamin B_{12}-producing bacteria. In the loch with the lowest vitamin B_{12} concentration the algae were mainly non-vitamin B_{12} requiring cyanobacteria whereas in Forfar, the loch with the highest vitamin B_{12} concentration, small, fast growing green algae and diatoms which may have been vitamin B_{12} auxotrophs were dominant. Hagedorn (1969), and Ohwada and Taga (1972, 1973) have also shown correlations between maximum production of vitamins and algal blooms or between the vertical profile of chlorophyll and vitamin B_{12} concentrations, indicating some sort of relationship between the two and the bacteria which may have produced the vitamins.

Other specific positive responses between microscopic algae and bacteria

Schwabe and Mollenhauer (1967) observed that axenic cultures of *Nostoc sphaericum* developed only as minute colonies but on the addition of the naturally associated bacteria, the normal morphology was re-established. Bacteria isolated from *Nostoc punctiforme*, however, only partially induced normal colonies of *N. sphaericum*. Whitton (1973) suggested that in culture, *Nostoc* spp. colonies rarely resembled those in the field and concluded that such bacterial influences merited further investigation.

The readdition of a *Corynebacterium* sp. to axenic cultures of male and female *Oedogonium cardiacum* by Machlis (1973) restored its ability to form reproductive organs, a facility much reduced by the earlier elimination of the *Corynebacterium* sp. Surprisingly, *Pseudomonas putida* elicited a greater development of sex organs plus longer vegetative filaments compared with axenic *Oedogonium* sp. cultures or those containing *Corynebacterium* sp.

Stegeman and Hoober (1975) reported that *Chlamydomonas reinhardi* caused a bacterium to produce a specific protein, an effect which was probably mediated via a factor synthesized in the light by the alga, excreted and utilized by the bacterium.

Bunt (1961) found that *Caulobacter* sp. stimulated nitrogen fixation by *Nostoc* sp. and Bjälfve (1962) noted similar effects of some bacteria on *Nostoc calcicola*. A specific attachment of bacteria to the polar region of the heterocysts of *Aphanizomenon flos-aquae* and *Anabaena circinalis* was shown by Paerl (1976) using SEM and autoradiographic techniques (Fig. 1), who later (Paerl, 1978) differentiated two types of cyanobacterial-bacterial associations. Firstly, in the mucus surrounding the heterocysts and the vegetative cells of *Anabaena spiroides*, clumped bacteria were located which could metabolize glucose and produce reducing conditions (demonstrated by NBT formazan crystal deposition). High oxygen concentrations increased mucus production by the *Anabaena* sp. thereby stimulating colonization by bacteria attaching at right angles to the filaments. Secondly, morphologically distinct bacterial rods were found only on the heterocysts of *Anabaena oscillariodes*, *A. circinalis* and *A. spiroides*. Filtered (3 μm pore size Nuclepore filters) lake water yielded bacteria which colonized only the heterocysts of axenic *A. oscillaroides* with a statistically significant polar localization. The motile bacteria moved towards the region between a heterocyst and vegetative cell (Paerl and Kellar, 1978) possibly being chemotactically attracted; flagella activity soon ceased and firm attachment ensued.

The uptake of [^{3}H]-labelled organic substrates by the heterocyst specific bacteria and the deposition of NBT formazan crystals in the bacterial colonies (Paerl, 1978) suggested the production of reducing conditions near the heterocysts. Addition of bacteria to axenic *A. oscillariodes* greatly stimulated its nitrogen fixation (and the bacteria alone did not reduce acetylene). At high oxygen concentrations the bacteria-*Anabaena* sp. association showed less inhibition of nitrogen fixation and faster recovery from oxygen effects than did axenic "algal" cultures. Possibly the bacterial release of carbon dioxide stimulated "algal" photosynthesis and so, indirectly, nitrogen fixation. Paerl was, however, unable to show significant stimulation of nitrogenase activity by carbon dioxide enrichment. Therefore, a direct effect of the bacteria via their removal of oxygen, to protect the oxygen sensitive heterocyst nitrogenase at the heterocyst poles adjacent to the oxygen evolving vegetative cells, seems more probable. Paerl has thus demonstrated a symbiotic association whereby bacteria are nourished by extracellular products of the cyanobacteria, and the bacteria reciprocate by protecting the heterocystous nitrogenase system from oxygen.

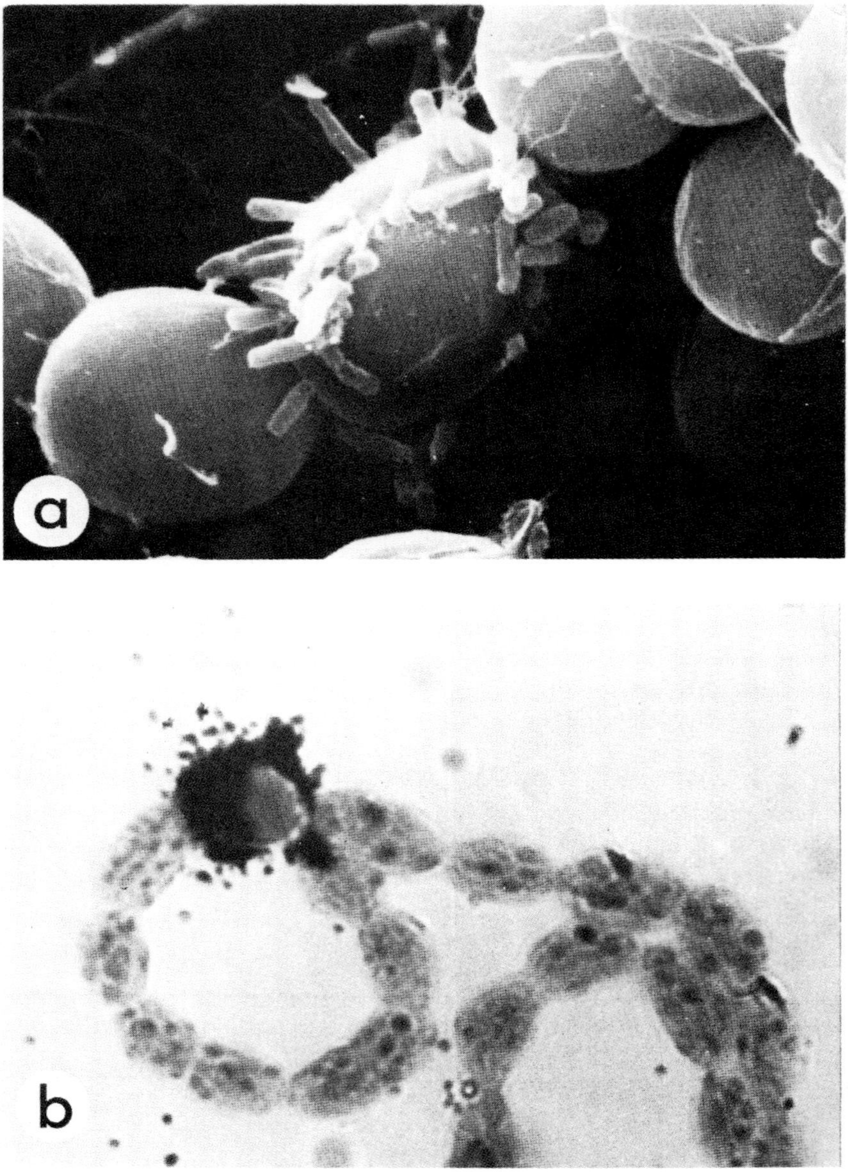

Fig. 1. (a) Specific attachment of bacteria to the heterocysts of *A. circinalis* freshly collected from L. Rotoaira. (b) Autoradiograph showing attached bacteria labelled with 2-^{3}H alanine after a 3-hour incubation with the isotope. Exposed silver grains are black here. (From Paerl (1978), *Microbial Ecology* **4**, 215–231 courtesy of Springer-Verlag, New York, Inc.)

Lange (1967) observed that under conditions of carbon dioxide limitation, the addition of organic carbon stimulated the growth of cyanobacteria in mixed culture with bacteria via carbon dioxide production and that the availability of metabolizable organic carbon compounds, such as sucrose and certain aliphatic compounds (Lange, 1970, 1971), could be important factors in the occurrence of algal blooms. King (1970) and Kerr *et al.* (1972) also suggested that carbon dioxide supply may be important in this context. Schindler (1971) disagreed and Goldman *et al.* (1972) also suggested that the conditions of high nutrient and low carbon dioxide concentration employed by Lange were not comparable to natural conditions and that carbon will rarely, if ever, be limiting to algal growth in nature. Lange (1973) stressed that only when carbon dioxide was limiting would carbon matter utilization and supply of carbon dioxide by bacteria be important in eutrophication. The sheath of cyanobacteria also increased on the addition of organic matter (Lange, 1973, 1976) and organisms with large and fluffy sheaths have been observed in natural productive waters. It was suggested that the sheath constituted a microenvironment of increased nutrient concentration around the "alga" thus, if organic carbon became limiting and bacteria associated with the "algae" then utilized the sheath substance making it thinner, "algal" growth might be reduced. Tison and Lingg (1979) demonstrated that the yield of two cyanobacteria and two green algae was increased when nutrients were not limiting, by the production of carbon dioxide and/or co-factors by bacteria utilizing glucose added to the cultures. The photosynthesising algae were able to furnish the major part of the oxygen supply to the bacteria. However, Lin (1972) found that although bacteria and glucose together increased the growth of *Chlorella pyrenoidosa*, single additions of glucose, bacteria and bicarbonate had no effect, thus indicating that bacterial carbon dioxide production ws probably not implicated directly. The production of carbon dioxide by bacteria could also offset the rise in pH caused by algal photosynthesis and so influence the availability and transport into the algae of substances whose solubility is pH dependent (Caldwell, 1977). The production of chelators, such as hydroxamate (Murphy *et al.*, 1976), by cyanobacteria or their associated bacteria may account for the existence of cyanobacteria in habitats with a high pH, where iron, for example, would otherwise be insoluble and unavailable (Caldwell, 1977).

The association of a *Zooglea* sp. with *Anabaena flos-aquae* was deemed to be obligate by Caldwell and Caldwell (1978b) as the bacterium was able to assimilate but not respire [^{14}C]-PDOC produced by the cyanobacterium, but would not grow without the *Anabaena* sp. in liquid culture, although it would do so on the equivalent agar medium. The occurrence of bacteria in unpolluted mountain streams was more prevalent in the epilithic situation than in stream water and a close physical relationship existed between attached epilithic algae and the bacteria; the algae providing a surface for bacterial colonization (Geesey *et al.*, 1978). McFeters *et al.* (1978a,b) found that high numbers of coliform bacteria in an unpolluted mountain stream were associated with a well-developed algal matt community of species of *Chlorella*, *Gloeocapsa*, and *Stigonema*. Pure cultures of coliforms incorporated [^{14}C]-PDOC and were able to grow in culture media from which previously growing *Chlorella* sp. had been removed by centrifugation, as would bacteria from the stream. The observation that *Legionella pneumophila* (Tison *et al.*, 1980) also grew in the presence of cyanobacteria suggests that a range of bacteria might opportunistically survive with algae and hence be brought unwittingly into the laboratory. Therefore, it is prudent to treat non-axenic algal cultures as potential sources of infection in the laboratory.

The algal-bacterial symbiosis in aerobic sewage-oxidation ponds.
The treatment of sewage in aerobic oxidation ponds is probably a form of symbiosis involving algae and bacteria (Oswald *et al.*, 1953a,b). The organic waste is converted by aerobic bacteria into carbon dioxide, ammonia and other inorganic nutrients, utilizing oxygen produced by photosynthesising algae which themselves assimilate the nutrients and carbon dioxide produced by the bacteria (Oswald and Golueke, 1968) (Fig. 2). However, Abeliovich and Weisman (1978) concluded that the *Scenedesmus* sp. was able to obtain 15% of its carbon directly from glucose and that the bacteria played only a minor role in biological oxygen demand (BOD) reduction, being mainly involved in the degradation of biopolymers, producing substrates for algal consumption.

Originally, the ponds were about 1.5 m deep and required 20 d retention periods for breakdown of the waste (McGarry, 1971) but the development of shallow (20 to 45 cm deep) high rate oxidation ponds (Oswald and Golueke, 1968; McGarry 1971) allowed better light penetration throughout the system and hence greater photosynthetic oxygen input. This resulted in shorter retention periods of 1.5 to 3 d and an effluent with a BOD less than 20 mg l^{-1}. The algal protein harvested from these ponds has been successfully fed to chickens and other farm animals and with areal production (production per unit area of land) rates 127 times that of soya bean and 560 times that of rice (McGarry, 1971); it seems a useful method of both sewage treatment and protein production in locations where the climate is appropriate and land costs not excessive. Soeder (1980) has discussed the potential for mass culti-

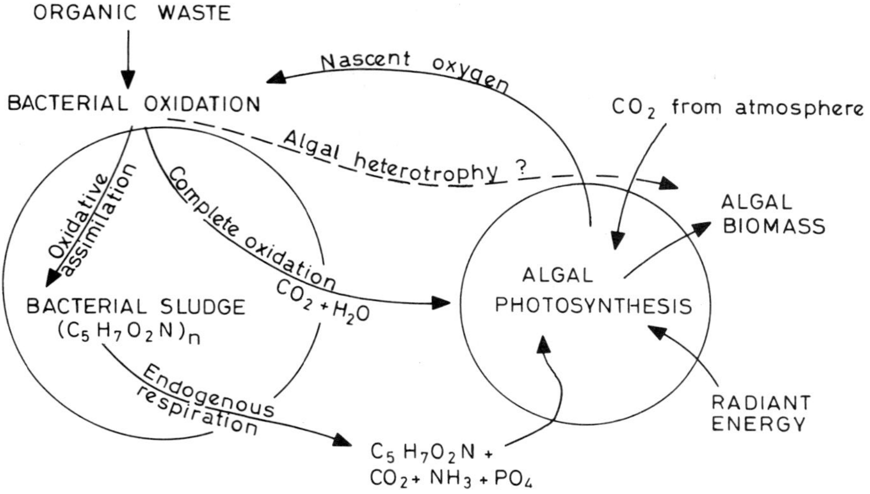

Fig. 2. Interactions of algae and bacteria.

vation of algae and concluded that the treatment of liquid wastes with algal/bacterial systems was the most promising microalgal technology.

The groups of algae found in all the ponds studied by Silva and Papenfuss (1953) were the *Volvocales*, *Chlorococcales*, and *Euglenales*. According to Allen (1955) *Chlorella* was the principal genus in the active oxidation phase of a pond, with *Scenedesmus* sp. as the sub-dominant population. When oxidation was almost complete, a mixed flora with *Scenedesmus* and *Chlamydomonas* as principal genera, succeeded the *Chlorella*. Dor (1976) suggested that the dominance of the *Chlorella* and *Scenedesmus* genera was due to their production of antibiotics which controlled the bacterial generation of nutrients. Benson-Evans and Williams (1975) have also listed the algae of oxidation ponds and the literature surveyed in its compilation. They concluded that the four most constant and cosmopolitan algal genera were *Chlorella*, *Scenedesmus*, *Chlamydomonas*, and *Euglena*. *Chlorella* species appeared to be the dominant algae when oxidation ponds were operating maximally (Isaac and Lodge, 1958); *Chlamydomonas* species were considered by Oswald (1961) to be the least desirable and Kott and Ingerman (1966) regard the genus as indicative of recurrent anaerobic conditions, whereas *Euglena* species were regarded as an indicator of good conditions. This was interesting as the feebly photosynthetic *Chlamydomonas mundana* apparently utilized acetate produced by bacteria in anaerobic waste stabilization lagoons (Eppley and MaciasR, 1963). The bacteria involved in these aerobic ponds have not been identified with certainty (Ganapati, 1975) but they may be similar to those in activated sludge or trickling filter populations. In laboratory experiments, Ganapati (1975) found a slight variation in bacterial flora associated with the different algae employed, but amongst the genera identified were *Pseudomonas*, *Flavobacterium*, *Alcaligenes*, *Aerobacter*, *Comamonas*, *Achromobacter* and *Zooglea*; that is, genera which can be detected amongst the flora of biological sewage treatment plants.

As well as utilizing carbon dioxide and inorganic nutrients released by bacterial activity, Ganapati (1975) considered that the algae could also increase the area of submerged surfaces where nutrients may concentrate and function as resting places for bacteria; furthermore the algae also aided the processes of autoflocculation and bioflocculation important in the first phase in the purification of settled, raw sewage.

Drawing on the theories of Eckenfelder and Weston (1956) and Eckenfelder and O'Connor (1961), Ganapati (1975) proposed that the bacteria utilized and oxidized the organic material in a three stage process; firstly, complete oxidation, resulting in the production of carbon dioxide, water and energy; secondly, cellular synthesis and oxidative assimilation of the remaining matter, with newly formed bacterial cells as a major end product; and thirdly, endogenous metabolism, where the bacteria derive their energy via metabolism of their storage products and cell material. If the last stage is significant, little accumulation of bacterial sludge would occur.

After an initial accumulation of aerobic sludge in oxidation ponds the volume remains fairly constant (Oswald, 1960), corresponding to the small amount of sludge accumulation observed by Ganapati (1975) in his experiments. Ganapati (1975) suggested that the mechanism of purification in these oxidation ponds was similar to the extended aeration modification of the activated sludge process, where the objective is to oxidize the biological solids produced in the second or synthesis phase noted previously, and that the constant volume of sludge consisted of inert organic matter and active sludge.

Ganapati (1975) has demonstrated a direct correlation between the formation of algal biomass and the COD consumed and hence bacterial activity in laboratory experiments. Using a number of assumptions, it was calculated that there was usually an excess of algal oxygen production over that needed for the bacterial oxidation of sewage, whereas insufficient carbon dioxide was generated by the bacteria to support algal photosynthesis. It was suggested that this deficiency was made good by carbon dioxide originating from the atmosphere and the bicarbonate-carbonate system in the aqueous phase. Thermodynamic considerations have led Oron *et al.* (1979) to suggest that 300 bacterial units may be required to supply sufficient carbon dioxide for each alga. Under operating conditions the ratio of algae to bacteria (in terms of numbers) was estimated at 1 : 250 and therefore it was considered that an alternative source of carbon dioxide may be required to keep the system in thermodynamic equilibrium. As adequate treatment was achieved in their ponds at a ratio of 1 : 100, it is presumed that the extra carbon dioxide was supplied via aeration.

Similar processes probably occur where *Chlorella vulgaris* has been employed for the purification of pig slurry (Garett and Allen, 1976; Allen and Garett, 1977). The bacterial flora of the slurry was altered by the addition of *Chlorella* sp. and the bacteria made phosphorus available to the algae in the early stages of the process. Differing light intensities employed in the process resulted in variations of the bacterial flora, but these variations were *not* a direct effect of the light. It was also interesting to note that different populations of bacteria were obtained if the plates used to isolate them were incubated at 4°C as opposed to 22°C and 37°C.

Bacterial endosymbionts of algae

Rod-shaped bacteria with an endonuclear location have been observed in several species belonging to the Euglenophyceae (Leedale, 1969). The bacteria possessed a mucilaginous sheath outside the cell membrane and there appeared to be a degree of synchrony of cell division between these bacteria and their algal hosts. Endogenous bacteria have been observed in the cytoplasm of *Bryopsis hypnoides* (Burr and West, 1970) in all vegetative parts, and even in the gametes, and Dodge (1973) has reported the presence of bacteria in the cytoplasm of the Dinophycean algae *Amphidinum herdmania* and *Katodinium glandulum.* Rod-shaped bacterial endosymbionts about 1.65 μm long were observed in *Volvox carteri* located in the cytoplasm near the periphery of somatic cells, gonidia and sperm, and frequently close to cell organelles but never within them (Kochert and Olsen, 1970; Lee and Kochert, 1976). The symbionts were not transferred when infected and non-infected algae were grown together in asexual cultures, but transfer was observed on induction of sexual reproduction and all the progeny carried the endosymbiont.

Rod-shaped intracellular bacteria were also found in 5 to 10% of the cells of *Gonyostomum semen* (Heywood, 1978), but the exact nature of this relationship seems uncertain at present. Rod shaped intracellular bacteria were recently observed in *Pleurocapsa minor* (Wujek, 1979), distributed throughout all regions of the cytoplasm.

3.1.2 Antagonistic Interactions

Antibacterial activity of phytoplankton

Pratt *et al.* (1944) first showed that a growth inhibiting substance, chlorellin, which appears to consist of peroxides of unsaturated fatty acids, was found in old cultures of *Chlorella* species. Spoehr *et al.* (1949) and Scutt (1964) confirmed these findings and Steemann-Nielsen (1955) suggested that similar substances reduced bacterial growth in light bottles used in photosynthesis experiments. Between 20 to 25% of Mediterranean diatoms may produce inhibitors (Aubert and Gauthier, 1967), the chemical nature of which is variable: acrylic acid (Sieburth 1960, 1961), fatty acids (Pesanda, 1972), nucleosides (Aubert *et al.*, 1970), peptides or gelbstoff (Berland *et al.*, 1972c), polysaccharides derived from chlorophyll degradation (Jorgensen, 1962) have all been suggested. Gauthier *et al.* (1978) found that *Asterionella japonica* and *Chaetoceros lauderi* produced a similar lipid antibiotic, the activity of which was increased by light. When the two diatoms were grown with *Prorocentrum micans* they contained larger amounts of antibiotic and lower quantities of carotenoid pigments. It was hypothesized that the dinoflagellate produced an extracellular substance which inhibited pigment synthesis by the diatoms with a consequent increase of the photoactivation of the lipid antibiotics *in vivo*.

These algal substances were generally most active against Gram-positive bacteria but some were also active against Gram-negative organisms or anaerobes (Chróst, 1972, 1975a,b). Chróst *et al.* (1975) have also described the inhibition of bacteria by freshwater algae and Chróst and Siuda (1978) found a negative relationship between bacterial heterotrophic activity and chlorophyll *a* concentration in the photic zone, whereas a positive relationship existed between them in the meta- and hypolimnion of the lakes investigated. This difference was thought to be the result of light activated bacterial inhibitors being released by algae in the photic zone, although no such compound was identified. In batch cultures of *Anabaena cylindrica*, two different strains of *Pseudomonas* were dominant, one in the early and the other with later stages of growth. These bacteria could utilize PDOC released by the "alga" but in the maximum population phase of growth an increased accumulation of extracellular products, presumably not well utilized by the bacteria, was noted. Peptide-like compounds with infra-red absorption spectra similar to substances from *Trichodesmium* sp. (Ramamurthy, 1973) formed part of the algal extracellular material.

The green alga *Chlorococcum humicolum* was shown to have a broad spectrum antibiotic substance (Pande and Gupta, 1977) and ethanol extracts of 14 out of 24 species of green algae exhibited significant inhibitory activity against either *Staphylococcus aureus* or *Bacillus subtilis* or both (Debro and Ward, 1979).

Moebus (1972) found that seasonal changes in the antibacterial activity of seawater against *Escherichia coli*, *Staphylococcus aureus* and *Serratia marinoruba* was correlated with the breakdown of diatom blooms, although some increase was also noted during the active growth of various algae.

Acrylic acid, produced by *Phaeocystis pouchetii* (Sieburth, 1968; Guillard and Hellebust, 1971), was probably the cause of the sterile gut of penguins feeding on euphasids which in turn had been grazing on *Phaeocystis* sp. Similarly, sea gulls, feeding on fish whose gut contents comprised 80 to 90% *Trichodesmium erythraeum*, had gastro-intestinal contents possessing antibacterial activity. Thus some algal antibiotics may exert an effect along the food chain.

Sieburth (1968) attributed the relatively bacteria-free state of many marine diatoms to an acid microzone around the cells caused by organic acid excretion and suggested that the perpendicular attachment of the few bacteria which were stuck to the algal cells was a response to ameliorate this effect. However, Sjobald and Mitchell (1979) found that glycollic acid, acrylic acid and dimethyl sulphide acted as attractants for *Vibrio alginolyticus* at concentrations of 1×10^{-4} M, 1×10^{-5} M, and 1×10^{-3} M, respectively. As acrylate only reduced bacterial viability at 1×10^{-1} M, it was concluded that instead of repelling bacteria at concentrations likely to be found in the sea, these substances contribute to the development of a phycosphere.

Rieper (1976) found in Schlei fjord that a spring bloom of *Chlorella* species, maximum population phase cultures of which would inhibit *Escherichia coli*, was accompanied by red bacteria whose growth it could enhance. Whereas the summer growth of *Microcystis* sp. was accompanied by white bacteria which suppressed the saprophytic bacteria of the Schlei.

Anti-algal effects of bacteria

The ability of bacteria to elaborate compounds exhibiting anti-algal activity has been well established (Berland *et al.*, 1972a,b); for example, *Bacillus* sp. (Reim *et al.*, 1974) and *Pseudomonas fluorescens* (Shiaris and Morrison, 1976) were shown to produce substances inhibitory to certain cyanobacteria. A complex situation was revealed by Delucca and McCracken (1977); *Pseudomonas* sp., *Xanthomonas* sp., and a *Flavobacterium* sp. all stimulated *Oscillatoria* sp. and other algae responded differently to each of the bacteria and to mixtures of them (Tables 3 and 4).

Table 3. Responses of algal cultures to bacteria isolated from *Oscillatoria* species (after Delucca and McCracken, 1977).

	Bacteria		
Algae	*Pseudomonas sp.*	*Flavobacterium* sp.	*Xanthomonas* sp.
Oscillatoria sp.	Stimulation; thick growth around disc	Stimulatory	Stimulatory
Chlorella pyrenoidosa	Inhibition followed by 2° growth	No response	Stimulatory; growth heavy on disc
Chlamydomonas reinholdt	Stimulatory; dark ring around disc	Stimulatory; dark growth around disc	Inhibition; small zone around disc
Anabaena flos-aquae	Stimulatory; dense ring around disc	No response	No response
Euglena gracilis "Z"	Inhibition; some 2° growth	Inhibition; small zone around disc	Inhibition; no mat growth
Haematococcus locustriis	No response	Stimulatory; spotting on disc	Stimulatory; slight darker growth on disc
Botrydiopsis arhiza	Inhibition; small zone—fine mat growth	Stimulatory; spotting on disc	Stimulatory; growth heavy on bottom of disc

Chlorella vulgaris, which was sensitive to hydroxylamine, was inhibited by an *Arthrobacter* sp., capable of releasing hydroxylamine during the oxidation of ammonium and other reduced nitrogenous compounds (Berger *et al.*, 1979). A *Pseudomonas* sp. isolated by Baker and Herson (1978) would not grow in the culture medium employed in the absence of *Thallasiosira pseudonanna*, but inhibited the growth of the alga when they were cultured together. Such inhibition probably results from proteinatious material excreted by the bacterium. However, Aubert (1978) questioned the relevance of anti-algal substances *in situ* because of the low concentrations likely to be involved.

Table 4. Responses of algal cultures to mixed populations of bacteria (after Delucca and McCracken, 1977).

Algae	Bacteria		
	Pseudomonas sp. + *Flavobacterium* sp.	*Pseudomonas* sp. + *Xanthomonas* sp.	*Pseudomonas* sp. + *Xanthomonas* sp. + *Flavobacterium* sp.
Oscillatoria sp.	Stimulatory; dark growth around disc	Stimulatory	Stimulatory
Chlorella pyrenoidosa	Inhibition followed by 2° growth	Inhibition followed by 2° growth	Inhibition followed by 2° growth
Chlamydomonas reinholdt	Inhibition	Inhibition	Inhibition
Anabaena flos-aquae	Stimulatory; heavy spots of growth on disc	Stimulatory	Stimulatory
Euglena gracilis "Z"	Stimulatory; slow growth	Slight stimulation	Stimulation; dark growth on disc
Haematococcus locustriis	No response	No response	No response
Botrydiopsis arhiza	No response	No response	No response

Microbial pathogens of algae have attracted recent attention and the topic has been reviewed by Stewart and Daft (1976, 1977). Safferman and Morris (1962, 1963) found an Actinomycete which was capable of lysing primarily cyanobacteria but also some green algae. Bershova *et al.* (1968) found that the bacteria antagonistic to cyanobacteria exceeded the numbers of satellite bacteria which stimulated "algal" growth. *Cellvibrio* species have been shown to excrete low molecular weight substances active against a range of cyanobacteria, with the effects appearing like those of penicillin in so far as cell wall synthesis was prevented (Granhall and Berg, 1972). *Anabaena*, *Phormidium*, and *Nostoc* species have been lysed by *Flexibacter flexilis* (Gromov *et al.*, 1972) and the flexibacteria may be an important group in this context (Stewart and Daft, 1976).

Fruiting myxobacteria were considered to have caused the death of cyanobacteria (Geitler, 1924) and more recently Stewart and Brown (1970) have demonstrated that four fruiting myxobacteria could lyse *Nostoc muscorum*.

The important lysing agents in the lochs studied by Stewart *et al.* (1977) were non-fruiting myxobacteria with high guanine + cytosine ratios. These affected both green algae and cyanobacteria (Daft *et al.*, 1975) and in Forfar loch a sewage works was an important source of these bacteria, as was treated sewage in Eglwys Nunydd reservoir in South Wales. In this latter reservoir and in Thrieply loch, the chlorophyll *a* concentration and the numbers of lysing bacteria were directly correlated (Stewart and Daft, 1976) and a similar seasonal variation of algae and lytic bacteria occurred in Monikie reservoir (Daft *et al.*, 1973).

Most of these myxobacteria needed contact with the host to ensure lysis (Stewart and Daft, 1977) although one isolate produced a soluble "lysozyme-like" enzyme (Stewart and Brown, 1971). Myxobacteria of the CP-1 type (Daft and Stewart, 1971, 1973; Daft *et al.*, 1973) were grown separately from other organisms and it was suggested that they coexisted in lakes with the algae. When conditions favouring bacterial growth or alternatively inhibitory to algal growth occurred, the balance was upset resulting in large scale algal lysis (Daft *et al.*, 1975).

Schnepf *et al.* (1974) have described a *Spirillum* sp. which penetrates the cell wall of *Scenedesmus* sp. and grows intracellularly.

The extracellular lysis of *Phormidium luridum* can be achieved by *Bdellovibrio bacteriovorus* or by a heat resistant factor in the *Bdellovibrio* sp. culture medium supernatant which was also effective against *Synechococcus* sp. (Burnham *et al.*, 1976; Burnham and Sun, 1977). *Bdellovibrio* sp. caused a 75% inhibition of photosynthesis within 4 h in *Phormidium* sp., and by 48 h intrathylakoidal vesicles formed prior to cell wall breakdown and cell lysis. The pattern was different in *Synechococcus* sp. but also suggested that the *Bdellovibrio* sp. secretions stimulated an autolytic mechanism in these cyanobacteria.

A *Bdellovibrio*-like parasitic bacterium was found infecting unicellular cyanobacteria in two coral reef sponges, *Neofibularia crata* and *Jaspis stelifera* (Wilkinson, 1979). The bacterium was located between the cell wall and cytoplasmic membrane of the "algae" and appeared to be similar to previously described *Bdellovibrio* sp.

Gromov and Mamkaeva (1972) isolated a bacterium which was considered to be a *Bdellovibrio* sp., capable of destroying *Chlorella vulgaris*. Coder and Starr (1978) argued that there was a lack of relationship with *Bdellovibrio* sp. as the life cycle of this bacterium was different, the flagellum unsheathed and because it was extracellularly located, even though its DNA base composition was similar to *Bdellovibrio bacteriovorus*. This "chlorellavorous bacterium" would only grow in the presence of live *Chlorella* sp. cells and not in the presence of other organisms tested and seemed to have a very limited host range.

Bacteria similar in structure to some actinomycetes, but lacking a cell wall or a boundary between them and the host cytoplasm, were found in *Euglena* sp. (Gerola and Bassi, 1978). There were no signs of host cytoplasmic reaction round the bacterium so it is probably premature to describe this as a case of parasitism.

Cyanobacteria can also produce substances which inhibit eukaryotic algae (Vance, 1965) and Lefèvre *et al.* (1950, 1952) found that water samples from blooms of species of *Aphanizomenon*, *Oscillatoria*, *Anabaena* and *Microcystis* were inhibitory to a variety of test algae. Lam and Sylvester (1979) also found that *Microcystis aeruginosa* inhibited *Chlorella* species probably via the production of extracellular material. The apparent inhibition of eukaryotic algae by some cyanobacteria could also be due to the production of hydroxamate chelators by the cyanobacteria or associated bacteria. Hydroxamate was shown by Murphy *et al.* (1976) to inhibit pure cultures of *Scenedesmus* sp. but have no effect on *Anabaena* sp.

Finally, although laboratory studies may indicate potential antagonistic reactions between organisms, such a potential is not always realized in nature. For example, *Bacillus polymyxa*, a common epiphyte on cyanobacteria, produces polymyxin B which is active against such hosts. This may be explained in part by the production of extracellular products by the alga which increase its resistance to the antibiotic, as was the case with *Anabaena cylindrica* (Whitton, 1965, 1967).

Competition for organic carbon and nitrogen

Parsons and Strickland (1962) suggested that the uptake of organic substances by natural plankton populations obeyed simple Michaelis-Menten kinetics. This was confirmed for freshwater populations at low concentrations of glucose and acetate (<0.1 mg l^{-1}) (Wright and Hobbie, 1965, 1966), whereas at concentrations between 0.5 to 5 mg l^{-1} uptake exhibited a direct dependence on substrate concentration. The uptake at low concentrations was said to occur via active uptake and on the basis of selective filtration and other evidence was attributed to the bacteria. The uptake at higher concentrations was ascribed to diffusion into the algae. Since then it has largely been taken for granted in much of the ecological literature that the uptake of organic solutes by algae occurs by diffusion and that algae cannot compete with bacteria for available organic substrates at the concentrations normally present in aquatic ecosystems.

However, recent physiological studies have revealed that diatoms, green algae and cyanobacteria possess well-defined active uptake systems. The best documented is that of *Chlorella* sp. (Tanner *et al.*, 1970, 1977; Komor and Tanner, 1971, 1974) where uptake is by an inducible proton symport system. Hellebust and Lewin (1977) have listed the uptake systems described in a range of diatoms for amino acids, organic acids, glucose and galactose. It was concluded that the few species of predominantly planktonic, centric diatoms which possessed heterotrophic capabilities were isolated from habitats rich in DOC, whereas amongst pennate diatoms which are mainly benthic or epiphytic and exist in a DOC rich environment, the capacity for heterotrophy is fairly common. Raven (1980) has compiled an extensive review of the whole topic of nutrient transport in micro-algae, but was still unable to draw a firm conclusion regarding the competitive heterotrophic capabilities of algae in comparison with specialist microscopic heterotrophs.

Wangersky (1978) has suggested that the experimental procedures at present employed to assess heterotrophic activity in a water sample are heavily biased in favour of the bacteria. He also proposed that mixotrophy could be the primary mode of organic substrate assimilation by algae and, therefore, experiments should be carried out in the light rather than the dark. As some diatom uptake systems require 12 h dark incubation for full activation and can be inactivated by high light intensities (Hellebust, 1970, 1971; Lewin and Hellebust, 1970, 1975, 1976; Jolley *et al.*, 1976), then Wangersky's suggestion is not universally applicable. Lack of awareness of the light inactivation of certain diatom uptake systems could account for the failure to detect uptake by natural diatom assemblages in some investigations. It may also be ecologically significant (Hellebust and Lewin, 1977) that such uptake systems do not require high external substrate concentrations for their induction, and in fact uptake activity can be lowered if glucose is present during the dark induction period (Jolley *et al.*, 1976).

Navicula muralis exhibited a lower Michaelis-Menten constant (K_m) for glucose uptake than did an associated *Flavobacterium* sp., indicating a greater affinity for glucose by the diatom. However, the recorded $K_m = 1.5 \times 10^{-5}$ M seemed too high to allow effective removal of glucose at natural substrate concentrations (Jolley *et al.*, 1976). The *Navicula* sp. also grew heterotrophically in the presence and absence of the *Flavobacterium* sp. at concentrations of glucose as low as 10 μM and at 1 mM glucose, the addition of the *Flavobacterium* sp. to heterotrophic cultures of the diatom reduced the doubling time of the alga in the dark from 22 to 12 h (Jolley and Jones, 1977). Bratten and Jones (1979) have also found that three bacterial isolates and *Dactylococcus inequalis* could stimulate the uptake of [^{14}C]-glucose by *Chlorella vulgaris* when incubated in separate dialysis sacs in the same apparatus. This indicates the difficulty in translating results obtained with single pure cultures in the laboratory to the situation pertaining in the field.

Saks *et al.* (1976) detected both enhancement and inhibition on adding a variety of organic substrates to saltmarsh epiphytic algae. Darley *et al.* (1979) found that populations of pennate diatoms from a saltmarsh could both take up and respire acetate, lactate and glucose at 1 μM concentrations and estimated that they could obtain 1% of their carbon heterotrophically if several usable substrates were each available at 1 μM concentrations.

In 1 h incubations in the light 70% of the total uptake of glucose and mannose by *Cylindrotheca* sp. and *Aeromonas* sp. was achieved by the diatom at substrate concentrations between 1 to 10 μM (Saks and Kahn, 1979). The diatom was also capable of competing with the bacterium for amino acids which accords with observations of Hellebust (1970) that the K_m for arginine transport by *Melosira nummuloides* compared very favourably with those of a number of bacterial systems reported by other workers. Allen (1971a) noted a *Cyclotella* sp. was able to take up glucose via a system showing first order kinetics at concentrations generally dominated by bacterial uptake.

Therefore, there is some indication from laboratory studies that a few algae can compete with a limited range of bacteria for some organic substrates. Field studies have not usually revealed a confirmatory picture. Using an autoradiographic procedure, Munro and Brock (1968) showed that bacteria, rather than algae, in a community attached to sand grains were responsible for [^{3}H]-acetate uptake. Fry and Ramsay (1977) also found that bacteria and not algae were labelled in autoradiographs of epiphytic populations. It is surprising that no label was incorporated into algae, even from respired [^{14}C]-carbon dioxide, since field studies (G. M. Bratten and A. K. Jones, unpublished observations), have shown that [^{14}C]-supplied to *Scirpus maritimus* as [^{14}C]-carbon dioxide can find its way to pure cultures of algae suspended in dialysis sacs near the macrophyte via photosynthesis by the macrophyte and excretion of PDOC into the water. This is similar to results of Allen (1971a) who detected [^{14}C]-transfer via excreted photosynthate from *Najas flexilis* to bacteria and algae in the laboratory. In uptake experiments it was shown that a mixture of a diatom, a green alga and a bacterium took up substrate poorly whereas in mixtures of one alga with one bacterium, uptake activities greater than the sum of the individual cultures was shown, thus emphasizing the complex interactions that may be taking place in natural communities.

According to Hoppe (1976) exposure of autoradiographs for only 3 to 7 d probably accounted for the failure of Munro and Brock (1968), Paerl and Goldman (1972) and, extending his argument, Hollibaugh (1976), to detect algal uptake of organic substrates. Using 14 d exposure periods for autoradiography, Hoppe (1976) detected weak labelling of algae from a Kiel fjord sample incubated with a [^{3}H]-labelled amino acid mixture. Pollingher and Berman (1976) and Saunders (1972) also demonstrated heterotrophic algal uptake employing autoradiography. Autoradiographic techniques are not straightforward in their application and some controversy exists as to the best procedure to employ (Paerl and Stull, 1979; Knoechel and Kalff, 1979). Common and dominant algae in samples from the Western Baltic Sea exhibited slight heterotrophic uptake of a [^{3}H]-labelled amino acid mixture supplied at near *in situ* concentrations (Hoppe, 1978). Their activity was very low, however, and perhaps negligible in quantitative terms although some advantage may have been gained by the algae after acquiring such nutrients.

Following size fractionation of lake water samples, Allen (1971b) showed that organisms between 3 to 8 μm were responsible for most of the "active" uptake of glucose and acetate, whilst the organisms in the bacterial size class were unimportant. He concluded that nannoplanktonic algae with surface area to volume ratios similar to those of bacteria were competing with bacteria for organic substrates. However, the possibility that bacteria were attached to the larger size organisms should not be discounted. Hansen and Wiebe (1977) examining [^{14}C]-glucose uptake state that although Wiebe and Pomeroy (1972) demonstrated that

> few recognisable bacteria were visible microscopically on estuarine particulate matter . . . we found that >70% of the activity was associated with particles between 14 and 180 μm; thus, as with coastal waters most of the metabolically active bacteria were associated with particles.

With their lack of microscopic evidence for the location of bacteria on the particles between 14 to 180 μm, and as there is evidence for uptake by algae in this size range, their conclusions cannot be accepted without qualification.

Thus the picture remains unresolved. There is still very little firm evidence that algae are significant competitors with bacteria for organic substrates in aquatic ecosystems. What evidence there is suggests that algae probably utilize their facultative heterotrophic capabilities for survival or slow growth in the dark or at low light intensities and that organic nitrogen compounds can serve as supplementary nitrogen sources.

Competition for phosphorus

Both bacteria and algae may be important in removing phosphorus from the aquatic environment (Harris, 1957; Lean, 1973) but some suggest that bacteria are the primary agents (Johannes, 1964; Correll *et al.*, 1975). Faust and Correll (1976), employing selective filtration, found that bacteria in an estuary would take up [^{32}P]-phosphate at high rates throughout the year, whereas algal uptake was high in the summer when there was a peak in algal biomass. The rate of phosphate uptake also appeared to vary with the species of algae present. They were unable to detect competition for phosphate at low concentrations whereas Rhee (1972) showed that the growth of *Scenedesmus* sp. was severely limited by a *Pseudomonas* sp., after external phosphate was exhausted in the culture medium. On incubating algae and bacteria with [^{33}P]-phosphate, the bacteria were labelled after 2 min whereas the specific activity per unit of biomass in the *Anabaena* sp. was only 30% of that in bacteria after 10 min incubation (Paerl and Lean, 1976). However, in lake samples, algal biomass greatly exceeded bacterial biomass, consequently the algae were quantitatively more significant in removal of phosphate from lake water. Both algae and bacteria excreted large amounts of phosphorus as fibrillar and amorphous colloidal particles, so this could be a method of retaining the phosphorus in the water mass ready for re-utilization when required.

When a *Flavobacterium* sp. was inoculated into a steady state chemostat culture of *Ankistrodesmus braunii*, the numbers of algae initially declined as the *Flavobacterium* sp. grew rapidly after which steady state was obtained with both organisms present. Rapid wash-out of the alga occurred on addition of *Pseudomonas* sp. to the axenic culture in a similar fashion. When *Pseudomonas* sp. was inoculated into the steady state culture of *Ankistrodesmus*/*Flavobacterium* mixture, the alga washed out of the system first, then the *Flavobacterium* sp. and finally the *Pseudomonas* sp. The limiting nutrient for the alga in the chemostat was phosphate at 2 μM; however, by transiently increasing the phosphate to 200 μM, *Ankistrodesmus* sp. increased in numbers, followed by both bacteria when there was sufficient organic carbon supplied by the enlarged algal population. These results indicated a competition for phosphate between the alga and the bacteria and that the rate of bacterial growth was limited by the rate at which the *Ankistrodesmus* sp. produced organic substrates (Mayfield and Innis, 1978).

Shuter (1978) noted that the subsistence quotas for phosphorus for the cyanobacteria *Anacystis* sp. and *Anabaena* sp. were much nearer those of bacteria than were those of eukaryotic algae. This may in part explain why competition for phosphate at low substrate concentrations is more likely to be detected between eukaryotic algae and bacteria rather than between cyanobacteria and their prokaryotic relatives. It could also account for the observation of Lam and Silvester (1979) that *Anabaena* sp. appeared to inhibit *Chorella* sp. by out-competing it for available phosphate in an experiment where the cultures were separated in the arms of a U-tube by a membrane filter.

4. BACTERIA AND SEAWEEDS

4.1 The Distribution of the Bacterial Flora

The scanning electron microscope has revealed distinctive epibacterial floras on seaweeds (Fig. 3) (Sieburth, 1976, 1979). In *Ulva lactuca* the bacteria are usually attached perpendicular to the thallus and in *Monostroma* only one type of bacteria was observed, growing preponderantly along the thallus edge.

Sieburth (1968) found no definite differences between the bacterial flora of the red alga *Polysiphonia* and the brown *Sargassum*. The genus *Vibrio* was dominant and little difference was detected between the flora living on active and inactive segments of those seaweeds. This similarity was partly attributable to the possible action of inhibitory polyphenols extracted when the algae were ground with water to obtain the bacteria for culturing. On *Sargassum*, a reduction by two to three orders of magnitude in the numbers of bacteria was detected on the tips of the fronds. Mitchell and Cundell (1977) and Cundell *et al.* (1977), using the scanning electron microscope (SEM), observed that only microcolonies of yeasts were present on the tips of *Ascophyllum nodosum*, the main stipe above the hold-fast was covered by a lawn of end-attached bacterial rods and population diversity was greater in the internodal region representing the fourth year of growth, where a dense lawn of end-attached bacteria was overlaid by filamentous bacteria. Sieburth *et al.*

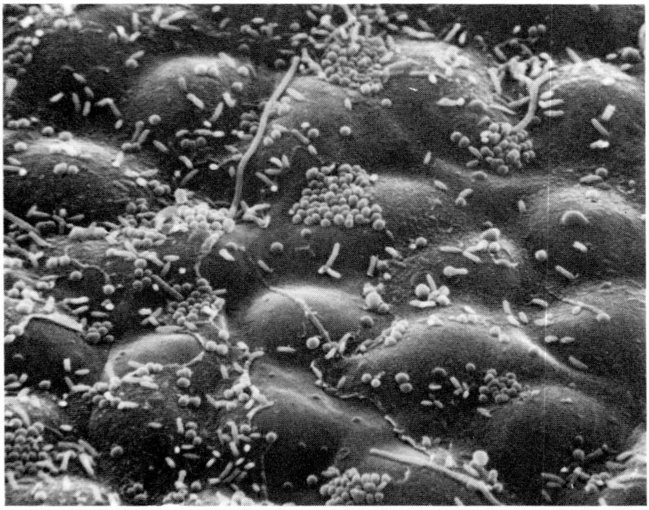

Fig. 3. The green alga *Ulva lactuca* (sea lettuce) which is covered by a variety of rod-shaped and coccoid bacteria, but is free of diatoms. ×1900. (Courtesy of Sieburth (1975) "Microbial Seascapes", University Park Press, Baltimore, USA.)

(1974) found that although the surfaces of *Ascophyllum* were relatively free of bacteria when it was actively growing in spring, microzones with surface slime containing adhering microbes could be observed. Also attached were short filaments of *Leucothrix mucor* which, with pennate diatoms and yeasts, formed dense microzonal areas in winter when *Ascophyllum* could be relatively dormant.

Similarly, the red alga *Polysiphonia lanosa*, an epiphyte on *Ascophyllum*, initially appeared relatively free of an epiflora in summer. More critical investigation revealed localized growths of bacterial cells, filaments of *Leucothrix mucor*, yeasts and pennate diatoms, particularly between bifurcations of the thallus and in the regions between the pericentral cells. In winter the density of this microflora increased markedly and the filaments of *Leucothrix* largely coalesced to form thick bundles.

Leucothrix species were found to be most abundant on red algae by Bland and Brock (1973), particularly on *Bangia* and to a lesser extent *Porphyra*. This coincided with observations on Marshall Island (Johnson *et al.*, 1971) where 13 of 22 species of red algae, 3 of 6 species of brown algae, 2 of 11 species of green algae and 3 of 9 cyanobacterial species had considerable *Leucothrix* sp. populations.

Laycock (1974) obtained 4200 bacterial isolates from *Laminaria longicruris*. Only one was a Gram-positive rod, all the rest were Gram-negative motile rods and the 1,500 of these which were examined were all polarly flagellate. The population was divided into three groups:

(i) glucose oxidizing (2 types of *Pseudomonas* sp., I and II);
(ii) glucose inert (*Pseudomonas* I, II, and IV) and;
(iii) glucose fermenting (*Vibrio* sp.).

Yellow pigmented isolates were classified as *Flavobacterium* sp. and tan pigmented isolates resembled the *Pseudomonas putrefaciens* group found on fish (Laycock and Regier, 1970). In winter, psychrophilic bacteria which were predominantly laminarin hydrolysing were associated with the *Laminaria*. They were most abundant on the eroded frond tips and their abundance varied inversely with water temperature. Associated with the decaying algal frond in summer was a population of mesophilic isolates capable of hydrolysing mannitol, protein, and alginate.

Shiba *et al.* (1979) recently isolated from *Enteromorpha linza*, *Porphyra* sp. and *Sargassum*, aerobic bacteria containing bacteriochlorophyll *a*. These bacteria constituted between 0.9 to 1.1% of the aerobic heterotrophic bacteria on the seaweed and between 1.2 to 6.3% of the population on beach sand.

4.2 Factors Affecting Bacterial Abundance and Distribution on Seaweeds

Antibacterial activity by seaweeds has been demonstrated by a number of workers including Pratt *et al.* (1951), Chesters and Stott (1956), Roos (1957), Allen and Dawson (1960), and Burkholder *et al.* (1960). Sieburth (1964, 1968) has reviewed much of this work and, more recently, Hornsey and Hide (1974, 1976a,b) have surveyed a large number of seaweeds for antibacterial activity using, as did many of the earlier workers, organisms such as *Staphylococcus aureus* and *Escherichia coli* amongst their test bacteria. Although they demonstrated variation of antibacterial activity both on a seasonal basis and within individual algal thalli, it is difficult to relate their work precisely to variations in the natural microflora, as the spectrum of antibiotic activity as revealed by their test organisms could possibly bear little relation to the responses of bacteria likely to reside on seaweeds. However, the circumstantial evidence that the antibacterial substances produced by seaweeds affects bacterial abundance and distribution seems strong. Phenolic substances seem to be those most strongly implicated.

Sieburth (1968, 1976, 1979), and Sieburth *et al.* (1974) have suggested that polyphenols inhibit colonization seasonally in *Ascophyllum* and the absence of bacteria from its apical tips was accounted for by the secretion of tannin (polyphenol) like compounds, desiccation during exposure to air and active meristematic growth (Mitchell and Cundell, 1977; Cundell *et al.*, 1977). Tannic acid may reach concentrations of 1 to 2 mg l^{-1} in seawater around *Ascophyllum* and such concentrations have elicited negative chemotactic responses from motile marine bacteria. Mitchell and Cundell (1977) proposed, therefore, that negative chemotaxis could be a powerful deterrent to the colonization of surfaces by marine bacteria.

Similarly, in *Fucus vesciculosus* inhibitory polyphenols were found at the distal end of frond tips (Craigie and McLachlan, 1964), and the similar regions in *Sargassum natans* were only free of fouling organisms when young and containing tannin-like substances (Sieburth and Conover, 1965).

The failure of *Leucothrix mucor* to colonize some algae may also be partly related to algal production of antimicrobial compounds; for example, *Desmarestia* has sulphuric acid in its cell sap at concentrations up to 0.44 M (Eppley and Bovell, 1958). Bland and Brock (1973) related this feature to the rare occurrence of the epiphytes on this alga. They also suggested that the failure of *Leucothrix* sp. to colonize some algae was because these organisms did not produce the carbon compounds that *Leucothrix* sp. could utilize. The preferential attachment of *Leucothrix* sp. to *Bangia fuscopurpurea* rather than plastic was related to the production of mannans and mannose in the algal mucilage, as mannose, along with glutamate and peptone, were good carbon sources for the bacterium. Bland and Brock (1973) suggested that the predominance of *Leucothrix* in the epiphyton of *Bangia* and *Porphyra*, located at the top of the shore, was also accounted for by the water retention of the algae and the resistance to desiccation of *Leucothrix* sp.

Even though an inverse relationship was detected between numbers of psychrophilic bacteria growing on the tips of *Laminaria* and temperature, Laycock (1974) suggested that availability of assimilable nutrients regulated bacterial abundance since isolated psychrophiles grew more rapidly at 10°C than at 3°C. The peak of algal growth occurred during low temperature periods and it was suggested that if healthy cells at the tips of the *Laminaria* fronds were disrupted, then nutrients, such as amino acids, would be released and used by the bacteria. Such substances were not considered to be as readily available when the frond was senescent later in the year.

4.3 Effects of Bacteria on the Growth and Morphology of Seaweeds

After being rendered axenic, cultures of *Ulva lactuca* and *Ulva taeniata* only exhibited their normal morphology in one seawater sample enriched with nitrogen, phosphorus, trace metals, and vitamins plus adenine and kinetin (Provasoli, 1958). Other samples of seawater with similar additions only supported the growth of pin cushion-like thalli rather than the normal distromatic blade. Foyn (1960), however, reported that *Ulva* has been cultured in a bacterized state in Erd-Schreiber medium since 1934. However, single additions of 40 bacterial isolates to cultures of *Ulva* caused a vigorous growth of hollow, tubular filaments virtually indistinguishable from *Enteromorpha*; none induced the usual *Ulva* morphology (Provasoli and Pintner, 1972). Contact with the alga was apparently necessary to exert the effect as supernatants of the bacterial cultures were inactive.

Enteromorpha linza and *E. compressa* also assumed pin cushion-like growth stages after a period in axenic culture (Fries, 1975). Bacterially contaminated cultures could be easily recognized by their formation of tubular fronds. The addition of indole acetic acid, adenine and kinetin, in the same combination and at the same concentrations which induced frond formation in *Ulva* (Provasoli, 1958), to filamentous stages of *Enteromorpha linza*, failed to alter its morphology. Even though Chandramohan (1971) has shown that bacteria living on *Enteromorpha* could convert tryptophan into indole acetic acid, Fries (1975) concluded that bacteria exert their effect via agents other than indole acetic acid.

Normal morphology could be restored to axenic cultures of *Monostroma oxyspermum* by the addition of a marine bacterium, or a marine yeast or by their culture supernatants, either filter sterilized or autoclaved. The filter sterilized supernatants of cultures of several axenic red algae and the brown substances produced by axenic *Sphacelaria* or freshly collected *Fucus* were also active. However, none of these filtrates produced a normal morphology when added to axenic cultures of *Ulva* (Provasoli and Pintner, 1964).

Axenic cultures of *Polysiphonia urceolata* and *Dasya pedicellata* also exhibited an abnormal morphology, but normal morphology was restored to *Polysiphonia* by the addition of 27 out of the 200 bacterial isolates tested individually. None of these, even in mixtures of 10 clones at a time, restored the appearance of *Dasya* to its usual form. However, specific bacteria seem to be needed for each species of seaweed, supernatants active for *Monostroma* did not affect *Ulva* and *Polysiphonia*. The only apparent polyvalent bacterial isolates were probably mixed cultures, as re-streaking resulted in several species specific clones.

However, Bonneau (1977) has succeeded in obtaining an axenic clone of *Ulva lactuca* which retained its blade morphology. From the same axenic zoospore population plants were grown ranging in structure from typical *Ulva* through to typical *Enteromorpha* morphology and even some individuals which were *Ulva* like in one area of the blade and *Enteromorpha* like in another. As these results were achieved employing defined media and axenic cultures, then it is possible that large variations in morphology of some of the previously mentioned algae could be independent of the effect of the external microflora. Alternatively, the harsh methods employed to obtain axenic *Ulva* may have influenced Bonneau's results, although he gave reasons to discount this possibility.

5. CONCLUSIONS

Interactions between mixtures of algae and bacteria under laboratory conditions have been described as stimulatory, antagonistic, and competitive. The term "phycosphere" has evolved to denote a zone around algae where algal extracellular products influence these relationships. Most bacteria associated with algae are Gram-negative, possibly partially due to the exclusion of Gram-positive organisms by algal antibiotics. Each alga, therefore, has a moderately specific bacterial flora in its phycosphere. There is some evidence of close coupling between PDOC production by some phytoplankton and its utilization by associated bacteria. Evidence for the direct attachment of many marine bacteria to healthy marine phytoplankton is meagre but close associations may still exist. Freshwater cyanobacterial and green algal colonies and filamentous cyanobacteria are often well colonized, whereas some healthy freshwater diatoms appear virtually bacteria-free (Maximum population phase cultures of bacterized algae are often heavily colonized). The bacteria colonizing larger seaweeds seem to avoid zones of possible antibiotic activity and favour areas of nutrient availability.

The ecological significance of the laboratory observations is debatable. There are contradictory reports concerning correlations between algal populations assessed numerically and numbers of associated bacteria, or their activities. Whether bacteria primarily utilize PDOC or the dead and decaying remains of the algae is also highly controversial. The possible reasons for such discrepancies may be methodological due to the use of different bacterial media and cultural conditions, leading to the selective growth of different bacteria. Very probably many bacteria likely to respond to specific algal extracellular products have yet to be cultured. Unfortunately, direct counts of both algae and bacteria may not reflect the most active, or even viable, organisms. Finally, the range of radioactive substrates employed to assess heterotrophic activity is usually limited and possibly not metabolized by the bacteria able to respond specifically to algal PDOC. Other controversial opinions concerning algal-bacterial interactions reflect biological variations. The effective concentrations of substrates, antibiotics and other metabolites, in the phycosphere is as dependent upon factors, such as water stability and turbulence, as upon the rates of production or utilization by the organisms concerned. The concentration factor is often very important in both qualitative and quantitative respects; for example, acrylic acid may be an antibiotic at high concentration but it elicited a positive chemotactic response from *Vibrio alginolyticus* at low concentrations. Algal production of PDOC also varies with the phase of growth and with environmental conditions. For example, high light intensity and low carbon dioxide concentration favour glycollate excretion, and in a stable water column, glycollate utilizing bacteria might show a very different distribution compared with that of the total heterotrophic flora (which might even be inhibited by the algal bloom). This difference would be detected only if samples were taken at the correct location, and if suitable bacterial media and culture conditions were employed. In lakes, beside factors such as temperature and pH, there are always allochthonous sources of organic carbon which easily obscure additional specific relationships between total bacterial numbers and algae. Variations in the specific algal components of the bloom are also important because some healthy algae are known to have very few bacteria associated with them whereas others are abundantly colonized. In the latter case, during active growth of a monospecific algal bloom, one might detect a very different bacterial flora re both numbers and species compared with the bacteria utilizing the decaying algal remains at the end of the bloom. The latter bacterial flora might probably resemble more closely the total heterotrophic population and also those bacteria living on the decaying remains of algae which are never heavily colonized. A totally different picture would obtain if the products released on autolysis of the algal cells were antibiotic in action at the concentrations likely to be reached in the environment. Thus contradictory conclusions in the literature may reflect both differences in laboratory techniques and real differences in the biological situation.

To advance soundly in this controversial area it will be necessary to standardize laboratory procedures as far as possible and also to identify the specific organisms active in the associations. Further exploitation of techniques, such as continuous cultures, using both known mixtures of specific microorganisms and also natural communities of algae and bacteria under conditions more closely pertaining to their natural habitats, are obvious promising approaches in the laboratory (see Parkes, pp. 45–102). In the field studies the production of PDOC of high specific activity coupled with the use of autoradiography, selective filtration and selective inhibitors, must be further refined and utilized to facilitate the more precise identification of those components of the microbiota which are actively functioning in the ecosystem. More meaningful correlations between specific algae and bacteria may then be possible as well as more accurate assessment of the overall importance of their interactions. Such an understanding should have significant predictive value for those concerned with the management of our water resources.

ACKNOWLEDGMENTS

I would like to thank Muriel Rhodes-Roberts for reading a draft of this article and making a number of helpful suggestions, and Bob Cannon for his comments.

REFERENCES

Abeliovich, A. and Weisman, D. (1978). Role of heterotrophic nutrition in growth of the alga *Scenedesmus obliquus* in high-rate oxidation ponds. *Applied and Environmental Microbiology* **35**, 32–37.

Al-Hasan, R. H., Coughlan, S. J., Pant, A., and Fogg, G. E. (1975). Seasonal variations in phytoplankton and glycollate concentrations in the Menai Straits, Anglesey. *Journal of the Marine Biological Association of the U.K.* **55**, 557–565.

Allen, H. L. (1971a). Primary productivity, chemo-organotrophy, and nutritional interactions of epiphytic algae and bacteria on macrophytes in the littoral of a lake. *Ecological Monographs* **41**, 97–127.

Allen, H. L. (1971b). Dissolved organic carbon utilization in size-fractionated algal and bacterial communities. *International Revue Der Gesammten Hydrobiologie* **56**, 731–749.

Allen, M. D. B. and Garrett, M. K. (1977). Bacteriological changes occurring during the culture of algae in the liquid phase of animal slurry. *Journal of Applied Bacteriology* **42**, 27–43.

Allen, M. B. (1955). General features of algal growth in sewage oxidation ponds. *State Water Pollution Control Board, Sacramento, California, Publication No. 15.*

Allen, M. B. and Dawson, E. Y. (1960). Production of antibacterial substances by benthic tropical marine algae. *Journal of Bacteriology* **79**, 459–460.

Anderson, G. C. and Zeutschel, R. P. (1970). Release of dissolved organic matter by marine phytoplankton in coastal and offshore areas of the Northeast Pacific Ocean. *Limnology and Oceanography* **15**, 402–407.

Aubert, M. (1978). Telemediateurs et Rapports Inter-espèces dans le Domaine des Micro-organismes Marins: Les Substances organiques naturelles dissoutes dans l'eau de mer. *In* "Colloque du Groupement pour l'avancement de la Biochemie Marine", pp. 174–199. (Ceccaldi, H. J., ed.) Centre National de la Recherche Scientifique, Paris.

Aubert, M. and Gauthier, M. J. (1967). Origine et nature des substances antibiotiques présentes dans le milieu marin. 8° Partie: Étude systématique de l'action antibactérienne d'espèces phytoplanctoniques vis-à-vis de germes telluriques aérobies. *Revue Internationale D'Oceanographie médicale* **5**, 63–71.

Aubert, M., Pesando, D., and Gauthier, M. J. (1970). Phénomènes d'antibiose d'origine phytoplanctonique en milieu marin. Substances antibactériennes produites par une Diatomee, *Asterionella japonica* (Cleve). *Revue Internationale D'Oceanographie médicale* **18–19**, 69–76.

Baker, K. H. and Herson, D. S. (1978). Interactions between the diatom *Thallasiosira pseudonanna* and an associated pseudomonad in a mariculture system. *Applied and Environmental Microbiology* **35**, 791–796.

Baumann, P., Baumann, L., and Mandel, M. (1971). Taxonomy of marine bacteria: the Genus *Beneckea*. *Journal of Bacteriology* **107**, 268–294.

Baumann, L., Baumann, P., Mandel, M., and Alden, R. D. (1972). Taxonomy of aerobic marine eubacteria. *Journal of Bacteriology* **110**, 402–429.

Bednarz, T. (1972). Attempts at elimination of bacteria from algal cultures by means of antibiotics. *Acta Microbiologia Polonica, Series B* **4**, 165–169.

Bell, W. and Mitchell, R. (1972). Chemotactic and growth responses of marine bacteria to algal extracellular products. *The Biological Bulletin—Marine Biological Laboratory* **143**, 265–277.

Bell, W. H., Lang, J. M., and Mitchell, R. (1974). Selective stimulation of marine bacteria by algal extracellular products. *Limnology and Oceanography* **19**, 833–839.

Benson-Evans, K. and Williams, P. F. (1975). Algae and bryophytes. *In* "Ecological Aspects of Used Water Treatment", pp. 153–282. (Curds, C. R. and Hawkes, H. A., eds.) Academic Press, London and New York.

Berger, P. S., Rho, J., and Gunner, H. B. (1979). Bacterial suppression of *Chlorella* by hydroxylamine production. *Water Research* **13**, 267–273.

Berland, B. R. and Maestrini, S. Y. (1969). Study of bacteria associated with marine algae in culture. II. Action of antibiotic substances. *Marine Biology* **3**, 334–335.

Berland, B. R., Bianchi, M. G., and Maestrini, S. Y. (1969). Etude des bactéries associeés aux algues marines en culture. I. Détermination préliminaire des espèces. *Marine Biology* **2**, 350–355.
Berland, B. R., Bonin, D. J., and Maestrini, S. Y. (1970). Study of bacteria associated with marine algae in culture. III. Organic substrates supporting growth. *Marine Biology* **5**, 68–76.
Berland, B. R., Bonin, D. J., and Maestrini, S. Y. (1972a). Are some bacteria toxic for marine algae? *Marine Biology* **12**, 189–193.
Berland, B. R., Bonin, D. J., and Maestrini, S. Y. (1972b). Etude des relations algues-bactéries du milieu marin. Possibilité d'inhibition des algues par les bactéries. *Thetys* **4**, 339–348.
Berland, B. R., Bonin, D. J., Cornu, A. L., Maestrini, S. Y., and Marino, J. P. (1972c). The antibacterial substances of the marine alga: *Stichochrysis immobilis* (Chrysophyta). *Journal of Phycology* **8**, 383–392.
Berland, B. R., Bonin, D. J., Fiola, M., and Maestrini, S. Y. (1978). Importance des vitamines en mer. Consommation et production par les algues et les bacteries. *In* "Les Substances Organiques Naturelles Dissoutes dans l'eau de Mer", pp. 121–146. (Ceccaldi, H. J., ed.) Centre National de la Recherche Scientifique, Paris.
Berman, T. and Holm-Hansen, O. (1974). Release of photoassimilated carbon as dissolved organic matter by marine phytoplankton. *Marine Biology* **28**, 305–310.
Bershova, O. I., Kopteva, Zh, P., and Tantsyurenko, E. V. (1968). The interrelations between the blue-green algae—The causative agents of water "bloom" and bacteria. "Tsvetenie" Vody. Kiev, pp. 159–171. (Topachevsky, A. V., ed.) (English translation, pp. 100–117, British Library, Boston Spa).
Bjälfve, G. (1962). Nitrogen fixation in cultures of algae and other microorganisms. *Physiologia Plantarum* **15**, 122–129.
Bland, J. A. and Brock, T. D. (1973). The marine bacterium *Leucothrix mucor* as an algal epiphyte. *Marine Biology* **23**, 283–292.
Bratten, G. M. and Jones, A. K. (1979). Interactions between algae and bacteria isolated from the epiphyton of *Scirpus maritimus*. *British Phycological Journal* **14**, 120.
Bonneau, . R. (1977). Polymorphic behaviour of *Ulva lactuca* (Chlorophyta) in axenic culture. I. Occurrence of *Enteromorpha*-like plants in haploid clones. *Journal of Phycology* **13**, 133–140.
Bonotto, S. (1976). Multicellular plants. *In* "Marine Ecology", Vol. 3, pp. 467–529. (Kinne, O., ed.) John Wiley and Sons, London.
Bunt, J. S. (1961). Nitrogen-fixing blue-green algae in Australian rice soils. *Nature, London* **192**, 479–480.
Burkholder, P. R. (1963). Some nutritional relationships among microbes of sea sediments and waters. *In* "Marine Microbiology", pp. 133–150. (Oppenheimer, C. H., ed.) Charles C. Thomas, Springfield, Illinois.

Burkholder, P. R. and Burkholder, L. M. (1956). Vitamin B_{12} in suspended solids and marsh muds collected along the coast of Georgia. *Limnology and Oceanography* **1**, 202–208.

Burkholder, P. R., Burkholder, L, M., and Almodovar, L. R. (1960). Antibiotic activity of some marine algae of Puerto Rico. *Botanica Marina* **2**, 149–156.

Burnham, J. C., Stetak, T., and Locher, G. (1976). Extracellular lysis of the blue-green alga *Phormidium luridum* by *Bdellovibrio bacteriovorus*. *Journal of Phycology* **12**, 306–313.

Burnham, J. C. and Sun, D. (1977). Electron microscope observations on the interaction of *Bdellovibrio bacteriovorus* with *Phormidium luridum* and *Synechococcus* sp. (Cyanophyceae). *Journal of Phycology* **13**, 203–208.

Burr, F. A. and West, J. A. (1970). Light and electron microscope observations on the vegatative and reproductive structures of *Bryopsis hypnoides*. *Phycologia* **9**, 17–37.

Caldwell, D. E. (1977). The planktonic microflora of lakes. *In* "C.R.C. Critical Reviews in Microbiology", Vol. 5, pp. 305–370. (Laskin, A. and Le Chevalier, H., eds.) C.R.C. Press Inc., Cleveland, Ohio.

Caldwell, D. E. and Caldwell, S. J. (1978a). Bacteriological characterization of phytoplankton cell surfaces. *In* "The Aquatic Environment. Environmental Biogeochemistry and Geomicrobiology", Vol. 1, pp. 101–107. (Krumbein, W. E. ed.) Ann Arbor Publishers.

Caldwell, D. E. and Caldwell, S. J. (1978b). A *Zoogloea* sp. associated with blooms of *Anabaena flos-aquae*. *Canadian Journal of Microbiology* **24**, 922–931.

Carlucci, A. F. and Bowes, P. M. (1970a). Production of vitamin B_{12}, thiamine, and biotin by phytoplankton. *Journal of Phycology* **6**, 351–357.

Carlucci, A. F. and Bowes, P. M. (1970b). Vitamin production and utilization by phytoplankton in mixed culture. *Journal of Phycology* **6**, 393–400.

Carlucci, A. F. (1974). Production and utilization of dissolved vitamins by marine phytoplankton. *In* "Effects of the Ocean Environment on Microbial Activities", pp. 449–456. (Colwell, R. R. and Morita, R. Y., eds.) University Park Press, Baltimore.

Carmichael, W. W. and Gorham, P. R. (1977). Factors influencing the toxicity and animal susceptibility of *Anabaena flos-aquae* (Cyanophyta) blooms. *Journal of Phycology* **13**, 97–101.

Chan, E. C. S. and McManus, E. A. (1967). Development of a method for the total count of marine bacteria on algae. *Canadian Journal of Microbiology* **13**, 295–301.

Chandramohan, D. (1971). Indole acetic acid synthesis in sea. *Proceedings of the Indian Academy of Science, Series B* **73**, 105–109.

Chesters, C. G. C. and Stott, J. A. (1956). The production of antibiotic substances by seaweeds. Proceedings of the 2nd International Seaweed Symposium, pp. 49–54. Pergamon Press, New York.

Chróst, R. J. (1972). Growth of bacteria in *Chlorella vulgaris* cultures. *Acta Microbiologica Polonica, Series B* **4**, 171–174.

Chróst, R. J. (1975a). Inhibitors produced by algae as an ecological factor affecting bacteria in water ecosystems. I. Dependence between phytop-

lankton and bacteria development. *Acta Microbiologica Polonica, Series B* **7**, 125–133.
Chróst, R. J. (1975b). Inhibitors produced by algae as an ecological factor affecting bacteria in water ecosystems. II. Antibacterial activity of algae during blooms. *Acta Microbiologica Polonica*, Series B **7,** 167–176.
Chróst, R. J. (1978). The estimation of extracellular release by phytoplankton and heterotrophic activity of aquatic bacteria. *Acta Microbiologica Polonica, Series B* **27**, 139–146.
Chróst, R. J. and Brzeska, D. (1978). Extracellular release of organic products and growth of bacteria on *Anabaena cylindrica* (blue-green alga) culture. *Acta Microbiologica Polonica, Series B* **27**, 287–295.
Chróst, R. J., Owczarek, J., and Matusiak, K. (1975). Degradation of urea by bacteria and algae in mass cultures. *Acta Microbiologica Polonica, Series B* **7**, 231–236.
Chróst, R. J. and Siuda, W. (1978). Some factors affecting the heterotrophic activity of bacteria in lake. *Acta Microbiologica Polonica, Series B* **27**, 129–138.
Coder, D. M. and Starr, M. P. (1978). Antagonistic association of the chlorellavorus bacterium ("Bdellovibrio" chlorellavorus) with *Chlorella vulgaris. Current Microbiology* **1**, 59–64.
Collins, V. G., Jones, J. G., Hendrie, M. S., Shewan, J. M., Wynn-Williams, D. D., and Rhodes, M. E. (1973). Sampling and estimation of bacterial populations in the aquatic environment. *In* "Sampling—Microbiological Monitoring of Environments", pp. 77–110. (Board, R. G. and Lovelock, D. W., eds.) Academic Press, London and New York.
Correll, D. L., Faust, M. A., and Severn, D. F. (1975). Phosphorus flux and cycling in estuaries. *In* "Estuarine Research: Chemistry and Biology", Vol. 1, pp. 108–316. (Cronin, L. E., ed.) Academic Press, London and New York.
Coughlan, S. J. and Al-Hasan, R. H. (1977). Studies of uptake and turnover of glycollic acid in the Menai Straits, North Wales. *Journal of Ecology* **65**, 731–746.
Coveney, M. F., Cronberg, G., Enell, M., Larsson, K., and Olofsson, L. (1977). Phytoplankton, zooplankton and bacteria—standing crop and production relationships in a eutrophic lake. *Oikos* **29**, 5–21.
Craigie, J. S. and McLachlan, J. (1964). Excretion of colored ultra-violet-absorbing substances by marine algae. *Canadian Journal of Botany* **42**, 23–33.
Cundell, A. M., Sleeter, T. D., and Mitchell, R. 1977). Microbial populations associated with the surface of the brown alga *Ascophyllum nodosum. Microbial Ecology* **4**, 81–91.
Daft, M. J. and Stewart, W. D. P. (1971). Bacterial pathogens of freshwater blue-green algae. *New Phytologist* **70**, 819–829.
Daft, M. J. and Stewart, W. D. P. (1973). Light and electron-microscope observations on algal lysis by bacterium CP-1. *New Phytologist* **72**, 799–808.
Daft, M. J., McCord, S., and Stewart, W. D. P. (1973). The occurrence of blue-green algae and lytic bacteria at a waterworks in Scotland. *Water Treatment and Examination* **22**, 114–124.

Daft, M. J., McCord, S., and Stewart, W. D. P. (1975). Ecological studies on algal-lysing bacteria in fresh waters. *Freshwater Biology* **5**, 577–596.

Darley, W. M., Ohlman, C. T., and Wimpee, B. B. (1979). Utilization of dissolved organic carbon by natural populations of epibenthic salt marsh diatoms. *Journal of Phycology* **15**, 1–5.

Debro, L. H. and Ward, H. B. (1979). Antibacterial activity of freshwater green algae. *Planta Medica* **36**, 375–378.

Delucca, R. and McCracken, M. D. (1977). Observations on interactions between naturally-collected bacteria and several species of algae. *Hydrobiologia* **55**, 71–75.

Derenbach, J. B. and Williams, P. J. LeB. (1974). Autotrophic and bacterial production: fractionation of plankton populations by differentiated filtration of samples from the English Channel. *Marine Biology 25*, 263–269.

Dhevendaran, K., Santhaman, R., and Selvakumar, N. (1978). Bacteria associated with plankton of Porto Novo Region. *Current Science* **47**, 745–747.

Dodge, J. D. (1973). Symbiosis. *In* "The Fine Structure of Algal Cells", pp. 209–213. (Dodge, J. D, ed.) Academic Press, London and New York.

Dor, I. (1976). Why *Chlorella* and *Scenedesmus* are the dominant algae in the waste waters over the world: a hypothesis. Proceedings of the 7th Scientific Conference of the Israel Ecological Society, Tel Aviv, pp. 13–25.

Droop, M. R. (1962). Organic micronutrients. *In* "Physiology and Biochemistry of Algae", pp. 141–154. (Lewin, R. A., ed.) Academic Press, London and New York.

Droop, M. R. (1968). Vitamin B_{12} and marine ecology. IV. The kinetics of uptake, growth and inhibition in *Monochrysis lutheri*. *Journal of the Marine Biological Association of the U.K.* **48**, 689–733.

Droop, M. R. (1969). Algae. *In* "Methods in Microbiology", Vol. 3B, pp. 269–313. (Norris, J. R. and Ribbons, D. W., eds.) Academic Press, London and New York.

Droop, M. R. and Elson, K. G. R. (1966). Are pelagic diatoms free from bacteria? *Nature, London* **211**, 1096–1097.

Droop, M. R. (1967). A procedure for routine purification of algal cultures with antibiotics. *British Phycological Bulletin* **3**, 295–297.

Duursma, E. K. (1963). The production of dissolved organic matter in the sea, as related to the primary gross production of organic matter. *Netherlands Journal of Sea Research* **2**, 85–94.

Eckenfelder, W. S. and O'Connor, D. J. (1961). "Ecological Waste Treatment". Pergamon Press, New York.

Eckenfelder, W. W. and Weston, R. P. (1956). Kinetics of biological oxidation. *Biological Treatment of Sewage and Industrial Waste* **1**, 18–34.

Englemann, T. W. (1884). Intersuchungen über die quantitativen Beziehungen zwischen Absorption des Lichtes und Assimilation in Pflanzenzellen. *Botanische Zeitung* **42**, 81–93.

Eppley, R. W. and Bovell, C. R. (1958). Sulfuric acid in *Desmarestia*. *The Biological Bulletin—Marine Biological Laboratory* **115**, 101–106.

Eppley, R. W. and Macias, R. F. M. (1963). Role of the alga *Chlamydomonas mundana* in anaerobic waste stabilization lagoons. *Limnology and Oceanography* **8**, 411–416.

Faust, M. A. and Correll, D. L. (1976). Comparison of bacterial and algal utilization of orphophosphate in the estuarine environment. *Marine Biology* **34**, 151–162.

Fogg, G. E. (1962). Extracellular products. *In* "Physiology and Biochemistry of Algae", pp. 475–489. (Lewin, R. A., ed.) Academic Press, London and New York.

Fogg, G. E. (1966). The extracellular products of algae. *Oceanography and Marine Biology Annual Review* **4**, 195–212.

Fogg, G. E. (1971). Extracellular products of algae in freshwater. *Archiv für Hydrobiologie* **5**, 1–25.

Fogg, G. E. (1977). Excretion of organic matter by phytoplankton. *Limnology and Oceanography* **22**, 576–577.

Fogg, G. E. and Nalewajko, C. (1964). Glycollic acid as an extracellular product of phytoplankton. *Verhandlungen Internationale Vereinigung für theoretische und angewundle limnologie* **15**, 806–881.

Fogg, G. E. and Watt, W. D. (1965). The kinetics of release of extracellular products of photosynthesis by phytoplankton. *Memorie—Istituto Italiano Di Idrobiologia Marco De Marchi* **18** (Supplement), 165–174.

Foyn, B. (1960). Sex-linked inheritance in *Ulva*. *The Biological Bulletin* **118**, 407–411.

Fries, L. (1963). On the cultivation of axenic red algae. *Physiologia Plantarum* **16**, 695–708.

Fries, L. (1975). Some observations on the morphology of *Enteromorpha linza* (L). J. Ag. and *Enteromorpha compressa* (L). Grev. in axenic culture. *Botanica Marina* **18**, 251–253.

Fry, J. C. and Ramsay, A. J. (1977). Changes in the activity of epiphytic bacteria of *Elodea canadensis* and *Chara vulgaris* following treatment with the herbicide, paraquat. *Limnology and Oceanography* **22**, 556–561.

Fry, J. C. and Humphrey, N. C. B. (1978). Techniques for the study of bacteria epiphytic on aquatic maerophytes. *In* "Techniques for the Study of Mixed Populations", pp. 1–29. (Lovelock, D. W. and Davies, R., eds.) Academic Press, London and New York.

Ganapati, S. V. (1975). Biochemical studies of algal-bacterial synthesis in high-rate oxidation ponds with varying detention periods and algae. *Archiv für Hydrobiologie* **76**, 302–367.

Garrett, M. K. and Allen, M. D. B. (1976). Photosynthetic purification of the liquid phase of animal slurry. *Environmental Pollution* **10**, 127–139.

Gauthier, M. J., Bernard, P., and Aubert, M. (1978). Modification of the antibiotic activity of two marine diatoms, *Asterionella japonica* (Cleve) and *Chaetoceros lauderi* (Ralfs), by the dinoflagellate *Prorocentrum micans* (Ehrenberg). *Journal of Experimental Marine Biology and Ecology* **33**, 37–50.

Geitler, L. (1924). Über *Polyangium parasiticum* n. sp., eine submerse, parasitische Myxobacteriaceae. *Archiv für Protistenkunde* **50**, 67–88.

Gerola, F. M. and Bassi, M. (1978). A case of parasitism in *Euglena*. *Journal of Submicroscopic Cytology* **10**, 261–263.

Geesey, G. G., Mutch, R., and Costerton, J. W. (1978). Sessile bacteria: an important component of the microbial population in small mountain streams. *Limnology and Oceanography* **23**, 1214–1223.

Godlewska-Lipowa, W. A. and Jablonska, I. (1972). Spatial differentiation abundance of bacteria in the water of Mikolajske lake. *Polish Journal of Ecology* **29**, 367–371.

Goldman, C. R., Gerletti, M., Javornicky, P., Melchiorri-Santolin, U., and De Amezaga, E. (1968). Primary productivity bacteria, phyto and zooplankton in Lake Maggiore. Correlations and relationships with ecological factors. *Memorie Istituto Italiano Di Idrobiologia Marco De Marchi* **23**, 49–127.

Goldman, J. C., Poriella, D. B., Middlebrooks, E. J., and Toerien, D. F. (1972). The effect of carbon on algal growth—its relationship to eutrophication. *Water Research* **6**, 637–679.

Granhall, U. and Berg, B. (1972). Antimicrobial effects of *Cellvibrio* on blue-green algae. *Archiv für Mikrobiologie* **84**, 234–242.

Gromov, B. V. and Mamkaeva, K. A. (1972). Electron microscope examination of *Bdellovibrio chlorellavorus* parasitism on cells of the green alga *Chlorella vulgaris*. *Tsitologiya* **14**, 256–260.

Gromov, B. V., Ivanov, O. G., Mamkaeva, K. A., and Avilova, I. A. (1972). A flexibacterium that lyses blue-green algae. *Microbiology* **41**, 952–956. (*Mikrobiologija* **41**, 1074–1079).

Guillard, R. R. L. and Hellebust, J. A. (1971). Growth and the production of extracellular substances by two strains of *Phaeocystis poucheti*. *Journal of Phycology* **7**, 330–338.

Guillard, R. R. L. and Wangersky, P. J. (1958). The production of extracellular carbohydrates by some marine flagellates. *Limnology and Oceanography* **3**, 449–454.

Gunderson, K. (1976). Cultivation of microorganisms. Bacteria. *In* "Marine Ecology", Vol. 3, pp. 301–335. (Kinne, O., ed.) John Wiley, London.

Guseva, K. A. (1967). Interrelation of phytoplankton and saprophyte bacteria in a reservoir. *Zoologicheeskiy Institute Trudy Problemnykth; Tematiches Kikh Soveshchaniy* (*National Technical Information Service AD 650–805*) **1**, 34–38.

Hagedorn, V. H. (1969). The vertical distribution of thiamine bacteria. *Berichte der Deutschen Botanischen Gesellschaft* **82**, 223–234.

Hanson, R. B. and Wiebe, W. J. (1977). Heterotrophic activity associated with particle size factions in a *Spartina alterniflora* saltmarsh estuary, Sapelo Island, Georgia, USA and the Continental Shelf waters. *Marine Biology* **42**, 321–330.

Harris, E. (1957). Radiophosphorous metabolism in zooplankton and in microorganisms. *Canadian Journal of Zoology* **35**, 769–782.

Hellebust, J. A. (1967). Excretion of organic compounds by cultured and natural populations of marine phytoplankton. *In* "Estuaries", pp. 361–366. (Lauff, G. H., ed.) American Association for the Advancement of Science. Washington, DC.

Hellebust, J. A. (1970). Uptake and assimilation of organic substances by marine phytoplankton. *In* "Organic Matter in Natural Waters", pp. 225–256. (Hood, D. W., ed.) University of Alaska: Institute of Marine Science Occasional Publication.

Hellebust, J. A. (1971). Glucose uptake by *Cyclotella cryptica*: dark induction and light inactivation of transport system. *Journal of Phycology* **7**, 345–349.

Hellebust, J. A. (1974). Extracellular products. *In* "Algal Physiology and Biochemistry", pp. 838–863. (Stewart, W. D. P., ed.) Blackwell Scientific Publications, Oxford.

Hellebust, J. A. and Lewin, J. (1977). Heterotrophic nutrition. *In* "Biology of Diatoms", pp. 169–197. (Werner, D., ed.) Blackwell Scientific Publications, Oxford.

Heywood, P. (1978). Intracellular bacteria in *Gonyostomum semen* (Chloromonadophycaae). *Journal of Phycology* **14**, 121–122.

Hollibaugh, J. T. (1976). The biological degradation of arginine and glutamic acid in seawater in relation to the growth of phytoplankton. *Marine Biology* **36**, 303–312.

Hoppe, H-G. (1976). Determination and properties of actively metabolizing heterotrophic bacteria in the sea, investigated by means of micro-autoradiography. *Marine Biology* **36**, 291–302.

Hoppe, H-G. (1978). Relations between active bacteria and heterotrophic potential in the sea. *Netherlands Journal of Sea Research* **12**, 78–98.

Hoppe, H-G. and Horstmann, U. (1981). Aspects of blue-green algae and bacteria production in Swedish Baltic Coastal waters. *Marine Biological Letters*, in press.

Hornsey, I. S. and Hide, D. (1974). The production of antimicrobial compounds by British marine algae. I. Antibiotic producing marine algae. *British Phycological Journal* **9**, 353–361.

Hornsey, I. S. and Hide, D. (1976). The production of antimicrobial compounds by British marine algae. II. Seasonal variation in production of antibiotics. *British Phycological Journal* **11**, 63–67.

Hornsey, I. S. and Hide, D. (1976b). The production of antimicrobial compounds by British marine algae. III. Distribution of antimicrobial activity within the algal thallus. *British Phycological Journal* **11**, 175–181.

Isaac, P. C. G. and Lodge, M. (1958). Algae and sewage treatment. *In* "New Biology", Vol. 25, pp. 85–97. (Johnson, M. L., Abercrombie, M., and Fogg, G. E., eds.) Penguin Books, Harmondsworth.

Iturriaga, R. and Hoppe, H-G. (1977). Observations of heterotrophic activity on photoassimilated organic matter. *Marine Biology* **40**, 101–108.

Johannes, R. G. (1964). Uptake and release of dissolved organic phosphorus by representatives of a coastal ecosystem. *Limnology and Oceanography* **9**, 224–234.

Johnson, P. W., Sieburth, J. McN., Sastry, A., Arnold, C. R., and Doty, M. S. (1971). *Leucothrix mucor* infestation of benthic crustacea, fish eggs and tropical algae. *Limnology and Oceanography* **16**, 962–969.

Jolley, E. T., Jones, A. K., and Hellebust, J. A. (1976). A description of glucose uptake in *Navicula pelliculosa* (Breb) Hilse, including a brief comparison with an associated *Flavobacterium* sp. *Archives of Microbiology* **109**, 127–133.

Jolley, E. T. and Jones, A.K (1977). The interaction between *Navicula muralis* Grunow and an associated species of *Flavobacterium*. *British Phycological Journal* **12**, 315–328.

Jones, A. K., Rhodes, M. E., and Evans, S. C. (1973). The use of antibiotics to obtain axenic cultures of algae. *British Phycological Journal* **8**, 185–196.

Jones, J. G. (1972). Studies on freshwater bacteria; association with algae and alkaline phosphatase activity. *Journal of Ecology* **60**, 59–75.

Jones, J. G. (1977a). The effect of environmental factors on estimated viable and total populations of planktonic bacteria in lakes and experimental enclosures. *Freshwater Biology* **7**, 67–91.

Jones, J. G. (1977b). The study of aquatic microbial communities. *In* "Aquatic Microbiology", pp. 1–30. (Skinner, F. A. and Shewan, J. M., eds.) Academic Press, London and New York.

Jørgensen, E. G. (1962). Antibiotic substances from cells and culture solutions of unicellular algae with special reference to some chlorophyll derivatives. *Physiologia Plantarum* **15**, 530–545.

Kerr, P. C., Brockway, D. L., Paris, D. F., and Barnett, J. T. (1972). The interrelation of carbon and phosphorus in regulating heterotrophic and autotrophic populations in an aquatic ecosystem, Shriners Pond. *In* "Nutrients and Eutrophication", pp. 41–62. (Likens, G. E., ed.) American Society of Limnology and Oceanography.

King, D. L. (1970). The role of carbon in eutrophication. *Journal of the Water Pollution Control Federation* **12**, 2035–2051.

Kinne, O. (1976). Marine Ecology, Vol. 3. John Wiley, London.

Knoechel, R. and Kalff, J. (1979). The advantages and disadvantages of grain density and track autoradiography. *Limnology and Oceanography* **24**, 1170–1171.

Kochert, G. and Olson, L. W. (1970). Endosymbiotic bacteria in *Volvox carteri*. *Transactions of the American Microscopical Society* **89**, 475–478.

Kogure, K., Simidu, U., and Taga, N. (1979). Effect of *Skeletonema costatum* (Grev) Cleve on the growth of marine bacteria. *Journal of Experimental Marine Biology and Ecology* **36**, 201–215.

Komor, E. and Tanner, W. (1971). Characterization of the active hexose transport system of *Chlorella vulgaris*. *Biochimica et Biophysica Acta (Amsterdam)* **241**, 170–179.

Komor, E. and Tanner, W. (1974). The nature of the energy metabolite responsible for sugar accumulation in *Chlorella vulgaris*. *Zeitschrift für Pflanzenphysiologie* **71**, 115–128.

Kott, Y. and Ingerman, R. (1966). The biochemical dynamics of waste stabilization ponds. *Air and Water Pollution International Journal* **10**, 603–609.

Lam, C. W. Y. and Silvester, W. B. (1979). Growth interactions among blue-green (*Anabaena oscillarioides*, *Microcystis aeruginosa*) and green (*Chlorella* sp.) algae. *Hydrobiologia* **63**, 135–143.

Lancelot, C. (1979). Cross excretion rates of natural marine phytoplankton and heterotrophic uptake of excreted products in the Southern North Sea, as determined by short-term kinetics. *Marine Ecology—Progress Series* **1**, 179–186.
Lange, W. (1967). Effect of carbohydrates on the symbiotic growth of planktonic blue-green algae with bacteria. *Nature, London* **215**, 1277–1278.
Lange, W. (1970). Cyanophyta-bacteria systems: effects of carbon compounds or phosphate on algal growth at low nutrient concentrations. *Journal of Phycology* **6**, 230–234.
Lange, W. (1971). Enhancement of algal growth in Cyanophyta-bacteria systems by carbonaceous compounds. *Canadian Journal of Microbiology* **17**, 303–314.
Lange, W. (1973). Bacteria-assimilable organic compounds, phosphate, and enhanced growth of bacteria-associated blue-green algae. *Journal of Phycology* **9**, 507–509.
Lange, W. (1976). Speculations on a possible essential function of the gelatinous sheath of blue-green algae. *Canadian Journal of Microbiology* **22**, 1181–1185.
Larsson, U. and Hagström, A. (1979). Phytoplankton exudate release as an energy source for the growth of pelagic bacteria. *Marine Biology* **52**, 199–206.
Laycock, R. A. (1974). The detrital food chain based on seaweeds. I. Bacteria associated with the surface of *Laminaria* fronds. *Marine Biology* **25**, 223–231.
Laycock, R. A. and Regier, L. W. (1970). Pseudomonads and achromobacters in the spoilage of irradiated haddock of different pre-irradiation quality. *Applied Microbiology* **20**, 333–341.
Lean, D. R. S. (1973). Phosphorus dynamics in lake water. *Science* **179**, 678–680.
Lee, W. S. and Kochert, G. (1976). Bacterial endosymbionts in *Volvox carteri* (Chlorophyceae). *Journal of Phycology* **12**, 195–197.
Leedale, G. (1969). Observations on endonuclear bacteria in euglenoid flagellates. *Öesterreichisches Botanisches Zeitschrift* **116**, 279–294.
Lefèvre, M., Jakob, H., and Nisbet, M. (1950). Sur la sécretion, par certaines Cyanophytes de substances algostatiques dans les collections d'eau naturelles. *Comptes Rendus Academie des Seances Paris* **230**, 2226–2227.
Lefèvre, M., Jakob, H., and Nisbet, M. (1952). Auto- et hetéroantagonisme chez les algues d'eau douce. *Annales de la Station Centrale de Hydrobiologie Appliquee* **4**, 197 pp.
Lewin, J. and Hellebust, J. A. (1970). Heterotrophic nutrition of the marine pennate diatom *Cylindrotheca fusiformis*. *Canadian Journal of Microbiology* **16**, 1123–1119.
Lewin, J. and Hellebust, J. A. (1975). Heterotrophic nutrition of the marine pennate diatom *Navicula pavillardi* Hustedt. *Canadian Journal of Microbiology* **21**, 1335–1342.
Lewin, J. and Hellebust, J. A. (1976). Heterotrophic nutrition of the marine pennate diatom *Nitzschia angularis*, var. *affinis* (Grun) Perag. *Marine Biology* **36**, 313–320.

Lin, C. K. J. (1972). Phytoplankton succession in a eutrophic lake with special reference of blue-green algal blooms. *Hydrobiologia* **39**, 321–334.
Lucas, C. E. (1947). The ecological effects of external metabolites. *Biological Reviews* **22**, 270–295.
Machlis, L. (1973). The effects of bacteria on the growth and reproduction of *Oedogonium cardiacum*. *Journal of Phycology* **9**, 342–344.
Mague, T. H., Freiberg, E., Hughes, D. J., and Morris, I. (1980). Extracellular release of carbon by marine phytoplankton; a physiological approach. *Limnology and Oceanography* **25**, 262–279.
Makisimova, E. A. (1973). Annual dynamics of the vertical distribution of heterotrophic bacteria in the Southern Baikal. *Mikrobiologuja* **42**, 469–474.
Marker, A. F. H. (1965). Extracellular carbohydrate liberation in the flagellates *Isochrysis galbana* and *Prymnesium parvum*. *Journal of the Marine Biological Association of the U.K.* **45**, 755–772.
Martin, Y. P. and Bianchi, M. A. (1980). Structure, diversity and catabolic potentialities of aerobic heterotrophic bacterial populations associated with continuous cultures of natural marine phytoplankton. *Microbial Ecology* **5**, 265–279.
Mayfield, C. I. and Innis, W. E. (1978). Interactions between freshwater bacteria and *Ankistrodesmus braunii* in batch and continuous culture. *Microbial Ecology* **4**, 331–344.
McFeters, G. A., Stuart, S. A., and Olson, S. A. (1978a). Growth of heterotrophic bacteria and algal extracellular products in oligotrophic waters. *Applied and Environmental Microbiology* **35**, 383–391.
McFeters, G. A., Stuart, S. A., and Olson, S. B. (1978b). Interactions of algae and heterotrophic bacteria in an oligotrophic stream. *In* "Proceedings in Life Sciences. Microbial Ecology", pp. 57–61. (Loutit, M. W. and Miles, J. A. R., eds.) Springer-Verlag, Berlin.
McGarry, M. G. (1971). Water and protein reclamation from sewage. *Process Biochemistry* **6**, 50–53.
Mikhaylenko, L. Ye. and Kulikova, I. Ya. (1973). Interdependence of bacteria and blue-green algae. *Hydrobiological Journal* **9**, 32–38.
Mitchell, R. A. and Cundell, A. M. (1977). The role of microorganisms in marine fouling and boring processes. Technical Report No. 3, US Office of Naval Research Contract NOOO14–76–c–0042 NR–104–967.
Moebus, K. (1972). Seasonal changes in antibacterial activity of North Sea water. *Marine Biology* **13**, 1–13.
Munro, A. L. S. and Brock, T. D. (1968). Distinction between bacterial and algal utilization of soluble substances in the sea. *Journal of General Microbiology* **51**, 35–42.
Murphy, T. P., Lean, D. R. S., and Nalewajko, C. (1976). Blue-green algae: their excretion of iron-selective chelators enables them to dominate other algae. *Science* **192**, 900–902.
Nalewajko, C., Chowdhuri, N., and Fogg, G. E. (1963). Excretion of glycollic acid and the growth of planktonic *Chlorella*. *In* "Studies on Microalgae and

Photosynthetic Bacteria", pp. 171–183. University of Tokyo Press.
Nalewajko, C., Dunstall, T. G., and Shear, H. (1976). Kinetics of extracellular release in axenic algae and in mixed algal-bacterial cultures. Significance in estimation of total (gross) phytoplankton excretion rates. *Journal of Phycology* **12**, 1–5.
Nalewajko, C. and Lean, D. R. S. (1972). Growth and excretion in planktonic algae and bacteria. *Journal of Phycology* **8**, 361–366.
Nalewajko, C. and Marin, L. (1969). Extracellular production in relation to growth of four planktonic algae and of phytoplankton populations from Lake Ontario. *Canadian Journal of Botany* **47**, 405–413.
Ohwada, K. and Taga, N. (1972). Vitamin B_{12}, thiamine and biotin in Lake Sagami. *Limnology and Oceanography* **17**, 315–320.
Ohwada, D. and Taga, N. (1973). Seasonal cycles of Vitamin B_{12}, thiamine and biotin in Lake Sagami. Patterns of their distribution and ecological significance. *Internationale Revue Der Gesammten Hydrobiologie* **58**, 851–871.
Oppenheimer, C. and Zobell, C. E. (1952). The growth and viability of sixty-three species of marine bacteria as influenced by hydrostatic pressure. *Journal of Marine Research* **11**, 10–18.
Oron, G., Shelef, G., Levi, A., Meydan, A., and Azov, Y. (1979). Algae/bacteria ratio in high rate ponds used for waste treatment. *Applied and Environmental Microbiology* **38**, 570–576.
Oswald, W. J. (1960). Fundamental factors in stabilization pond design. *Sewage and Industrial Wastes* **29**, 437–455.
Oswald, W. J. (1961). Research and installation experiences in California. *In* "Waste Stabilization Lagoons", pp. 33–39. US Public Health Service Publication No. 872.
Oswald, W. J. and Golueke, C. G. (1968). Harvesting and processing of waste-grown microalgae. *In* "Algae, Man and Environment", pp. 371–389. (Jackson, D. F., ed.) Syracuse University Press, New York.
Oswald, W. J., Gotaas, H. B., Ludwig, H. F., and Lynch, V. (1953a). Algae symbiosis in oxidation ponds. II. Growth characteristics of *Chlorella pyrenoidosa* culture in sewage. *Sewage and Industrial Wastes* **25**, 26–37.
Oswald, W. J., Gotaas, H. B., Ludwig, H. F., and Lynch, V. (1953b). Algae symbiosis in oxidation ponds. III. Photosynthetic oxygenation. *Sewage and Industrial Wastes* **25**, 692–705.
Overbeck, J. (1967). Zur Bakteriologie der Süsswassersees—Ergebruisse und Probleme. *Das Gas und Wasserfach* **44**, 1258–1260 (cited in J. G. Jones, 1972).
Overbeck, J. (1968). Prinzipielles zum Vorkommen der Bakterien im See. *Mitteilungen internationale Veirinigung für theoretische und angewandte Limnologie* **14**, 134–144.
Paerl, H. W. (1976). Specific associations of the blue-green algae *Anabaena* and *Aphanizomenon* with bacteria in freshwater blooms. *Journal of Phycology* **12**, 431–435.
Paerl, H. W. (1978). Role of heterotrophic bacteria in promoting N_2 fixation by *Anabaena* in aquatic habitats. *Microbial Ecology* **4**, 215–231.

Paerl, H. W. and Goldman, C. R. (1972). Heterotrophic assays in the detection of water masses at Lake Tahoe, California. *Limnology and Oceanography* **17**, 145–148.

Paerl, H. W. and Kellar, P. E. (1978). Significance of bacterial-*Anabaena* (Cyanopyceae) associations with respect to N_2 fixation in freshwater. *Journal of Phycology* **14**, 254–260.

Paerl, H. W. and Lean, D. R. S. (1976). Visual observations of phosphorous movement between algae, bacteria and abiotic particles in lake waters. *Journal of the Fisheries Research Board of Canada* **33**, 2805–2813.

Paerl, H. W. and Stall, E. A. (1979). In defense of grain density autoradiography. *Limnology and Oceanography* **24**, 1166–1169.

Pande, B. N. and Gupta, A. B. (1977). Antibiotic properties of *Chlorococcum humicolum* (Naeg) Rabenh. (Chlorophyceae). *Phycologia* **16**, 439–441.

Parsons, T. R. and Strickland, J. D H. (1962). On the production of particulate organic carbon by heterotrophic processes in sea water. *Deep-Sea Research* **8**, 211–222.

Pesanda, D. (1972). Étude chimique et structurale d'une substance lipidique antibiotique produite par une diatomée marine: *Asterionella japonica*. *Revue Internationale D'Oceanographie Medicale* **25**, 46–69.

Pollingher, U. and Berman, T. (1976). Autoradiographic screening for potential heterotrophs in natural algal populations of Lake Kinneret. *Microbial Ecology* **2**, 252–260.

Pratt, R., Daniels, T. C., Eiler, J. J., Gunnison, J. B., Kummler, W. D., Oneto, J. F., Spoehr, H. A., Hardin, G. J., Milner, H. W., Smith, J. H. C., and Strain, H. H. (1944). *Chlorellin*, an antibacterial substance from *Chlorella*. *Science* **49**, 351–352.

Pratt, R., Mautner, R. H., Gardner, G. M., Sha, Y., and Dufrenoy, J. (1951). Report on the antibiotic activity of seaweed extracts. *Journal of the American Pharmaceutical Association, Science Edition* **40**, 575–579.

Provasoli, L. (1958). Nutrition and ecology of protozoa and algae. *Annual Review of Microbiology* **12**, 279–308.

Provasoli, L. (1963). Growing marine seaweeds. *Proceedings of the International Seaweed Symposium* **4**, 9–17.

Provasoli, L. (1971). Nutritional relationships in marine organisms. *In* "Fertility of the Sea", pp. 369–382. (Costlow, J. D., ed.) Gordon and Breach Science Publishers, New York.

Provasoli, L. and Carlucci, A. F. (1974). Vitamins and growth regulators. *In* "Algal Physiology and Biochemistry", pp. 741–787. (Stewart, W. D. P., ed.) Blackwell Scientific Publications, Oxford.

Provasoli, L. and Pintner, I. J. (1964). Symbiotic relationships between microorganisms and seaweeds (Abst.). *American Journal of Botany* **51**, 681.

Provasoli, L. and Pintner, I. J. (1972). Effects of bacteria on seaweed morphology. *Journal of Phycology* **8**, p. 10 (Supplement).

Ramamurthy, V. D. (1973). Infra-red spectral analysis of antibacterial substance isolated from *Trichodesmium erythraeum* (marine blue-green alga). *Hydrobiologia* **41**, 247–250.

Raven, J. A. (1980). Nutrient transport in microalgae. *In* "Advances in Microbial Physiology", vol. 21, pp. 48–226.
Reim, R. L., Shane, M. S., and Cannon, R. E. (1974). The characterization of a *Bacillus* capable of blue-green bactericidal activity. *Canadian Journal of Microbiology* **20**, 981–986.
Rhee, G. Y. (1972). Competition between an alga and an aquatic bacterium for phosphate. *Limnology and Oceanography* **17**, 505–514.
Rieper, M. (1976). Investigations on the relationships between algal blooms and bacterial populations in the Schlei Fjord (Western Baltic Sea). *Helgolander wiss. Meeresunters* **28**, 1–18.
Roos, H. (1957). Untersuchungen über Vorkommen antimikrobieller Substanzen in Meeresalgen. *Kieler Meeresforschungen* **13**, 41–58.
Safferman, R. S. and Morris, M. E. (1962). Evaluation of natural products for algicidal properties. *Applied Microbiology* **10**, 289–292.
Safferman, R. S. and Morris, M. E. (1963). The antagonistic effects of Actinomycetes on algae found in waste stabilization ponds. *Bacteriological Proceedings*, p. 14.
Saks, N. M., Stone, R. J., and Lee, J. J. (1976). Autotrophic and heterotrophic nutritional budget of salt marsh epiphytic algae. *Journal of Phycology* **12**, 443–448.
Saks, N. M. and Kahn, E. G. (1979). Substrate competition between a salt marsh diatom and a bacterial population. *Journal of Phycology* **15**, 17–21.
Saunders, G. W. (1972). Potential heterotrophy in a natural population of *Oscillatoria agarhii* var. *isothrix* Skuja. *Limnology and Oceanography* **17**, 704–711.
Saunders, G. W. and Storch, T. A. (1971). Coupled oscillatory control mechanism in a planktonic system. *Nature, New Biology* **230**, 58–60.
Schindler, D. W. (1971). Carbon, nitrogen and phosphorus and the eutrophication of freshwater lakes. *Journal of Phycology* **7**, 321–329.
Schwabe, G. H. and Mollenhauer, R. (1967). Über den Einfluss der Begleitbakterien auf das Lagerbild von *Nostoc sphaericum*. *Nova Hedwigia* **13**, 77–80.
Schnepf, E., Hegewald, E., and Soeder, C. J. (1974). Electron microscopic observations on parasites of *Scenedesmus* mass cultures. IV. Bacteria. *Archives of Microbiology* **98**, 133–145.
Scutt, J. E. (1964). Autoinhibitor production by *Chlorella vulgaris*. *American Journal of Botany* **51**, 581–584.
Sharp, J. H. (1977). Excretion of organic matter by marine phytoplankton: Do healthy cells do it? *Limnology and Oceanography* **22**, 381–399.
Shah, N. M. and Wright, R. T. (1974). The Occurrence of glycollic acid in coastal sea water. *Marine Biology* **24**, 122–124.
Shiaris, M. P. and Morrison, S. M. (1976). The inhibition of blue-green algae by *Pseudomonas fluorescens*. *Abstract 76th Annual Meeting of the American Society for Microbiology*, p. 180.
Shiba, T., Simidu, U., and Taga, N. (1979). Distribution of aerobic bacteria which contain bacteriochlorophyll *a*. *Applied and Environmental Microbiology* **38**, 43–45.
Shuter, B. J. (1978). Size dependence of phosphorus and nitrogen substance

quotas in unicellular organisms. *Limnology and Oceanography* **23**, 1248–1255.

Sieburth, J. McN. (1960). Acrylic acid, an 'antibiotic' principle in *Phaeocystis* blooms in Antarctic waters. *Science* **132**, 676–677.

Sieburth, J. McN. (1961). Antibiotic properties of acrylic acid, a factor in the gastro-intestinal antibiosis of polar marine animals. *Journal of Bacteriology* **82**, 72–79.

Sieburth, J. McN. (1964). Antibacterial substances produced by marine algae. *Developments in Industrial Microbiology* **5**, 124–134.

Sieburth, J. McN. (1968). The influence of algal antibiotics in the ecology of marine microorganisms. *In* "Advances in Microbiology of the Sea", pp. 63–94. (Droop, M. R. and Ferguson-Wood, E. J. F., eds.) Academic Press, London and New York.

Sieburth, J. McN. (1976). Bacterial substrates and productivity in marine ecosystems. *Annual Review of Ecology and Systematics* **7**, 259–285.

Sieburth, J. McN. (1979). "Sea Microbes". Oxford University Press, New York.

Sieburth, J. McN., Brooks, R. D., Gessner, R. V., Thomas, C. D., and Tootle, J. L. (1974). Microbial colonization of marine plant surfaces as observed by scanning electron microscopy. *In* "Effects of the Ocean Environment on Microbial Activities", pp. 418–432. (Colwell, R. R. and Morita, R. Y., eds.) University Park Press, Baltimore.

Sieburth, J. McN. and Conover, J. T. (1965). *Sargassum* tannin, an antibiotic that retards fouling. *Nature, London* **208**, 52–53.

Silva, P. C. and Papenfuss, G. F. (1953). A systematic study of the algae of sewage oxidation ponds. *California State Water Pollution Control Board Publication No. 7.*

Silvey, J. K. G. and Roach, A. W. (1964). Investigation of the microbiological cycle in surface waters. *Journal of the American Water Works Association* **56**, 60–71.

Simidu, U., Asturio, K., and Kaneko, E. (1971). Bacterial flora of phyto- and zoo-plankton in the inshore water of Japan. *Canadian Journal of Microbiology* **17**, 1157–1160.

Simidu, U., Kaneko, E., and Taga, N. (1977). Microbiological studies of Tokyo Bay. *Microbial Ecology* **3**, 173–191.

Sjobald, R. D. and Mitchell, R. (1979). Chemotactic responses of *Vibrio alginolyticus* to algal extracellular products. *Canadian Journal of Microbiology* **25**, 964–967.

Smith, D. F. and Higgins, H. W. (1978). An interspecies regulatory control of dissolved organic carbon production and incorporation by microheterotrophs. *In* "Proceedings in Life Sciences, Microbial Ecology", pp. 34–39. (Loutit, M. W. and Miles, J. A. R., eds.) Springer-Verlag, Berlin.

Soeder, C. J. (1980). Massive cultivation of microalgae: results and prospects. *Hydrobiologia* **72**, 197–209.

Sorokin, Yu. I. (1969). Seasonal dynamics of plankton productivity in shore and pelagic areas of Rybinsk reservoir. *Byulletin Instituta biologii Vodokhranilishcha* **3**, 7–10.

Sorokin, Yu. I. (1972). Biological productivity of the Rybinsk reservoir. *In* "Productivity Problems of Inland Waters", pp. 493–503. (Kajak, Z., ed.) Polish Scientific Publishers, Warszawa Krakov.

Sorokin, Yu. I. (1977). The heterotrophic phase of plankton succession in the Japan Sea. *Marine Biology* **41**, 107–117.

Sorokin, Yu. I. and Paveljeva, E. B. (1972). The quantitative characteristics of the pelagic ecosystem of Dalnee lake (Kamchatka). *Hydrobiologia* **40**, 519–552.

Spencer, C. P. (1952). On the use of antibiotics for isolating bacteria-free cultures of marine phytoplankton organisms. *Journal of the Marine Biological Association of the U.K.* **31**, 97–106.

Spoehr, H. A., Smith, J. H. C., Strain, H. H., Milner, H. W., and Hardin, G. J. (1949). Fatty acid antibacterials from plants. *Carnegie Institution of Washington Publications* **586**.

Steemann-Nielsen, E. (1952). The use of radio-active carbon [^{14}C] for measuring organic production in the sea. *Journal du Conseil—Conseil Permanent International Pour l'exploration de la Mer* **18**, 117–140.

Steemann-Nielsen, E. (1955). The production of antibiotics by plankton algae and its effect upon bacterial activities in the sea. *In* "Deep Sea Research Special Suplement to Volume 3, Papers in Marine Biology and Oceanography", pp. 281–286. Pergamon Press, Oxford.

Stegeman, W. J. and Hoober, J. K. (1975). Induction of synthesis of bacterial protein by excretory product of the alga *Chlamydomonas reinhardii* y−1. *Nature, London* **257**, 244–246.

Stein, J. R. (1973). Handbook of phycological methods. Culture methods and growth measurements. Cambridge University Press, London.

Stewart, J. R. and Brown, R. M. (1970). Killing of green and blue-green algae by a non-fruiting myxobacterium *Cytophaga* N−J. *Bacteriological Proceedings*, p. 18.

Stewart, J. R. and Brown, R. M. (1971). Algicidal non-fruiting myxobacteria with high G+C ratios. *Archiv für Mikrobiologie* **80**, 176–190.

Stewart, W. D. P. and Daft, M. J. (1976). Algal lysing agents of freshwater habitats. *In* "Microbiology in Agriculture, Fisheries and Food", pp. 63–90. (Skinner, F. A. and Carr, J. G., eds.) Academic Press, London and New York.

Stewart, W. D. P. and Daft, M. J. (1977). Microbial pathogens of cyanophycean blooms. *In* "Advances in Aquatic Microbiology", Vol. 1, pp. 177–218. (Droop, M. R. and Jannasch, J. W., eds.) Academic Press, London and New York.

Stewart, W. D. P., Sinada, F., Christofi, N., and Daft, M. J. (1977). Primary production and microbial activity in Scottish fresh water habitats. *In* "Aquatic Microbiology", pp. 31–54. (Skinner, F. A. and Shewin, J. M., eds.) Academic Press, London and New York.

Straskrabova, V. (1975). Seasonal variations in the production and biomass of bacterial plankton in the Kilicava reservoir and their relation to the production of algae. *Folia Microbiologia* **20**, 76.

Taga, N. and Matsuda, O. (1974). Bacterial populations attached to plankton and detritus in seawater. *In* "Effects of the Ocean Environments on Microbial Activities", pp. 433–447. (Colwell, R. R. and Morita, R. Y., eds.) University Park Press, Baltimore.

Tanaka, N., Nakanishi, M., and Kadota, H. (1974). Nutritional interrelation between bacteria and phytoplankton in a pelagic ecosystem. *In* "Effect of the Ocean Environment on Microbial Activities", pp. 495–509. (Colwell, R. R. and Morita, R. Y., eds.) University Park Press, Baltimore.

Tanner, W., Grunes, R., and Kandler, O. (1970). Specificity and turnover of inducible hexose uptake systems of *Chlorella. Zeitschrift für Pflanzenphysiologie* **62**, 376–386.

Tanner, W., Komor, E., Fenzil, F., and Decker, M. (1977). Sugar-Proton cotransport systems. *In* "Regulation of Cell Membrane Activities in Plants", pp. 79–90. (Marre, E. and Ciferri, O., eds.) Elsevier/North Holland Biomedical Press, Amsterdam.

Thomas, J. P. (1971). Release of dissolved organic matter from natural populations of marine phytoplankton. *Marine Biology* **11**, 311–323.

Tilzer, M. M. (1972). Dynamics and productivity of phytoplankton and pelagic bacteria in a high mountain lake. *Archiv für Hydrobiologie, Supplement* **40**, 201–273.

Tison, D. L. and Lingg, A. J. (1979). Dissolved organic matter utilization and oxygen uptake in algal-bacterial microcosms. *Canadian Journal of Microbiology* **25**, 1315–1320.

Tison, D. L., Pope, D. H., Cherry, W. B., and Fliermans, C. B. (1980). Growth of *Legionella pneumophila* in association with blue-green algae (Cyanobacteria). *Applied and Environmental Microbiology* **39**, 456–459.

Ukeles, R. and Bishop, J. (1975). Enhancement of phytoplankton growth by marine bacteria. *Journal of Phycology* **11**, 142–149.

Ukeles, R. (1976). Cultivation of Plants. Unicellular Plants. *In* "Marine Ecology", Vol. 3, pp. 367–466. (Kinne, O., ed.) John Wiley, London.

Vance, B. D. (1965). Composition and succession of Cyanophycean water blooms. *Journal of Phycology* **1**, 81–86.

Wagner, F. S. (1969). Composition of the dissolved organic compounds in seawater: A review. *Contributions in Marine Science* **14**, 115–153.

Wangersky, P. J. (1978). Production of Organic Matter. *In* "Marine Ecology", vol. 4, pp. 115–220. (Kinne, O., ed.) John Wiley, Chichester.

Watt, W. D. (1966). Release of dissolved organic material from the cells of phytoplankton populations. *Proceedings of the Royal Society of London, Series B* **164**, 521–551.

Whitton, B. A. (1965). Extracellular products of blue-green algae. *Journal of General Microbiology* **40**, 1–11.

Whitton, B. A. (1967). Studies on the toxicity of polymixin B to blue-green algae. *Canadian Journal of Microbiology* **13**, 987–993.

Whitton, B. A. (1973). Interactions with other organisms. *In* "The Biology of Blue-green Algae", pp. 415–433. (Carr, N. G. and Whitton, B. A., eds.) Blackwell Scientific Publications, Oxford.

Wiebe, W. J. and Pomeroy, L. R. (1972). Microorganisms and their association with aggregates and detritus in the sea: a microscopic study. *Memorie—Istituto Italiano Di Idrobiologia Marco De Marchi, Supplement* **29**, 325–352.
Wiebe, W. J. and Smith, D. F. (1977). Direct measurement of dissolved organic carbon release by phytoplankton and incorporation by microheterotrophes. *Marine Biology* **42**, 213–223.
Wilkinson, C. R. (1979). *Bdellovibrio*-like parasites of Cyanobacteria symbiotic in marine sponges. *Archives of Microbiology* **123**, 101–103.
Williams, P. J. LeB. and Yentsch, C. S. (1976). An examination of photosynthetic production, excretion of photosynthetic products, and heterotrophic utilization of dissolved organic compounds with reference to results from a coastal sub-tropical sea. *Marine Biology* **35**, 31–40.
Wright, R. T. (1970). Glycollic acid uptake by planktonic bacteria. *In* "Organic Matter in Natural Waters", pp. 521–536. (Hood, D. W., ed.) Institute of Marine Science, Alaska.
Wright, R. T. (1974). Mineralization of organic solutes by heterotrophic bacteria. *In* "Effects of the Ocean Environment on Microbial Activities", pp. 546–565. (Colwell, R. R. and Morita, R. Y., eds.) University Park Press, Baltimore.
Wright, R. T. (1975). Studies on glycollic acid metabolism by freshwater bacteria. *Limnology and Oceanography* **20**, 626–633.
Wright, R. T. and Hobbie, J. E. (1965). The uptake of organic solutes in lake water. *Limnology and Oceanography* **10**, 22–28.
Wright, R. T. and Hobbie, J. E. (1966). The use of glucose and acetate by bacteria and algae in aquatic ecosystems. *Ecology* **47**, 447–464.
Wright, R. T. and Shah, N. M. (1975). The trophic role of glycollic acid in coastal seawater. I. Heterotrophic metabolism in seawater and bacterial cultures. *Marine Biology* **33**, 175–183.
Wright, R. T. and Shah, N. M. (1977). The trophic role of glycollic acid in coastal seawater. II. Seasonal changes in concentration and heterotrophic use in Ipswich Bay, Massachusetts, USA. *Marine Biology* **43**, 257–263.
Wujek, D. E. (1979). Intracellular bacteria in the blue-green alga *Pleurocapsa minor*. *Transactions of the American Microscopical Society* **98**, 143–145.
Zagallo, A. C. (1953). Oxidative metabolism of some bacteria and blue-green algae alone and in association. *Agronomia Lusitania* **15**, 315–345.
Zobell, C. E. (1946). "Marine Microbiology". Chronica Botanica Co., Waltham, Massachusetts.

7

Freshwater Protozoan Communities

J. Cairns, Jr.

1. INTRODUCTION

Protozoan communities are ecologically fascinating systems that provide a magnificent opportunity to investigators interested in testing various ecological hypotheses. In many instances, these communities fit models developed for higher organisms. The small size of protozoan communities and the speed with which events occur are simultaneously advantageous and frustrating. In a small space and a short time, research can be accomplished that would take many years for larger organisms. However, rapid movement of individuals and turnover of species make individual interactions difficult or impossible to detect and quantify, unless the components are studied in isolation from the remainder of the community.

Although the literature on freshwater protozoan communities is not large, it is still too large to cover adequately in a single chapter. The majority of readers probably do not have extensive first hand research experience in this area but may wish to carry out such studies or at least be aware of their significance. As a consequence, the objectives of this chapter are:

(1) to illustrate the utility of protozoan communities for both theoretical and applied studies;
(2) to provide references so that the reader can easily acquire additional information on protozoan communities, and
(3) to acquaint the reader with the use of artificial substrates which are now commonly used in a variety of studies of protozoan and other microbial communities.

Much of the early literature on ecology has been summarized by Noland and Gojdics (1967), a good general text is Sleigh (1973), and one of the important older papers on protozoan community structure is Picken (1937). Unfortunately, studies of spatial relationships within a protozoan community are rare and such information is badly needed. Picken (1937) was one of the first to note that an assemblage of protozoans is a complex community of herbivores, carnivores, omnivores, and detritus feeders which form a closed social structure. He also analysed food chains in protozoan communities that had notably different microbial associated communities of diatoms, cyanobacteria, and bacteria. Much attention has been given to the protozoans in sewage treatment systems where they have important roles to fulfill (Lackey, 1925; Barker, 1946; Curds, 1966, 1973) and to the degradation of organic compounds in natural waters (Bick, 1971). The passive dispersal of protozoans which is important in both colonization and succession has been studied by Gislen (1948), Schlichting (1961, 1964, 1969), Maguire (1963), Milliger *et al.* (1971), and Blanchard and Parker (1977). Succession has been studied by Woodruff (1912), Eddy (1928), Unger (1931), Cooke (1967), and Yongue and Cairns (1971). Ecological factors affecting protozoan distribution have been studied by Hausman (1917), Stout (1956, 1974), Kitching (1957), and Webb (1961). Interactions with other microorganisms have been studied by Hairston *et al.* (1968) and competition between protozoan species by Evans (1958). Maguire (1977) studied colonization processes with particular attention to changes in the autotroph to heterotroph ratio during this process. Brooks and Dodson (1965), Hrbacek (1977) and others have shown that predation may alter community composition of the plankton, including protozoan component. The importance of protozoans as grazers of natural bacterial populations has been discussed by

Bick (1958) and Fenchel (1977). A good summary of the literature on the importance of surface films may be found in Parker and Barson (1970). Ruggiu (1969) discusses the benthic ciliates in the profundal of Lake Orta, Northern Italy. Other studies of benthic protozoans include Moore (1939), Cole (1955), and Goulder (1974). Wang (1928) discusses the seasonal distribution of protozoans.

Various functions, e.g. respiration and production, have been studied by Ganf (1974) and Finlay (1978). Quantitative assessment methods have been developed by Sramek-Husek (1958), Borkott (1975), and Finley *et al.* (1979). Colonization of artificial, uninhabited substrates by microorganisms has been discussed by Butcher (1946), Cooke (1956), Grzenda and Brehmer (1960), and Spoon and Burbanck (1967).

In studies of natural communities, theoretical possibilities must be distinguished from what can be measured. Many years ago the poet Francis Thompson wrote ". . . Thou canst not stir a flower without troubling of a star." Later Hardin (1969) illustrated the validity of this hypothesis by lifting a flower in a vase and pointing out that he does indeed "trouble a star" because Newton's Law states "every body attracts every other body with a force that . . ." and so on. As Hardin lifted the flower, literally every star in the universe, even those beyond the reach of the most powerful telescopes, had its position and motion altered by virtue of the law of universal gravitation. However, although the validity of this assumption is recognized, it is ignored because it is practically of no importance. Thus, some theoretical effects are quantitatively beyond our ability to measure them and, therefore, of no operational utility. Similarly, although every alteration in environmental quality or community composition triggers a chain of events within the community, these are often beyond our capacity to measure.

The long evolutional history of protozoan communities, the high probability of cosmopolitan distribution for many of these species, and their relative morphological stability (keys produced by Kahl (1930) and Pascher (1913–1927) to mention only two that are still as effective as they were when they were written) raise the tantalizing possibility that assemblages of protozoan species may have a reasonably constant structure as well as synergistic, functional relationships. All of the statements just made about protozoans are also probably equally valid for diatoms. In addition, aggregations of diatom species have the advantage of being more readily preserved and, therefore, provide a greater opportunity for detailed counts of both the species diversity and the number of individuals per species (evenness). Such analyses are much more difficult for protozoans since preservation of a community consisting of a large number of species (many in extremely low numbers mixed with the inevitable detritus) is virtually impossible. Some of these problems have been discussed in detail by Cairns (1965) and will not be repeated here. However, since the similarities just noted exist between protozoans and diatoms, it is advantageous to use information generated with diatoms when comparable information for protozoans is exceedingly difficult to obtain. One of the most fascinating of these data sets is the curves originally proposed by Preston (1948, 1962) and confirmed by Patrick (1949) with diatoms, illustrated in Fig. 1. Patrick, frequently in association with various colleagues, has published a number of papers on this subject; many are discussed in Patrick (1977). The fascinating implication of this distribution which occurs in a wide variety of temperature zone streams at all seasons, is that not only is the species richness remarkably constant (considering the number of potential colonizing species) but also the numbers of individuals per species are arrayed in a predictable fashion. In short, a very high probability exists that a certain number of species with two to four individuals per species as well as a certain number with four to eight individuals per species

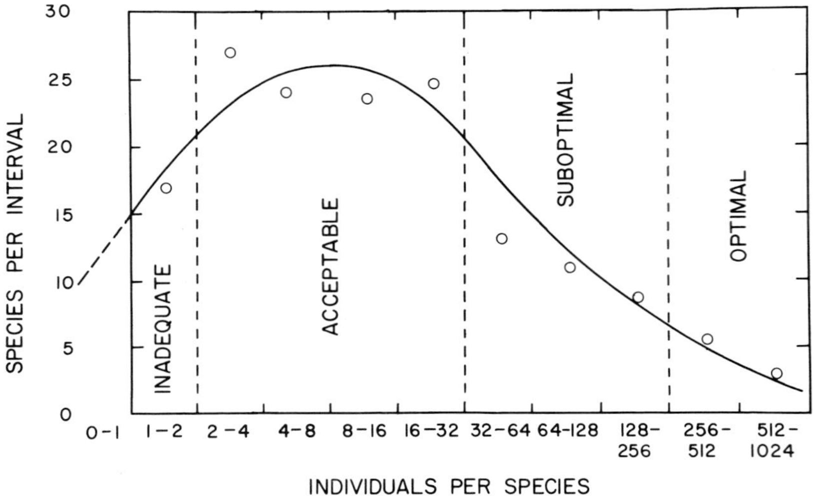

Fig. 1. Model for the truncate normal curve distribution hypothesized by Preston (1948), confirmed by Patrick *et al.* (1954), and modified by Cairns (1971).

and so on will occur. Protozoans quite likely follow the same distributional system because crude estimates of abundance show that low density species are by far the most numerous (Cairns *et al.*, 1969) and high density species are few. This follows in a very general way from the more detailed analysis possible with diatoms. The fact that protozoan communities appear to be organized in distributional patterns, similar to those of diatoms and even birds, e.g. the MacArthur-Wilson model, shows that such communities deserve careful attention and study.

Cairns and Yongue (1977) offered two hypotheses (not mutually exclusive) for the truncate normal curve found by Patrick and her co-workers. First, the chances of finding optimal conditions by organisms which are passively transported are slim but, if they do, they will flourish (i.e. low number of species—high number of individuals per species). The chances of finding sub-optimal conditions are greater, but the species that do so will not flourish and so on. The second hypothesis assumes a limited number of functional roles with any one of an array of species filling the role at a particular time (Fig. 2).

A necessary caveat at the conclusion of this introduction is that protozoans should not be considered in isolation from other members of the aquatic microbial community, such as algae, bacteria, and fungi, as well as the smaller metazoans, such as rotifers and gastrotrichs.

2. SIMILARITIES OF PROTOZOAN COMMUNITIES TO THOSE OF HIGHER ORGANISMS

A fallacious belief is widely held that protozoan communities have characteristics quite different from communities of higher organisms. A few illustrations will demonstrate that protozoan communities are structured by the same principles as those that shape communities of higher organisms.

2.1 Species—Area Curves

A species–area curve is developed when a comparatively homogeneous area (in terms of the habitat) is sampled starting with a relatively small area and increasing the area geometrically. This usually provides a clear demonstration that:

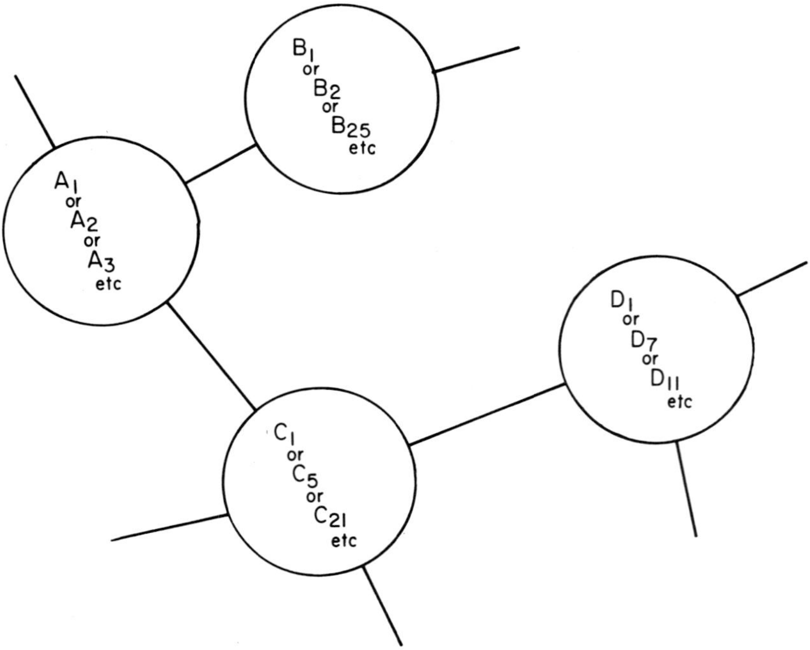

Fig. 2. Role A might be temporarily filled by a species designated A_1 or, as far as species temporarily filling role B is concerned, A_7. Thus community structure would be relatively stable despite species replacing each other because the "role relationships" remain constant. Division of role A among species might be determined by differential tolerances to pH, temperature, etc. The number of roles would, of course, greatly exceed those depicted in this simplified diagram (from Cairns, 1977).

(1) when the initial area is small, increasing the size of the area sampled results in a marked increase in the number of species found;

(2) but further increases generally result in a diminished number of species in proportion to the increase in area sampled.

A comparable curve for protozoans is given in Fig. 3 (Cairns and Ruthven, 1970).

2.2 Colonization—The MacArthur-Wilson Model

Simberloff (1974) has pointed out that "any patch of habitat isolated

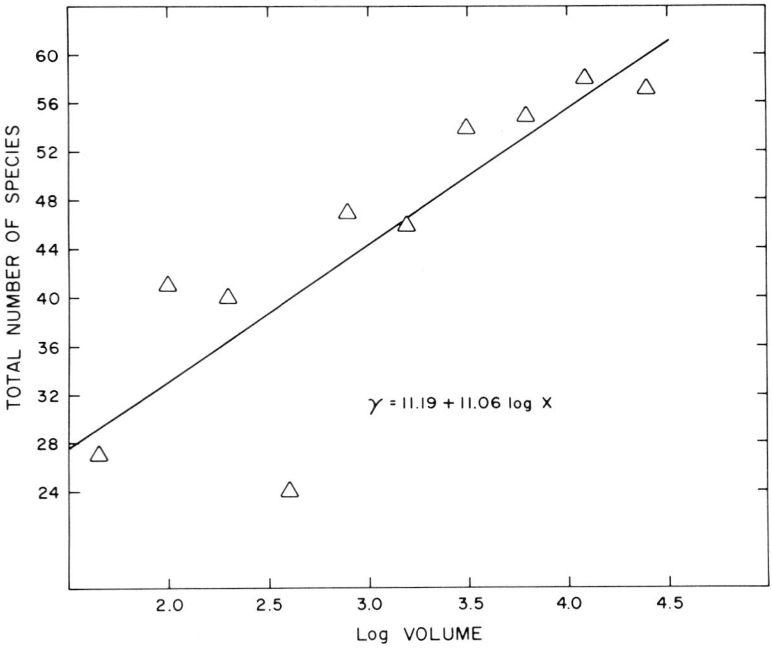

Fig. 3. Relationship between average number of species of freshwater protozoans and log substrate size.

from similar habitat by different, relatively inhospitable terrain traversed only with difficulty by organisms of the habitat patch may be considered an island". Thus a rock in a river, a mud flat, submerged log, or artificial substrate may be considered an ecological island (i.e., most micro-habitats).

Initial island colonization is a non-interactive process primarily influenced by the dispersal capacities and extinction potentials of the colonizing organisms (MacArthur and Wilson, 1963). With the establishment of an equilibrium (asymptotic) species number (Figs. 4 and 5), interactive processes, such as competition and predation, take precedence in determining the island's biotic composition. The island assemblage soon manifests the characteristics of an integrated community capable of maintaining a degree of autonomy with respect to the

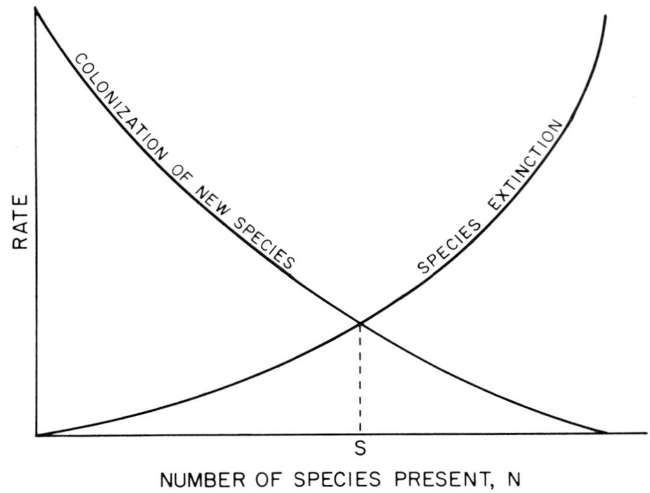

Fig. 4. Basic MacArthur-Wilson model with the equilibrium point(s) of a relatively constant number of species being reached when the colonization rate equals the extinction rate (from MacArthur-Wilson, 1963).

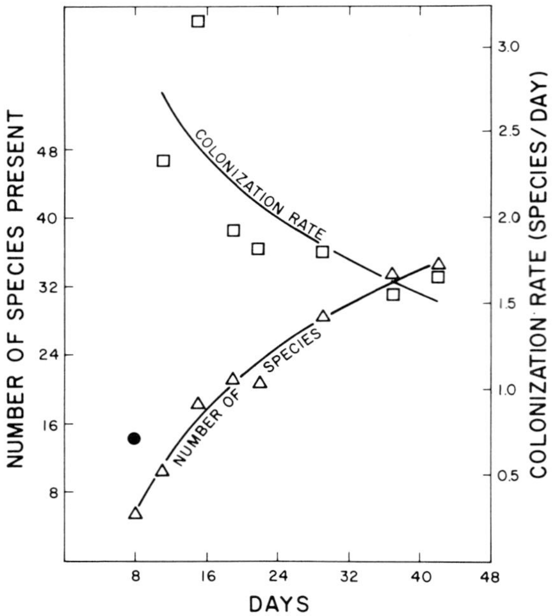

Fig. 5. Colonization of artificial substrates by Protozoa. Notice that the form of the curves is consistent with the predictions of the MacArthur-Wilson equilibrium model. The colonization rate declines as the number of species rises (from Cairns *et al.*, 1969).

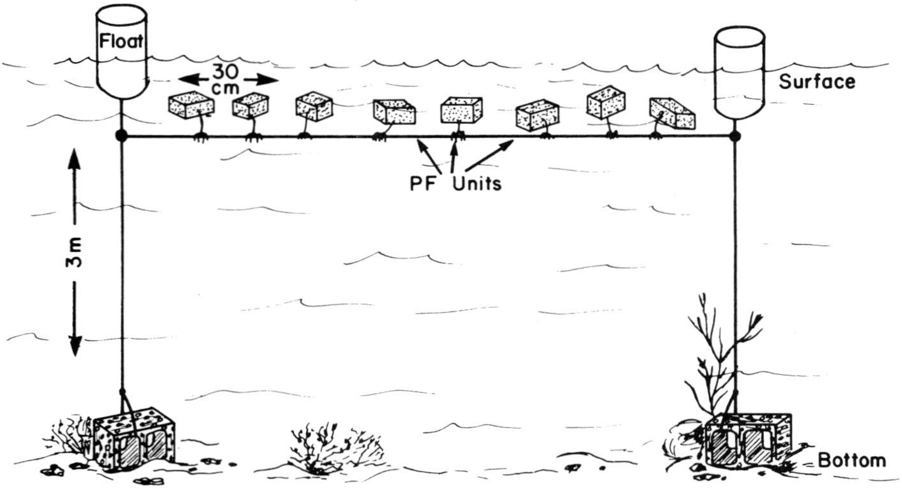

Fig. 6. Typical artificial substrate placement for surface water sampling in a lake or pond (from Cairns *et al.*, 1979).

surrounding environment. Thus, during the non-interactive phase of colonization, artificial substrates or uninhabited natural substrates (e.g. a rock that falls into a stream) act as sampling devices which passively collect organisms from the natural community (Fig. 6). After establishing an equilibrium, the substrate ceases to function directly as a sampling device and the associated species assemblage begins to evolve its own characteristic composition. If an artificial substrate is used as a species sampling device, it should be retrieved from the environment just before or soon after it acquires an equilibrium species number. The appropriate immersion time varies under different environmental conditions.

The question of the influence of time on colonization became especially important following an earlier study (Cairns *et al.*, 1976a) which indicated that location of substrates in a lake was of little importance in determining the outcome of colonization.

The chemical-physical upheaval associated with lake overturn (and presumably other episodic events, such as floods) apparently disrupts the stability of protozoan communities in terms of species richness (Cairns *et al.*, 1976b). It is possible that ecological space again becomes available with the breakdown of competitive mechanisms and non-interactive colonization dynamics take precedence in determining the species' numbers. All species are qualitatively equated with respect to their capacities to occupy the newly available space. Species already inhabiting a particular substrate have an advantage over pioneer occupants. Occupation of the remaining space is determined by a random selection from the available species pool (Gilroy, 1975). Rapid changes in composition have also been observed to occur concomitantly with the spring circulation and other catastrophic perturbations (Jassby and Goldman, 1974).

The disruption of competitive interactions also profoundly affects the species assemblages of the littoral and sublittoral sediments. Unlike the pelagic zone, the primary mechanisms of passive dispersal between the various areas of the sediments include differential sedimentation and turbulent mixing. The latter is probably the only important force affecting dispersal in the open water and is equivalent at all depths, particularly following overturn. Sedimentation, however, does tend to deposit materials (and presumably species) differently within the littoral and sublittoral sediments (Davis and Brubaker, 1973). Therefore, if potential colonists from other areas of the lake are also deposited in greater abundances in these zones, sediment-associated protozoan communities would be expected to exhibit quantitative differences in species numbers, but become somewhat more similar to all other areas after the overturn phase. This appears to be the case.

Of the sediment assemblages, those at the 5 m location most resemble those of the pelagic zone. In fact, they are normally included within the overall compositional cluster for the open water. The implication from pre-overturn that the 5 m location exhibits characteristics of both the pelagic and benthic environments is confirmed. Finally, these conclusions imply that the profundal sediments are inundated with a large number of species, primarily from the pelagic zone which had been previously excluded.

Drastic environmental perturbation, such as lake overturn, causes the partial or complete breakdown of interactive mechanisms (Cairns *et al.*, 1976b). Simple functions of non-interactive colonization assume primary importance in determining the nature of particular assemblages.

Henebry and Cairns (1980a) studied the colonization of artificial islands in a closed laboratory ecosystem using epicentres (source of species) colonized in natural systems. After seven days, the islands with the smallest area or closest to the epicentre had the highest species number. The latter fits the MacArthur and Wilson (1963) model but the former appears contradictory. However, the islands had not reached equilibrium which was the main point of the original hypothesis. Islands exposed to epicentres of intermediate maturity (half-way to equilibrium) had significantly greater species richness than islands tested with mature (at MacArthur-Wilson equilibrium) epicentres. Evidence gathered in this study strongly suggests that the kinds of species present during different periods of colonization are responsible for differences in species richness on islands exposed to epicentres of different maturities. In addition Henebry and Cairns (1980b) supported the hypothesis that colonization rates onto artificial islands were influenced by the maturity of source pools of species and the proportion of pioneer species in the source pool communities. Plafkin *et al.* (1980) found the acquisition of a stable equilibrium number to be more rapid for artificial islands drawing colonists from a species pool in a natural system (mostly lakes) stressed by organic enrichment.

2.3 Succession and Seasonal Change

Succession in protozoan communities is a well established phenomenon (Patrick *et al.*, 1967). Changes in dominance and composition may result from changes in organic loading (Fig. 7, McKinney and Gram,

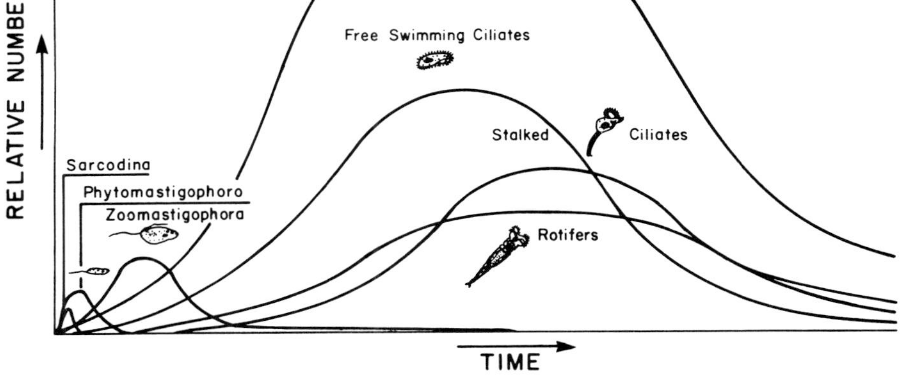

Fig. 7. Relative predominance of microorganisms in activated sludge systems.

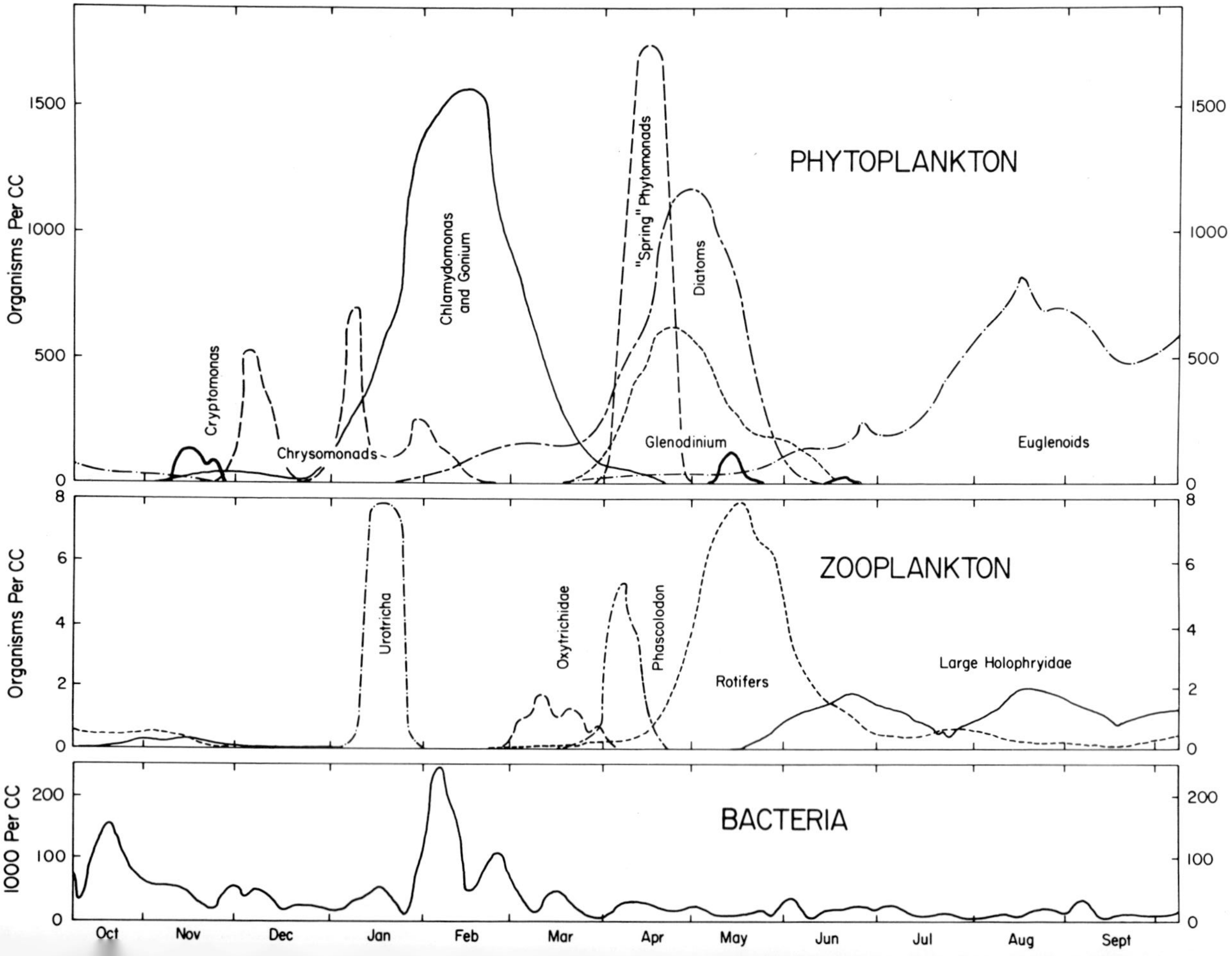
PHYTOPLANKTON
Organisms Per CC
1500
1000
500
0
Cryptomonas
Chrysomonads
Chlamydomonas and Gonium
"Spring" Phytomonads
Diatoms
Glenodinium
Euglenoids
ZOOPLANKTON
Organisms Per CC
8
6
4
2
0
Urotricha
Oxytrichidae
Phascolodon
Rotifers
Large Holophryidae
BACTERIA
1000 Per CC
200
100
0
Oct
Nov
Dec
Jan
Feb
Mar
Apr
May
Jun
Jul
Aug
Sept

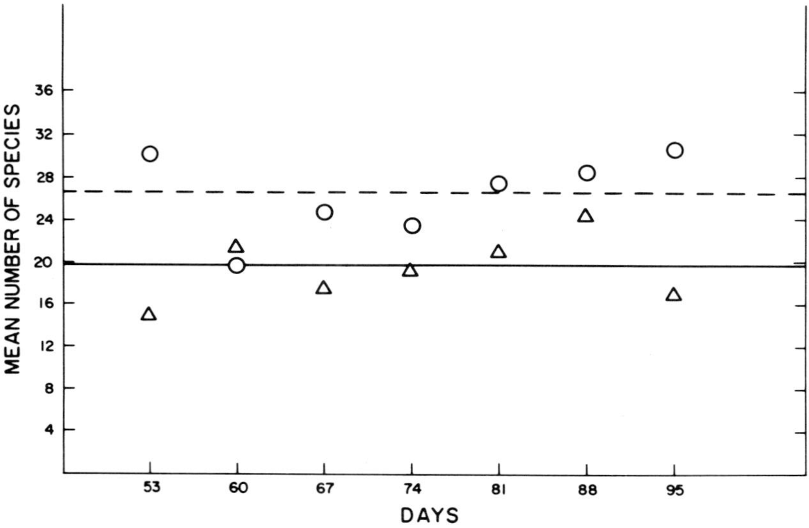

Fig. 9. Weekly mean number of species oscillating about the mean number of species for all substrates which were not previously squeezed (i.e. harvested) (△) and weekly mean number of species oscillating about the mean number of species for all substrates which were squeezed once previously (○).

1967) or in the planktonic community in a small pond as a consequence of seasonal change (Fig. 8, Bamforth, 1958). Flagellates make up the major portion of the early pioneer community becoming established on bare substrates and reaching equilibrium much earlier than the other taxonomic groups (Yongue and Cairns, 1978). However, the number of species or substrates may typically oscillate about a mean once the MacArthur-Wilson equilibrium point has been reached (Cairns *et al.*, 1971a). A segment of a long-term study from the paper just cited illustrates this point (Fig. 9).

Fig. 8. Seasonal variation and succession of dominant organisms in a small artificial pond. The curves are smoothed by 15-point moving averages (after Bamforth, 1958).

3. ANALYSIS OF NATURAL POPULATIONS

3.1 Analysis of Protozoan Communities

Often problems arise in identifying species in protozoan communities before the community becomes seriously altered through reproduction, death, and encystment. There is a difference between obtaining the correct Latin name on each species and accurately determining how many taxons are represented. Of course, precision in identification is always desirable but when precision in identifying one species beyond any reasonable doubt means that the community composition of other species is altered before other members of the community are identified, it is not a desirable practice for effective community analysis. That is, precision in identification of a few "difficult" species may mean loss of precision in analysing the community structure. In case this is taken as a license for inaccurate taxonomy, I hasten to add that it is possible occasionally to have the best of both worlds by a combination of preparatory identifications and the use of technniques based on the principles of aeroplane recognition. The method consists of spending a week or so with a more leisurely identification of the species characteristic of the sampling area just before the main analysis begins. Although successional processes and species turnover does occur, many of the species identified in the more leisurely and, therefore, more precise preparatory investigation, will be found for quite a few days thereafter, and perhaps even for the entire period of the investigation. As a consequence, a substantial percentage of the identifications are confirmations of previous identifications and the number of totally new identifications are normally a minority.

Although the practice of fractionating a microbial community into the commonly accepted taxonomic subcomponents, such as protozoans or diatoms, has been deplored, it is necessary to discuss protozoan communities, primarily in a restricted taxonomic sense, by ignoring interactions with other microbial taxonomic groups since this is the standard procedure adopted. It is also worth noting that higher organisms, such as pulmonate snails, can markedly affect attached microbial community structures by selective cropping of species (R. Patrick, personal communication). For planktonic microbial communities, the classic work of Brooks and Dodson (1965) has shown that higher organisms, such as the alewife (an American fish), also affect community structure significantly by particle size discrimination in planktonic feeding. The study of subdivisions of a true microbial community, such as a protozoan community, is not without scientific justification, however, since predictions can be made based on this limited taxonomic array that can be verified and are remarkably consistent.

There is also justification for limiting the scope of a study by the use of the definition due to Whittaker (1975), that a community is "a system of organisms living together and linked together by their effects on one another and their responses to the environment they share". Whittaker (1975) defined an ecosystem as "a community and its environment treated together as a functional system of complementary relationships and transfer and circulation of energy and matter". Patrick (1949) discussed means of assessing the effects of pollution on aquatic communities using sections of streams from bank to bank and several hundred feet along the stream, by collecting samples from all of the common habitats within the system. The well-established species in the samples are identified and the results from all the samples combined to determine the community composition of each of the major groups of aquatic organisms, including algae and protozoans. Since the original study was concerned with pollution effects, breaking the system into discrete study areas seemed a satisfactory procedure to delimit a community. Alternatively, river microbial communities might be viewed as a continuum since it is theoretically possible that the actions of organisms on one bank of a river might have some effect upon a microorganism on the other bank. It is unlikely, however, that such effects could be demonstrated. A protozoan community thus defined (i.e. from bank to bank in a river) provides a useful means of detecting gross pollution effects (Cairns, 1965) but not the more subtle effects. The amount of work involved is substantial since frequently six or more subsamples require identification. One also has to assess the weight given to the subsamples, according to the percentage of the total habitat they are presumed to represent, or whether to measure qualitative changes in species richness without regard to the percentage of habitat represented by the sample from which the determinations were made. This in turn depends on whether interest centres on the qualitative or quantitative degree of the environmental impact under study.

3.2 Response to Changes in Water Quality

Interpretation of the response of protozoan communities to changes in water quality is enhanced by obtaining information on other taxonomic groups of aquatic organisms from the same area and collected at the same time. This is illustrated clearly by the histograms from Patrick (1949) who was one of the first to show the utility of an array of taxonomic groups in analysing pollution effects in freshwater ecosystems (Fig. 10). Each of the histogram columns represents a different number of species, despite the fact that they are nearly the same height in the healthy station. For example, the histogram for fish might represent approximately 35 species at 100%, which would be the average value for a number of healthy stations on unpolluted streams. The insect column that had 100% might represent over 80 species of aquatic insect larvae established at the same number of stations on unpolluted streams. Therefore, the 100% value represents the typical species richness (i.e., number of species present) in unpolluted systems. Note that for the semi-healthy station, the percentage of protozoan species remained about the same as that found in the healthy station, while the percentage of fish species declined substantially. However, the pollution tolerant worms rose by a substantial percentage. For the polluted station the changes were more dramatic: both the pollution tolerant algae and rotifers, and pollution tolerant worms rose significantly. In the case of the pollution tolerant algae and rotifers there was a doubling in numbers compared to the healthy station and the tolerant worms increased by 50%. The fish disappeared entirely and the taxonomically higher animals (rotifers, clams, insecta, and crustacea) almost disappeared. Finally, in the very polluted station, the percentage of aquatic organisms compared to the healthy station was greatly diminished.

It appears that pollution, particularly organic pollution, favours the lower organisms and, for some species, this is true. However, the effect displayed in the histogram series (Fig. 10) was probably due to the result of reduced predation on the lower organisms which permitted the expansion of certain populations, coupled with a direct benefit for some species from the substance causing the pollution. That the loss of the predator species is at least partly responsible for the increase in the columns representing the lower organisms is supported by the progression exhibited from the healthy to the polluted station. Note the rather dramatic loss in the very polluted station of all species, indicating that the pollutants were detrimental to almost all forms of life (except for a relatively few species) when at a high concentration. Analyses of the histograms shows that the number of protozoan species alone may not provide an early indication of pollutional effects as well as for some of the other aquatic organisms (e.g. fish).

In situations, such as the one just mentioned, where the number of species remains constant, there may be a shift in the numbers of individuals per species so that the individuals of a few species could become exceedingly abundant (this would change the diversity index without changing the species richness). A shift in dominance from

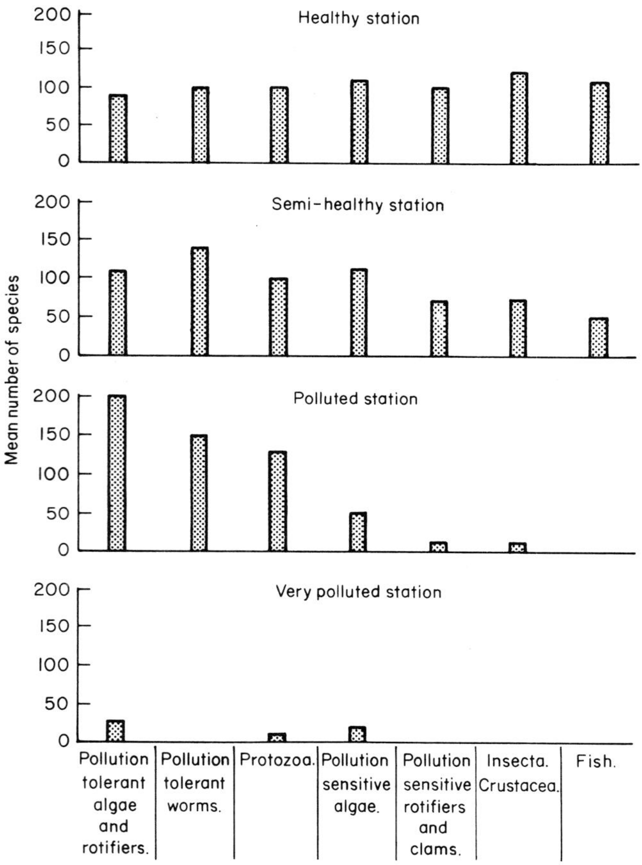

Fig. 10. Histograms illustrating the response of an aquatic community to pollution (from Patrick, 1949).

flagellates to ciliates or some other shift in species composition might occur. Additionally, there might be a shift in species function (e.g. from primarily autotrophic to primarily heterotrophic). Thus, even when using only a single taxonomic group, such as protozoan communities, the pollution investigator would do well to obtain sufficient data so that more than one analysis can be made. If time and financial circumstances permit additional evidence beyond that furnished by the microbial communities, such evidence should be obtained.

3.3 Sampling Natural Communities

The most important problem in sampling natural systems is defining the boundaries of the systems being sampled. For pollution studies in a river or a lake, sampling usually occurs along a gradient from high to low concentration of the offending material. Once the nature of the gradient has been established (and this is not always a simple task), the next consideration is to determine that the types of habitats sampled at each of the locations along the gradient are ecologically comparable to the reference (control) area. Very often a contaminant, such as heated waste water or a hazardous chemical, has thermal or concentration gradients that require sampling in areas where the habitats differ from those found in the reference station or from one part of the gradient to another. In such situations, the comparison may be facilitated by the use of artificial substrates (e.g. Cairns *et al.*, 1979). This ensures that the microhabitats are structurally comparable and that species from this type of habitat are comparable. Artificial substrates can be used in situations where normally only plankton are found because suitable invading species (i.e. those usually associated with substrates) are present in most freshwater situations. Whether this is true or not for marine environments outside the coastal zone has not been determined. Duplication of natural communities by artificial substates does not necessarily occur (although this is usually the result), but determining whether there is a biological stress due to pollution or some other effect differentially exerted from one sampling area to another is essential. If it is assumed that an equal colonization opportunity exists in each area, and that the primary variable is the presence or absence of the pollutant or stress, then comparisons in community structure and/or function that display differences should be considered valid. In pollution assessments, extrapolation from the response of a few organisms to many is usually the case and testing of all the species possibly at risk is rare. It is not possible to test all the organisms that could conceivably be affected. Consequently, testing an array of species through the use of community analysis assumes that the response range will be adequately displayed.

3.4 Sampling Natural Substrates

When analysing protozoan communities, particularly those that are difficult to preserve (most freshwater systems as already discussed), there has to be a choice of how to distribute the analytical time before the fresh sample is seriously altered through death, reproduction, or encystment. A sample from each of the major substrate habitats (i.e. surfaces of mud, rock, submerged vegetation) is examined separately. By coupling this information with the relative proportion of the various microhabitats within an aquatic ecosystem, a very detailed and realistic apportioning of the species distribution within the system and their abundance in the different habitat types can be obtained. The same physical structure (e.g. submerged vegetation or wood) in shady and sunlit situations as well as at different depths and in different current velocities should be analysed. Unfortunately, if one collected as many samples as should be necessary to evaluate properly all the differences just mentioned, the samples would be far too numerous to investigate before serious alteration in community composition began. The errors when the samples deteriorate are far greater than if only enough samples were collected to analyse before undergoing marked alteration. The recording of species occurring at different time intervals (the result of the rapid turnover that occurs in some microbial communities) introduces a serious distortion. Alternatively, composite results may be prepared from a series of samples. Compositing might reduce the apparent abundance of the extremely common species found in certain habitats but not found elsewhere, to the point where they might be below the threshold density that makes recognition possible. Normally it is generally preferable to compromise by taking a series of samples from the most common major habitats (i.e. surfaces of rocks, submerged vegetation or mud) and ignoring some of the finer distinctions previously mentioned. Since these are usually baseline studies carried out before, say, an industry begins discharging wastes, the type of community inhabiting the area in a general way is determined. By comparing the post treatment samples with the pretreatment samples, any major deleterious effects that occur may be determined.

The problem of how many species to take in characterizing the planktonic community is somewhat similar, although, of course, the life histories and ecological requirements of planktonic species are different. Planktonic species have an advantage in that many are comparatively easy to preserve and there is generally less detritus to interfere with sample examination.

4. THE USE OF PROTOZOANS IN THE ASSESSMENT OF WATER QUALITY

Protozoans provide a useful means of assessing water quality but they are not frequently used. Some of the advantages of using protozoans for this purpose are as follows:

(1) They require relatively small containers compared to those needed for fishes and macroinvertebrates.

(2) Some microbial species are no more difficult to handle than rainbow trout and other commonly used test species.

(3) They grow comparatively rapidly so that the effects of potential pollutants on reproduction, growth, metabolism and other properties may be readily tested over several generations without waiting months or years for results.

(4) They can be maintained in synthetic media under defined conditions so that assays are completely reproducible—a situation which is expensive for the large volumes of fluid required for higher organisms and difficult to achieve.

(5) Many species have both sexual and asexual reproduction stages—thus clonal uniformity is available as well as the unique characteristics associated with sexual reproduction.

(6) Unicellular organisms are in more intimate contact with their environment and often have shorter response times than higher organisms.

(7) Since most free-living microorganisms have a cosmopolitan distribution and are likely to be found wherever the natural conditions are appropriate, the same species may be used as an indicator organism on different continents and help to reduce the conflicts due to differing results with different species.

(8) Since many species can often be kept in stock culture at a slow rate of growth, it is possible to keep a collection more easily and in less space than is possible for higher organisms.

(9) The differences in tolerance to various waste materials among fishes, invertebrates, and microbial species is not as great as is generally supposed. Patrick *et al.* (1968) have shown that diatoms are sometimes more sensitive than fishes, sometimes less, and sometimes quite comparable, but rarely are there orders of magnitude differences in response. Thus, fish tolerance is no more or less representative of the entire aquatic community than a microbial or macroinvertebrate species.

(10) Protozoans, together with other microbial species, constitute the major portion of the biomass of many aquatic systems. Therefore, in terms of weight per unit area (or volume), they are frequently a dominant portion of aquatic ecosystems.

(11) A collecting permit usually is not required!

We are now entering an era where the hazard evaluation of toxic substances, biological monitoring, and environmental quality control systems are essential for the maintenance of the quality of life in industrial societies. In addition, such systems are necessary for the protection of human health. Even for situations where the sole intent is the protection of human health, the use of biological early-warning systems involving surrogate organisms will almost certainly become increasingly common in the future (Miller, 1977). As a consequence of widespread use, more careful attention will probably be given to the cost of generating the information. Cost will be partially dependent upon the space requirements of the organisms used in the various types of biological assessments. It is, therefore, highly probable that the use of protozoans for this purpose will increase (Cairns, 1979).

Protozoans may be used as indicators of water quality in three major ways:

(1) in surveys of rivers, streams, lakes, and other bodies of water which are carried out preferably by a team of specialists working with organisms ranging from bacteria to fishes;
(2) in laboratory bioassays designed to determine the effects of various changes in water quality; and
(3) in laboratory microecosystems, artificial streams, and the like which are designed to fill the void between the single species bioassay and the complex, highly variable natural systems (e.g. Cooke, 1977).

4.1 Use of Protozoans in Field Studies of Aquatic Ecosystems

The use of protozoan communities in the assessment of water quality of rivers, streams, lakes, estuaries and oceans has some serious drawbacks and some marked advantages which have been discussed at length by Cairns (1974).

Protozoan communities may be used in a variety of ways for assessments.

(1) They may be used as indicator species as exemplified by Bick (1971).

(2) They may be counted to give the number of species and the number of individuals per species, and this data can be analysed by cluster analyses, principal component analyses, ordination, and other procedures. Either natural substrates (Lackey, 1944) or artificial substrates (Spoon and Burbanck, 1967) may be used. Artificial substrates offer some advantages in this type of sampling. Firstly, the substrate may be positioned in the best locations relative to a waste discharge rather than depending on locating the natural habitat. Secondly, the species found are usually the same as those on natural substrates, which is expected since this is where the invaders of the artificial substrate

originate. Thirdly, the sampling process is simplified since composite results from a variety of substrates of dissimilar composition and location, in the proportion in which they occur in nature, do not have to be determined since the proportionality is to a certain extent taken care of by the invasion and subsequent colonization processes.

Examination of microbial communities is normally not done on artificial substrates during the early stages of the colonization process and often not until they are presumed to have reached a MacArthur-Wilson equilibrium condition. However, there is a strong possibility that the colonization process itself may be more informative than the equilibrium condition. This is demonstrated by Fig. 11 from a study of the protozoan communities on artificial substrates for six stations at Smith Mountain Lake, Virginia (Cairns *et al.*, 1979). The arm of the reservoir containing sampling stations 1 and 2 receives sewage and heavy metals from the city of Roanoke, Virginia, just above station 1 (Roanoke River Basin Comprehensive Water Resources Plan, 1975). This produces a well-defined eutrophic gradient with the arm of the reservoir containing station 1 representing the organically enriched situation with some water quality recovery at station 2 and a more marked recovery at the confluence of the two arms represented by stations 3 and 4. Stations 5 and 6 are on an unpolluted section of the reservoir. Although the equilibrium species number (day 21) is quite similar for all stations, the early colonization species number (days 1 and 3) is much higher for the polluted stations (1 and 2) than at the unpolluted stations. Smith Mountain Lake is not a badly polluted lake: if it were, this would be reflected in the species richness at equilibrium. The fact that it is in a threshold condition in certain areas and that the differences between stations under these circumstances are best determined during colonization occurring before equilibrium than after is worth further attention.

Ecological perturbations, such as organic pollution in an aquatic environment, generally produce certain predictable changes in community structure. Species with low tolerances are eliminated, while those species best suited for survival in enriched habitats become excessively dominant. Fig. 12a illustrates two hypothetical distributions of species into abundance classes. Stress distorts the normal distributi n U by eliminating many low to moderately abundant species and in reasing the number of high abundance species. The net result is that in a stressed situation (distribution S) a greater proportion of the total species are present in high abundance. Such changes in community structure must affect the colonization dynamics of initially barren islands drawing colonists from the perturbed system.

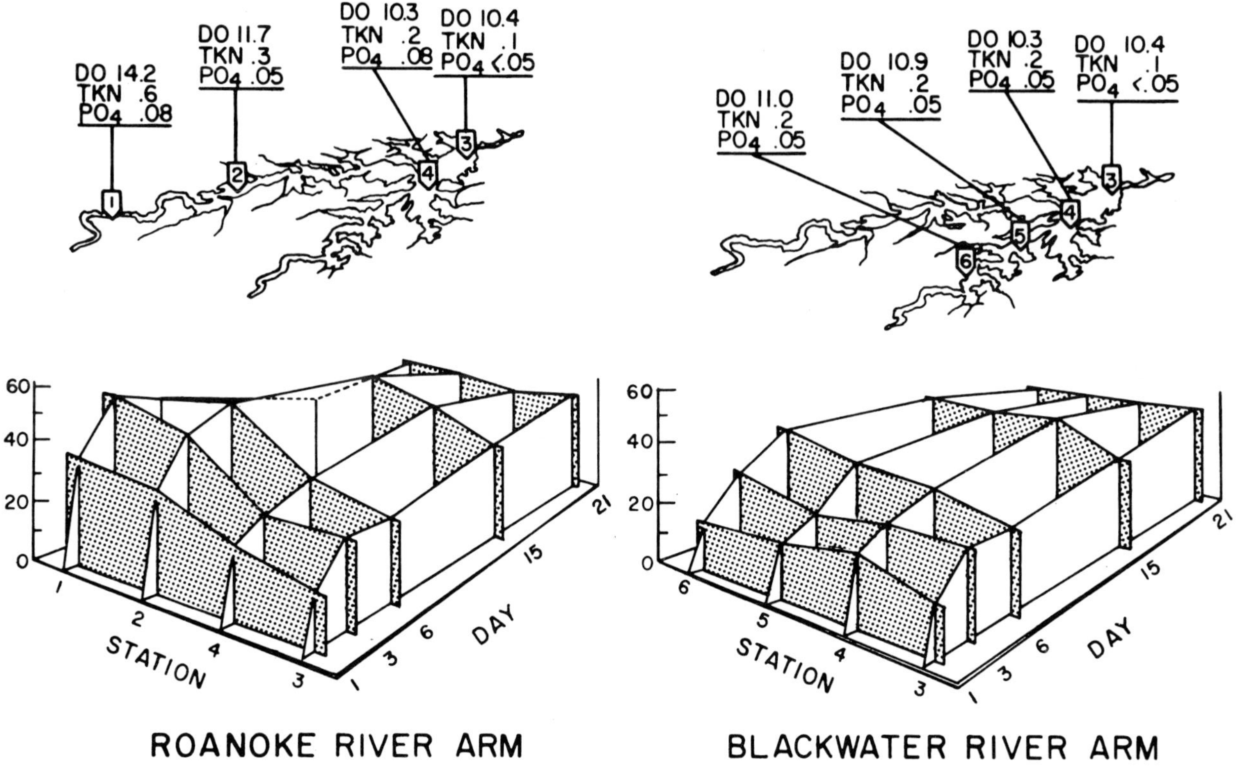

Fig. 11. Mean colonization curves showing the protozoan colonization of artificial substrates in Smith Mountain Lake (a reservoir) Virginia (after Cairns *et al*., 1979). See text for details of stations 1 to 6. DO = dissolved oxygen concentration; TKN = the nitrogen concentration; PO_4 = the phosphate concentration.

There are four reasons to expect that the establishment of a stable equilibrium number would be more rapid for uncolonized islands (in this case polyurethane (PF) islands) drawing colonists from a stressed environment.

(1) In organically enriched areas, colonization is rapid because the PF islands supply immediately suitable habitats for immigrant heterotrophs. Under comparatively oligotrophic circumstances, however, colonization is slower because preparation of the substrate is first required before many protozoan species can invade successfully.

Gilroy (1975) noted that in defaunation experiments (Wilson and Simberloff, 1968) "non-ideal" islands (i.e. those which did not reach an equilibrium during the first year of the monitoring period and did not fit the theoretical model) had suffered considerable damage during the defaunation procedure. Recolonization was very slow because the suitability of these islands as habitats had been significantly diminished.

Foam islands at the cleaner Smith Mountain Lake sampling stations 3 and 4 were also comparatively slow to accumulate species. The initial phases of colonization at stations 3 and 4 were adequately described by the non-interactive model (Table 1, days 1 to 6), but the apparent acquisition of equilibrium by day 6 and the subsequent increase in species numbers suggested that habitat islands were not immediately suitable for certain components of the community. Because habitat islands at the cleaner stations are not subject to the large and immediate influx of organic nutrients which abound at more polluted stations, colonization and habitat preparation by early invaders was probably first required to modify these habitats for subsequent successful invasion by other heterotrophic protozoa.

(2) A stressed community is commonly dominated by r-selected opportunistic species. These species are particularly well adapted to the pioneer stages of colonization when resources are plentiful and competitive pressures are minimal (Luckinbill, 1979). Barren islands drawing propagules from this type of pool should accumulate species more readily than islands drawing from a more complex source.

Opler *et al.* (1975) have observed a situation analogous to this in the tropical lowland forests of Costa Rica. Recolonization of clear-cut plots drawing propagules from severely perturbed source areas was extremely rapid. This was also attributed to the relatively large numbers of pioneer species within this pool. Plots surrounded by more mature forest, however, exhibited much slower increases in species richness within an equivalent time frame.

(3) Since island community establishment is considered an essentially non-interactive process (Simberloff, 1969), it can be simply

Table 1. Nonlinear regression analysis of model $S = \hat{S}_{eq}(1 - e^{-GT})$ in 6 Smith Mountain Lake stations. Lack of Fit (L.O.F.) F and α level attained are presented ($\alpha[F] < 0.01$ is required for decision level). Estimates of the parameters are given where model was adequate ($t_{99\%}$ is time to reach 99% of $\hat{S}_{eq}$ the equilibrium species number).*

		Station 1		Station 2	Station 3	Station 4	Station 5	Station 6
			w/o 15†					
All data	F	6.97	0.622	1.403	27.34	45.78	4.98	1.078
	$\alpha(F)$	0.0086	>0.50	>0.75	≪0.001	≪0.001	=0.023	>0.75
	$\hat{S}_{eq}$	52.04	55.12	58.72	LOF	LOF	48.26	48.47
	G	1.64	1.42	0.89			0.31	0.40
	$t_{99\%}$	2.81	3.24	5.17	(no equilibrium)		14.85	11.51
Days 1 to 15	F	—	—	—	12.92	86.36	—	—
	$\alpha(F)$	—	—	—	0.004	≪0.001	—	—
	$\hat{S}_{eq}$	—	—	—	—	—	—	—
	G	—	—	—	LOF	LOF	—	—
	$t_{99\%}$	—	—	—	—	—	—	—
Days 1 to 6	F	—	—	—	7.43	11.84	—	—
	$\alpha(F)$	—	—	—	0.037	0.017	—	—
	$\hat{S}_{eq}$	—	—	—	32.29	35.59	—	—
	G	—	—	—	1.99	1.14	—	—
	$t_{99\%}$	—	—	—	2.31	4.04	—	—

* Reprinted with permission from Cairns *et al.* (1979).
† Without Day 15.

viewed as a sampling phenomenon whose dynamics are a function of the source species distribution.

If the dispersal capacities of the organisms are similar which is likely to be the case for passively dispersed protozoa in a benthic system (Kuhn and Plafkin, 1977), then the likelihood of any particular species invading the substrate is directly related to its abundance: the greater a species' abundance, the greater its likelihood of colonizing. If samples are taken at an early time (T1, Fig. 12b), a greater proportion of the total species pool would be sampled from the stressed versus the unstressed distribution. As the sample time increases (T2, Fig. 12c), the likelihood of a substrate receiving new species from low abundance classes is increased. Eventually, as the sample time is expanded further, a point of diminishing return is reached where continued exposure draws in fewer and fewer new species. This point is reached sooner in the stressed situation than in the unstressed where there is a greater proportion of low abundance species (T3, Fig. 12d).

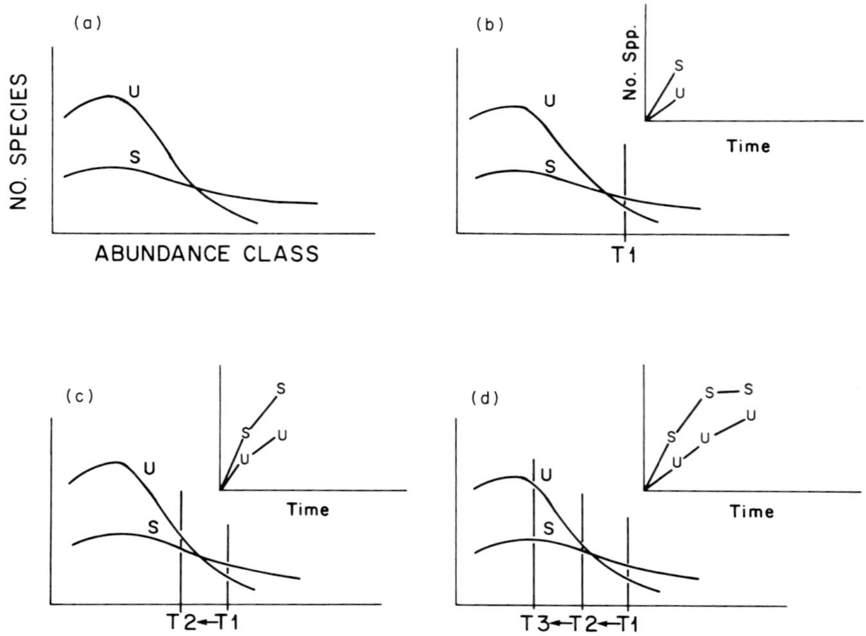

Fig. 12. Distortion of relative abundance structure by stress (from Cairns *et al.*, 1979).

(4) Another direct result of organic pollution is to increase the carrying capacity of the habitat in question for heterotrophic organisms. Fertilization can be visualized as causing a displacement of the mode of the effected species pool distribution to the right (Fig. 13). This displacement effectively increases the number of high abundance species at enriched locations, potentially speeding substrate colonization even more than the distortions of relative abundance structure which have just been considered.

The protozoan colonization of barren habitat appears to be a direct function of the characteristics surrounding natural protozoan community. The process reflects characteristics of both the composition and productivity of the source pool.

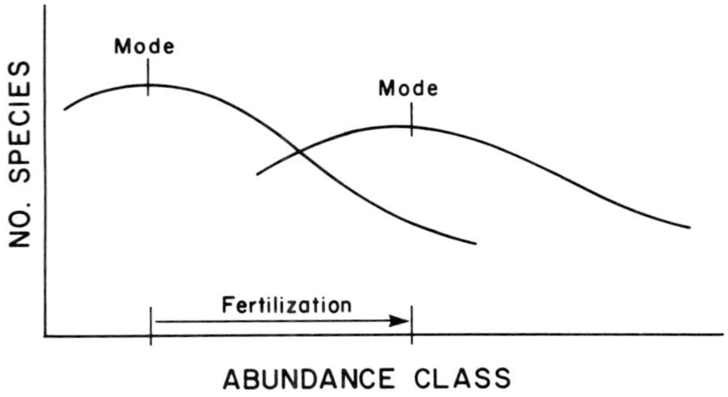

Fig. 13. Fertilization shifts the mode of the relative abundance distribution to the right, reflecting the increased carrying capacity of the enriched habitat. A hypothetical distortion of the curve's shape in response to pollutional stress is also illustrated (axes as in Fig. 9). (From Cairns *et al.*, 1979.)

4.2 Laboratory Bioassys

A variety of laboratory bioassay techniques using protozoans to determine the toxicity of chemical substances has been developed (Bovee, 1975; Bringmann and Kuhn, 1959; Butzel *et al.*, 1960; Gross, 1962; Gross and Jahn, 1962; Bick, 1971; Mitchell, 1972; Maloney and Palmer, 1956). It is worth noting that some Protista organisms may be referred to as either algae or protozoans. Some of the bioassays involved protozoan communities (Cairns and Plafkin, 1971); others involve single species (Yongue *et al.*, 1979); some involve inhibition (Kostyaev, 1973); others depend on direct lethal effects or the disappearance of members of a community (Cairns *et al.*, 1971b). Schultz and Dumont

(1977) exposed the protozoan *Tetrahymena pyriformis* to phenol and found that concentrations of less than 75 mg l^{-1} alter cell motility, shape, and contractile vacuole activity. After 3 min exposure to less than 10 mg phenol l^{-1} reduced the oxygen uptake with a concommitant increase in electron density of the mitochondrial matrics. Alterations in mucocysts, pellicle, and glycogen were also observed. Bringmann and Kuhn (1959) and Epstein *et al.* (1963) also devised tests for the protozoan *Microragma*.

A comparison of the protozoan response to phenol with some other commonly used aquatic test organisms demonstrates that the protozoan response is within the range of the response of other organisms. For example, Alekseev and Antipin (1976) found an LC50 value of 320 μg phenol ml^{-1} for *Physa fontinalis* and Bringmann and Kuhn (1977) report an LC50 for *Daphnia magna* of 31 μg phenol ml^{-1}. Additional evidence of the response relationship may be found in the literature (Buikema *et al.*, 1979) and a general discussion may be found in Hutner (1964) and Hutner *et al.* (1965).

The paucity of toxicity tests on protozoan communities forces estimation of the concentrations that will not impair community integrity. There are a number of drawbacks to this approach.

(1) Interactions between and among species are ignored.

(2) As the level of organization increases (i.e. species, community, ecosystem), properties emerge that were not apparent at lower levels (e.g. energy flow).

(3) Detoxification is more likely in a complex system as is disappearance into an environmental sink (e.g. sediments). Cairns *et al.* (1980) found that a sublethal dose of copper sulphate significantly decreased the colonization rate of uncolonized substrates in association with both mature (at MacArthur-Wilson equilibrium) and immature communities (not at MacArthur-Wilson equilibrium). The effects were more pronounced when immature communities were involved.

4.3 Microcosms

There is a vast chasm between single species laboratory tests in which variables are controlled and the highly variable complex natural ecosystems where they are not. Clearly, a system of intermediate complexity with more control over variables is quite useful for a variety of research endeavours, including the verification of predictive models developed either from single species laboratory or natural system data. Some of these are "species defined" gnotobiotic systems (e.g. Taub,

1969; Crow and Taub, 1979; Taub and Crow, 1981; Taub *et al.*, 1981), but many are not. Microcosm is defined as "a little world: a miniature universe." It is, in the context of this chapter, a patch of an ecosystem lacking many of its important characteristics but, if one has chosen the patch carefully, it exhibits the characteristics one wishes to study reasonably well (see Parkes, pp. 45–102). The recent literature on methods in this field has been summarized by Cooke (1977) and Salt (1971). Although investigations with microcosms are not yet widely established, it appears to be a very promising approach.

5. THE FUTURE OF RESEARCH ON PROTOZOAN COMMUNITIES

Although there is certainly no evidence of widespread interest in protozoan community structure, there are good reasons for believing that such interest will develop in the near future. A broad interest presently exists among biological scientists in the ways in which communities function and are structured. This interest seems to be growing quickly and some fundamental questions about community organization now seem to be resolvable given computer technology as an assistance in handling the masses of data necessary. The decline in federal and state support of university and private institute research, coupled with increased difficulties both in transportation and political unrest of getting to remote locations from the home base, make it likely that studies of the larger vertebrates and vascular plant communities may be partly replaced by studies of microbial communities where the hypothesis being tested is ammenable to such use. There is no intent to imply that hypotheses involving microbial communities for their own sake are not desirable, but rather that microbial community research may be a suitable surrogate for research on communities of larger organisms for which funding may not be as readily available as it has been in the recent past. Even if research with the larger systems is contemplated, it might well be advantageous to carry out some screening studies with smaller systems to define more precisely the parameters to be measured and the questions to be asked.

Three very important and essentially unanswered questions regarding the response of communities to stress illustrate this point.

(1) Are the single species toxicity tests useful for predicting responses of entire communities? Does an application factor derived from tests carried out on a single species actually protect an entire community?

(2) Do communities in different areas made up of different aggregations of species respond in a similar way to an identical concentration of a toxicant? In this instance one might have to allow for differences in water quality that affect toxicity.

(3) Do communities of different maturity respond similarly to identical concentrations of a toxicant?

The primary assumption stimulating question (1) is that the commonly used test species, such as the rainbow trout, the white rat, the guinea pig, the bluegill, and the rabbit, will furnish evidence from which extrapolations can be made that will protect all other organisms not tested. They are thought to furnish a reasonable representation of the range of biological response. Differences between the customary test species and those of other species not included in the testing procedure are thought to be adequately predicted through the intelligent and knowledgeable use of existing information. Questions (2) and (3) are fairly straightforward.

Protozoan communities have been useful in:

(1) Testing hypotheses in theoretical ecology.
(2) Biological monitoring of pollutional effects.
(3) Toxicity testing.
(4) Evaluating water quality.

It is curious that they are not used more frequently considering the many advantages mentioned in the text. There are only two major drawbacks:

(i) Skill in identification is not easily acquired, and
(ii) Preservation usually distorts the community structure and there is a loss of some taxonomic characters even for the more durable species.

These seem comparatively minor when one considers the many benefits.

REFERENCES

Alekseev, V. A. and Antipin, B. N. (1976). Toxicological characteristics and symptom complex of the acute phenol poisoning of certain freshwater crustaceans and mollusks, *Gidrobiologicheskii Zhurnal* **12**, 37–44 (Russ.); *Chemical Abstracts* **85**, 117–120.

Bamforth, S. S. (1958). Ecological studies of the planktonic Protozoa of a small artificial pond. *Limnology and Oceanography* **3**, 398–412.

Barker, A. N. (1946). The ecology and function of Protozoa in sewage purification. *Annals of Applied Biology* **33**, 314–325.

Bick, H. (1958). Okologische Untersuchungen an Ciliaten fallaubreicher Kleingewasser. *Archiv fuer Hydrobiologie* **54**, 506–542.
Bick, H. (1971). The potentialities of ciliated Protozoa in the biological assessment of water pollution levels. *In* "Proceedings of the International Symposium on Identification and Measurement of Environmental Pollutants," pp. 305–309, Ottawa, Canada.
Blanchard, D. C. and Parker, B. C. (1977). The freshwater to air transfer of microorganisms and organic matter. *In* "Aquatic Microbial Communities" (J. Cairns, Jr, ed.), pp. 625–658. Garland Publishing Inc., New York.
Borkott, H. (1975). A method for quantitative isolation and preparation of particle-free suspensions of bacteriophagous ciliates from different substrates for electronic counting. *Archiv fuer Protistenkunde* **117**, 261–268.
Bovee, E. C. (1975). "Effects of Certain Chemical Pollutants on Aquatic Animals," Office of Water Research and Technology, PB-241, 336 pp.
Bringmann, G. and Kuhn, R. (1959). Water toxicology studies with protozoans as test organisms. *Gesundheitswesen-Ingenieur* **80**, 239–242.
Bringmann, G. and Kuhn, R. (1977). Befund der Schadwirkung wasser gefahrdender Stoffe gegen *Daphnia magna* (Damaging and action of water pollutants to *Daphnia magna*). *Zeitschrift fur Wasser und Abwasser Forschung* **19**, 161–166.
Brooks, J. L. and Dodson, S. T. (1965). Predation, body size and composition of plankton. *Science* **150**(3692), 28–35.
Buikema, A. L., McGinniss, M. J., and Cairns, J., Jr. (1979). Phenolics in aquatic ecosystems: A selected review of recent literature. *Marine Environmental Research* **2**(2), 87–181.
Butcher, R. W. (1946). The biological detection of pollution. *Journal and Proceedings of the Institute of Sewage Purification* **2**, 92–97.
Butzel, H. H., Brown, L. H., and Martin, W. B., Jr. (1960). Effects of detergents upon electromigration of *Paramecium aurelia. Physiological Zoology* **33**, 39.
Cairns, J., Jr. (1965). The environmental requirements of freshwater Protozoa. *In* "Biological Problems in Water Pollution, Third Seminar, 1962", pp. 48–52. U.S. Public Health Service Publ. No. 999-WP-25.
Cairns, J., Jr. (1971). "Factors Affecting the Number of Species in Freshwater Protozoa Communities." Research Division Monograph 3, Virginia Polytechnic Institute and State University, Blacksburg, Va.
Cairns, J., Jr. (1974). Protozoans (Protozoa). *In* "Pollution Ecology of Freshwater invertebrates" (C. W. Hart, Jr. and S. L. Fuller, eds) pp. 1–28. Academic Press, New York and London.
Cairns, J., Jr. (1977). "Aquatic Microbial Communities". Garland Publishing Inc., New York.
Cairns, J., Jr. (1979). A strategy for use of protozoans in the evaluation of hazardous substances. *In* "Biological Indicators of Water Quality", (S. James, ed.) pp. 6-1 to 6-17. University of Newcastle upon Tyne, UK.
Cairns, J., Jr., Beamer, T., Churchill, S., and Ruthven, J. A. (1971b). Response of protozoans to detergent-enzymes. *Hydrobiologia* **38**(2), 193–205.

Cairns, J., Jr., Dahlberg, M. L., Dickson, K. L., Smith, N., and Waller, W. T. (1969). The relationship of freshwater protozoan communities to the MacArthur-Wilson equilibrium model. *American Naturalist* **103**(933), 439–454.

Cairns, J., Jr., Dickson, K. L., and Yongue, W. H., Jr. (1971a). The consequences of nonselective periodic removal of portions of freshwater protozoan communities. *Transactions of the American Microscopical Society* **90**(1), 71–80.

Cairns, J., Jr., Hart, K. M., and Henebry, M. S. (1980). The effects of a sublethal dose of copper sulfate on the colonization rate of freshwater protozoan communities. *American Midland Naturalist* **104**, 93–101.

Cairns, J., Jr., Kaesler, R. L., Kuhn, D. L., Plafkin, J. L., and Yongue, W. H., Jr. (1976b). The influence of natural perturbation upon protozoan communities inhabiting artificial substrates. *Transactions of the American Microscopical Society* **95**(4), 646–653.

Cairns, J., Jr., Kuhn, D. L., and Plafkin, J. L. (1979). Protozoan colonization of artificial substrates. *In* "Methods and Measurements of Attached Microcommunities: A Review", (R. L. Weitzel, ed.) pp. 34–57. American Society for Testing and Materials, Philadelphia.

Cairns, J., Jr. and Plafkin, J. L. (1971). Response of protozoan communities exposed to chlorine stress. *Archiv fuer Protistenkunde* **117**, 47–53.

Cairns, J., Jr. and Ruthven, J. A. (1970). Artificial microhabitat size and the number of colonizing protozoan species. *Transactions of the American Microscopical Society* **89**(1), 100–109.

Cairns, J., Jr. and Yongue, W. H., Jr. (1977). Factors affecting the number of species in freshwater protozoan communities. *In* "Aquatic Microbial Communities" (J. Cairns, Jr., ed.) pp. 257–303. Garland Publishing Inc., New York.

Cairns, J., Jr., Yongue, W. H., Jr., and Kaesler, R. L. (1976a). Qualitative differences in protozoan colonization of artificial substrates. *Hydrobiologia* **51**(3), 233–237.

Cole, G. A. (1955). An ecological study of the microbenthic fauna of two Minnesota lakes. *American Midland Naturalist* **53**(1), 213–230.

Cooke, G. D. (1967). The pattern of autotrophic succession in laboratory microcosms. *BioScience* **17**, 717–721.

Cooke, G. D. (1977). Experimental aquatic laboratory ecosystems and communities. *In* "Aquatic Microbial Communities" (J. Cairns, Jr., ed.) pp. 59–103. Garland Publishing Inc., New York.

Cooke, W. B. (1956). Colonization of artificial bare areas by microorganisms. *Botanical Review* **22**, 613–638.

Crow, M. E. and Taub, F. B. (1979). Designing a microcosm bioassay to detect ecosystem level effects. *International Journal of Environmental Studies* **13**, 141–147.

Curds, C. R. (1966). An ecological study of the ciliated Protozoa in activated sludge. *Oikos* **15**, 282–289.

Curds, C. R. (1973). The role of Protozoa in the activated sludge process. *American Zoologist* **13**, 161–169.

Davis, M. B. and Brubaker, L. B. (1973). Differential sedimentation of pollen grains in lakes. *Limnology and Oceanography* **18**, 635–646.
Eddy, S. (1928). Succession of Protozoa in cultures under controlled conditions. *Transactions of the American Microscopical Society* **147**, 283–319.
Epstein, S. S., Small, M., Koplan, J., Jones, H., Mantel, N., and Hutner, S. H. (1963). A photodynamic bioassay of benzo(a)pyrene using *Paramecium caudatum*. *Journal of the National Cancer Institute* **31**, 163–168.
Evans, F. R. (1958). Competition for food between two carnivorous ciliates. *Transactions of the American Microscopical Society* **77**, 390–395.
Fenchel, T. (1977). The significance of bactivorous Protozoa in the microbial community of detrital particles. *In* "Aquatic Microbial Communities" (J. Cairns, Jr., ed.) pp. 529–544. Garland Publishing Inc., New York.
Finlay, B. J. (1978). Community production and respiration by ciliated Protozoa in the benthos of a small eutrophic lock. *Freshwater Biology* **8**, 327–341.
Finlay, B. J., Laybourn, J., and Strachan, I. (1979). A technique for the enumeration of benthic ciliated Protozoa. *Oecologia* **39**, 375–377.
Ganf, G. G. (1974). Community respiration of equatorial plankton. *Oecologia* **15**, 17–32.
Gilroy, D. (1975). The determination of the rate constants of island colonization. *Ecology* **56**(4), 915–923.
Gislen, T. (1948). Aerial plankton and its conditions of life. *Biological Review* **23**, 109–126.
Goulder, R. (1974). The seasonal and spatial distribution of some benthic ciliated Protozoa in Esthwaite Water. *Freshwater Biology* **4**, 127–147.
Gross, J. A. (1962). Cellular responses to thermal and photo stress. II. Chlorotic Euglenoides and Tetrahymena. *Journal of Protozoology* **9**(4), 415–418.
Gross, J. A. and Jahn, T. L. (1962). Cellular responses to thermal and photo stress. I. Euglena and Chlamydomonas. *Journal of Protozoology* **9**, 340–346.
Grzenda, A. R. and Brehmer, M. L. (1960). A quantitative method for the collection and measurement of stream periphyton. *Limnology and Oceanography* **5**(2), 190–194.
Hairsten, N. H. G., Allan, J. D., Colwell, R. K., Futuyama, D. J., Howell, J., Lubin, M. D., Mathias, J., and Vandermeer, J. H. (1968). The relationship between species diversity and stability: an experimental approach with Protozoa and bacteria. *Ecology* **49**, 1091–1101.
Hardin, G. (1969). Not peace, but ecology. *In* "Diversity and Stability in Ecological Systems", (G. M. Woodwell and H. H. Smith, eds) pp. 151–158. Brookhaven National Laboratory, Springfield, Virginia.
Hausman, L. A. (1917). Observations on the ecology of the Protozoa. *American Naturalist* **51**, 157–172.
Henebry, M. S. and Cairns, J., Jr. (1980a). The effect of island size, distance, and epicenter maturity on the colonization process in freshwater protozoan communities. *American Midland Naturalist* **104**, 80–92.
Henebry, M. S. and Cairns, J., Jr. (1980b). The effect of source pool maturity on the process of island colonization: An experimental approach with protozoan communities. *Oikos* **35**, 107–114.

Hrbacek, J. (1977). Competition and predation in relation to species composition of freshwater zooplankton, mainly *Cladocera*. *In* "Aquatic Microbial Communities". (J. Cairns, Jr., ed.) pp. 305–353. Garland Publishing Inc., New York.

Hutner, S. H. (1964). Protozoa as toxicological tools. *Journal of Protozoology* **11**(1), 1–6.

Hutner, S. H., Baker, H., Aaronsen, S., and Zahalsky, A. C. (1965). Bacteria-Protozoa as toxicological indicators in purifying waters. *In* "Transactions of the Third Seminar on Biological Problems in Water Pollution", (C. Tarzwell, ed.). PHS Publ. No. 999-WP-25.

Jassby, A. D. and Goldman, C. R. (1974). A quantitative measure of succession rate and its application to the phytoplankton of lakes. *American Naturalist* **108**, 688–693.

Kahl, A. (1930). Wimperitier oder Ciliata (Infusionia). *In* "Die Tierwelt Deutschlands, Teil 18", (F. Dahl, ed.). I. G. Fischer, Jena.

Kitching, J. A. (1957). Some factors in the life of free-living Protozoa. *In* "Microbial Ecology", (R. E. O. Williams and C. C. Spicer, eds). Society of General Microbiology, London.

Kostyaev, V. Ya. (1973). Effect of phenol on algae. *Trudy Institute Biologii* Vnutrennikh Vod Akademii Nauk SSR **24**(27), 93–113.

Kuhn, D. L. and Plafkin, J. L. (1977). The influence of organic pollution on the dynamics of artificial island colonization by Protozoa. *Bulletin of the Ecological Society of America* **58**, 14.

Lackey, J. B. (1925). The fauna of Imhoff tanks. *Bulletin of the New Jersey Experimental Station* **417**, 1–39.

Lackey, J. B. (1944). Stream microbiology. *In* "Stream Sanitation", (E. B. Phelps, ed.) pp. 227–263. John Wiley & Sons, New York.

Luckinbill, L. S. (1979). Selection in /K continuum in experimental populations of Protozoa. *American Naturalist* **113**(3), 427–437.

MacArthur, R. and Wilson, E. O. (1963). An equilibrium theory of insular zoogeography. *Evolution* **17**, 373–387.

Maguire, B., Jr. (1963). The passive dispersal of small aquatic organisms and their colonization of insolated bodies of water. *Ecological Monographs* **33**, 161–185.

Maguire, B., Jr. (1977). Community structure of protozoan and algae with particular emphasis on recently colonized bodies of water. *In* "Aquatic Microbial Communities", (J. Cairns, Jr., ed.) pp. 355–397. Garland Publishing Inc., New York.

Maloney, T. E. and Palmer, C. M. (1956). Toxicity of six chemical compounds to 30 cultures of algae. *Water and Sewage Works* **103**, 509–513.

McKinney, R. E. and Gram, A. (1967). Protozoa and activated sludge. *In* "Biology of Water Pollution: A Collection of Selected Papers on Stream Pollution Waste Water and Waste Treatment", (L. E. Keup, W. M. Ingram, K. M. Mackenthun, eds) pp. 252–262. U.S. Department of the Interior, Federal Water Pollution Control Administration, Washington, D.C.

Miller, W. F. (1977). The development of the WRC water quality monitor using fish. *In* "Proceedings of the Practical Aspects of Water Quality Monitoring Systems". Water Research Centre, Stevenage Laboratory, UK.

Milliger, L. E., Stewart, K. W., and Silvey, J. K. G. (1971). The passive dispersal of viable algae, protozoans, and fungi by aquatic and terrestrial Coleoptera. *Annals of the Entomological Society of America* **64**, 36–45.

Mitchell, R., ed. (1972). "Water Pollution Microbiology". John Wiley & Sons, New York.

Moore, G. M. (1939). A limnological investigation of the microscopic benthic fauna of Douglas Lake, Michigan. *Ecological Monographs* **9**, 537–582.

Noland, J. L. and Gojdics, M. (1967). Ecology of free-living Protozoa. *In* "Research in Protozoology", (T. Chen, ed.) pp. 215–266.

Opler, R. A., Baker, H. G., and Frankie, G. W. (1975). Recovery of tropical lowland forest ecosystems. *In* "Recovery and Restoration of Damaged Ecosystems", (J. Cairns, Jr., ed.) pp. 372–421. University Press of Virginia, Charlottesville.

Parker, B. and Barson, G. (1970). Biological and chemical significance of surface microlayers in aquatic ecosystems. *Bioscience* **20**(1), 87–93.

Pascher, A. (1913–1927). Flagellates. *In* "Die susswasser-flora Deutschlands, Ostereichs, und der Schweiz". J. G. Fischer, Jena.

Patrick, R. (1949). A proposed measure of stream conditions, based on a survey of the Conestoga Basin, Lancaster County, Pennsylvania. *Proceedings of the Academy of Natural Science of Philadelphia* **101**, 277–341.

Patrick, R. (1977). Diatom communities. *In* "Aquatic Microbial Communities", (J. Cairns, Jr., ed.) pp. 139–160. Garland Publishing Inc., New York.

Patrick, R., Cairns, J., Jr., and Roback, S. S. (1967). An ecosystematic study of the flora and fauna of the Savannah River. *Proceedings of the Academy of Natural Sciences of Philadelphia* **118**(5), 109–407.

Patrick, R., Cairns, J., Jr., and Scheier, A. (1968). The relative sensitivity of diatoms, snails, and fish to twenty common constituents of industrial wastes. *Progressive Fish-Culturist* **30**(3), 137–140.

Patrick, R., Hohn, M. H., and Wallace, J. H. (1954). A new method for determining the pattern of the diatom flora. *Notulae Naturae* (Philadelphia) **259**, 1–12.

Picken, L. E. R. (1937). The structure of some protozoan communities. *Journal of Ecology* **25**(2), 324–368.

Plafkin, J. L., Kuhn, D. L., Cairns, J., Jr., and Yongue, W. H., Jr. (1980). Protozoan species accrual on artificial islands in differing lentic and wetlands systems. *Hydrobiologia* **75**, 161–178.

Preston, F. W. (1948). The commonness, and rarity, of species. *Ecology* **29**, 254–283.

Preston, F. W. (1962). The cannonical distribution of commonness and rarity. *Ecology*, part I, 43, 185–215; part II, 43, 410–432.

Roanoke River Basin Comprehensive Water Resources Plan. (1975). "Plan-

ning Bulletin 247A, Vol. V-A(1)". Commonwealth of Virginia State Water Control Board, Bureau of Water Control Management.
Ruggiu, D. (1969). Benthic ciliates in the profundal of Lake Orta (Northern Italy). *Verhandlungen Internationale Vereinigung fur Theoretische und Angewandte Limnologie* **17**, 255–258.
Salt, G. W. (1971). The role of laboratory experimentation in ecological research. *In* "The Structural and Function of Fresh-Water Microbial Communities", (J. Cairns, Jr., ed.) pp. 87–100. Research Division Monograph No. 3, Virginia Polytechnic Institute and State University, Blacksburg, Virginia.
Schlichting, H. E., Jr. (1961). Viable species of algae and Protozoa in the atmosphere. *Lloydia* **24**(2), 81–88.
Schlichting, H. E., Jr. (1964). Meteorological conditions affecting the dispersal of airborne algae and Protozoa. *Lloydia* **17**, 64–78.
Schlichting, H. E., Jr. (1969). The importance of airborne algae and Protozoa. *Journal of the Air Pollution Control Association* **19**, 946–951.
Shultz, T. W. and Dumont, J. N. (1977). Cytotoxicity of synthetic fuel product on *Tetrahymena pyriformis*. I. Phenol. *Journal of Protozoology* **24**(1), 164–172.
Simberloff, D. S. (1969). Experimental zoogeography of islands. A model for insular colonization. *Ecology* **50**, 296–314.
Simberloff, D. S. (1974). Equilibrium theory of island biogeography and ecology. *Annual Review of Ecology and Systematics* **5**, 161–182.
Sleigh, M. (1973). "The Biology of Protozoa". Arnold, London.
Spoon, D. M. and Burbanck, W. D. (1967). A new method for collecting sessile ciliates in plastic petri dishes with tight-fitting lids. *Journal of Protozoology* **14**, 735–744.
Sramek-Husek, R. (1958). Die Rolle der Ciliatennanalyse bei der biologischen Kentrolle von Flussverunreinigungen. *Verhandlungen Internationale Vereinigung Limnologie* **13**, 626–645.
Stout, J. D. (1956). Reaction of ciliates to environmental factor. *Ecology* **37**, 178–191.
Stout, J. D. (1974). Protozoa. *In* "Biology of Plant Litter Decomposition, Vol. 12", (C. H. Dickinson and G. J. F. Pugh, eds) pp. 385–420. Academic Press, New York and London.
Taub, F. B. (1969). A biological model of a freshwater community: a gnotobiotic ecosystem. *Limnology and Oceanography* **14**(1), 136–142.
Taub, F. B. and Crow, M. E. (1981). Synthesizing aquatic microcosms. *In* "Microcosms in Ecological Research". Savannah River Laboratory, Georgia.
Taub, F. B., Crow, M. E., and Hartman, H. J. (1981). Response of aquatic microcosms to acute mortality. *In* "Microcosms in Ecological Research". Savannah River Laboratory, Georgia.
Unger, W. B. (1931). The protozoan sequence in five plant infusions. *Transactions of the American Microscopical Society* **50**, 144–153.
Wang, C. C. (1928). Ecological studies of the seasonal distribution of Protozoa in a freshwater pond. *Journal of Morphology* **46**(2), 431–478.

Webb, M. C. (1961). The effects of thermal stratification on the distribution of benthic Protozoa in Esthwaite. *Water Journal of Animal Ecology* **30**, 137–152.
Whittaker, R. H. (1975). "Communities and Ecosystems 2nd ed." Macmillan Publishing Co., Inc., New York.
Wilson, E. O. and Simberloff, D. S. (1968). Experimental zoogeography of islands. Defaunation and monitoring techniques. *Ecology* **50**, 267–277.
Woodruff, L. L. (1912). The origin and sequence of the protozoan fauna of hay infusions. *Journal of Experimental Zoology* **12**, 205–264.
Yongue, W. H., Jr., Berrent, B. L., and Cairns, J., Jr. (1979). Survival of *Euglena gracilis* to sublethal temperature and hexavalent. *Journal of Protozoology* **26**(1), 122–125.
Yongue, W. H., Jr. and Cairns, J., Jr. (1971). Colonization and succession of freshwater protozoans in polyurethane foam suspended in a small pond in North Carolina. *Notulae Naturae (Philadelphia)* **433**, 1–13.
Yongue, W. H., Jr. and Cairns, J., Jr. (1978). The role of flagellates in pioneer colonization of artificial substrates. *Polskie Archiwum Hydrobiologii* **25**(4), 787–801.

8

Genetic Interactions among Microbial Communities

D. C. Reanney, W. P. Roberts, and W. J. Kelly

1. INTRODUCTION

The pioneering studies of Koch and others which established bacteriology as a *bona fide* science owed much of their success to the use of pure cultures of microorganisms. In so doing a paradox was created which persists to this day. In order to bring reproducibility into experimental results it is necessary to insulate test organisms from the very sources of variability that sustain natural evolution. Thus, the analytical approach which has made the study of microbiology so fruitful must, in a sense, be regarded as the enemy of an integrated appreciation of microbial ecology.

These abstract considerations can be given substance by considering an issue that lies at the core of the present article, namely the presence of extrachromosomal DNA in bacteria. For example strains of *Bacillus* sp., freshly isolated from the environment, have been reported to contain multiple bands of covalently closed, circular DNA (Bernhard *et al.*, 1978; González and Carlton, 1980; A. J. Radford and D. C. Reanney, unpublished observations). However, in an extensive study of 38 *Bacillus* sp. cultures held by the American Type Culture Collection, Lovett and Bramucci (1975) could detect extrachromosomal DNA in only five cases. This suggested, but did not prove, that repeated passage on nutrient media selected for organisms which had lost genetic features characteristic of the wild-type organisms.

Similar problems arise from the use of artificially constructed strains of bacteria in the study of bacterial genetics. Most of the mutants which have been indispensable tools of the geneticist are selectively disadvantaged with respect to their wild-type counterparts. Even when deliberately imposed mutation and selection are absent, an organism may adapt irreversibly to the unnatural environment provided by the laboratory. A well-known example is provided by *Escherichia coli*, the K12 subline of which has been passaged so often under laboratory conditions that it can no longer colonize its original habitat, the mammalian gastro-intestinal tract (Gorbach, 1978).

This review considers the genetic interactions which enable communities of microorganisms to respond adaptively to change. Figure 1 forms our starting point by setting out in the form of a genetic "circuit diagram" some of the known pathways of DNA exchange among microorganisms. Each link in the circuit is the result of a single pairwise interaction carried out under laboratory conditions, and whether a common gene could diffuse among all the links in this network in the natural environments concerned, remains to be demonstrated.

2. MECHANISMS OF GENETIC EXCHANGE

There is now very strong evidence that most strains of bacteria in nature contain, in addition to the chromosome, one or more species of covalently closed, circular DNA molecules. Many of these plasmids are relatively small (size range: 1 to 15 Md) and often occur in multiple copies (copy numbers for plasmids vary from 1 to about 50).

Plasmids which are unable *per se* to move from one cell to another are called non-conjugative plasmids (Novick *et al.*, 1976). The cell-to-cell movement of bacterial genes is made possible by the existence of various vector systems. One of the best studied mechanisms is a form of mating mediated by conjugative plasmids (Novick *et al.*, 1976). Conjugative plasmids can also mobilize non-conjugative plasmids in either of two ways. Firstly, the larger replicon may recombine physically with

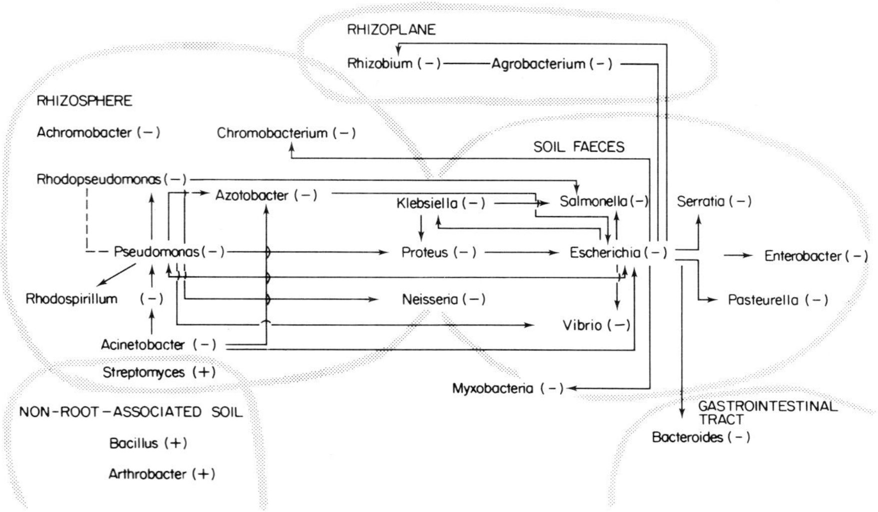

Fig. 1. Known pathways of intergeneric genetic exchange among bacteria in several ecosystems. The *rhizoplane* is the surface of plant roots and the *rhizosphere* the zone surrounding plant. Enteric organisms have been grouped under the heading "soil faeces" since the huge populations of bacteria present both in the faeces and in rhizosphere soil have many opportunities to exchange genes when faecal matter is reintroduced into the soil environment. Most of the DNA transfers shown have been carried out under laboratory conditions. There is no certainty that the same transfers occur in nature. The transfers listed were all mediated by conjugation and transduction. No attempt has been made to include transfers made possible by transformation. (−) and (+) refer to the reaction of the taxa shown to the Gram stain. Details of these transfers and a tabulation of primary source material are given in Reanney (1977). (From Reanney, 1977.)

the smaller to give a cointegrate structure whose molecular weight is often the sum of the two contributing units. In this way the small plasmid can exploit the transfer functions of the larger one. Secondly, small plasmids can move across the pilus or bridge which connects two mating cells without necessarily recombining with the conjugative plasmid (H. W. Smith, 1977). Conjugation is discussed in detail by Rowbury (1977), Clark and Warren (1979), and Broda (1979).

A second vector system is provided by temperate bacteriophages. Transducing phages can incorporate into their capsids genes derived from any DNA species resident in the infected cell. Transfer in this case is purely passive with the viral head inadvertently acting as a transfer capsule for non-viral genes. The topic of transduction has been extensively reviewed (Ozeki and Ikeda, 1968; Barksdale and Arden, 1974; Low and Porter, 1978).

A third system is transformation, i.e. the uptake of homospecific and heterospecific naked DNA by competent cells, and has been reviewed by Notani and Setlow (1974), and Dubnau (1976).

In addition to these major, well-documented systems of DNA transfer, several examples exist of poorly understood processes which may represent either novel phenomena or known mechanisms acting in concert. Thus *Streptococcus pneumoniae* exhibits a process of phage-associated gene transfer known as pseudotransduction (Porter *et al.*, 1979). Phage PG24 packages host genes in particles that resemble normal phages. However, the transfer of these genes into recipient cells requires the development of competence. A different variation is seen in *Staphylococcus aureus*. In mixed cultures transfer is dependent not only on the lysogenic donor but also on cell-to-cell contact (Lacey, 1975b) and virtually no transducing particles have been detected in supernatants from donors.

Any cell which has accepted heterologous DNA is said to be genetically recombinant and the fate of the introduced DNA depends on its character. Plasmids which successfully establish themselves in a new cell as a result of a transfer event retain in the recipient cell the property of autonomous replication which they possessed in the donor. This means that these genetic units may be stably maintained throughout subsequent generations without physically recombining with any resident species of host DNA (Table 1). By contrast other DNA species or fragments must normally enter into a physical association with resident replicons to become a permanent component of the hereditary apparatus of the receiving cell. Two types of physical recombination are distinguished. Integrative events which require extensive, pre-existing nucleotide homology between the incoming and the established DNA molecules and which require a functional *rec* system, take place through a process of generalized recombination. Those which require little or no homology and which can occur in the absence of a functional *rec* system occur through a process of quantal recombination. The replacement of a mutant gene, e.g. gal^- by its wild-type allele gal^+, is an example of generalized recombination. The integration of phage λ into the chromosome of *Escherichia coli* K12 at a definite, pre-programmed point, is a classic example of quantal recombination. By means of this site-specific interaction two replicons of differing size and sequence can be united into one molecule. Because of its ability to generate radically new genetic structures through single crossover events quantal recombination is now believed to be the prime accelerator of evolutionary change in bacteria (Cohen, 1976; Reanney, 1976, 1977; Kleckner, 1977; Fox, 1978; Davey and Reanney, 1980).

Table 1. Mechanisms that result in genetic interchange in bacteria

Mechanism	Examples	References
Intracellular		
Transposition of defined DNA sections	Insertion sequences and transposons	Many
Plasmid–plasmid interactions	Formation of cointegrate plasmids with molecular weights equal to that of the recombining molecules	Nugent and Hedges (1979)
Plasmid–phage interactions	Formation of hybrids with phage P1 in *Escherichia coli*	Iida and Arber (1977) Mise and Nakaya (1977)
	Formation of hybrids with phage λ in *E. coli*	Dempsey *et al.* (1978)
	Formation of hybrids with phage P22 in *Salmonella typhimurium*	Mise and Nakaya (1977)
	Formation of hybrids with phage S1 in *Staphylococcus aureus*	Inoue *et al.* (1976)
Plasmid–chromosome interactions	Integration of the F plasmid in *E. coli*	Davidson *et al.* (1975)
	Formation of F^1 plasmids in *E. coli*	Low (1972)
	Formation of R^1 plasmids in *Pseudomonas aeruginosa*	Holloway (1978)
Phage–chromosome interactions	Integration of temperate phages, e.g. phage λ in *E. coli*	Gottesman and Weisberg (1971)
Phage–phage interactions	Recombination between phage λ and phage P22 in *E. coli*	Botstein and Herskowitz (1974)
Intercellular—Recombination dependent		
Transformation by chromosomal DNA		
Transduction of chromosomal DNA		
Integration of temperate phages, such as λ		
Conjugal transfer of chromosomal DNA		
Intercellular—Recombination independent		
Transformation by plasmid DNA		
Transduction of plasmid DNA		
Infection by temperate phages, such as P1		
Conjugal transfer of plasmid DNA		

It is difficult to give an impression of the multiplicity of interacting mechanisms which endow superficially simple cells, such as bacteria, with their formidable adaptive flexibility. Some conjugative plasmids can, in a single conjugal event, effect the transfer of their genes between members of such widely separated families as the Enterobacteriaceae and the *Pseudomonadaceae*. In other cases, transfer may occur in a stepping stone fashion. For example, the plasmid pMG1, normally found in *Pseudomonas aeruginosa* cannot transfer into *E. coli*. However, if *P. aeruginosa* cells containing pMG1 are incubated in the presence of *Aeromonas salmonicida* containing the *E. coli* plasmid R6K and recombinant cells are then incubated with *E. coli*, genes can pass from *P. aeruginosa* into *E. coli* by way of the intermediate organism (Olsen and Wright, 1976). Sequential transfers of this type may occur frequently in nature.

A hypothetical scenario may suffice to give some insight into the mechanism(s) which allow rapid rearrangement and dissemination of DNA sequences among microbial populations. Consider a given gene *X* on a non-conjugative plasmid in species of soil *Enterobacter*. Let us assume that this gene, flanked by inverted repeat sequences, constitutes an authentic transposon (Kleckner, 1977; Cohen and Shapiro, 1980) able to replicate *in situ* and translocate from one genetic locus to others. A conjugative plasmid now enters the *Enterobacter* cell from a different species in the same ecosystem, say, *Klebsiella* sp. Translocation of *X* from the resident non-conjugative plasmid to the incoming conjugative plasmid will occur under suitable circumstances. As a result gene *X* acquires the expanded host range and cell-to-cell communicability of the vector replicon. During successive transfers into other ecologically grouped species gene *X* may transpose into:

(1) other conjugative plasmids and hence acquire other transfer capacities and efficiencies;
(2) non-conjugative plasmids in which case gene *X* may tend to become a quasi-stable part of the genetic heritage of a variety of cells;
(3) the chromosomes of receiving organisms in which case the identity of the cell may eventually be altered in a more permanent way.

The common theme in these genetic processes is transferability. Indeed this coupling of inter-molecular transferability (made possible by various DNA to DNA interactions) and inter-cell transferability (made possible by effective vector systems) lies at the core of the remarkable adaptive plasticity of bacterial ecosystems. It seems possible that, under appropriate selection, a given nucleotide sequence can move between members of any two taxonomic groups, no matter how widely separated the terminal donor and receptor cells may be in terms of genetical or physiological properties (Reanney, 1977).

One point should be stressed. Doolittle and Sapienza (1980) and Orgel and Crick (1980) have suggested that defined tracts of DNA which possess the ability to replicate *in situ* and spread among DNA species should be thought of as selfish DNA, i.e. they need have no function other than their own survival within the genome. Many conjugative plasmids are equally selfish since, by spreading their survival options across many taxonomic baskets, such plasmids ensure that their genes will survive even if whole species of sensitive hosts become extinct. This new perspective does not invalidate conventional theories which assign to plasmids and phages a central role in the adaptive genetics of the bacterial cell. What such ideas do is to serve as a necessary reminder that any evolutionary advantage conferred on a bacterium by the acquisition of plasmids may be an accidental by-product of the survival strategy of the plasmid's own DNA.

3. POTENTIAL BARRIERS TO PROCESSES OF GENE FLOW AMONG MEMBERS OF MICROBIAL COMMUNITIES

3.1 Genetic Barriers

While the establishment and survival of heterologous DNA in an acceptor cell depends on a complex interplay of many factors, some of the barriers which impede the colonization of organism *B* by genes from organism *A* are well documented. It has been observed that introduced DNA is often degraded in particular strains of bacteria. Such cells evidently possess a mechanism which effectively distinguishes between self DNA and alien DNA. The molecular basis of this distinction has been clarified by the work of Arber and others (Arber and Linn, 1969; Boyer, 1971) and forms the basis of the phenomenon of restriction and modification (res/mod). Organisms containing a functional restriction system manufacture enzymes which selectively recognize short target sequences in duplex DNA and produce cleavages in both strands. These sequences are usually palindromes. Self DNA is protected from the degradative action of these nucleases by the enzymatic addition of methyl groups within the sensitive sequences which so modify the target sequence that the nuclease no longer cleaves. Any introduced DNA which contains this target sequence but lacks the earmarking methyl group is automatically at risk. Res/mod systems are found on bacterial chromosomes, plasmids, and phages (Roberts, 1976). Each system operates independently, i.e., the res/mod system encoded by the chromosome of a given organism normally exhibits a specificity different from that (those) encoded by resident plasmids.

Res/mod systems appear to constitute molecular guardians of the genetic individuality of all replicons which are stable residents of a common organism although it also seems likely that one class of restriction nucleases participates in "natural genetic engineering" (Reanney, 1976; Chang and Cohen, 1978). One of the more surprising aspects of plasmid biology is the extent to which DNA can move among heterologous organisms, even when antagonistic res/mod systems are present. It is possible that propagation of some plasmids occurs so fast that at least a few copies of the incoming replicon acquire the protective label before they are cleaved by pre-existing nucleases. Once a single plasmid has been suitably modified, epidemic spread to other sensitive cells in the population may readily occur.

The res/mod system of a cell is directed against foreign DNA in general irrespective of the nature or origin of the DNA. If one considers only plasmid DNA a more selective barrier can be seen to exist. Infection of a plasmid-containing organism by an isogenic or related plasmid usually results in the displacement of one or the other unit. This phenomenon, often referred to as plasmid incompatibility (Datta, 1979), is analogous in some respects to the superinfection immunity which prevents related phages from establishing common lysogeny in a single cell. The molecular basis of plasmid incompatibility is still debated (Novick and Hoppenstaedt, 1978). Plasmids of the Enterobacteriaceae are currently classified into 17 incompatibility (Inc) groups (Datta, 1979) and members of a common Inc group may carry different genes and still be unable to coexist jointly in a given cell.

Linked to the issue of the establishment of a heterologous DNA species in a new cell is the problem of the adequate expression of the genes it encodes. Many R plasmids inhibit the transfer functions of the classical sex factor F. A series of elegant experiments (Willetts, 1977, 1978; Achtman and Skurray, 1977; Clark and Warren, 1979) has shown that this so-called fertility inhibition is due to the inhibition of transcription of specific genes in the F plasmid. The result is that the protein which constitutes the F-pilus is not manufactured and the transfer capacity of F is diminished or abolished. Similar inhibitory systems operate among many groups of plasmids (Olsen and Shipley, 1973, 1975). Fertility inhibition is important when considering potential gene flow among different bacteria as repression of the transfer functions of any conjugative plasmid lowers the overall intercell communicability of plasmid-encoded genes.

In considering the colonization of host *B* by a plasmid from host *A* it is important to assess the effect of the new genetic background on the stability of the plasmid. Most well-studied R plasmids of clinical origin are normally stable in standard strains of *Escherichia coli*, *Shigella* spp., and *Klebsiella* spp. However, in the absence of antibiotic selection, certain types of plasmid are highly unstable in hosts, such as *Salmonella typhimurium*. For example, Watanabe and Ogata (1970) suggested that only 4 to 10 passages on unselective media may be needed to eliminate an entire pattern of multiple resistance from cultures of *S. typhimurium*. Whether the plasmid itself is eliminated is uncertain as some evidence (Falkow, 1975) suggests that the transfer genes of the plasmid may be retained even though associated resistance markers are lost.

Taxonomy may also affect the degree to which an introduced gene is expressed in its new genetic background. For example, an aminoglycoside resistance gene found in *Bacillus circulans* is more efficiently expressed in *Escherichia coli* after cloning and transfer than in its original host (Courvalin *et al.*, 1977). Our impression from the available literature is that barriers to gene transfer operate more at the level of entry and DNA synthesis than at the level of transcription and translation.

3.2 Entry Barriers

The first and primary barrier to the exchange of genes between heterologous cells lies at the cell surface, in the structure of the cell wall or the cell membrane. Considering the importance of this issue, surprisingly little is known of the molecular factors which influence the entry of non-homologous DNA. Some conjugative plasmids can only colonize organisms which are close evolutionary relatives. Thus Inc F11 plasmids can successfully infect only enteric bacteria (Bukhari *et al.*, 1977). While post-entry barriers may also occur, it seems likely that enteric organisms possess some surface feature(s) lacking in other bacteria. By contrast the IncP plasmid RP1 has an extraordinary host range since Gram-negative organisms as far apart taxonomically as *Escherichia coli* and *Rhizobium* sp. can maintain this plasmid (Reanney, 1977). In this case the requirements for effective pair or aggregate formation appear to be low, although post-entry factors must also contribute to the promiscuity of this replicon. In general, bacteriophages have much more stringent specificities than conjugative plasmids. Many phages are species and even strain specific and the molecular basis of some phage host ranges is partly understood. The DNA-containing phage PRD1 only infects organisms which harbour plasmids belonging to Inc groups P, W, and N (Bradley and Cohen, 1977). Presumably these plasmids specify some surface feature(s) necessary for the effective adsorption and/or penetration of the phage. RNA-containing phages, such as MS2 and r17, adsorb specifically to the pili specified by IncF plasmids (Lawn *et al.*, 1967) while phage PRR1 adsorbs to the surface appendage specified by IncP plasmids (Olsen and Thomas, 1973).

The ribophage example illustrates the point that polynucleotide exchange between different organisms may be helped or hindered by the presence of indigenous extrachromosomal units resident in cells of either partner. Many, perhaps most, plasmids possess gene(s) which specify a surface receptor that prevents the transfer of identical or similar plasmids. This phenomenon of entry exclusion should not be confused with incompatibility as the former operates at the cell surface while the latter affects post-entry events. It has been suggested that entry exclusion prevents pair formation between donor cells (Falkow, 1975).

One well-studied mechanism whose primary barrier lies at the cell surface is transformation. The ability of organisms to bind and take up exogenous DNA, termed competence, is a distinct physiological state, often induced by growth conditions. Bacteria, such as *Bacillus subtilis*, *Neisseria gonorrhoeae*, *Haemophilus influenzae*, and *Streptococcus pneumoniae*, can acquire competence under natural conditions. The specificity of transformation may differ sharply from system to system: thus while some species of *Bacillus* and *Streptococcus* can take up heterospecific DNA, *Haemophilus* species have been shown to possess an efficient mechanism which restricts the successful uptake to homospecific DNA. The *Haemophilus influenzae* genome contains about 600 copies of a base sequence which determines uptake (Sisco and Smith, 1979). This uptake sequence is apparently recognized and bound by a specific membrane protein found only in competent cells (Deich and Smith, 1980). This protein appears to act in different fashion to a surface-located endonuclease in *S. pneumoniae* in which only one strand of transforming DNA is drawn into competent cells (Lacks, 1977). *N. gonorrhoeae* is believed to have two distinct surface receptors, one of which binds all types of DNA while the other selects only homologous or related DNA (Saunders, 1979). This example suggests that while DNA exchange among gonococci is normally restricted to cells of the same species, the presence of saturating amounts of heterologous DNA may on occasions permit the entry of such DNA.

These systems appear better suited to the uptake of chromosomal DNA than plasmid DNA (Low and Porter, 1978). In an ecological context it is interesting to note that several species excrete DNA (Catlin, 1956).

3.3 Temperature Barriers

In few instances is the medical bias of modern bacteriology more evident than in the conditions used to detect conjugal transfer of plasmid DNA between a freshly isolated donor cell and a standard recipient. Especially noticeable is the fact that the temperature normally used to detect transfer is that of a homiothermic mammal. As has been pointed out elsewhere (Reanney, 1978), warm-blooded animals probably did not evolve until the Triassic period so that plasmids adapted to transfer optimally at 37°C are unlikely to be representative of the vast majority of conjugative elements in the bacterial flora.

At the time of writing rather few systematic studies on the temperature preference of conjugative plasmids have been carried out. Kelly and Reanney (1982) detected conjugative plasmids in several soil bacteria, including an *Enterobacter* organism and two species of *Pseudomonas*. For the resident *Enterobacter* sp., plasmid transfer to *Escherichia coli* at 28°C was 1000 times more efficient than transfer at 37°C under similar conditions. The markers carried by these transmissible plasmids, especially the linked resistances to tellurium and mercuric ions, suggested that the plasmids belonged to the IncH2 class. Plasmids in this group (Taylor and Levine, 1980) displayed a temperature sensitive (*ts*) phenotype even when resident in clinically derived organisms adapted to growth at 37°C, e.g., enteric species of *Klebsiella*. The optimal temperature for transfer reflected not only the molecular character of the plasmid but also the biology of the interacting organisms. In the case of plasmids resident in the soil *Enterobacter* spp. referred to above transfer at 12.5°C was detected only when *Erwinia herbicolor* was used as recipient (Kelly and Reanney, 1982).

An interesting study by Cooke (1978), suggested that even R^+ bacteria, normally regarded as part of the enteric flora, can transfer effectively at environmental temperatures. Cooke (1978) found that R^+ coliforms taken, not from the caecum but from filter-feeding shellfish in a New Zealand lake, transferred at 15°C with efficiencies as high as 1×10^{-2}. For several plasmids a drop in temperature from 35 to 15°C—such as might occur on going from colon to environment—increased the transfer frequencies to levels reminiscent of derepressed plasmids. Under identical conditions the classical R factor, R1, transferred with frequencies of 4×10^{-6} at 15°C. It is also worth noting that the highly promiscuous R plasmid RP1, while transferring optimally at 37°C, transferred at 15°C with efficiencies of up to 1×10^{-5} depending on the recipient cell and the assay conditions (W. J. Kelly and D. C. Reanney, unpublished observations).

A well-studied instance of thermosensitive transfer occurs with the Ti plasmid found in gall-causing strains of *Agrobacterium tumefaciens*. Efficient transfer occurred in the range 22 to 29°C but transfers at 32°C were 100 times less efficient and at 37°C transfer was undetectable (Tempe *et al*., 1977; Hooykass *et al*., 1979) despite the fact that both donor and recipient grew at these temperatures. It is significant that many strains of *Agrobacterium* sp. show high-frequency loss of the Ti plasmid when grown at 37°C. While it has not been established that the incubation temperature was the chief cause of the loss of the plasmid-associated phenotype, it does not seem unreasonable to suggest that plasmids may readily be discarded under conditions where they confer no necessary survival advantage, as may be the case if the growth temperature of the host cell is in any sense unnatural. Heat is a well-known curing agent for organisms adapted to optimal growth at 37°C, but the term high temperature is purely relative. A temperature of 37°C may well be high for soil bacteria which, for the most part, propagate in the temperature range 5 to 25°C. On the other hand, high temperatures *per se* are not incompatible with the stable maintenance of plasmids as circular DNA species have been observed in thermophilic strains of *Bacillus stearothermophilus* which propagates at temperatures of 55°C (Bingham *et al*., 1979).

In this context it is of interest that both transposons Tn3 (Kretschmer and Cohen, 1979) and Tn951 (Tn*lac*) (Cornelis, 1980) have been reported to show a temperature optimum for transposition of 30°C and that transposition decreased rapidly above this temperature. This lends credence to the idea that these transposons have diffused into their present hosts from organisms normally resident outside the clinical environment.

3.4 Ecological Barriers

Physical contact between donor and acceptor cells is a *sine qua non* if genes are to move among bacterial populations by conjugal mechanisms. Also transformation is likely to be ineffective unless the interacting cells are in close physical proximity. It follows that population densities are likely to be important factors in determining the frequency and extent of gene transfers in the environment.

These considerations can be illustrated by considering the possible extent of gene transfer in dry soil devoid of vegetation and in the rhizosphere. The microbial population of dry soil is restricted to bacteria which are good survivors under adverse conditions and to taxa, such as *Bacillus*, which can manufacture spores. Facile exchange of DNA under these circumstances seems unlikely. By contrast cell counts in the neighbourhood of plant roots can exceed 1×10^9 organisms per gram rhizosphere soil (Carpenter, 1972). The diversity of the rhizosphere community is impressive including such taxa as *Pseudomonas*, *Agrobacterium*, *Achromobacter*, *Azotobacter*, and *Rhizobium*. Similarly the mammalian gastrointestinal tract constitutes a natural chemostat which harbours enormous microbial populations: the human colon contains up to 1×10^{11} bacteria (g wet weight colon content)$^{-1}$ (Savage, 1977) and it is well known that microbial cells make up about 20 to 40% of the bulk of normal faeces. In both the colon and the rhizosphere the metazoan host provides a supply of nutrients and dead cell material as well as a sequestered environment; both ecosystems contain a characteristic and relatively constant microflora.

In the light of these ecological considerations many of the intergeneric transfers noted in Fig. 1 seem possible. Thus, the transfer of R genes between *Escherichia coli* and unrelated organisms, such as *Bacteroides fragilis* and *Fusobacterium* spp. (Burt and Woods, 1976) do not seem strange when it is realized that the majority of the colon flora is made up of obligate anaerobes. Likewise the transfer of genes between *E. coli* and Myxobacteria (Kaiser and Dworkin, 1975; Parish, 1975) can be rationalized by noting that myxobacteria often grow on faecal matter. These ecological patterns infer that the degree to which selection opens up, and sustains, pathways for the cell-to-cell movement of genes may depend more on ecological proximity than on evolutionary relatedness. Accordingly it is observed that, in the laboratory at least, the IncP plasmid RP1 can colonize members of most rhizosphere taxa and most enteric genera (Reanney, 1977). The overlap between the two ecosystems is not unexpected as large faecal bacteria populations are continually reintroduced into the densely populated rhizosphere zone (Fig. 1).

A second ecological factor which is likely to influence transfer possibilities is stability. Mating in liquid suspension probably represents a poor model for genetic interactions in nature since, in natural habitats, the bulk of the bacterial biomass is probably concentrated at liquid/surface interfaces. Chilton *et al.* (1976) observed that the co-transfer of plasmid RP1 and virulence-associated plasmids of Agrobacteria species occurred optimally on solid media. Haas and Holloway (1976) reported that little chromosome transfer could be detected for the P group plasmid R68–45 of *Pseudomonas aeruginosa* in liquid medium. Transfer of the Ti plasmid of *Agrobacterium tumefaciens* was inefficient in liquid medium but occurred very readily on membrane filters under appropriate conditions (Kerr *et al.*, 1977). These preferences may reflect an in-vitro memory on the part of the mating cells of their natural interactive situations on root surfaces or soil particles (Reanney, 1978).

The number of ecological factors which influence gene transfer in microbial communities is vast and complex. Moreover, just as a successful transfer *in vitro* of DNA does not guarantee a similarly effective process *in vivo* so failure to observe transfer *in vitro* does not mean that the organisms concerned cannot directly exchange genes in nature. A well-known example of the latter situation is provided by the Ti plasmid of *Agrobacterium tumefaciens*. For many years transfer of the Ti plasmid could only be demonstrated in crown gall tissue (Kerr, 1969; 1971) and all attempts to demonstrate transfer using standard bacterial conjugation methods failed. Self-transfer of the Ti plasmid under defined conditions was only achieved when two unusual amino acids, octopine and nopaline, were added to the system (Kerr *et al.*, 1977; Gentello *et al.*, 1977; Petit *et al.*, 1978). The Ti plasmid codes for the utilization of these amino acids by the bacterium (Bomhoff *et al.*, 1976; Montoya *et al.*, 1977). Transfer of a part of the Ti plasmid to the plant cell leads to crown gall disease and the production of these compounds in plant tissue (Chilton *et al.*, 1977; Merlo *et al.*, 1980; Yang *et al.*, 1980; Nuti *et al.*, 1980).

4. KNOWN GENETIC INTERACTIONS AMONG BACTERIA IN NATURE

4.1 Bacteria Associated with Animals

4.1.1 Gram-negative Enteric Bacteria

The normal habitat of these organisms is the gastrointestinal tract of animals, in particular the caecum and colon (Savage, 1977). The enterobacteria constitute only a minor part of the microbial flora of these regions which are dominated numerically by a small number of anaerobic species (Brown, 1977). Although plasmid transfer can be readily demonstrated under laboratory conditions, transfer in the gastrointestinal tract has proved much more difficult to establish unequivocally and has been relatively little studied. Most of the investigations known have used antibiotic resistance markers for selection and some of these have been reviewed previously (Lacey, 1975a).

Transfer has been demonstrated in animals but only under artificial conditions where the intestinal flora was either not yet established, as in germ-free mice (Reed *et al.*, 1969) and newly hatched chickens (Walton, 1966) or had been modified by pretreatment with antibiotics (Kasuya, 1964, Guinée, 1970) or starvation (Smith, 1975).

H. W. Smith (1970, 1971a,b) demonstrated gene transfer from *Escherichia coli* to *Salmonella typhimurium* in chickens with a normal flora. These studies emphasize the importance of using strains which transfer readily *in vitro* and can colonize the intestinal tract well, if transfer is to be detected. *S. typhimurium* of phage type 29, a strain implicated in severe disease outbreaks amongst intensively reared calves in Great Britain, and frequently found to contain R-factors in nature (Anderson, 1968), was an especially good recipient. Antibiotic supplemented feeds greatly enhanced the transfer rate provided that the donor organism harboured a resistance determinant for the antibiotic that was being supplied. Comparable results have also been achieved (Nivas *et al.*, 1976) using one-day old turkey poults as the test animal. Transfer *in vivo* in chickens was also reported (Smith, 1972) in experiments showing the decreased virulence of R-factor carrying *S. typhimurium* strains.

A different experimental approach was used by Smith (1975) and was significant in that in all cases transfer was readily detected in the absence of any selective antibiotic treatment. R factor transfer from *Escherichia coli* strains to either *E. coli* (Smith, 1975; 1976) or *Salmonella* sp. (M. G. Smith, 1977a) recipients occurred when the donor and recipient organisms were inoculated in low numbers into the rumen of sheep. Detection of the transfer was dependent upon the animal being deprived of food for 24 to 48 h, thereby altering the rumen environment sufficiently to allow the introduced organisms to multiply. Later experiments (M. G. Smith, 1977b) were successful in detecting transfer from *E. coli* introduced into the rumen to *E. coli* recipients implanted in the lower gastrointestinal tract, and again a period of starvation was necessary if transfer was to be detected. Whether non-ruminant animals would behave in the same way has not been adequately tested.

Successful transfer experiments have been conducted using calves and pigs (Smith, 1970) and Gyles *et al.* (1978) observed the exchange of a plasmid determining antibiotic resistance and enterotoxin production when newly weaned pigs were given large oral doses of donor and recipient *Escherichia coli*. Such studies suggest that R factor transfer *in vivo* can occur in animals and that the imposition of stress, such as antibiotic treatment or starvation, can facilitate detection of the process. Consequently, the possibility of plasmid transfer in the human gastrointestinal tract, which is of considerable interest, has been the subject of similar studies (Watanabe, 1963; Smith, 1969; Burton *et al.*, 1974). Lacey (1975a) has reviewed the epidemiological evidence for R factor exchange *in vivo*.

In many of the above studies, the evidence for gene transfer is solely on the basis of resistance phenotypes, and because of this their validity in terms of a general phenomenon has been questioned (Walton, 1971; Richmond, 1973). Certain marker combinations are very common within the enterobacteria (Mitsuhashi, 1979) and as a result it is important to include a molecular characterization of the plasmids involved.

Molecular investigations have been used to demonstrate R plasmid transfer by Anderson *et al.* (1973b). Transfer of the antibiotic resistance plasmid R1 to both introduced and endogenous *Escherichia coli* recipients in the human gut was only detected following a course of treatment with the appropriate antibiotic (Anderson *et al.*, 1973a). Interestingly, some isolates contained plasmids which were one third larger than plasmid R1 and which had acquired a tetracycline (Tc) resistance determinant. Further investigation of the plasmid involved (R174) demonstrated that it was probably a result of recombination *in vivo* between the introduced plasmid R1 and an endogenous R factor, R157 (Ingram *et al.*, 1974). These two plasmids were compatible but shared considerable homology thus giving rise to a recombinant molecule containing almost all the DNA base sequences of the parental plasmids.

Transfer in the absence of antibiotic selection was not detected in the above study but by monitoring an individual's faecal flora for long periods Petrocheilou *et al.* (1976, 1977) have shown that this can occur. Molecular studies showed very similar plasmids in different *Escherichia coli* strains. However, in one instance no transfer could be detected for 7 months and therefore transfer in the gut, in the absence of selection, appears to be a very rare event. A similar conclusion was reached by E. S. Anderson (1975) who detected low frequency transfer of a hybrid plasmid only when very large numbers of potential donors were ingested.

The difficulties in demonstrating gene transfer *in vivo* have led to investigations into the existence of potentially inhibiting factors in the gastro-intestinal tract. While several compounds are known to prevent transfer *in vitro*, Ott *et al.* (1971) found no agents capable of inhibiting transfer *in vivo*, and suggested that the transfer process may be self-limiting. Dense suspensions of *Bacteroides fragilis* did inhibit transfer *in vitro* (J. D. Anderson, 1975) but lower numbers of cells of several bacterial species or the addition of bile salts were only moderately inhibitory. This suggested that antibiotic treatment may facilitate transfer both by reducing the anaerobic flora and imposing a selection pressure.

All the above investigations have used R factors to determine transfer because of the ease of selection, but in an interesting series of experiments Williams (1977) has shown that Col V plasmids can transfer from an ingested donor strain to bacteria resident in the gastro-intestinal tract. Col V production was observed after the ingested donor strain had been eliminated and plasmid transfer was confirmed by sedimentation and homology studies of plasmids isolated from the ingested donor and the resident recipient.

Why the transfer of Col V plasmids appears to take place more readily than R plasmid transfer is not known although it is possible that this may be a reflection of the survival capabilities of the plasmid-carrying bacteria. In this context it is noteworthy that Col V plasmids can enhance the survival ability of the bacterial strain in which they reside (Smith, 1974; Smith and Huggins, 1976). This increased survival ability has been ascribed to properties of increased iron uptake (Williams, 1979; Williams and George, 1979) and/or to a heightened resistance to the bacteriocidal effects of serum (Binns *et al.*, 1979).

It may be, therefore, that studies of R factor transfer alone provide a somewhat distorted picture of the prevalence of plasmid transfer *in vivo*, and we await with interest investigations with the growing numbers of plasmids reportedly involved in enhancing bacterial survival or pathogenicity.

4.1.2 *Pseudomonas aeruginosa*

The epidemiology of carbenicillin resistant strains of *Pseudomonas aeruginosa* provides further evidence of gene transfer *in vivo*. Roe *et al.* (1971) found that carbenicillin resistance could be readily transferred between strains of *Escherichia coli*, *P. aeruginosa*, *Proteus mirabilis* and *Klebsiella aerogenes* in mixed infections of mouse burns. The plasmid involved (R1822 = RP1) conferred five drug resistance phenotypes on the host strain and appeared to have been acquired by *P. aeruginosa* from *K. aerogenes* or some other enteric organism found in the mixed flora of burns. Richmond (1973) has reviewed the emergence of carbenicillin resistant *P. aeruginosa* in detail.

4.1.3 *Clostridium perfringens*

Although little is known of the gene transfer system of this organism, transfer *in vivo* has been reported (Brefort *et al.*, 1977). Donor and recipient cultures were orally administered to axenic mice and the transfer of tetracycline resistance was observed to occur readily, without selection. Interestingly, one strain arising from a mating *in vivo* behaved as a hybrid between the parental strains and was capable of promoting chromosomal gene transfer.

4.1.4 *Staphylococcus Species*

Gene transfer *in vivo* in *Staphylococcus aureus* was first demonstrated using mice in which staphylococcal kidney infection had been induced.

That transduction was possible in this system was shown by Jarolmen *et al.* (1965) who injected phage into mice with established infections and by Novick and Morse (1967)who used a high-frequency transducing element (Novick, 1967) to show that transfer could occur between strains. However, Lacey and Richmond (1974) have argued that this type of transfer, involving an element formed by the rare recombination of phage and plasmid DNA, is unlikely to occur in nature. In addition, Lacey (1971, 1972) has reported the transfer of chromosomal and plasmid-borne antibiotic resistance traits between strains in mixed culture in broth, and in milk products (Lacey and Didcock, 1973), without the use of specially constructed strains. Interspecific transfer between *S. aureus* and *S. epidermidis* has also been shown in mixed culture (Witte, 1977).

A natural staphylococcal habitat is the body surface (Marples, 1974). The presence of several strains at a single site, the high frequency of lysogeny amongst bacteria inhabiting the skin surface (Marples, 1974) and the abundance of staphylococcal plasmids (Lacey, 1975b) suggests that gene transfer in this context may occur quite readily. The possibility of the transfer of resistance plasmids on the skin surface has been shown by Lacey (1971) and by Lacey and Richmond (1974) using *Staphylococcus aureus*. This organism is not normally found on healthy skin (although it is common in the nose and throat—Noble and Sommerville, 1974) and plasmid transfer required a moist environment and a reduction in the anti-bacterial action of the skin. These conditions were produced experimentally by suspending the organisms in serum and covering the site of application to the skin, thereby simulating the conditions occurring in wounds and other infections (Lacey, 1975b). In this way transfer within 6 h at frequencies of 1×10^{-5} to 1×10^{-8} was detected (Lacey and Richmond, 1974).

More recently, Naidoo and Noble (1978) demonstrated the transmission of gentamicin resistance between strains, in the absence of antibiotic therapy and under conditions designed to simulate those found on a dermatological patient undergoing treatment. Interestingly, transfer occurred at a higher frequency on the skin surface than in culture medium. Both transduction and conjugation have been suggested as possible mechanisms of transfer (Naidoo and Noble, 1978; Witte, 1977). However, although phages are certainly implicated in most cases, precisely how the transfer occurs remains unknown (Lacey, 1975b).

The situation for *Staphylococcus epidermidis* is not as well studied, although this species also contains plasmids (Schaeffler, 1972; Groves, 1979; Olsen *et al.*, 1979) as well as transducing phages (Yu and

Baldwin, 1971). Gene exchange between *S. epidermidis* and *S. aureus* has been demonstrated *in vitro* (Yu and Baldwin, 1971; Witte, 1977). *S. epidermidis* may be an important reservoir of genetic information on body surfaces as it is more prevalent amongst the normal bacterial flora than is *S. aureus*. Many workers have postulated that saprophytic organisms act as reservoirs of resistance genes for enteric bacteria and an analogous situation may apply with the staphylococci (Rosendorf and Kayser, 1974). Gene transfer in the absence of selective pressure seems more likely to occur between staphylococci in their natural habitat than in the comparable situation with enterobacteria.

4.1.5 Streptococcus Species

Evidence is also available that pneumococci (*Streptococcus pneumoniae*) can exchange genes in nature and some of the original transformation studies (Griffith, 1928) were carried out in a living host, although heat-killed cells were used as the source of DNA. Prompted by the finding that transformation occurred between pneumococcal strains in mixed culture, Ottolenghi and MacLeod (1963) showed that transformation between strains, in which competence had been induced, could occur during the peritoneal infection of mice. Similarly, Conant and Sawyer (1967) detected transformation during mixed pneumococcal peritoneal and respiratory infections in mice, provided that both strains were able to grow. Transformation did not require competence to be induced prior to infection and still occurred if the donor and recipient cells were added 6 h apart.

The role which transformation might play in the natural adaptation of pneumococci is not known although the mechanism by which transformation occurs has been well studied.

Streptococcus pneumoniae (Smith *et al.*, 1980) and a number of other species of streptococci are reported to exchange genes *in vitro* by conjugation (Jacob and Hobbs, 1974; Le Blanc *et al.*, 1978; Kempler and McKay, 1979) or transduction (McKay *et al.*, 1973) and a phage-associated gene transfer system is documented for pneumococci (Porter *et al.*, 1979). Whether these processes are of any significance in nature is unknown but the habitats of several species would provide ample opportunities for these to take place.

4.1.6 Neisseria gonorrhoeae

Sarubbi and Sparling (1974) demonstrated the transfer of antibiotic resistance, presumably by transformation, during the broth cultivation of mixed strains of *Neisseria gonorrhoeae*. It was suggested that such a mechanism could effect gene transfer in nature, especially since *Neisseria* sp. commonly excrete DNA (Catlin, 1960) and mixed infections appear to occur commonly (Short *et al.*, 1977).

4.2. Bacteria Associated with Plants

Genetic systems in phytopathogenic bacteria have only recently begun to be intensely studied, primarily as an aid to understanding the virulence of these organisms. Many genera are now known either to contain plasmids or to be able to act as recipients for plasmids from other genera (Lacy and Leary, 1979) but few experiments have been conducted to determine if transfer occurs on or in plants.

Agrobacterium tumefaciens, the causative agent of crown gall, is undoubtedly the best studied of the phytopathogenic bacteria (Van Montagu and Schell, 1979). The transfer of pathogenicity between *Agrobacterium* species in tomato plants resulting in the conversion of a non-pathogenic species to virulence was first reported by Kerr (1969, 1971) and has been demonstrated in other studies using other plant species (Van Larebeke *et al.*, 1975; Watson *et al.*, 1975).

A related organism, *Agrobacterium rhizogenes*, causes hairy root disease of numerous higher plants and the infectivity of this species has recently been correlated with the presence of a large plasmid (Moore *et al.*, 1979). Using carrot root discs these authors have successfully detected transfer *in planta* to a non-infectious strain of *A. radiobacter*.

Other phytopathogens are less well studied but transfer *in planta* has been shown by using the wide host range R plasmid RP1. Transfer from *Escherichia coli* to *Pseudomonas glycinea* and *P. phaseolicola* and between these two *Pseudomonas* species occurred in watery suspensions in pods and on leaves of *Lima* bean plants (Lacy and Leary, 1975). Mating pair formation and gene exchange was rapid, thus avoiding any possible inhibiting effects of the host plant's hypersensitive reaction. Subsequent transfer of plasmid RP1 between *Erwinia chrysanthemi* strains was shown in maize plants (Lacy, 1978) and was unaffected by the presence of an inhibitor, to which the strains were sensitive, in the host tissue.

A significant point arising from these studies is that transfer frequencies were higher for transfers *in planta* than for the controls *in vitro*. The transfer frequency for *Erwinia chrysanthemi* was 100 times higher than the rate *in vitro* and constituted strong evidence that plasmid transfer was favoured by the conditions occurring within the host plant (Lacy, 1978). Lacy and Leary (1979) also reported that plasmid RP1 transferred from *Escherichia coli* to *Erwinia amylovora* and *Erwinia herbicola* in pear blossoms at high frequencies, and from *E. herbicola* to *Pseudomonas syringae* and *E. amylovora*.

The recent finding that genes for toxin production are plasmid-borne in phytopathogenic pseudomonads (Gonzalez and Vidaver, 1979; Gantotti *et al.*, 1979) indicates that gene transfer may be of great importance in the pathogenic relationship of bacteria with plants. In addition to an understanding of pathogenic relationships there is growing evidence that plasmids are implicated in the symbiotic relationship between rhizobia and leguminous plants (Brewin *et al.*, 1980) and it will be of considerable interest to determine if the exchange of these capabilities can occur *in planta* or in the rhizosphere soil.

Another recent study has documented gene exchange in a botanical environment in the course of an investigation into the taxonomy and antibiotic resistance of *Klebsiella* species isolated from various sources. In an attempt to determine if the transfer of antibiotic resistance occurred in nature, radish plants grown from seeds were exposed to donor and recipient *Klebsiella* sp. strains prior to germination and were used to simulate one of the habitats of these organisms (Talbot *et al.*, 1980). Transfer of resistance from clinical to non-clinical strains was detected with five of the 21 mating pairs and presumably occurred on or in close proximity to the plant surface. Although such a simulated environment may represent a somewhat artificial situation this report demonstrated that the potential exists for gene transfer to occur amongst introduced and endogenous organisms present in the microflora associated with plant roots.

4.3 Transfer in Soil

To date studies to determine if gene exchange can occur in soil have been very limited, even though it seems likely that such exchange can occur (Reanney, 1978). Although generalized transduction systems are known for common soil genera, such as *Bacillus* (Taylor and Thorne, 1963; Thorne, 1968; Lovett, 1972; Vary, 1979), *Pseudomonas* (Stanisich and Richmond, 1975) and *Rhizobium* (Kowalski, 1971; Buchanan-Wollaston, 1979), and while bacteriophages for some species are extremely common in soil (Hegazi and Jensen, 1973; Reanney and Teh, 1976), transduction has never been demonstrated in a model soil system.

Similarly, plasmids are known in many soil genera but the cell-to-cell transmission of these in soil by conjugation or by other mechanisms has not been demonstrated. Conjugal transfer of the F factor between *Escherichia coli* strains has been reported in soil (Weinberg and Stotzky, 1972) but follow-up studies involving bacteria better adapted to this environment and/or with characteristics likely to produce a survival advantage if transferred, are long overdue.

Therefore, it is to transformation that we must turn for the most compelling evidence of gene exchange in soil. Using spores of specially constructed *Bacillus subtilis* strains, Graham and Istock (1978, 1979) have demonstrated the exchange of genes in sterile soil. For undetermined reasons, some markers, for both drug resistance and auxotrophy, were favoured compared with others. After 8 d a single phenotype, with markers from both strains was dominant in all replicates, making up 79% of the total population. The parental strains were undetectable after one day. It must be emphasized that, although only a small fraction (1×10^{-6}) of the total population was examined, 149 of the possible 256 marker combinations were detected prior to the emergence of the dominant phenotype. These results suggested that a large number of possible combinations were created initially followed by rearrangement and elimination by selection.

It was assumed that transformation was the mechanism responsible as the *Bacillus subtilis* strains used lacked plasmids or transducing phages, and could be transformed by adding DNA to the soil. It is known that both growing cells (Ephrati-Elizur, 1968) and germinating spores (Borenstein and Ephrati-Elizur, 1969) release genetically active DNA and as DNase and competition with heterologous DNA did not affect the process, it is possible that the excreted DNA was protected by protein or membrane components (Streips and Young, 1974). It is also possible that these strains harboured defective phage particles, the frequent occurrence of which in bacilli, suggests a possible role in bacterial survival (Kelly and Reanney, 1978).

The use of sterile soil and of strains lacking plasmids and phage probably does not reflect the natural situation accurately. It does, however, raise interesting queries as to the extent of genetic exchange amongst the normal soil microflora and provides an excellent example of how an organism utilizes its genetic potential to the full in adaptive processes.

4.4 Transfer in Natural Waters

While liquid media are usually employed for bacterial gene transfer experiments, unsupplemented natural waters have not been used in these studies. However, it has been reported that bacteria can exchange genes in these environments. Grabow *et al.* (1975) detected the transfer of 3/10 R factors from *Escherichia coli* isolated from river water. Transfer occurred at low frequency in a dialysis bag containing river water immersed in the river, and also in river water incubated in the laboratory.

The observation that the proportion of transferable resistance increased during the sedimentation and chlorination steps associated with conventional sewage treatment has led Grabow *et al.* (1976) to suggest that R factor transfer may be of significance during these stages. However, the importance of gene transfer in sewage, or during the treatment process, awaits further elucidation.

The transfer of antibiotic resistance was detected (Cooke, 1978) using *Escherichia coli* K12 and coliforms isolated from water supplies as donors, in non-sterile, unfiltered river and sea water. Transfer occurred at low frequencies but over a range of temperatures.

Talbot *et al.* (1980) detected the transfer of antibiotic resistance between *Klebsiella* species in an aqueous suspension of sawdust in their attempts to simulate the natural habitats of these organisms.

5. EVIDENCE FOR THE MOVEMENT OF GENES AMONG MEMBERS OF MICROBIAL COMMUNITIES

The preceding section suggests that the cell-to-cell movement of genes, illustrated in Fig. 1, also occurs in natural ecosystems, although with differing efficiencies. Perhaps the most telling evidence for a widespread flux of genes among mixed populations of bacteria comes from the distribution of specific phenotypes (and their underlying nucleotide codes) in diverse groups of organisms.

The strongest evidence for the widespread dissemination of genes among microbial floras has come from the well-documented finding of antibiotic resistance genes among diverse organisms. One of the best studied examples is that of the enzyme TEM β lactamase which is responsible for ampicillin resistance in many taxa. The R gene(s) involved can be carried in a typical transposon (TnI) which has a molecular weight of 3 Md. β-Lactamases similar to that encoded by TnI have been found in naturally occurring plasmids of Inc groups FI, FII, N, X, O, I, C, and W (Heffron *et al.*, 1975) suggesting that the gene(s) can move very extensively by DNA to DNA interactions among diverse replicons. Some of these plasmids, e.g. those belonging to group W, have very wide host ranges and the acquisition of TnI by such transmissible plasmids would offer an efficient vector mechanism for its dissemination throughout multiple species of Gram-negative organisms. Accordingly the TEM β lactamase gene has been found distributed throughout a wide variety of species in different parts of the world (Table 2). The technique used to assess the identity of the enzyme—isoelectric focusing—is not of itself sufficiently accurate to guarantee that the enzymes under examination are homologous. Nonetheless the weight of available evidence suggests that many of the

Table 2. Efficiency of spread of translocating units for ampicillin resistance (TEM-type β Lactamase)

R plasmid	Incompatibility group	Species in which detected	Country of origin
R978b	FI	*Enterobacter cloacae*	U.S.A.
R452	FII	*Proteus morganii*	South Africa
R946a	FII	*Escherichia coli*	Indonesia
R820	Iα	*Salmonella typhi*	Mexico
R269N	N	*Shigella sonnei*	U.K.
R825	N	*Providencia* sp.	Canada
R40a	A-C complex	*Pseudomonas aeruginosa*	France
R935	A-C complex	*Serratia marcescens*	France
R1097	H	*Salmonella typhi*	Thailand
R472	L	*Serratia marcescens*	U.S.A.
R1P69	M	*Salmonella paratyphi B*	France
R842	P	*Proteus mirabilis*	U.K.
R934	P	*Serratia marcescens*	Japan
R1033	P	*Pseudomonas aeruginosa*	Spain
R826	S	*Serratia marcescens*	France
R394	T	*Proteus rettgeri*	South Africa
R7K	W	*Proteus rettgeri*	Greece
ϕAmp	Y	*Escherichia coli*	U.K.
R28K	Unknown	*Citrobacter*	Greece
R870	Unknown	*Proteus mirabilis*	Poland
RSF007	Not given	*Haemophilus influenzae*	Poland
	Not given	*Neisseria gonorrhoeae*	England

Taken with permission from: D. C. Reanney *Genetic Interaction and Gene Transfer*, Brookhaven Symposia in Biology: No. 29 (1977).

bacteria now harbouring a common TEM β lactamase phenotype acquired the gene(s) responsible by DNA transfer across taxonomic boundaries.

In this context it is interesting to note that preliminary amino acid sequence data on β lactamase indicated that enzymes from such taxonomically distant organisms as *Escherichia coli*, *Bacillus licheniformis* and *Staphylococcus aureus* contained homologous proteins (Ambler, 1979). Unless one makes the unlikely assumption that the common ancestor of all three taxa contained a lactamase gene, the presence of the gene in the contemporary organisms must be ascribed to DNA transfers at some stage in the evolutionary development of the cells concerned. Especially noteworthy is the fact that a homologous enzyme may be found in both Gram-positive and Gram-negative organisms. To date no known conjugative plasmids can infect both Gram-positive and Gram-negative bacteria except under highly artificial circumstances. However, the Gram divide seems unlikely to pose an impervious barrier to the exchange of genes, especially when selective pressures are strong. Benveniste and Davies (1973) have suggested that soil *Streptomyces* species constitute the ancestral reservoir of genes determining resistance to aminoglycoside antibiotics. Since *Streptomyces* species are Gram-positive the presence of genes conferring such resistance phenotypes in so many Gram-negative pathogens can only be explained, in terms of the Benveniste-Davies hypothesis, by DNA transfer. In like vein we note that the transfer of genes encoding resistance to the antibiotic butirosin from the Gram-positive producer *Bacillus circulans* to *E. coli* has been achieved *in vitro* (Courvalin *et al.*, 1977). It is our opinion that transfers between Gram-positive and negative organisms will eventually be demonstrated, perhaps by processes involving transformation.

6. CONCLUSION

This chapter began with Fig. 1 and we would like to conclude by bringing the reader back to Fig. 1, hopefully with a greater understanding of the issues involved. A bacterial cell, considered in isolation, has a relatively small genetic content and, hence, a limited adaptive potential. However, the range of known DNA exchange mechanisms and vectors suggests that bacteria in their normal niches have access to a large "library" of outside information which endows microbial populations with an extraordinary adaptive capability (Reanney, 1976, 1977). In our view then, a microbial community constitutes a genetically "open" system in which data originating in any one source can theoretically diffuse to any other point in the network. However, a complex variety of interacting constraints, few of which are constant, limit or restrain the pooling of these genetic data in nature.

ACKNOWLEDGEMENTS

We are grateful to Dr V. A. Stanisich for a critical reading of a draft form of this article.

REFERENCES

Achtman, M. and Skurray R. (1977). *In* "Microbial Interactions: Receptors and Recognition Series B, vol. 3", pp. 233–279. (J. L. Reissig, ed.) Chapman and Hall, London.

Ambler, R. P. (1979). News and Views. *Nature, London* **279**, 288–289.

Anderson, E. S. (1968). The ecology of transferable drug resistance in the Enterobacteriaceae. *Annual Review of Microbiology* **22**, 131–180.

Anderson, E. S. (1975). Viability of, and transfer of a plasmid from *E. coli* K12 in the human intestine. *Nature, London* **255**, 502–504.

Anderson, J. D. (1975). Factors that may prevent transfer of antibiotic resistance between gram-negative bacteria in the gut. *Journal of Medical Microbiology* **8**, 83–88.

Anderson, J. D., Gillespie, W. A., and Richmond, M. H. (1973a). Chemotherapy and antibiotic-resistance transfer between enterobacteria in the human gastro-intestinal tract. *Journal of Medical Microbiology* **6**, 461–473.

Anderson, J. D., Ingram, L. C., Richmond, M. H., and Wiedemann, B. (1973b). Studies on the nature of plasmids arising from conjugation in the human gastro-intestinal tract. *Journal of Medical Microbiology* **6**, 475–486.

Arber, W. and Linn, S. (1969). DNA modification and restriction. *Annual Review of Biochemistry* **38**, 467–500.

Barksdale, L. and Arden, S. B. (1974). Persisting bacteriophage infections, lysogeny and phage conversions. *Annual Review of Microbiology* **28**, 265–299.

Bernhard, K., Schrempf, H., and Goebel, W. (1978). Bacteriocin and antibiotic resistance plasmids in *Bacillus cereus* and *Bacillus subtilis*. *Journal of Bacteriology* **133**, 897–903.

Benveniste, R. and Davies, J. (1973). Aminoglycoside antibiotic-inactivating enzymes in Actinomycetes similar to those present in clinical isolates of antibiotic-resistant bacteria. *Proceedings of the National Academy of Science* (*U.S.A.*) **70**, 2276–2280.

Bingham, A. H. A., Bruton, C. J., and Atkinson, T. (1979). Isolation and partial characterization of four plasmids from antibiotic-resistant thermophilic bacilli. *Journal of General Microbiology* **114**, 401–408.

Binns, M. M., Davies, D. L., and Hardy, K. G. (1979). Cloned fragments of the plasmid ColV, I-K94 specifying virulence and serum resistance. *Nature, London* **279**, 778–781.

Bomhoff, G., Klapwiji, P. M., Kester, H. C. M., Schiperoort, R. A., Hernalsteens, J. P., and Schell, J. (1976). Octopine and nopaline synthesis and breakdown genetically controlled by a plasmid of *Agrobacterium tumefaciens*. *Molecular and General Genetics* **145**, 177–181.

Borenstein, S. and Ephrati-Elizur, E. (1969). Spontaneous release of DNA in sequential genetic order by *Bacillus subtilis*. *Journal of Molecular Biology* **45**, 137–152.

Botstein, D. and Herskowitz, I. (1974). Properties of hybrids between *Salmonella* phage P22 and coliphage λ. *Nature, London* **251**, 584–589.

Boyer, H. W. (1971). DNA restriction and modification mechanisms in bacteria. *Annual Review of Microbiology* **25**, 153–176.

Bradley, D. E. and Cohen, D. R. (1977). Adsorption of lipid containing bacteriophages PR4 and PRD1 to pili determined by a P-1 incompatibility group plasmid. *Journal of General Microbiology* **98**, 619–623.

Brefort, G., Magot, M., Ionesco, H., and Sebald, M. (1977). Characterisation and transferability of *Clostridium perfringens* plasmids. *Plasmid* **1**, 52–66.

Brewin, N. J., Beringer, J. E., Buchanan-Wollaston, A. V., Johnston, A. W. B., and Hirsch, P. R. (1980). Transfer of symbiotic genes with bacteriocinogenic plasmids in *Rhizobium leguminosarum*. *Journal of General Microbiology* **116**, 261–270.

Broda, P. (1979). "Plasmids". Freeman, Oxford.

Brown, J. P. (1977). Role of gut bacterial flora in nutrition and health: a review of recent advances in bacteriological techniques, metabolism, and factors affecting flora composition. *CRC Critical Reviews in Food Science and Nutrition* **8**, 229–336.

Buchanan-Wollaston, V. (1979). Generalised transduction in *Rhizobium leguminosarum*. *Journal of General Microbiology* **112**, 135–142.

Bukhari, A. I., Shapiro, J. A., and Adhya, S. L. (1977). *DNA Insertion Elements, Plasmids and Episomes*. Cold Spring Harbour Laboratory, New York.

Burt, S. J. and Woods, D. R. (1976). R factor transfer to obligate anaerobes from *Escherichia coli*. *Journal of General Microbiology* **93**, 405–409.

Burton, G. C., Hirsh, D. C., Blenden, D. C., and Zeigler, J. L (1974). The effects of tetracycline on the establishment of *Escherichia coli* of animal origin, and *in vivo* transfer of antibiotic resistance, in the intestinal tract of man. *In* "The Normal Microbial Flora of Man", pp. 241–253. (Skinner, F. R. and Carr, J. G., eds.) Academic Press, London and New York.

Carpenter, P. L. (1972). "Microbiology". Saunders, Philadelphia.

Catlin, B. W. (1956). The formation of extracellular deoxyribonucleic acid by various bacteria. *Bacteriological Proceedings* p. 123.

Catlin, B. W. (1960). Interspecific transformation of *Neisseria* by culture slime containing deoxyribonucleate. *Science* **131**, 608–610.

Chang, A. C. Y. and Cohen, S. N. (1978). Construction and characterization of amplifiable multicopy DNA cloning vehicles derived from the P15A cryptic miniplasmid. *Journal of Bacteriology* **134**, 1141–1156.

Chilton, M. D., Farrand, S. K., Levin, R. L., and Nester, E. W. (1976). RP4 promotion of transfer of a large *Agrobacterium* plasmid which confers virulence. *Genetics* **83**, 609–618.

Chilton, M. D., Drummond, M. H., Merlo, D. J., Schiahy, D., Montoya, A. L., Gordon, M. P., and Nester, E. W. (1977). Stable incorporation of plasmid DNA into higher plant cells: the molecular basis of crown gall tumorigenesis. *Cell* **11**, 263–271.

Clark, A. J. and Warren, G. J. (1979). Conjugal transmission of plasmids. *Annual Review of Genetics* **13**, 99–125.

Cohen, S. N. (1976). Transposable genetic elements and plasmid evolution. *Nature, London* **263**, 731–738.

Cohen, S. N. and Shapiro, J. H. (1980). Transposable genetic elements. *Scientific American* **242**, 36–45.

Conant, J. E. and Sawyer, W. D. (1967). Transformation during mixed pneumococcal infection of mice. *Journal of Bacteriology* **93**, 1869–1875.

Cooke, M. D. (1978). R-factor transfer in river and sea-water. Abstracts of New Zealand Microbiological Society Annual Conference, Nelson, New Zealand.

Cornelis, G. (1980). Transposition of Tn951 (Tn*lac*) and cointegrate formation are thermosensitive processes. *Journal of General Microbiology* **117**, 243–247.

Courvalin, P., Weisblum, B., and Davies, J. (1977). Aminoglycoside-modifying enzyme of an antibiotic-producing bacterium acts as a determinant of antibiotic resistance in *Escherichia coli*. *Proceedings of the National Academy of Sciences (U.S.A.)* **74**, 999–1003.

Datta, N. (1979). Plasmid classification: Incompatibility Grouping. *In* "Plasmids of Medical, Environmental and Commercial Importance", pp. 3–12. (Timmis, K. A. and Pühler, A., eds.) Elsevier/North Holland, Amsterdam.

Davey, R. B. and Reanney, D. C. (1980). Extrachromosomal genetic elements and the adaptive evolution of bacteria. *Evolutionary Biology* **13**, 113–147.

Davidson, N., Deonier, R. C., Hu, S., and Ohtsubo, E. (1975). Electron microscope heteroduplex studies of sequence relations among plasmids of *Escherichia coli*. X. Deoxyribonucleic acid sequence organisation of F and of F-primes, and the sequences invoved in Hfr formation. *In* "Microbiology 1974", pp. 56–65. (Schlessinger, D., ed.) American Society for Microbiology, Washington.

Deich, R. A. and Smith, H. U. (1980). Mechanism of homospecific DNA uptake in *Haemophilus influenzae* transformation. *Molecular and General Genetics* **177**, 369–374.

Dempsey, W. B., McIntire, S. A., Willets, N., Schottel, J., Kinscherf, T. G., Silver, S., and Shannon, W. A. (1978). Properties of lambda transducing bacteriophages carrying R100 plasmid DNA: mercury resistance genes. *Journal of Bacteriology* **136**, 1084–1093.

Doolittle, W. F. and Sapienza, C. (1980). Selfish genes, the phenotype paradigm and genome evolution. *Nature, London* **284**, 601–603.

Dubnau, D. (1976). Genetic transformation of *Bacillus subtilis*; a review with emphasis on the recombination mechanism. *In* "Microbiology 1976", pp. 14–27. (Schlessinger, D., ed.) American Society for Microbiology, Washington.

Ephrati-Elizur, E. (1968). Spontaneous transformation in *Bacillus subtilis*. *Genetical Research, Cambridge* **11**, 83–96.

Falkow, S. (1957). L forms of Proteus induced by filtrates of antagonistic strains. *Journal of Bacteriology* **73**, 443–444.

Falkow, S. (1975). "Infectious Multiple Drug Resistance". Pion Limited, London.

Fox, M. S. (1978). Some features of genetic recombination in prokaryotes. *Annual Review of Genetics* **12**, 47–68.

Gantotti, B. V., Patil, S. S., and Mandel, M. (1979). Apparent involvement of a plasmid in phaseotoxin production by *Pseudomonas phaseolicola. Applied and Environmental Microbiology* **37**, 511–516.

Gentello, C., Van Larebeke, N., Holsters, M., De Picker, A., Van Montagu, M., and Schell, J. (1977). Ti plasmids of *Agrobacterium* as conjugative plasmids. *Nature, London* **265**, 561–563.

González, J. M. and Carlton, B. C. (1980). Patterns of plasmid DNA in crystalliferous and acrystalliferous strains of *Bacillus thuringiensis. Plasmid* **3**, 92–98.

Gonzalez, C. F. and Vidaver, A. K. (1979). Syringomycin production and holcus spot disease of maize: plasmid associated properties in *Pseudomonas syringae. Current Microbiology* **2**, 75–80.

Gorbach, S. L. (1978). Recombinant DNA: an infectious disease perspective. *Journal of Infectious Diseases* **137**, 615–623.

Gottesman, M. E. and Weisberg, R. A. (1971). Prophage insertion and excision. *In* "The Bacteriophage Lambda", pp. 113–138. (Hershey, A. D., ed.) Cold Spring Harbor Laboratory.

Grabow, W. O. K., Prozesky, O. W., and Burger, J. S. (1975). Behaviour in a river and dam of coliform bacteria with transferable or non-transferable drug resistance. *Water Research* **9**, 777–782.

Grabow, W. O. K., Van Zyl, M., and Prozesky, O. W. (1976). Behaviour in conventional sewage purification processes of coliform bacteria with transferable or non-transferable drug-resistance. *Water Research* **10**, 717–723.

Graham, J. B. and Istock, C. A. (1978). Gene exchange in *Bacillus subtilis* in soil. *Molecular and General Genetics* **166**, 287–290.

Graham, J. B. and Istock, C. A. (1979). Gene exchange and natural selection cause *Bacillus subtilis* to evolve in soil culture. *Science* **204**, 637–639.

Griffith, F. (1928). The significance of pneumococcal types. *Journal of Hygiene* **27**, 113–159.

Groves, D. J. (1979). Interspecific relationships of antibiotic reistance in *Staphylococcus* sp: isolation and comparison of plasmids determining tetracycline resistance in *S. aureus* and *S. epidermidis. Canadian Journal of Microbiology* **25**, 1468–1475.

Guinée, P. A. M. (1970). Resistance transfer to the resident intestinal *Escherichia coli* of rats. *Journal of Bacteriology* **102**, 291–292.

Gyles, C., Falkow, S., and Rollins, L. (1978). In vivo transfer of an *Escherichia coli* enterotoxin plasmid possessing genes for drug resistance. *American Journal of Veterinary Research* **39**, 1438–1441.

Haas, D. and Holloway, B. W. (1976). R factor variants with enhanced sex factor activity in *Pseudomonas aeruginosa. Molecular and General Genetics* **144**, 243–251.

Heffron, F., Sublett, R., Hedges, R. W., Jacob, A., and Falkow, S. (1975). Origin of the TEM beta-lactamase gene found on plasmids. *Journal of Bacteriology* **122**, 250–256.

Hegazi, N. A. and Jensen, V. (1973). Studies of *Azotobacter* bacteriophages in Egyptian soils. *Soil Biology and Biochemistry* **5**, 231–243.

Holloway, B. W. (1978). Isolation and characterisation of an R′ plasmid in *Pseudomonas aeruginosa. Journal of Bacteriology* **133**, 1078–1082.

Hooykaas, P. J. J., Roobol, C., and Schilperoort, R. A. (1979). Regulation of the transfer of the Ti plasmid of *Agrobacterium tumefaciens*. *Journal of General Microbiology* **110**, 99–109.

Iida, S. and Arber, W. (1977). Plaque forming specialised transducing phage P1: isolation of P1CmSmSu, a precursor of P1Cm. *Molecular and General Genetics* **153**, 259–269.

Ingram, L. C., Anderson, J. D., Arrand, J. E., and Richmond, M. H. (1974). A probable example of R-factor recombination in the human gastro-intestinal tract. *Journal of Medical Microbiology* **7**, 251–257.

Inoue, S., Oshima, H., Saito, T., Okubo, T., and Mitsuhashi, S. (1976). Integration of the erythromycin(ero) resistance gene of a staphylococcal plasmid into S1*ppen.ion* phage. *Virology* **73**, 295–298.

Jacob, A. E. and Hobbs, S. J. (1974). Conjugal transfer of plasmid-borne multiple antibiotic resistance in *Streptococcus faecalis* var *zymogenes*. *Journal of Bacteriology* **117**, 360–372.

Jarolmen, H., Bondi, A., and Crowell, R. C. (1965). Transduction of *Staphylococcus aureus* to tetracycline resistance in vivo. *Journal of Bacteriology* **89**, 1286–1290.

Kaiser, D. and Dworkin, M. (1975). Gene transfer to a Myxobacterium by *Escherichia coli* phage P1. *Science* **187**, 653–654.

Kasuya, M. (1964). Transfer of drug resistance between enteric bacteria induced in the mouse intestine. *Journal of Bacteriology* **88**, 322–328.

Kelly, W. J. and Reanney, D. C. (1978). *In* "First International Symposium of Microbial Ecology", pp. 113–120. (Loutit, M. and Miles, J., eds.) Springer Verlag, Berlin.

Kelly, W. J. and Reanney, D. C. (1982). "Soil Biology and Biochemistry" (submitted).

Kemp, J. D. (1977). A new amino acid derivative present in crown gall tumor tissue. *Biochemical and Biophysical Research Communications* **74**, 862–868.

Kempler, G. M. and McKay, L. L. (1979). Genetic evidence for plasmid-linked lactose metabolism in *Streptococcus lactis* subsp. *diacetylactis*. *Applied and Environmental Microbiology* **37**, 1041–1043.

Kerr, A. (1969). Transfer of virulence between isolates of *Agrobacterium*. *Nature, London* **223**, 1175–1176.

Kerr, A. (1971). Acquisition of virulence by non-pathogenic isolates of *Agrobacterium radiobacter*. *Physiological Plant Pathology* **1**, 241–246.

Kerr, A., Manigault, P., and Tempe, P. (1977). Transfer of virulence *in vivo* and *in vitro* in *Agrobacterium*. *Nature, London* **265**, 560–561.

Kleckner, N. (1977). Translocatable elements in procaryotes. *Cell* **11**, 11–23.

Kowalski, M. (1971). Transduction in *Rhizobium meliloti*. *Plant and Soil, Special Volume*, 63–66.

Kretschmer, P. J. and Cohen, S. N. (1979). Effect of temperature on translocation frequency of the Tn3 element. *Journal of Bacteriology* **139**, 515–519.

Lacey, R. W. (1971). High-frequency transfer of neomycin resistance between naturally-occurring strains of *Staphylococcus aureus*. *Journal of Medical Microbiology* **4**, 73–84.

Lacey, R. W. (1972). Transfer of chromosomal genes between staphylococci in mixed cultures. *Journal of General Microbiology* **71**, 399–401.

Lacey, R. W. (1975a). A critical appraisal of the importance of R-factors in the Enterobacteriaceae in vivo. *Journal of Antimicrobial Chemotherapy* **1**, 25–37.

Lacey, R. W. (1975b). Antibiotic resistance plasmids of *Staphylococcus aureus* and their clinical importance. *Bacteriological Reviews* **39**, 1–32.

Lacey, R. W. and Didcock, R. (1973). Transfer of antibiotic resistance between strains of *Staphylococcus aureus* in milk products. *Lancet* **1**, 824.

Lacey, R. W. and Richmond, M. H. (1974). The genetic basis of antibiotic resistance in *S. aureus*: the importance of gene transfer in the evolution of this organism in the hospital environment. *Annals of the New York Academy of Sciences* **236**, 395–412.

Lacks, S. A. (1977). *In* "Microbial Interactions: Receptors and Recognition Series B", vol. 3, pp. 177–232. (Reissig, J. L., ed.) Chapman and Hall, London.

Lacy, G. H. (1978). Genetic studies with plasmid RP1 in *Erwinia chrysanthemi* strains pathogenic on maize. *Phytopathology* **68**, 1323–1330.

Lacey, G. H. and Leary, J. V. (1975). Transfer of antibiotic resistance plasmid RP1 into *Pseudomonas glycinea* and *Pseudomonas phaseolicola in vitro* and *in planta*. *Journal of General Microbiology* **88**, 49–57.

Lacey, G. H. and Leary, J. V. (1979). Genetic systems in phytopathogenic bacteria. *Annual Review of Phytopathology* **17**, 181–202.

Lawn, A. M., Meynell, E., Meynell, G. G., and Datta, N. (1967). Sex pili and the classification of sex factors in the Enterobacteriaceae. *Nature, London* **216**, 343–346.

LeBlanc, D. J., Hawley, R. J., Lee, L. N., and St. Martin, E. J. (1978). "Conjugal" transfer of plasmid DNA among oral streptococci. *Proceedings of the National Academy of Sciences (U.S.A.)* **75**, 3484–3487.

Lovett, P. S. (1972). PBP1: a flagella specific bacteriophage mediating transduction in *Bacillus pumilus*. *Virology* **47**, 743–752.

Lovett, P. S. and Bramucci, M. G. (1975). Plasmid deoxyribonucleic acid in *Bacillus subtilis* and *Bacillus pumilis*. *Journal of Bacteriology* **124**, 484–490.

Low, K. B. (1972). *Escherichia coli* K12 F-prime factors, old and new. *Bacteriological Reviews* **36**, 587–607.

Low, K. B. and Porter, D. D. (1978). Modes of gene transfer and recombination in bacteria. *Annual Review of Genetics* **12**, 249–287.

Marples, M. J. (1974). The normal microbial flora of the skin. *In* "The Normal Microbial Flora of Man", pp. 7–12. (Skinner, F. R. and Carr, J. G., eds.) Academic Press, London and New York.

McKay, L. L., Cords, B. R., and Baldwin, K. A. (1973). Transduction of lactose metabolism in *Streptococcus lactis* C2. *Journal of Bacteriology* **115**, 810–815.

Merlo, D. J., Nutter, R. C., Montoya, A. L., Garfinkel, D. J., Drummond, M. H., Chilton, M-D., Gordon, M. P., and Nester, E. W. (1980). The boundaries and copy numbers of Ti plasmid T-DNA vary in Crown Gall tumors. *Molecular and General Genetics* **177**, 637–643.

Mise, K. and Nakaya, R. (1977). Transduction of R plasmids by bacteriophages P1 and P22. Distinction between generalised and specialised transduction. *Molecular and General Genetics* **157**, 131–138.

Mitsuhashi, S. (1979). Drug resistance plasmids. *Molecular and Cellular Biochemistry* **26**, 135–181.

Montoya, A., Chilton, M. D., Gordon, M. P., Sciaky, D., and Nester, E. W. (1977). Octopine and nopaline metabolism in *Agrobacterium tumefaciens* and crown-gall tumor cells: role of plasmid genes. *Journal of Bacteriology* **129**, 101–107.

Moore, L., Warren, G., and Strobel, G. (1979). Involvement of a plasmid in the hairy root disease of plants caused by *Agrobacterium rhizogenes*. *Plasmid* **2**, 617–626.

Naidoo, J. and Noble, W. C. (1978). Transfer of gentamicin resistance between strains of *Staphylococcus aureus* on skin. *Journal of General Microbiology* **107**, 391–393.

Nivas, S. C., York, M. D., and Pomeroy, B. S. (1976). *In vivo* spread of infectious drug resistance in turkeys. *American Journal of Veterinary Research* **37**, 1211–1213.

Noble, W. C. and Somerville, D. A. (1974). *Microbiology of Human Skin*. Saunders, London.

Notani, N. K. and Setlow, J. K. (1974). Mechanism of bacterial transformation and transfection. *Progress in Nucleic Acids Research and Molecular Biology* **14**, 39–100.

Novick, R. P. (1967). Properties of a cryptic high-frequency transducing phage in *Staphylococcus aureus*. *Virology* **33**, 155–166.

Novick, R. P., Clowes, R. C., Cohen, S. N., Curtiss, R., Datta, N., and Falkow, S. (1976). Uniform nomenclature for bacterial plasmids. *Bacteriological Reviews* **40**, 168–189.

Novick, R. P. and Hoppenstaedt, F. C. (1978). On plasmid incompatibility. *Plasmid* **1**, 421–434.

Novick, R. P. and Morse, S. I. (1967). In vivo transmission of drug resistance factors between strains of *Staphylococcus aureus*. *Journal of Experimental Medicine* **125**, 45–59.

Nugent, M. E. and Hedges, R. W. (1979). Recombinant plasmids formed *in vivo* carrying and expressing two incompatibility regions. *Journal of General Microbiology* **114**, 467–470.

Nuti, M. P., Ledeboer, D. M., Durante, M., Nuti-Ronchi, V., and Schilperoort, R. A. (1980). Detection of Ti-plasmid sequences in infected tissues by *in situ* hybridisation. *Plant Science Letters* **18**, 1–6.

Olsen, R. H. and Thomas, D. (1973). Characteristics and purification of PRR1, an RNA phage specific for the broad host range *Pseudomonas* R1822 drug resistance plasmid. *Journal of Virology* **12**, 1560–1567.

Olsen, R. H. and Shipley, P. (1973). Host range and properties of the *Pseudomonas aeruginosa* R factor R1822. *Journal of Bacteriology* **113**, 772–780.

Olsen, R. H. and Shipley, P. L. (1975). RP1 properties and fertility inhibition among P, N, W and X incompatibility group plasmids. *Journal of Bacteriology* **123**, 28–35.

Olsen, R. H. and Wright, C. D. (1976). Interaction of *Pseudomonas* and *Enterobacteriaceae* plasmids in *Aeromonas salmonicida*. *Journal of Bacteriology* **128**, 228–234.

Olsen, W. C., Parisi, J. T., Totten, P. A., and Baldwin, J. N. (1979). Transduction of penicillinase production in *Staphylococcus epidermidis* and nature of the genetic determinant. *Canadian Journal of Microbiology* **25**, 508–511.

Orgel, L. E. and Crick, F. H. C. (1980). Selfish DNA: the ultimate parasite. *Nature, London* **284**, 604–607.

Ott, J. L., Short, L. J., and Holmes, D. H. (1971). In vitro and in vivo methodology for detecting and evaluating inhibitors of transfer of resistance factors. *Annals of the New York Academy of Sciences* **182**, 312–321.

Ottolenghi, E. and Macleod, C. M. (1963). Genetic transformation among living pneumococci in the mouse. *Proceedings of the National Academy of Sciences* (*U.S.A.*) **50**, 417–419.

Ozeki, H. and Ikeda, H. (1968). Transduction mechanisms. *Annual Review of Genetics* **2**, 245–278.

Parish, J. H. (1975). Transfer of drug resistance to Myxococcous from bacteria carrying drug resistance factors. *Journal of General Microbiology* **87**, 198–210.

Petit, A., Tempé, J., Kerr, A., Holsters, M., Van Montagu, M., and Schell, J. (1978). Substrate induction of conjugative activity of *Agrobacterium tumefaciens* Ti plasmids. *Nature, London* **271**, 570–571.

Petrocheilou, V., Grinsted, J., and Richmond, M. H. (1976). R-plasmid transfer in vivo in the absence of selection pressure. *Antimicrobial Agents and Chemotherapy* **10**, 753–761.

Petrocheilou, V., Richmond, M. H., and Bennett, P. M. (1977). Spread of a single plasmid clone to an untreated individual from a person receiving prolonged tetracycline therapy. *Antimicrobial Agents and Chemotherapy* **12**, 219–225.

Porter, R. D., Shoemaker, N. B., Rampe, G., and Guild, W. R. (1979). Bacteriophage-associated gene transfer in pneumococcus, transduction or pseudotransduction. *Journal of Bacteriology* **137**, 556–567.

Reanney, D. C. (1976). Extrachromosomal elements as possible agents of adaptation and development. *Bacteriological Reviews* **40**, 552–590.

Reanney, D. C. (1977). Genetic Interaction and Gene Transfer. Brookhaven Symposia in Biology No. 29.

Reanney, D. C. (1978). Coupled evolution: adaptive interactions among the genomes of plasmids, viruses, and cells. *In* "Aspects of Genetic Action and Evolution", pp. 1–68. (Bourne, G. H., Danielli, J. F., and Jeon, K. W., eds.) Academic Press, New York and London.

Reanney, D. C. and Teh, C. K. (1976). Mapping pathways of possible phage-mediated genetic interchange among soil bacilli. *Soil Biology and Biochemistry* **8**, 305–311.

Reed, N. D., Sieckmann, D. G., and Georgi, C. E. (1969). Transfer of infectious drug resistance in microbially defined mice. *Journal of Bacteriology* **100**, 22–26.

Richmond, M. H. (1973). Resistance factors and their ecological importance to bacteria and to man. *Progress in Nucleic Acids Research and Molecular Biology* **13**, 191–248.

Roberts, R. J. (1976). Restriction endonucleases. *CRC Critical Reviews of Biochemistry* **4**, 123–164.

Roe, E., Jones, R. J., and Lowbury, E. J. L. (1971). Transfer of antibiotic resistance between *Pseudomonas aeruginosa*, *Escherichia coli*, and other gram-negative bacilli in burns. *Lancet* **1**, 149–152.

Rosendorf, L. L. and Kayser, F. H. (1974). Transduction and plasmid deoxyribonucleic acid analysis in a multiply antibiotic-resistant strain of *Staphylococcus epidermidis*. *Journal of Bacteriology* **120**, 679–686.

Rowbury, R. J. (1977). Bacterial plasmids with particular reference to their replication and transfer properties. *Progress in Biophysics and Molecular Biology* **31**, 271–317.

Sarubbi, F. R. and Sparling, P. F. (1974). Transfer of antibiotic resistance in mixed cultures of *Neisseria gonorrhoeae*. *Journal of Infectious Diseases* **130**, 660–663.

Saunders, J. R. (1979). Specificity of DNA uptake in bacterial transformation. *Nature, London* **278**, 601–602.

Savage, D. C. (1977). Microbial ecology of the gastro-intestinal tract. *Annual Review of Microbiology* **31**, 107–133.

Schaeffler, S. (1972). Polyfunctional penicillinase plasmid in *Staphylococcus epidermidis*: bacteriophage restriction and modification mutants. *Journal of Bacteriology* **112**, 697–706.

Short, H. B., Ploscowe, V. B., Weiss, J. A., and Young, F. E. (1977). Rapid method for auxotyping multiple strains of *Neisseria gonorrhoeae*. *Journal of Clinical Microbiology* **6**, 244–248.

Sisco, K. L. and Smith, H. O. (1979). Sequence-specific DNA uptake in *Haemophilus* transformation. *Proceedings of the National Academy of Sciences (U.S.A.)* **76**, 972–976.

Smith, H. W. (1969). Transfer of antibiotic resistance from animal and human strains of *Escherichia coli* to resident *E. coli* in the alimentary tract of man. *Lancet* **1**, 1174–1176.

Smith, H. W. (1970). The transfer of antibiotic resistance between strains of enterobacteria in chickens, calves and pigs. *Journal of Medical Microbiology* **3**, 165–180.

Smith, H. W. (1971a). The effect of the use of antibacterial drugs on the emergence of drug-resistance bacteria in animals. *Advances in Veterinary Science and Comparative Medicine* **15**, 67–100.

Smith, H. W. (1971b). Observations on the *in vivo* transfer of R factors. *Annals of the New York Academy of Sciences* **182**, 80–90.

Smith, H. W. (1972). The effect on virulence of transferring R factors to *Salmonella typhimurium in vivo*. *Journal of Medical Microbiology* **5**, 451–458.

Smith, H. W. (1974). A search for transmissible pathogenic characters in invasive strains of *Escherichia coli*: the discovery of a plasmid-controlled toxin and a plasmid-controlled lethal character closely associated, or identical, with colicine V. *Journal of General Microbiology* **83**, 95–111.

Smith, H. W. (1977). Mobilisation of non-conjugative tetracycline, streptomycin, spectinomycin and sulphonamide resistance determinants of *Escherichia coli*. *Journal of General Microbiology* **100**, 189–196.

Smith, H. W. and Huggins, M. B. (1976). Further observations on the association of the colicine V plasmid of *Escherichia coli* with pathogenicity and with survival in the alimentary tract. *Journal of General Microbiology* **92**, 335–350.

Smith, M. D., Shoemaker, N. B., Burdett, V., and Guild, W. R. (1980). Transfer of plasmids by conjugation in *Streptococcus pneumoniae*. *Plasmid* **3**, 70–79.

Smith, M. G. (1975). *In vivo* transfer of R factors between *Escherichia coli* strains inoculated into the rumen of sheep. *Journal of Hygiene* **75**, 363–370.

Smith, M. G. (1976). R-factor transfer in vivo in sheep with *E. coli* K12. *Nature, London* **261**, 348.

Smith, M. G. (1977a). Transfer of R factors from *Escherichia coli* to Salmonellas in the rumen of sheep. *Journal of Medical Microbiology* **10**, 29–35.

Smith, M. G. (1977b). *In vivo* transfer of an R factor within the lower gastro-intestinal tract of sheep. *Journal of Hygiene* **79**, 259–268.

Stanisich, V. A. and Richmond, M. H. (1975). Gene transfer in the genus *Pseudomonas*. *In* "Genetics and Biochemistry of *Pseudomonas*", pp. 163–190. (Clarke, P. H. and Richmond, M. H., eds.) John Wiley, London.

Streips, U. N. and Young, F. E. (1974). Transformation in *Bacillus subtilis* using excreted DNA. *Molecular and General Genetics* **133**, 47–55.

Talbot, H. W., Yamamoto, D. K., Smith, M. W., and Seidler, R. J. (1980). Antibiotic resistance and its transfer among clinical and nonclinical *Klebsiella* strains in botanical environments. *Applied and Environmental Microbiology* **39**, 97–104.

Taylor, D. E. and Levine, J. G. (1980). Studies of temperature-sensitive transfer and maintenance of H incompatibility group plasmids. *Journal of General Microbiology* **116**, 475–484.

Taylor, M. J. and Thorne, C. B. (1963). Transduction of *Bacillus licheniformis* and *Bacillus subtilis* by each of two phages. *Journal of Bacteriology* **86**, 452–461.

Tempé, J., Petit, A., Holsters, M., Van Montagu, M., and Schell, J. (1977). Thermosensitive step associated with transfer of Ti-plasmid during conjugation: possible relation to transformation in crown gall. *Proceedings of the National Academy of Sciences* (*U.S.A.*) **74**, 2848–2849.

Thorne, C. B. (1968). Transduction in *Bacillus cereus* and *Bacillus anthracis*. *Bacteriological Reviews* **32**, 358–361.

Van Larebeke, N., Genetello, C., Schell, J., Schilperoort, R. A., Hermans, A. K., Hernalsteens, J. P., and Van Montagu, M. (1975). Acquisition of tumor-inducing ability by non-oncogenic agrobacteria as a result of plasmid transfer. *Nature, London* **255**, 742–743.

Van Montagu, M. and Schell, J. (1979). The plasmids of *Agrobacterium tumefa-*

ciens. *In* "Plasmids of Medical, Environmental and Commercial Importance", pp. 71–95. (Timmis, K. A. and Pühler, A., eds.) Elsevier/North Holland, Amsterdam.

Vary, P. S. (1979). Transduction in *Bacillus megaterium*. *Biochemical and Biophysical Research Communications* **88**, 1119–1124.

Walton, J. R. (1966). *In vivo* transfer of infectious drug resistance. *Nature, London* **211**, 312–313.

Walton, J. R. (1971). The public health implications of drug-resistant bacteria in farm animals. *Annals of the New York Academy of Sciences* **182**, 358–361.

Watanabe, T. (1963). Infective heredity of multiple drug resistance in bacteria. *Bacteriological Reviews* **27**, 87–115.

Watanabe, T. and Ogata, Y. (1970). Abortive transduction of resistance factor by bacteriophage P22 in *Salmonella typhimurium*. *Journal of Bacteriology* **102**, 596–597.

Watson, B., Currier, T. C., Gordon, M. P., Chilton, M-D., and Nester, E. W. (1975). Plasmid transfer required for virulence of *Agrobacterium tumefaciens*. *Journal of Bacteriology* **123**, 255–264.

Weinberg, S. R. and Stotzky, G. (1972). Conjugation and genetic recombination of *Escherichia coli* in soil. *Soil Biology and Biochemistry* **4**, 171–180.

Willetts, N. S. (1977). The transcriptional control of fertility in F-like plasmids. *Journal of Molecular Biology* **112**, 141–148.

Willetts, N. S. (1978). *In* "Microbiology 1978", pp. 211–213. (Schlessinger, D., ed.) American Society for Microbiology, Washington, D.C.

Williams, P. H. (1977). Plasmid transfer in the human alimentary tract. *FEMS Microbiology Letters* **2**, 91–95.

Williams, P. H. (1979). Novel iron uptake system specified by ColV plasmids: an important component in the virulence of invasive strains of *Escherichia coli*. *Infection and Immunity* **26**, 925–932.

Williams, P. H. and George, H. K. (1979). ColV plasmid-mediated iron uptake and the enhanced virulence of invasive strains of *Escherichia coli*. *In* "Plasmids of Medical, Environmental and Commercial Importance", pp. 161–172. (Timmis, K. N. and Pühler, A., eds.) Elsevier/North Holland, Amsterdam.

Witte, W. (1977). Transfer of drug-resistance-plasmids in mixed cultures of staphylococci. *Zentralblatt für Bakteriologie und Hygiene. Abteilung* **1**, *Originale A* **237**, 147–159.

Yang, F., Montoya, A. L., Merlo, D. J., Drummond, M. H., Chilton, M. D., Nester, E. W., and Gordon, M. P. (1980). Foreign DNA sequences in crown gall teratomas and their fate during the loss of the tumorous traits. *Molecular and General Genetics* **177**, 707–714.

Yu, L. and Baldwin, J. N. (1971). Intraspecific transduction in *Staphylococcus epidermidis* and interspecific transduction between *Staphylococcus aureus* and *Staphylococcus epidermidis*. *Canadian Journal of Microbiology* **17**, 767–773.

9

Hydrogen Transfer in Microbial Communities

M. J. Wolin

1. INTRODUCTION

Hydrogen is a major product of the fermentation of organic matter in anaerobic microbial ecosystems, examples of which are the intestinal tract (rumen and large intestine), swamps, and anaerobic waste decomposition. Although large quantities of hydrogen are produced in these environments, it often does not accumulate. The microbial communities present contain populations of hydrogen-using microorganisms (hydrogenotrophs) as well as hydrogen-producing microorganisms (hydrogenogens). Methane-producing bacteria (methanogens) are an important group of hydrogenotrophs, obtaining energy for growth by using hydrogen to reduce carbon dioxide (bicarbonate) to methane:

$$4H_2 + HCO_3^- \rightarrow CH_4 + 2H_2O + OH^- \qquad (1)$$

Production and use of hydrogen often represent a classical food chain. Hydrogenogens produce hydrogen from substrates that provide them with energy and this hydrogen becomes an energy substrate for hydrogenotrophs. A special feature of the food chain is that hydrogenotrophs, by using hydrogen, can cause hydrogenogens to produce more hydrogen than they would produce in the absence of hydrogenotrophs. Some hydrogen-producing systems of hydrogenogens are normally inhibited by hydrogen and can only be fully expressed when hydrogen is rapidly removed from the environment. Certain hydrogenogens depend almost entirely on these inhibited systems for oxidation and metabolism of their organic energy source. This means that on such substrates they can only grow together with hydrogenotrophs. These syntrophic associations are very important in several anaerobic ecosystems.

Some hydrogenogens have hydrogen-producing systems that, while they are inhibited by hydrogen, are not essential for growth. Such organisms can catabolize organic substrates in the absence of a hydrogenotroph. When a hydrogenotroph is present, however, the inhibited systems are expressed. Not only does the organism dispose of substrate electrons in different ways in the absence and presence of hydrogenotrophs, but the products of substrate metabolism differ as well. This type of interaction profoundly influences the course of fermentation in several anaerobic ecosystems.

This chapter reviews the relationships between hydrogenogens and hydrogenotrophs, and discusses the significance of these relationships to fermentation in anaerobic microbial ecosystems.

2. SYNTROPHIC ASSOCIATIONS

2.1 Use of Primary Alcohols

Barker's isolation and characterization of *Methanobacillus omelianskii* significantly advanced the understanding of the origin of biologically produced methane (Barker, 1940, 1941). It was shown that the culture produced acetate and methane from ethanol and bicarbonate, with the stoichiometry represented by the equation:

$$2C_2H_5OH + HCO_3^- \rightarrow 2CH_3CO_2^- + CH_4 + H_2O + H^+ \qquad (2)$$

The stoichiometry strongly supported the hypothesis, originally suggested by van Niel, that methane is formed by the reduction of carbon dioxide. *M. omelianskii* oxidizes ethanol to acetate and uses the electrons derived from the oxidation to reduce carbon dioxide to methane. Although it is now known that some methanogens can produce methane directly from the methyl groups of acetate, methanol and methylamines, bioproduction from carbon dioxide is nevertheless extremely important.

Methanobacillus omelianskii was subsequently discovered to be a syntrophic association of two distinct species of bacteria (Bryant *et al.*, 1967). One, the methanogen, was called strain M.o.H. but has now been named *Methanobacterium bryantii* (Balch *et al.*, 1979). It derives its energy for growth from reduction of carbon dioxide by hydrogen (equation 1). It cannot use any other electron donor (including ethanol) to reduce carbon dioxide. The other species, which is not a methanogen, is called the S organism; it has never been formally named. Organism S oxidizes ethanol to acetate and hydrogen but barely grows by itself because of hydrogen accumulation which inhibits the oxidation reaction that provides energy for the organism. The two species grow well together with ethanol and carbon dioxide, the S organism provides the methanogen with the hydrogen it needs for growth and the methanogen removes the hydrogen that inhibits S organism's growth (Fig. 1). The reason that the S organism needs the methanogen to grow well can be understood by examining the biochemistry and thermodynamics of its production of hydrogen from ethanol. The biochemical steps and free-energy changes are given by:

$$C_2H_5OH + NAD^+ \rightarrow CH_3CHO + NADH + H^+; \quad \Delta G^{\circ\prime} = +23.8\ \text{kJ} \tag{3}$$

$$NADH + H^+ \rightarrow NAD^+ + H_2; \quad \Delta G^{\circ\prime} = +18.0\ \text{kJ} \tag{4}$$

$$CH_3CHO + H_2O \rightarrow CH_3CO_2^- + H_2 + H^+; \quad \Delta G^{\circ\prime} = -32.2\ \text{kJ} \tag{5}$$

Total
$$C_2H_5OH + H_2O \rightarrow CH_3CO_2^- + 2H_2 + H^+; \quad \Delta G^{\circ\prime} = +9.6\ \text{kJ} \tag{6}$$

The free-energy changes are for the standard state of one atmosphere of hydrogen. (See Thauer *et al.* (1977) for conventions, and Wolin (1976) for more detailed discussions of aspects of thermodynamics mentioned in this article). The overall free-energy change is positive, as are those for equations (3) and (4). If hydrogen is not removed, oxidation of ethanol to acetate is inhibited because the reverse reactions that use

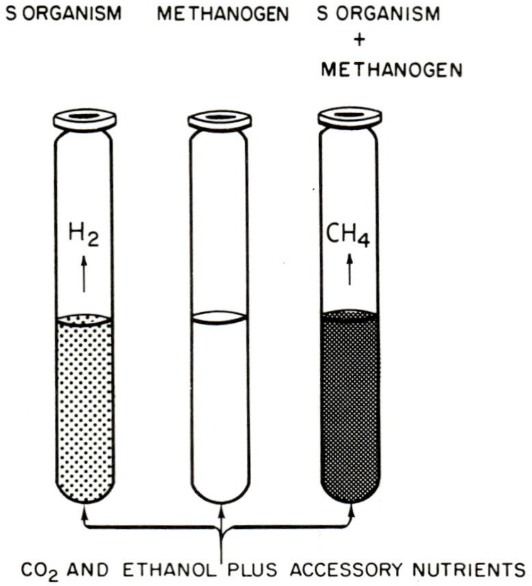

Fig. 1. Growth of the S organism, the methanogen, and S plus methanogen in an ethanol-containing medium with a carbon dioxide atmosphere. S organism alone grows poorly and produces hydrogen; the methanogen does not grow; and growth of the S organism plus the methanogen is extensive and results in the production of methane.

hydrogen are favoured. The continuous removal of hydrogen pulls the reactions in the direction of acetate and hydrogen production. If the partial pressure is 1×10^{-4} rather than one atmosphere, the $\Delta G^{\circ\prime}$ for equation (6) is -35.9 kJ.

The step most sensitive to the hydrogen partial pressure is the oxidation of NADH to NAD^+ and hydrogen. This is one of the most important steps in any of the interactions between hydrogenotrophs and hydrogenogens that have hydrogen-inhibited systems for producing hydrogen. Enzymes necessary for the production of hydrogen from NADH are not uncommon in the hydrogenogens of anaerobic ecosystems. The pathway of hydrogen production in the S organism emphasizes the central role of ferredoxin as a carrier of electrons from electron donors to protons, which become reduced to hydrogen (Fig. 2) (Reddy *et al.*, 1972b,c).

The S organism can grow with propanol or butanol instead of ethanol and oxidizes the alcohols to their corresponding acids (Reddy *et al.*, 1972a). As with ethanol, growth of the S organism requires carbon dioxide and the presence of a methanogen. The S organism is peculiar

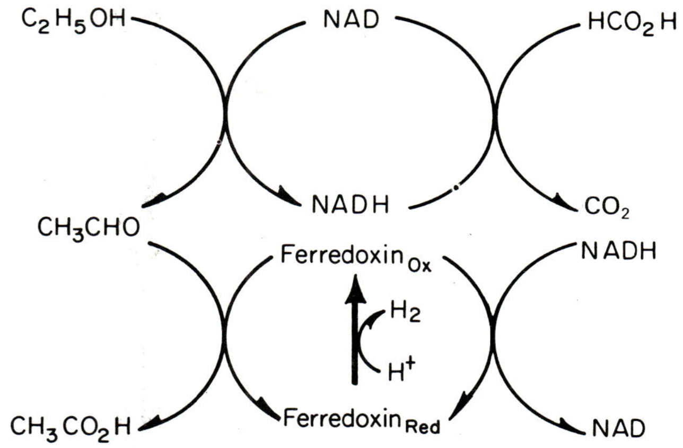

Fig. 2. Hydrogen pathway for the S organism. Ox = oxidized, Red = reduced.

in its ability to produce hydrogen from formic acid with a NAD^+-dependent system (Reddy *et al.*, 1972b). It cannot grow on formic acid but given ethanol and carbon dioxide it can grow and feed hydrogen from formic acid and ethanol to a methanogen.

Members of the genus *Desulfovibrio* can grow with ethanol, carbon dioxide, and a methanogen with the production of acetate and methane (Bryant *et al.*, 1977). Like the S organism, the *Desulfovibrio* species grow poorly by themselves; in contrast to it, however, they can use sulphate as an electron acceptor for the oxidation of ethanol. They can grow alone by oxidizing ethanol to acetate and by reducing sulphate to sulphide or, if a methanogen is present to use the hydrogen produced, they can grow by oxidizing ethanol to acetate and hydrogen. Both the *Desulfovibrio* species and the S organism presumably form ATP when acetaldehyde is oxidized to acetate, by analogy with the known reactions in *Clostridium kluyveri* which forms acetyl-CoA and acetyl phosphate from acetaldehyde. However, attempts to show ATP or acetyl phosphate formation from ethanol by the S organism were not successful (Reddy *et al.*, 1972b).

2.2 Use of Lactic Acid

As *Desulfovibrio* species can grow with ethanol and sulphate or with ethanol, carbon dioxide, and a methanogen, they also can grow with lactic acid and sulphate or lactic acid plus carbon dioxide and a methanogen (Bryant *et al.*, 1977). Although the enzymology has not been investigated, the system appears to be analogous to the S organism's ethanol system. Oxidation of lactic to pyruvic acid is probably NAD^+-dependent and the production of hydrogen from NADH depends on hydrogen removal by the methanogen. Subsequent oxidation of pyruvic acid to acetic acid and carbon dioxide may or may not involve NAD^+. In any case, the oxidations can be performed without a methanogen if sulphate is present and the desulfovibrio can use its capability of reducing sulphate to sulphide.

2.3 Hydrogenotrophs

Although the hydrogenotrophs discussed thus far have been methanogens, other hydrogenotrophs can also participate in syntrophic and other associations with hydrogenogens. Hydrogenotrophs by definition obtain energy for growth by using hydrogen to reduce an electron acceptor which may be carbon dioxide, sulphate, nitrate, oxygen, or fumarate (Table 1).

As well as the methanogens' reduction of carbon dioxide to methane, carbon dioxide is reduced by other bacteria to acetic acid (Balch *et al.*, 1977).

Certain *Desulfovibrio* species obtain energy by reducing sulphate to sulphide (Badziong *et al.*, 1978). *Desulfovibrio* species display a broad range of electron-transport reactions. As already mentioned, they can grow with organic electron donors and either sulphate or carbon dioxide plus a methanogen. Their ability to reduce sulphate with hydrogen allows them to grow and interact with other organisms that produce hydrogen from organic substrates not dissimilated by the sulphate-reducing bacteria.

Nitrate is reduced by some hydrogenotrophs to nitrous oxide and nitrogen or ammonia, depending on the species (Yoshinari, 1980).

Oxygen is a well-known acceptor for certain hydrogenotrophs. Although these organisms normally grow in aerobic environments with hydrogen that has diffused in from anaerobic environments, it is possible that they interact with anaerobic ecosystems at the anaerobic: aerobic zone interface.

Fumarate which appears to be the most common of the organic acceptors is reduced with hydrogen to succinate (Macy *et al.*, 1976; Wolin *et al.*, 1961).

2.4 Fatty Acids

Butyrate and propionate are important products of the fermentation of

Table 1. Reactions which provide energy for growth of hydrogenotrophs.

Reaction	Illustrative Species
$4H_2 + CO_2 \rightarrow CH_4 + 2H_2O$	*Methanobacterium bryantii*
$4H_2 + 2CO_2 \rightarrow CH_3CO_2H + 2H_2O$	*Acetobacterium woodii*
$4H_2 + SO_4^{2-} \rightarrow S^{2-} + 4H_2O$	*Desulfovibrio desulfuricans*
$5H_2 + 2NO_3^- + 2H^+ \rightarrow N_2 + 6H_2O$	*Paracoccus denitrificans*
$H_2 + NO_3^- \rightarrow NO_2^- + H_2O$	*Vibrio succinogenes**
$4H_2 + NO_3^- + 2H^+ \rightarrow NH_4^+ + 3H_2O$	*Vibrio succinogenes*
$H_2 + N_2O \rightarrow N_2 + H_2O$	*Vibrio succinogenes*
$2H_2 + O_2 \rightarrow 2H_2O$	*Paracoccus denitrificans*
H_2 + fumaric acid $\rightarrow$ succinic acid	*Vibrio succinogenes*

* Renamed *Wolinella succinogenes* (Tanner *et al.*, 1981).

organic substrates by populations of anaerobic microbes. In systems with long turnover times, these acids can be completely converted to methane. The equations for the conversions are:

$$2CH_3CH_2CH_2CO_2H + 2H_2O \rightarrow 5CH_4 + 3CO_2 \quad (7)$$

$$4CH_3CH_2CO_2H + 2H_2O \rightarrow 7CH_4 + 5CO_2. \quad (8)$$

Radioisotope studies indicated that both acids are oxidized to acetate with the reduction of carbon dioxide to methane (Stadtman and Barker, 1951). Acetic acid is subsequently converted to methane and carbon dioxide. The partial reactions can be represented as follows:

$$2CH_3CH_2CH_2COOH + 4H_2O \rightarrow 4CH_3COOH + 8(H) \quad (9)$$

$$8(H) + CO_2 \rightarrow CH_4 + 2H_2O \quad (10)$$

$$4CH_3CO_2H \rightarrow 4CH_4 + 4CO_2 \quad (11)$$

and

$$4CH_3CH_2COOH + 8H_2O \rightarrow 4CH_3COOH + 4CO_2 + 24(H) \quad (12)$$

$$24(H) + 3CO_2 \rightarrow 3CH_4 + 6H_2O \quad (13)$$

$$4CH_3CO_2H \rightarrow 4CH_4 + 4CO_2 \quad (14)$$

Recent studies (McInerny *et al.*, 1979; Boone and Bryant, 1980) demonstrated that the conversion of either fatty acid to acetate and methane is accomplished through syntrophic association of a specific bacterium (which oxidizes the fatty acid to hydrogen and acetate) with a hydrogenotroph (either a methanogen or *Desulfovibrio* species, which uses the hydrogen to produce methane or sulphide). One species oxidizes butyric, caproic, and caprylic acids to acetic acid plus hydrogen, as well as valeric and heptanoic acids to propionic and acetic acids

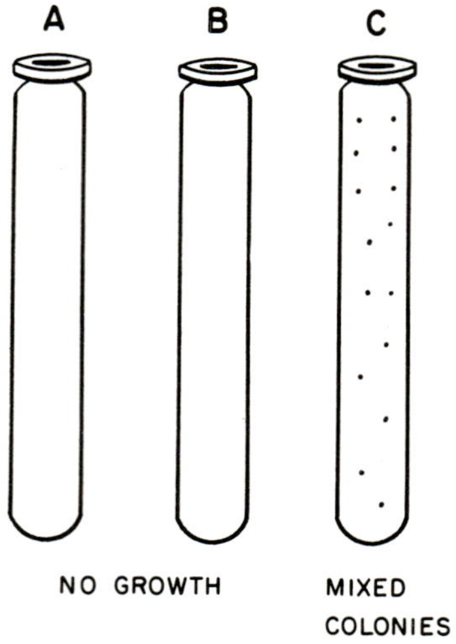

Fig. 3. Isolation of bacteria which use butyrate or propionate in syntrophic association with methanogens. A is inoculated with a dilution from an enrichment culture that produces methane from butyrate or propionate; B is inoculated with a large number of cells of a methanogen; and C = A + B. The agar medium contains the volatile fatty acid, accessory nutrients and a carbon dioxide atmosphere.

plus hydrogen. The other species oxidizes propionic acid to acetic acid and hydrogen. The fatty-acid oxidizer cannot grow at all without the hydrogenotroph, in contrast to the S organism and the *Desulfovibrio* species which can grow to a limited extent on ethanol without a hydrogenotroph. To purify the fatty-acid oxidizer, it is necessary to obtain isolated colonies on a lawn of a hydrogenotroph and to pick the mixed colonies (Fig. 3). Hydrogen accumulation in the absence of a lawn is probably sufficient to prevent any visible growth. Ethanol oxidizers, on the other hand, can grow sufficiently to produce visible colonies before hydrogen accumulation inhibits growth.

The thermodynamic barrier to hydrogen production from butyrate, propionate, or any other short- or long-chain fatty acid (except acetic acid) is much greater than the barrier to production from the corresponding alcohols, but it apparently can also be overcome by the removal of hydrogen (Table 2). The biochemical mechanisms used by the fatty acid-oxidizing species have not been established. Comparative biochemistry suggests, however, that for fatty acids with chain lengths of four or more carbon atoms, a sequence of activation of the acid with ATP and coenzyme A, oxidation to an unsaturated acyl-CoA, hydration to a hydroxyacyl-CoA, oxidation to a β-ketoacyl-CoA and, finally, cleavage to acetyl-CoA and an acyl-CoA residue.

The major thermodynamic barrier to hydrogen production is the oxidation of the saturated to the corresponding unsaturated acid. This can be seen by comparing the free energy changes for the oxidation of butyric, crotonic, and β-hydroxybutyric acids (Table 2). Wherever it has been studied, the oxidation of acyl-CoA esters of saturated acids to the unsaturated esters is catalyzed by a flavin-dependent enzyme. The oxidation of a reduced flavin enzyme to hydrogen plus the oxidized enzyme may be the step in the fatty-acid syntrophic systems that requires the maintenance of extremely low hydrogen partial pressures by the hydrogenotroph.

Pathways for oxidation of propionic acid might involve:

(1) prior conversion to succinic acid, followed by oxidation to fumaric acid and then the sequence of malic to oxalacetic to pyruvic to acetic acids; or
(2) oxidation to acrylic acid, followed by the sequence of lactic to pyruvic to acetic acids.

Both pathways require the oxidation of a saturated to an unsaturated acid, i.e. succinic to fumaric acid and propionic to acrylic acid. Again, through analogies from comparative biochemistry, these oxidations would require flavin enzymes and extremely low hydrogen partial pressures to produce hydrogen.

Degradation of long-chain fatty acids to methane in anaerobic ecosystems probably requires syntrophic associations similar to those that have been found for butyric and propionic acids. Even-numbered carbon chains would be cleaved to acetic acid and methane, and odd-numbered chains to acetic and propionic acids and methane, followed

Table 2. Free-energy changes for conversion of analogous alcohols and acids to acetic acid and hydrogen.

Reaction	$\Delta G^{o'}$ (kJ)
Propanol → propionic acid	+12.1
Propionic acid → acetic acid	+76.1
Butanol → butyric acid	+16.3
Butyric acid → acetic acid	+48.1
Crotonic acid → acetic acid	−26.9
β-hydroxybutyric acid → acetic acid	−35.1

by conversion of propionic acid to acetic acid and methane. Acetic acid, an intermediate in all of these fermentations, is cleaved to methane and carbon dioxide by methanogens that produce methane from the methyl group of acetic acid.

2.5 Syntrophism and Anaerobic Respiration

The syntrophic electron-transport reactions that have been discussed are special cases of anaerobic respiration of organic compounds. An organism that uses anaerobic respiration to dissimilate an organic compound requires an additional compound to serve as an electron acceptor for oxidation of the substrate. However, the organism cannot generate from the substrate, intermediates that can be used as electron acceptors in its oxidation. This may be due to the nature of the substrate itself to the physiological characteristics of the organism. It is difficult to visualize how an organism could oxidize ethanol and at the same time generate a thermodynamically and biochemically viable electron acceptor from ethanol. Organisms like propionibacteria can dissimilate lactic acid without an electron acceptor, while organisms like desulfovibrios cannot. *Desulfovibrio* species can dissimilate both ethanol and lactic acid by anaerobic respiration if provided with the electron acceptor sulphate. Some *Desulfovibrio* species can also dissimilate the substrates by reducing protons (which are ubiquitous electron acceptors) to hydrogen. When thermodynamics preclude continuous reduction of protons to hydrogen, a hydrogenotroph must be included in a syntrophic system to scavenge the hydrogen.

The examples of specific syntrophic associations that have been discussed are based on what has been demonstrated with a few species of hydrogenogens and hydrogenotrophs, but they may represent only the tip of the iceberg. For example, it is possible that bacteria exist which can oxidize glucose to acetic acid, hydrogen, and carbon dioxide in syntrophic association with hydrogenotrophs but which cannot ferment glucose alone.

3. HYDROGENOTROPHS AND CARBOHYDRATE FERMENTATION

In syntrophic associations, oxidation of organic substrates which supply energy to a hydrogenogenic species is critically dependent on hydrogen use by a hydrogenotroph. Removal of hydrogen by hydrogenotrophs can also significantly influence the kinds of products formed by fermentative species. By definition, fermentation does not require the addition of an electron-acceptor compound to oxidize an organic substrate. Oxidations of fermentation intermediates are coupled to the reduction of substrate-produced intermediates.

3.1 Formation of Hydrogen

Until recently the hydrogen produced from carbohydrate fermentation was generally assumed to come directly or indirectly from the cleavage of pyruvate. Pyruvate can be cleaved directly to acetyl-CoA, hydrogen, and carbon dioxide by some bacteria. Others cleave pyruvate to acetyl-CoA and formate and then convert formate to hydrogen and carbon dioxide. The free-energy change for the formation of the fermentation products acetic acid, hydrogen, and carbon dioxide by either of these cleavage pathways (equation 15) is −47.3 kJ. Production of hydrogen from pyruvate should be unaffected by the hydrogen partial pressure at ordinary pressures if thermodynamic factors are the only consideration.

$$CH_3COCO_2^- + 2H_2O \rightarrow CH_3CO_2^- + HCO_3^- + H_2 + H^+ \quad (15)$$

The possibility that oxidation of NADH to NAD^+ and hydrogen is a source of hydrogen produced from carbohydrates has usually been ignored. Interpretation of several classical fermentation balances in terms of well established biochemical pathways strongly implicates NADH, as well as pyruvate, as an electron source for the reduction of protons to hydrogen. Tables 3 and 4 illustrate this for the fermentation of glucose by *Clostridium butyricum* (Donker, 1926). Two oxidations are known to occur in the fermentation: the oxidation of glyceraldehyde-3-

Table 3. Clostridium butyricum fermentation of glucose (after Donker, 1926).

Product	mmol product per 100 mmol glucose fermented
Carbon dioxide	195.5
Hydrogen	233.0
Acetic acid	42.6
Butyric acid	75.3
% Carbon recovered	97.0
Oxidation/reduction index	1.02
C_1/C_2 balance	1.01

Table 4. Clostridium butyricum fermentation. Equations based on data in Table 3.

	$97C_6H_{12}O_6 + 194NAD^+ \rightarrow 194CH_3COCO_2H + 194NADH + 194H^+$
	$194CH_3COCO_2H \rightarrow 194CH_3CO_2H + 194H_2 + 194CO_2$
	$150CH_3CO_2H + 150NADH + 150H^+ \rightarrow 75CH_3CH_2CH_2CO_2H + 150NAD^+$
	$44NADH + 44H^+ \rightarrow 44H_2 + 44NAD^+$
Total	$97C_6H_{12}O_6 \rightarrow 75CH_3CH_2CH_2CO_2H + 44CH_3CH_2H + 238H_2 + 194CO_2$

phosphate coupled to NAD^+ reduction, and the oxidation of pyruvic acid coupled to proton reduction. All of the calculated NADH formed by oxidation cannot be used in the terminal reduction steps that form butyrate (Table 4); some of it must be used to reduce protons to hydrogen and the equations of Table 4 closely fit the experimental values of Table 3. Experimental studies confirm that oxidation of NADH to NAD^+ and hydrogen is a source of hydrogen in clostridial carbohydrate fermentations (Jungermann *et al.*, 1973).

One possible reason why NADH has been overlooked as a source of hydrogen in fermentations is that the thermodynamics of the reaction (see equation 4), i.e. the conversion of NADH to NAD^+ and hydrogen under standard conditions, shows a large positive change in free energy. As has already been discussed, the free energy is changed when the hydrogen is removed by a hydrogenotroph. A simple thermodynamic argument, however, is insufficient to explain effects of hydrogen on oxidation of the NADH of glycolysis to NAD^+ and hydrogen. For the hypothetical fermentation of equation (16), the free-energy change is -206 kJ, despite the fact that the only method for oxidation of NADH leads to the formation of NAD^+ and hydrogen:

$$C_6H_{12}O_6 \rightarrow 2CH_3CO_2^- + 4H_2 + 2HCO_3^- + 4H^+ \qquad (16)$$

In theory, the negative free-energy changes of the exergonic reactions of glycolysis should be sufficient to drive the endergonic oxidation of NADH to NAD^+ and hydrogen. However, there are no organisms known that use the process in equation (16) for obtaining energy for growth, except in the presence of a hydrogenotroph. This is probably because mechanisms do not exist for using ATP to drive the endergonic oxidation of NADH to NAD^+ and hydrogen. Therefore, in the presence of the appropriate enzymes and small amounts of hydrogen, almost all of the NAD^+ of the cell would be converted to NADH. Glycolysis would be blocked, e.g. in the Embden-Meyerhof-Parnas (EMP) scheme glyceraldehyde-3-phosphate would not be oxidized. The equilibrium ratio of NAD^+ to NADH at pH 7.0 and one atmosphere of hydrogen is $7 \times 10^{-4}:1$. A hydrogenotroph increases this ratio by removing hydrogen (and thus promoting oxidation of NADH to NAD^+) if the appropriate enzymes are available.

3.2 Growth of the S Organism on Pyruvic Acid

Direct experimental evidence for fermentation shifts caused by hydrogenotrophs came from studies of pyruvic acid fermentation by the S organism (Reddy *et al.*, 1972a). Pyruvic acid is the only organic substrate, of many tested, that supports good growth of the S organism alone; otherwise the S organism relies on alcohols and syntrophic growth with methanogens. The major products of pyruvic acid fermentation are acetic acid and ethanol and little hydrogen is formed. However, if the S organism is grown with a methanogen, acetic acid is the major two-carbon product and little ethanol is produced from pyruvic acid. Methane is produced in the mixed culture in amounts far greater than expected on the basis of the hydrogen produced by the S organism alone. The fermentation data for the S organism and the S organism plus *Methanobrevibacter smithii* are given in Table 5. The amount of hydrogen needed for the production of the 24 mol methane (100 mol pyruvic acid)$^{-1}$ is 4×24 (or 96) mol, i.e. about 32 times that produced by the S

Table 5. Fermentation of pyruvic acid by S organism without and with a methanogen. (Data from Reddy *et al.*, 1972a). Diacetyl, acetoin and 2,3-butanediol were also found in amounts less than 2 mol (100 mol pyruvic acid)$^{-1}$. Calculated formic acid as mol two-carbon minus one-carbon products

	Mol (100 mol pyruvic acid)$^{-1}$	
Product	S organism	S organism plus *Methanobrevibacter smithii*
Ethanol	34	3
Acetic acid	61	92
Hydrogen	3	0
Carbon dioxide	66	59
Formic acid	29	12
Methane	0	24

Table 6. Equations for fermentation of pyruvic acid by the S organism without and with a methanogen

$$100CH_3COCO_2H + 100CoASH \rightarrow 100CH_3COSCoA + 100HCO_2H$$

A. The S organism alone

$$66HCO_2H + 66NAD^+ \rightarrow 66CO_2 + 66NADH + 66H^+$$
$$66NADH + 66H^+ + 33CH_3COSCoA \rightarrow 33C_2H_5OH + 66NAD^+ + 33CoASH$$
$$67CH_3COSCoA + 67H_2O \rightarrow 67CH_3CO_2H + 67CoASH$$

Total: $100CH_3COCO_2H + 67H_2O \rightarrow 67CH_3CO_2H + 33C_2H_5OH + 34HCO_2H + 66CO_2$

B. The S organism plus methanogen

$$90HCO_2H + 90NAD^+ \rightarrow 90CO_2 + 90NADH + 90H^+$$
$$90NADH + 90H^+ \rightarrow 90H_2 + 90NAD^+$$
$$90H_2 + 22.5CO_2 \rightarrow 22.5CH_4 + 45H_2O$$
$$100CH_3COSCoA + 100H_2O \rightarrow 100CH_3CO_2H + 100CoASH$$

Total: $100CH_3COCO_2H + 65H_2O \rightarrow 100CH_3CO_2H + 10HCO_2H + 22.5CH_4 + 67.5CO_2$

organism alone. The data of Table 5 can be rationalized by the equations of Table 6. These equations assume that formic acid accumulates when the S organism is grown alone and ignore the minor amounts of other products. Although formic acid was not measured, the S organism contains an NAD^+-linked formic dehydrogenase and formic acid would account for the considerable missing one-carbon in the fermentation balance of the S organism alone as already pointed out by Reddy *et al.* (1972a). The equations of Table 6 agree reasonably well with the data of Table 5 as far as accounting for the stoichiometry of:

(1) the conversion of pyruvic acid to ethanol, acetic acid, and one-carbon products in the single culture; and
(2) the conversion of pyruvic acid to acetic acid, methane and carbon dioxide in the mixed system.

It is thermodynamically feasible for the S organism to obtain energy for growth by converting pyruvic acid to acetic acid, hydrogen, and carbon dioxide directly, but the organism appears unable to avoid making the reduced product ethanol when it is grown by itself. This is probably due to the rapid achievement of equilibrium between hydrogen and the pyridine nucleotide system. Since the biochemistry of pyruvic acid metabolism by the S organism has not yet been investigated, it is not certain if pyruvic acid is cleaved to acetyl-CoA and formic acid or if its oxidation is coupled to pyridine nucleotide reduction (Table 6). Maintenance of the pyridine nucleotide pool in the reduced state by hydrogen reduction would block oxidation of pyruvic as well as formic acid. Besides the one given in Table 6, mechanisms can be written for pyruvic acid catabolism which directly involve pyridine nucleotides (Wolin, 1976). The explanation for ethanol formation, however, is basically the same; regardless of how the electrons are removed from pyruvic acid, they remove NAD^+ from the cell pool by reduction to NADH and a sufficient NAD^+ concentration is no longer available for some essential cell function. Ethanol formation, therefore, becomes necessary for re-oxidation of NADH. When the methanogen is added, direct oxidation of NADH to NAD^+ and hydrogen can be expressed, due to the hydrogen removal, and more carbon flows from pyruvic to acetic acid because ethanol formation is no longer necessary and is not favoured in the mixed system. Figure 4 shows the alternative pathways for oxidation of NADH.

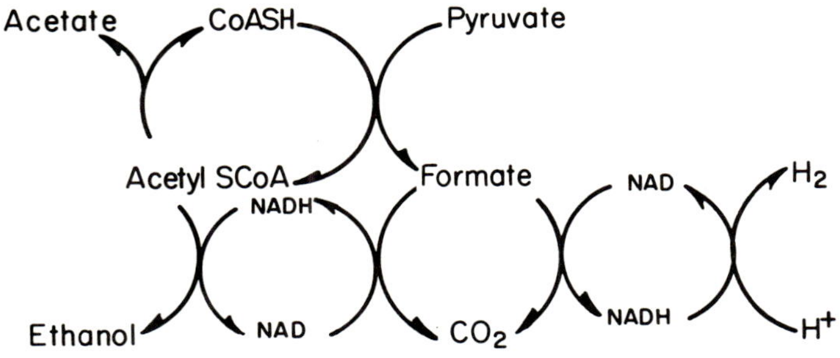

Fig. 4. Pyruvate fermentation and alternate pathways for NADH oxidation by S organism.

3.3 Growth of *Ruminococcus albus* on glucose

Ruminococcus albus is one of the major cellulolytic species of the rumen microbial ecosystem (Iannotti *et al.*, 1973). When grown on limiting concentrations of glucose in a chemostat, it produces ethanol, acetate, hydrogen, and carbon dioxide. Batch culture fermentation results in the same products plus formic acid from glucose, cellobiose, or cellulose. Formation of ethanol is the major electron sink for the re-oxidation of NADH formed during glycolysis.

To determine the effect of hydrogen removal on the carbohydrate fermentation, *Ruminococcus albus* was grown in a glucose-limited chemostat

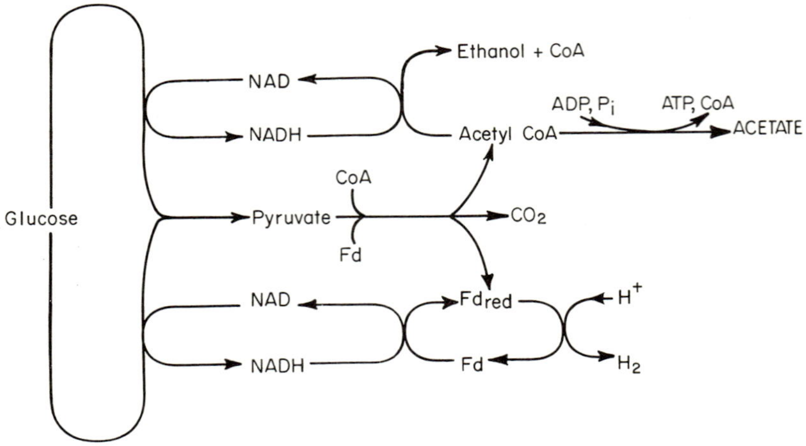

Fig. 5. Glucose fermentation and alternate pathways for NADH oxidation by *Ruminococcus albus*. (Glass *et al.*, 1977; reprinted by permission of the American Society of Microbiology.)

with the hydrogenotroph *Vibrio succinogenes*. *V. succinogenes* derives energy for growth by using hydrogen to reduce fumaric to succinic acid. Although the medium used for growing the mixed culture contained fumaric acid, *V. succinogenes* did not grow unless *R. albus* also grew and supplied hydrogen to *V. succinogenes*. In continuous-flow culture, both organisms grew well, producing significant amounts of succinic acid which were formed by *V. succinogenes*. No ethanol was produced from glucose and the only organic product of glucose fermentation was acetic acid. Similar results were obtained with a methanogen as the hydrogenotroph, except that carbon dioxide was reduced to methane instead of reducing fumaric acid to succinic acid (M. J. Wolin, unpublished results). Figure 5 shows a schematic representation of the differences in the fermentation of *R. albus* without and with a hydrogenotroph. The enzymological basis of the scheme has been established (Miller and Wolin, 1973; Glass *et al.*, 1977). All the results are consistent with the explanation that lowering the concentration of hydrogen by the hydrogenotroph favours proton reduction by NADH rather than ethanol formation.

The fermentation of *Clostridium thermocellum* is similar to that of *Ruminococcus albus*. Cellulose ferments to ethanol, acetate, hydrogen, and carbon dioxide. Co-fermentation with a methanogen has the same affect on the *C. thermocellum* fermentation as on the *R. albus* fermentation, i.e. ethanol formation decreases and acetate increases (Weimer and Zeikus, 1977).

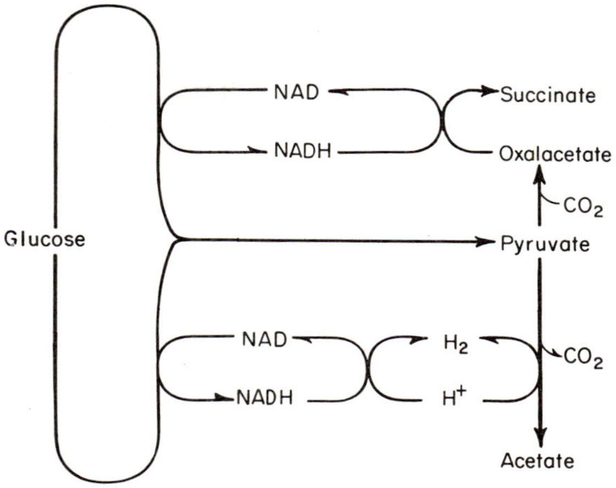

Fig. 6. Glucose fermentation and alternate pathways for NADH oxidation by *Ruminococcus flavefaciens*.

3.4 Growth of *Ruminococcus flavefaciens* on Cellulose

Ruminococcus flavefaciens is another major cellulolytic species of the rumen microbial ecosystem (Latham and Wolin, 1977). It ferments cellulose to acetic and succinic acids and small amounts of formic acid and hydrogen. Formation of succinic acid is the major electron sink for NADH oxidation. When *R. flavefaciens* is grown on cellulose with a methanogen, much larger amounts of methane are formed than are expected from the amounts of hydrogen and formic acid produced by the *Ruminococcus* species alone. The methanogen strain reduces carbon dioxide to methane with formic acid as well as hydrogen. There is a large decrease in the amount of succinic acid and a large increase in the amount of acetic acid produced by the combined culture, compared to *R. flavefaciens* alone. The results strongly suggest that the use of hydrogen favours the flow of electrons from NADH of the *R. flavefaciens* fermentation to protons for hydrogen formation, rather than to oxalacetic and fumaric acids to form succinic acid (Fig. 6).

3.5 Growth of *Selenomonas ruminantium* on Glucose and Lactic Acid

Selenomonas ruminantium, whilst also a major rumen species, is not cellulolytic. Starch, sugars, glycerol, or lactic acid is fermented, depending

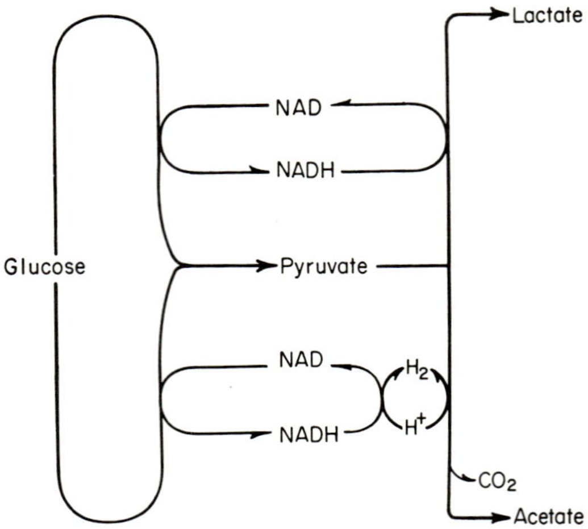

Fig. 7. Glucose fermentation and alternate pathways for NADH oxidation by *Selenomonas ruminantium*.

on the strain, to varying proportions of lactic, propionic, and acetic acids and carbon dioxide (Chen and Wolin, 1977). Some strains were recently found to produce trace amounts of hydrogen which was not previously detected because when the fermentation products were first described, sensitive gas chromatographic procedures for the detection of hydrogen were not available. The effects of growing hydrogen-producing strains of *S. ruminantium* with a methanogen are very dramatic. Large amounts of methane are formed in the mixed cultures and the almost cryptic hydrogen production of *S. ruminantium* is well expressed in the mixed cultures (Scheifinger *et al*., 1975). One strain of *S. ruminantium* produces almost 100 times the amount of hydrogen (calculated from the methane produced) when grown with a methanogen. These results emphasize a need for caution in concluding, on the basis of measurements of mono-culture fermentations, that an organism is not capable of producing significant amounts of hydrogen. Its hydrogen-producing system may be very sensitive to inhibition by low hydrogen partial pressures, so that production may be significant only in the presence of a hydrogenotroph.

When grown alone, the selenomonad strain rapidly forms large amounts of lactic acid from glucose, and from this lactic acid it slowly forms propionic and acetic acids and carbon dioxide. The short-chain fatty acids are ultimately the major acid products. In short-term fermentations with a methanogen present, large amounts of methane and

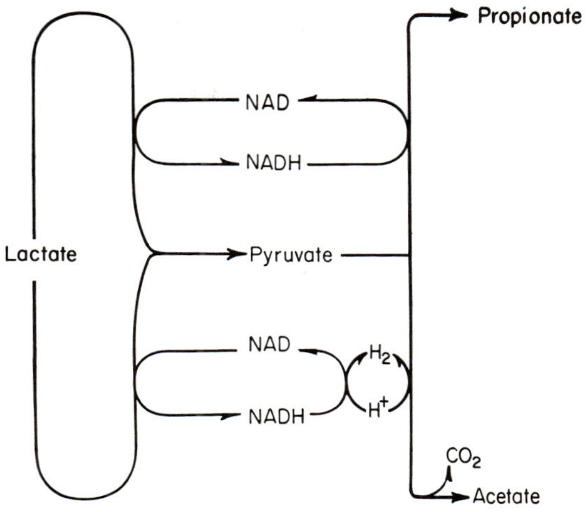

Fig. 8. Lactate fermentation and alternate pathways for NADH oxidation by *Selenomonas ruminantium*.

acetic acid, but little lactic and propionic acids, are formed. When lactic acid is the source of carbon and energy, propionic and acetic acids are the major acids produced by the selenomonad alone. With a methanogen, the amount of propionic acid formed decreases substantially, and acetic acid and methane are the major products. Figures 7 and 8 outline the biochemical basis of the fermentation shift and emphasize the effect of the removal of hydrogen on the production of hydrogen from NADH.

3.6 Effect on Homo-acetic Acid Fermentation

Acetobacterium woodii is a hydrogenotroph that can derive energy for growth not only by using hydrogen to reduce carbon dioxide to acetic acid, but also from the fermentation of fructose to acetic acid (Winter and Wolfe, 1980). The production of three mol acetic acid from one mol hexose is called a homo-acetic acid fermentation. The biochemistry of the fermentation has been elucidated mainly through studies of *Clostridium thermoaceticum* (Ljungdahl and Wood, 1969). A general representation of the pathway of formation of acetic acid is given by:

$$C_6H_{12}O_6 + 2NAD^+ \rightarrow 2CH_3COCOOH + 2NADH + 2H^+ \quad (17)$$

$$2CH_3COCOOH \rightarrow 2CH_3COOH + 4(H) + 2CO_2 \quad (18)$$

$$2NADH + 2H^+ + 4(H) + 2CO_2 \rightarrow CH_3COOH + 2NAD^+ \quad (19)$$

Two mol acetic acid are produced by the cleavage of pyruvic acid and one mol by reduction of carbon dioxide. The carbon dioxide reduction is used for regeneration of NAD^+ to sustain the fermentation. In the presence of a methanogen, *A. woodii* produces only 2 mol acetic acid from 1 mol fructose, and the methanogen forms 1 mol methane. In the mixed culture, the eight hydrogen-atom equivalents used to reduce carbon dioxide to acetic acid in the absence of the methanogen are transferred to the methanogen. *A. woodii* apparently has a mechanism for producing hydrogen from all of the hydrogen equivalents and the methanogen uses the hydrogen (Fig. 9). The interaction is unusual: the methanogen is only doing what *A. woodii* is known to be able to do alone (i.e. use hydrogen to reduce carbon dioxide) but apparently the methanogen is more efficient at the process.

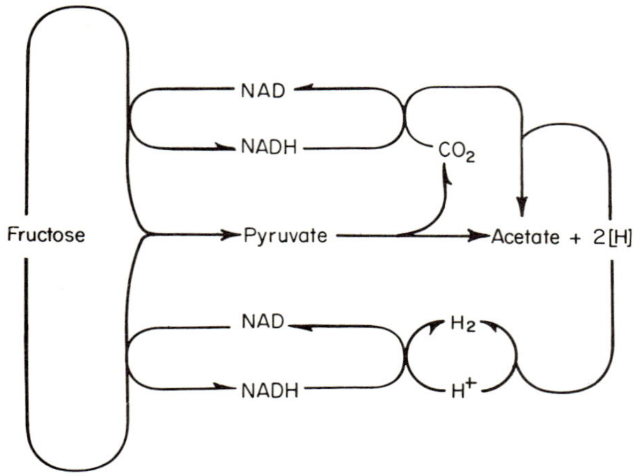

Fig. 9. Fructose fermentation and alternate pathways for NADH oxidation by *Acetobacterium woodii*.

3.7 Advantages of Interactions

Non-syntrophic interactions are obviously beneficial to hydrogenotrophs which obtain hydrogen and use it as a source of energy. The carbohydrate-fermenting hydrogenogens may benefit by obtaining more ATP from substrate in the presence than in the absence of hydrogenotrophs. The mixed culture dissimilation is a type of anaerobic respiration rather than a fermentation. Instead of reducing a fermentation intermediate to an electron-sink organic product, hydrogenotrophs use the electrons to reduce an added electron acceptor. Except for *Acetobacterium woodii*, all the interactions discussed result in more conversion of pyruvic to acetic acid than to an electron-sink product (ethanol or lactic, propionic, or succinic acids). Formation of acetic acid from pyruvic acid is generally associated with ATP formation by substrate phosphorylation, specifically by production of acetic acid and ATP from acetyl phosphate. None of the pathways to electron-sink production leads to ATP formation by substrate phosphorylation. For *A. woodii*, ATP is expended during the reduction of carbon dioxide to acetic acid because formation of the intermediate formyltetrahydrofolic acid requires ATP.

3.8 Speculation about Hydrogenotrophs and Other Fermentations

The examples of the effects of hydrogenotrophs on fermentation were taken from the results of experiments specifically designed to demonstrate the change in fermentation products caused by the presence of a hydrogenotroph. *Vibrio succinogenes* and methanogens are the only hydrogenotrophs used so far for these experiments, but there is no reason to believe that other types of hydrogenotrophs cannot participate in similar interactions.

Several reports also illustrate the effect of hydrogen partial pressure on specific fermentations (although they do not examine the effects of mixed culture growth with hydrogenotrophs). The studies of Kubowitz (1934) suggest that the addition of one atmosphere of hydrogen to cultures of *Clostridium butyricum* causes the normal butyric acid fermentation to shift to a lactate fermentation. Since the pyruvate cleavage system of a similar organism, *C. pasteurianum*, is not significantly inhibited by hydrogen (Mortenson and Wilson, 1951), the effect of hydrogen on the $NADH/NAD^+$ ratios in *C. butyricum* may shift the pathway for NADH re-oxidation from production of butyric acid and hydrogen to lactic acid production.

The oxidations of hypoxanthine to xanthine and hydrogen, and of xanthine to uric acid and hydrogen by *Veillonella alcalescens* (= *Micrococcus lactilyticus*) are reversible ($\Delta G^{\circ\prime}$ of 8.4 and 10.5 kJ, respectively). Whitely and Douglas (1951) showed that a hydrogen atmosphere inhibits the fermentation of hypoxanthine or xanthine by resting cell suspensions. Xanthine is reduced to hypoxanthine which accumulates and is not fermented (whereas it is in the absence of hydrogen). Growth on hypoxanthine or xanthine occurs under a hydrogen atmosphere. Whether hydrogenotrophs can influence the rates of these fermentations and whether the organisms can serve as hydrogenotrophs in the presence of uric acid or xanthine is not known. Production of hydrogen from hypoxanthine or xanthine by *V. alcalescens* is mediated by ferredoxin (Valentine and Wolfe, 1963).

Production of hydrogen from NADH by clostridia has already been discussed but, except for *Clostridium thermocellum*, there have been no studies of the interactions with hydrogenotrophs. Some clostridia can reduce nitrate, causing a shift from butyric to acetic acid production (Ishimoto *et al.*, 1974). Whether similar shifts could be effected by hydrogenotrophs is not known. The influence of the hydrogen partial pressure on fermentation of glycine by *Peptostreptococcus* (*Diplococcus*) *glycinophilus* to acetic acid, carbon dioxide, hydrogen, and ammonia is similar to the effect of hydrogenotrophs on the homoacetic acid fermentation of *Acetobacterium woodii*; decreasing the partial pressure favours formation of hydrogen and carbon dioxide over acetic acid production (Cardon and Barker, 1947).

Many fermentations and possible substrates for syntrophic association have not been examined from the point of view of interactions between hydrogenogens and hydrogenotrophs. In addition to glycine fermentation, amino acid fermentation and the important special case of anaerobic respiration known as the Stickland reaction (in which one amino acid is an electron donor and another is an electron acceptor), should be interesting subjects for investigation.

4. ECOSYSTEMS

4.1 Intestinal-tract Ecosystems

4.1.1 The Rumen

The elegant studies of Hungate and co-workers provided substantial evidence that the interactions between hydrogenogens and hydrogenotrophs significantly influence the course of the fermentation of plant polymers by the microbial community of the rumen. Acetic, propionic, and butyric acids plus methane and carbon dioxide are the major products of the fermentation; the acids are the major sources of carbon and energy for maintenance and growth of ruminants. Although ethanol and lactate are significant products of pure culture fermentations, by important populations of the community, ethanol is not normally produced or used in the rumen and lactate is produced only if the animals are fed high-grain diets (Jayasuria and Hungate,

1959; Moomaw and Hungate, 1963). Interactions between rumen hydrogenogens that form ethanol and lactate (*Ruminococcus albus* and *Selenomonas ruminantium*, respectively) and rumen hydrogenotrophic methanogens have already been discussed. Hungate's studies also showed that hydrogen is produced and used at high rates in the rumen ecosystem, with the net result of a low hydrogen partial pressure of 3×10^{-4} atmospheres (Hungate, 1967). The studies of the ecosystem itself and the pure culture studies based on the ecosystem observations, provide strong evidence that the interactions between hydrogenotrophs and methanogens profoundly influence the course of rumen fermentation.

The effect of the interactions on fermentation may be more complicated than is indicated by the relationships between ethanol and lactate formation and hydrogen use. Hydrogenotrophs can also influence the relative amounts of propionic, succinic, and acetic acids formed by important carbohydrate-fermenting hydrogenogens of the rumen. Succinic acid is a major precursor of propionic acid in the rumen (Blackburn and Hungate, 1963). Interactions between propionate- and succinate-producing species and methanogens already discussed indicate that methanogens potentially influence the proportions of propionic and acetic acid ultimately produced from rumen fermentation. The relative amounts of these acids significantly influence ruminant nutrition. The antibiotic monensin is extensively used as a feed additive for steers raised on high-grain diets in feedlots. It causes decreased methane and increased propionate production in the rumen and it increases the efficiency of feed utilization by the animals, probably due to the change in fermentation pattern. Monensin selects against populations that produce the major amounts of acetic and butyric acids and hydrogen in the rumen, while it selects for those that produce more propionic acid (Chen and Wolin, 1979). Methane production is probably reduced by diminished production of the methanogens' substrate for growth (hydrogen) rather than by direct inhibition of them. Reduction of methanogenesis could influence the relative amounts of propionic and acetic acids formed by the propiogenic community selected for by monensin, because the methanogens would drain fewer electrons away from the production of propionic acid.

Another potential complication, not yet studied extensively, is the possibility that populations which produce both hydrogen and propionic acid can also use hydrogen. The highest propionic acid : hexose ratio that can be formed is 4 : 3 (1.5 hexose $\rightarrow$ 2 propionic acid + acetic acid + carbon dioxide) but it could be increased to 2 : 1 by use of hydrogen (hexose + hydrogen $\rightarrow$ 2 propionic acid). In theory, a propionic acid-forming organism able to produce or take up hydrogen could shift from one extreme, i.e. of producing no propionic but only acetic acid from hexose (if all of the electrons of glycolytic oxidations are used to reduce protons to hydrogen), to the other extreme, i.e. of a homopropionic acid fermentation (if the organism can use all of the glycolytic electrons plus an extra pair (from hydrogen) for the reduction reactions involved in producing propionic acid from pyruvic acid).

Whether these complex ramifications of the interactions between hydrogenogens and hydrogenotrophs actually occur in the rumen or any other ecosystem is not known. They have been discussed because they should be considered when attempts are made to evaluate ecosystem fermentations on the basis of the fermentative abilities of single cultures examined under a limited number of laboratory conditions.

4.1.2 Large Intestine

Much less is known about the microbial communities of the large intestine of monogastric herbivores, omnivores, and carnivores. The communities that have been examined have products that are similar to those in the rumen, but the status of gas production is unclear. Approximately one third of human adults produce methane and the rest probably produce hydrogen (Levitt, 1969; Bond *et al.*, 1971). Whether methane is produced in the same portion of the large intestine as the volatile fatty acids is unknown, as are the dynamics of hydrogen production and methanogenesis. It is premature to speculate on the role of interactions between hydrogenogens and hydrogenotrophs in these ecosystems, except to say that the occurrence or absence of interactions no doubt influences the nature and amounts of the organic end products.

4.1.3 Anaerobic Waste Decomposition

The conventional perception of this fermentation is that polymer-fermenting populations form acetic, propionic and butyric acids, hydrogen, and carbon dioxide, and methanogenic species then form methane from these products. Almost certainly the only major precursors actually used by methanogens are acetic acid, hydrogen and carbon dioxide. The studies of Bryant and co-workers showed that if butyric and propionic acids are formed, they are converted to acetate and methane by the syntrophic interaction between specific hydrogenogens and methanogens (McInerny *et al.*, 1979; Boone and Bryant, 1980). Yet to be resolved is the extent of syntrophic metabolism of propionic and butyric acids versus the prevention of their formation by the influence of hydrogenotrophs on fermentations of hydrogenogens. If methanogens can cause syntrophic utilization of propionic and butyric acids, they can also prevent formation of the acids by fermenting populations that have the potential for producing them without methanogens. The observed accumulation of propionic and butyric acids when methanogenesis is interfered with or when hydrogen is added to waste fermentations could be due to interference with their conversion to methane and/or to a turning-on of their formation. Syntrophic metabolism of the acids and prevention of their production by some propionate- and butyrate-forming hydrogenogens can probably occur simultaneously. Wastes may contain significant amounts of propionic and butyric acids. The fermentation can also be engineered to operate in two stages: the volatile acids accumulate in the first stage and are converted to methane in the second stage (Ghosh *et al.*, 1975). Lipid decomposition would also yield at least propionic acid, if not butyric acid, during conversion to methane. Carbohydrates and proteins, however, could be decomposed with or without formation of propionic and butyric acids.

An additional feature of hydrogen metabolism that may be very important in the continuous bioconversion of wastes to methane and carbon dioxide is related to the influence of hydrogen on conversion of acetate to methane and carbon dioxide. This conversion can account for 50 to 70% of the methane formed, depending on the chemical composition of the feedstock. Mah and co-workers found that hydrogen represses or inhibits methane formation from acetate by acetate-using mixed or pure cultures obtained from waste digestion systems (Smith and Mah, 1978; Zinder and Mah, 1979). Thus any interference with the maintenance of a very low steady state of hydrogen by balanced production and use could have a profound cascading effect on the ecosystem. Through inhibition of its utilization, acetic acid would accumulate, as would propionic and butyric acids through either of the mechanisms discussed above. This accumulation of acids would lead to further perturbations of the system due to a drop in pH. There may be no other ecosystem that in practical performance is so closely regulated by the hydrogen partial pressure as the bioconversion of wastes to methane and carbon dioxide (Kasper and Wuhrmann, 1978).

4.1.4 Natural Aquatic Environments

Electron acceptors used for the decomposition of organic matter in aquatic ecosystems are oxygen, nitrate, sulphate, and carbon dioxide. The acceptors are listed in their favoured order of use. Anaerobic sediments, especially in marine environments, often contain large amounts of sulphate, and anaerobic respiration with sulphide production takes precedence over methane production from organic substrates (Winfrey and Zeikus, 1977; Abram and Newell, 1978a,b). One explanation of this is that sulphide *per se* inhibits methanogens (Cappenberg and Prins, 1974). A more likely possibility is suggested by the multi-functional roles that sulphate-reducing bacteria can have in these environments. On an organic substrate, such as ethanol or lactate, given the choice between reducing sulphate to sulphide and feeding hydrogen to methanogens for carbon dioxide reduction, sulphate reducers evidently prefer to reduce sulphate (Bryant *et al.*, 1977). Syntrophic utilization of butyric and propionic acids appears to proceed more readily when the hydrogenotroph is a sulphate reducer rather than a methanogen (McInerny *et al.*, 1979; Boone and Bryant, 1980). Thus when sulphate is present, its reduction by internal electron transport or by external electron transport mediated by gaseous hydrogen would appear to be kinetically favoured. When sulphate disappears, its reducers can assume the role of syntrophic producers of hydrogen that can be used by methanogens. Since oxidation of acetic acid to carbon dioxide coupled with sulphate reduction has been reported (Widdell and Pfennig, 1977), the terminal products of complete anaerobic bioconversion can be carbon dioxide and sulphide when sulphate is present, and methane and carbon dioxide when it is absent.

In principle, anaerobic decomposition in natural aquatic environments is similar to anaerobic decomposition of organic wastes (as discussed above, pp. 346–347). When the organic load is very large compared to the levels of oxygen, nitrate, and sulphate, the general features of the two types of ecosystems are essentially the same. All the paths lead to hydrogen, carbon dioxide, and acetate, and then to methane. As with waste decomposition little can be said at present about the relative significance of interactions between hydrogenotrophs and fermenting hydrogenogens.

Studies of sediments indicate that the rates of production of hydrogen and acetic acid from organic carbon—rather than the rates of conversion of hydrogen and acetic acid to methane—limit the rate of methanogenesis (Winfrey and Zeikus, 1979a,b). This suggests that there are sufficient methanogens to use hydrogen produced by fermentative microorganisms or by syntrophic associations if the methanogens and the populations that catabolize organic substrates other than acetic acid are in close proximity. The presence of sulphate is a complicating feature as discussed above, but in addition it complicates anaerobic decomposition of wastes to methane because sulphide production takes precedence when sulphate is introduced into waste decomposition ecosystems.

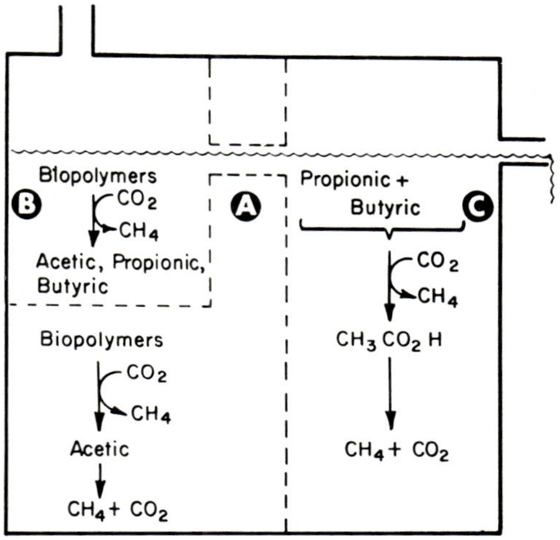

Fig. 10. Diagramatic representation of methanogenesis in anaerobic ecosystems.

The methanogenic systems can be generalized as in Fig. 10. If the dashed lines are ignored, part A of the figure represents the contents of a large fermentation tank where biopolymers are completely converted to methane and carbon dioxide. A system like A requires slow dilution rates and excellent interaction between hydrogenogens and hydrogenotrophs to prevent the accumulation of soluble organic acids. The dashed lines split the system into two subsystems, B and C. If the rate of input is the same as for system A, the dilution rate in B is much greater than for A. Production of methane from hydrogen and carbon dioxide is relatively rapid, but dilution is too fast to allow the use of volatile acids or to prevent their accumulation. System B resembles the rumen ecosystem. When the output of B flows into C, it enters a slow dilution rate phase, which permits the syntrophic utilization of propionic and butyric acids and the conversion of acetic acid to methane and carbon dioxide. A may be the sum of B and C, but this remains to be decided.

The portion below the liquid–gas interface can also be considered as a cross-section of an anaerobic sediment. However, the precise control of inputs and outputs, and the environments that are maintained in the rumen of a cow or in a well-managed waste treatment system, disappear. Although the model presented by A or B + C may apply in some instances, more disorder should reasonably be expected in sediments, wherein a variety of staged fermentation of biopolymers occur to form organic products, which may include alcohols as well as organic acids. Acetic acid would be one of these products and syntrophic methanogenesis might be a major mechanism for converting other products to acetic acid. The final stage would be production of methane and carbon dioxide. Acetic acid is known to be the major precursor of methane in sediments (Winfrey and Zeikus, 1979a), but the route from biopolymers to acetic acid may be tortuous.

4.1.5 General Comments About Ecosystems

Methane has been emphasized as an end product of the interactions between hydrogenogens and hydrogenotrophs, but hydrogenotrophs other than methanogens may play important roles when the appropriate electron acceptors are present. Recent isolations of sulphate reducers which obtain energy for growth by reducing sulphate with hydrogen (Badziong *et al.*, 1978) suggest that they may be important as hydrogenotrophs in interactions with hydrogenogens when sulphate is present in anaerobic ecosystems. Their role as well as that of other hydrogenotrophs in influencing hydrogen formation is not yet clear.

5. TECHNIQUES FOR STUDYING HYDROGEN TRANSFER IN MICROBIAL COMMUNITIES

An understanding of the hydrogen transfer process requires a broad spectrum of approaches. A listing of specific techniques is beyond the scope of this article, but many can be found in the references cited. The methods initially used define the physical, chemical, and biological characteristics of the microbial ecosystem. To understand relationships between hydrogenogens and hydrogenotrophs it is important to know the organic carbon inputs and outputs of the environment as well as the nature of the gaseous environments. Kinetic studies are particularly useful and studies with isotopically labelled substrates and presumptive intermediates help in assessing kinetic parameters. A necessary but often difficult task is the isolation and characterization of the major microbial populations of the community. Since the communities exist in anaerobic environments, particular emphasis must be given to sound anaerobic techniques for isolation and study of pure cultures (especially for methanogens and the bacteria that oxidize short-chain volatile acids to acetic acid and hydrogen). Gas analysis of pure cultures is obviously important, but surprisingly often characteristics of microorganisms are listed without any mention of the kinds of gases that they produce or use. Sensitive analyses of biologically important gases can easily be performed by gas chromatography.

Studies on mixtures of pure cultures of hydrogenogens and hydrogenotrophs are valuable to establish the details of syntrophic and fermentation interactions. This is not as easy a task as might be imagined. For fermentation interactions conditions must be established where the hydrogenotroph is always in high enough concentrations to interact with the hydrogenogen. If the hydrogenotroph has a slow growth rate while the hydrogenogen has a fast one, the latter will grow and produce its fermentation products before the former has a change to interact. Selection of the appropriate pair of organisms and the appropriate media is important. Chemostats are useful for regulating the growth rate of the hydrogenogen. Only by quantitative analysis of substrate utilization and product formation without and with a hydrogenotroph can the influence of the hydrogenotroph on fermentation by a hydrogenogen be established. For syntrophic associations growth itself can be used to monitor an interaction; however, the combinations of organisms must have compatible growth rates.

At the biochemical level it is important to elucidate the enzyme pathways used for catabolism of organic substrates, production of hydrogen by hydrogenogens, and use of hydrogen by hydrogenotrophs. Especially useful is the study of the kinetics and regulation of those enzyme reactions that are of particular importance in interactions between hydrogenogens and hydrogenotrophs. For example, the K_m and V_{max} for hydrogen utilization by whole cells or hydrogenases of hydrogenotrophs are of interest for interpreting their potential for hydrogen utilization. In evaluating potential for interaction with hydrogenogens, the K_m's for reduced pyridine nucleotide and hydrogen and the V_{max}'s of enzyme systems in hydrogenogens that produce hydrogen from reduced pyridine nucleotides should be examined.

6. SUMMARY

Hydrogenogens produce hydrogen gas in anaerobic ecosystems by coupling oxidation of organic substrates to reduction of protons; hydrogenotrophs use hydrogen to reduce inorganic and organic electron acceptors and thereby obtain energy for growth. Examples of inorganic acceptors are carbon dioxide and sulphate, which are reduced to methane and sulphide respectively. Fumarate is an organic acceptor that is reduced to succinate.

Important hydrogen-producing systems of hydrogenogens, although inhibited by hydrogen, are expressed in anaerobic communities because hydrogenotrophs can prevent the accumulation of hydrogen. Some hydrogenogens depend on interactions with hydrogenotrophs to oxidize and obtain energy from certain organic substrates. Examples of such syntrophic associations include those involved in the oxidation of ethanol and short-chain volatile fatty acids to acetic acid and hydrogen; they depend on the reduction of carbon dioxide to methane by methanogenic bacteria.

Hydrogenotrophs can also interact with fermentative hydrogenogens. By removing hydrogen the hydrogenotrophs permit some fermentative species to use protons as electron acceptors for oxidation of organic substrates, instead of the electron acceptors formed as fermentation intermediates. This causes a difference between the fermentation products formed by hydrogenogens in the absence and presence of hydrogenotrophs.

Syntrophic and fermentative interactions between hydrogenogens and hydrogenotrophs are significant in several important anaerobic microbial ecosystems. The role of these interactions in the community activities of intestinal tract ecosystems, systems that decompose organic wastes to methane, and anaerobic aquatic sediments are described in this chapter. Methods for studying interactions between hydrogenogens and hydrogenotrophs are briefly discussed.

REFERENCES

Abram, J. W. and Nedwell, D. B. (1978a). Inhibition of methanogenesis by sulfate reducing bacteria competing for transferred hydrogen. *Archives of Microbiology* **117**, 89–92.

Abram, J. W. and Nedwell, D. B. (1978b). Hydrogen as a substrate for methanogenesis and sulfate reduction in anaerobic saltmarsh sediment. *Archives of Microbiology* **117**, 93–97.

Badziong, W., Thauer, R. K., and Zeikus, J. G. (1978). Isolation and characterization of *Desulfovibrio* growing on hydrogen plus sulfate as the sole energy source. *Archives of Microbiology* **116**, 41–49.

Barker, H. A. (1940). Studies upon the methane fermentation. IV. The isolation and culture of *Methanobacillus omelianskii*. *Antonie van Leeuwenhoek* **6**, 201–220.

Barker, H. A. (1941). Studies on the methane fermentation. V. Biochemical activities of *Methanobacterium omelianskii*. *Journal of Biological Chemistry* **137**, 153–167.

Balch, W. E., Shoberth, S., Tanner, R. S., and Wolfe, R. S. (1977). *Acetobacterium*, a new genus of hydrogen-oxidizing, carbon dioxide-reducing, anaerobic bacteria. *International Journal of Systematic Bacteriology* **27**, 355–361.

Balch, W. E., Fox, G. E., Magrum, L. J., Woese, C. R., and Wolfe, R. S. (1979). Methanogens: reevaluation of a unique biological group. *Bacteriological Reviews* **43**, 260–296.

Blackburn, T. H. and Hungate, R. E. (1963). Succinic acid turnover and propionate production in the bovine rumen. *Applied Microbiology* **11**, 132–135.

Bond, J. H., Engel, R. R., and Levitt, M. D. (1971). Factors influencing pulmonary methane excretion in man. *Journal of Experimental Medicine* **133**, 572–588.

Boone, D. R. and Bryant, M. P. (1980). Propionate-degrading bacterium, *Syntrophobacter wolinii* sp. nov. gen. nov., from methanogenic ecosystems. *Applied and Environmental Microbiology* **40**, 626–632.

Bryant, M. P., Wolin, E. A., Wolin, M. J., and Wolfe, R. S. (1967). *Methanobacillus omelianskii*, a symbiotic association of two species of bacteria. *Archive für Mikrobiologie* **59**, 20–31.

Bryant, M. P., Campbell, L. L., Reddy, C. A. and Crabill, M. R. (1977). Growth of *Desulfovibrio* on lactate or ethanol media low in sulfate in association with H_2-utilizing methanogenic bacteria. *Applied and Environmental Microbiology* **33**, 1162–1169.

Cappenberg, T. E. and Prins, R. A. (1974). Inter-relationships between sulfate-reducing and methane-producing bacteria in bottom sediments of a fresh-water lake. III. Experiments with ^{14}C-labelled substrates. *Antonie van Leeuwenhoek* **40**, 457–469.

Cardon, B. P. and Barker, H. A. (1947). Amino acid fermentation by *Clostridium propionicum* and *Diplococcus glycinophilus*. *Archives of Biochemistry* **12**, 165–180.

Chen, M. and Wolin, M. J. (1979). Effect of monensin and lasalocid-sodium on the growth of methanogenic and rumen saccharolytic bacteria. *Applied and Environmental Microbiology* **38**, 72–77.

Chen, M. and Wolin, M. J. (1977). Influence of CH_4 production of *Methanobacterium ruminantium* on the fermentation of glucose and lactate by *Selenomonas ruminantium*. *Applied and Environmental Microbiology* **34**, 756–759.

Donker, H. J. L. (1926). Bijdrage tot de kennis der boterzuurbutulalcoholen acetongistingen. Dissertation. Delft.

Ghosh, S., Conrad, J. R., and Klass, D. L. (1975). Anaerobic acidogenesis of wastewater sludge. *Journal of Water Pollution Control Federation* **47**, 30–45.

Glass, T. L., Bryant, M. P., and Wolin, M. J. (1977). Partial purification of ferredoxin from *Ruminococcus albus* and its role in pyruvate metabolism and reduction of nicotinamide adenine dinucleotide by H_2. *Journal of Bacteriology* **131**, 463–472.

Hungate, R. E. (1967). Hydrogen as an intermediate in the rumen fermentation. *Archive für Mikrobiologie* **59**, 158–164.

Iannotti, E. L., Kafkewitz, D., Wolin, M. J., and Bryant, M. P. (1973). Glucose fermentation products of *Ruminococcus albus* grown in continuous culture with *Vibrio succinogenes*: Changes caused by interspecies transfer of H_2. *Journal of Bacteriology* **114**, 1231–1240.

Ishimoto, M., Umeyama, M., and Chiba, S. (1974). Alteration of fermentation products from butyrate to acetate by nitrate reduction in *Clostridium perfringens*. *Zeitschrift für Allgemeine Mikrobiologie* **11**, 115–121.

Jayasuria, G. C. N. and Hungate, R. E. (1959). Lactate conversions in the bovine rumen. *Archives of Biochemistry and Biophysics* **82**, 274–287.

Jungermann, K., Thauer, R. K., Leimenstoll, G., and Decker, K. (1973). Function of reduced pyridine nucleotide-ferredoxin oxidoreductases in saccharolytic *Clostridia*. *Biochimica et Biophysica Acta* **305**, 268–280.

Kasper, H. F. and Wuhrmann, K. (1978). Kinetic parameters and relative turnovers of some important catabolic reactions in digesting sludge. *Applied and Environmental Microbiology* **36**, 1–7.

Kubowitz, F. (1934). Über die hemmung der buttersäuregärung durch kohlenoxyd. *Biochemische Zeitschrift* **274**, 285–298.

Latham, M. J. and Wolin, M. J. (1977). Fermentation of cellulose by *Ruminococcus flavefaciens* in the presence and absence of *Methanobacterium ruminantium*. *Applied and Environmental Microbiology* **34**, 297–301.

Levitt, M. D. (1969). Production and excretion of hydrogen gas in man. *The New England Journal of Medicine* **281**, 122–127.

Ljungdahl, L. G. and Wood, H. G. (1969). Total synthesis of acetate from CO_2 by heterotrophic bacteria. *Annual Reviews of Microbiology* **23**, 515–538.

McInerny, M. I., Bryant, M. P., and Pfennig, N. (1979). Anaerobic bacterium that degrades fatty acids in syntrophic association with methanogens. *Archives of Microbiology* **122**, 129–135.

Macy, J., Kulla, H., and Gottschalk, G. (1976). H_2-dependent anaerobic growth of *Escherichia coli* on L-malate: succinate formation. *Journal of Bacteriology* **125**, 423–428.

Miller, T. L. and Wolin, M. J. (1973). Formation of hydrogen and formate by *Ruminococcus albus*. *Journal of Bacteriology* **116**, 836–846.

Moomaw, C. R. and Hungate, R. E. (1963). Ethanol conversion in the bovine rumen. *Journal of Bacteriology* **85**, 721–722.

Mortenson, L. E. and Wilson, P. W. (1951). Effect of molecular nitrogen and hydrogen on hydrogen evolution by *Clostridium pasteurianum*. *Journal of Bacteriology* **62**, 513–514.

Reddy, C. A., Bryant, M. P., and Wolin, M. J. (1972a). Characteristics of S organism isolated from *Methanobacillus omelianskii*. *Journal of Bacteriology* **109**, 539–545.

Reddy, C. A., Bryant, M. P., and Wolin, M. J. (1972b). Ferredoxin- and nicotinamide adenine dinucleotide-dependent H_2 production from ethanol and formate in extracts of S organism isolated from "Methanobacillus omelianskii". *Journal of Bacteriology* **110**, 126–132.

Reddy, C. A., Bryant, M. P., and Wolin, M. J. (1972c). The ferredoxin-dependent conversion of acetaldehyde to acetate and H_2 in extracts of S organism. *Journal of Bacteriology* **110**, 133–138.

Scheifinger, C. C., Linehan, B., and Wolin, M. J. (1975). H_2 production by *Selenomonas ruminantium* in the absence and presence of methanogenic bacteria. *Applied Microbiology* **29**, 480–483.

Smith, M. R. and Mah, R. A. (1978). Growth and methanogenesis by *Methanosarcina* strain 227 on acetate and methanol. *Applied and Environmental Microbiology* **36**, 870–879.

Stadtman, T. C. and Barker, H. A. (1951). Studies on the methane fermentation VII. Tracer experiments on fatty acid oxidation by methane bacteria. *Journal of Bacteriology* **61**, 67–80.

Tanner, A. C. R., Badger, S., Lai, C.-H., Listgarten, M. A., Visconti, R. A. and Socransky, S. S. (1981). *Wolinella* gen. nov., *Wolinella succinogenes* (*Vibrio succinogenes* Wolin *et al.*) comb. nov., and description of *Bacteroides gracilis* sp. nov., *Wolinella recta* sp. nov., *Campylobacter concisus* sp. nov., and *Eikenella corrodens* from humans with peridontal disease. *International Journal of Systematic Bacteriology* **31**, 432–445.

Thauer, R. K., Jungermann, K., and Decker, K. (1977). Energy conservation in chemotrophic anaerobic bacteria. *Bacteriological Reviews* **41**, 100–180.

Valentine, R. C. and Wolfe, R. S. (1963). Role of ferredoxin in the metabolism of molecular hydrogen. *Journal of Bacteriology* **85**, 1114–1120.

Weimer, P. J. and Zeikus, J. G. (1977). Fermentation of cellulose and cellobiose by *Clostridium thermocellum* in the absence and presence of *Methanobacterium thermoautotrophicum*. *Applied and Environmental Microbiology* **33**, 289–297.

Whitely, H. R. and Douglas, H. C. (1951). The fermentation of purines by *Micrococcus lactilyticus*. *Journal of Bacteriology* **61**, 605–616.

Widdel, F. and Pfennig, N. (1977). A new anaerobic sporing acetate-oxidizing, sulfate-reducing bacterium: *Desulfotomaculum* (emend.) *acetoxidans*. *Archives of Microbiology* **112**, 119–122.

Winfrey, M. R. and Zeikus, J. G. (1977). Effect of sulfate on carbon and electron flow during microbial methanogenesis in fresh-water sediments. *Applied and Environmental Microbiology* **33**, 275–281.

Winfrey, M. R. and Zeikus, J. G. (1979a). Anaerobic metabolism of immediate methane precursors in Lake Mendota. *Applied and Environmental Microbiology* **37**, 244–253.

Winfrey, M. R. and Zeikus, J. G. (1979b). Microbial methanogenesis and acetate metabolism in a meromictic lake. *Applied and Environmental Microbiology* **37**, 213–221.

Winter, J. U. and Wolfe, R. S. (1980). Methane formation from fructose by syntrophic associations of *Acetobacterium woodii* and different strains of methanogens. *Archives of Microbiology* **124**, 73–79.

Wolin, M. J., Wolin, E. A., and Jacobs, N. J. (1961). Cytochrome-producing anaerobic vibrio, *Vibrio succinogenes*, sp. n. *Journal of Bacteriology* **81**, 911–917.

Wolin, M. J. (1976). Interactions between H_2-producing and methane-producing species. *In* "Microbial formation and utilization of gases (H_2, CH_4, CO)", pp. 141-150. (H. G. Schlegel, G. Gottschalk, and N. Pfennig, eds.) Goltze, Göttingen.

Yoshinari, T. (1980). N_2O reduction by *Vibrio succinogenes*. *Applied and Environmental Microbiology* **39**, 81–84.
Zinder, S. H. and Mah, R. A. (1979). Isolation and characterization of a thermophilic strain of *Methanosarcina* unable to use H_2–CO_2 for methanogenesis. *Applied and Environmental Microbiology* **38**, 996–1008.

10

Microbial Interactions and Communities in Biotechnology

J. D. Linton and J. W. Drozd

1. INTRODUCTION

Traditional foods, such as sauerkraut, yoghurt, cheese, and soy sauce have been consumed by man ever since recorded history. However, only recently has it become apparent that these foods develop their highly prized nutritional and organoleptic qualities as a result of the metabolic activity of a host of different microorganisms. In certain cases, such as in soy sauce production, the fermentation may take up to 3 years and require the metabolic activity of a succession of different microorganisms (Young and Wood, 1974). These traditional foods and their associated mixed microbial population are consumed world-wide in large quantities annually. In spite of this, the deliberate use of defined mixed cultures in industry has only recently been considered seriously. Indeed, to date the "Symba Process" for single-cell protein production from potato waste is the only recent mixed culture system to have been used on an industrial scale.

Most of the research on the industrial use of mixed cultures has been confined to single-cell protein production from methane, methanol and longer chain alkanes. Recently, however, a number of new applications for mixed cultures have emerged. In this article the authors review the actual and potential use of mixed cultures in industry but traditional fermented foods, beverages and conventional waste treatment processes will not be discussed (see Steinkraus p. 407).

2. CLASSIFICATION OF MICROBIAL INTERACTIONS

The types of interaction found in mixed populations of microorganisms can be conveniently classified into a number of well-defined categories (Table 1) (Bungay and Bungay, 1968; Slater and Bull, 1978; see Bull and Slater pp. 13–44). In practice, however, interactions rarely separate into these discrete groups and more than one type of interaction often occurs simultaneously. Although a number of elegant academic studies have been carried out (Meers, 1973; Veldkamp and Jannasch, 1972; Jannasch and Mateles, 1974), the interactions that occur in the numerous mixed culture system of potential industrial interest remain largely uncharacterized.

Mixed cultures used in industry may be arbitrarily classified into three broad groups on the basis of their origin and function. In the first group, the mixed cultures are assembled by combining two or more species of microorganism of known metabolic function so that the combination performs a precise task that is not easily accomplished by a single species. The production of ethanol and single-cell protein (SCP) from starch and cellulosic materials, the production of SCP from natural gas and multi-step transformation of steroids are examples of this type of structured mixed cultures (see Section 3).

In the second group, the mixed cultures are usually isolated by a single enrichment procedure and the functions of all the component organisms are not essential to the process. Indeed, alternative processes using single species can usually be operated quite satisfactorily. Examples of this latter type of mixed culture are those proposed for the production of SCP from methanol and methane.

The third group of mixed cultures are selected *in situ* during the operation of various waste treatment systems where the component microbial species perform vital roles in removing the multi-carbon substrates present in these effluents. This third group is discussed elsewhere and is not examined in detail in this review.

Table 1. Types of interaction found in mixed populations of microorganisms (after Meers, 1973).

Type of Interaction	Description	Comments
Competition	Growth of two or more populations of microorganisms mutually exclusive as they strive for a common factor required for growth.	Usually leads to the establishment of a population of single species. No industrial application.
Neutralism	When the growth of two species has no effect on each other.	Examples of neutralism have not been documented to occur in nature but this type of interaction has been demonstrated in the laboratory.
Amensalism	This interactions occurs when one one species retards the growth of another species.	Inhibition of growth of one species by another can be as a result of a number of factors such as excretion of toxic products or change of culture environment. Does not lead to stable mixed cultures.
Mutualism	The growth of each species in the mixed culture is promoted by the other.	Interactions ranges from a symbiotic relationship where two species are totally dependent on one another to a loose cooperative relationship. Commonly found in mixed cultures used for single cell protein from methanol and methane.
Commensalism	The growth of one organism promotes the growth of a second species, but the growth of the latter does not affect the former.	Stable in continuous culture systems and commonly found in mixed cultures used for single cell protein production for methane and methanol.
Predation	When one organism totally engulfs and metabolizes a second species.	Very important in nature but there are no obvious industrial applications for this type of interaction, apart from the obvious predation of protozoa and rotifers on bacteria in effluent treatment.

2.1 Enrichment for Mixed Cultures

A detailed account of the methods used for the enrichment and isolation of mixed microbial cultures is given in Chapter 3 (see Parkes, pp. 45–102). Nevertheless, it is instructive to examine some of the procedures employed for the enrichment of mixed cultures suitable for single-cell protein production from methane and methanol.

Most of the mixed culture systems capable of utilizing methane and methanol as their sole source of carbon and energy were isolated by traditional batch culture techniques (Brown *et al.*, 1964; Vary and Johnson, 1967) and in certain cases these cultures were subjected to a further selection in continuous-flow culture (Cremieux *et al.*, 1977).

2.1.1 Enrichment in Batch Culture

The procedures used for batch culture enrichments are straightforward (Vary and Johnson, 1967; Whittenbury *et al.*, 1970; Johnson, 1972; Sukatch and Johnson, 1972; Cremieux *et al.*, 1977). Mineral salts medium containing the desired organic compound as the sole source of carbon and energy is inoculated with a sample that is likely to contain the wanted organism, and is incubated at the required temperature. Vigorous aeration of the enrichment is recommended because there is a high oxygen demand for growth on reduced C_1 compounds. However, if the enrichments are not aerated then the growth of these aerobic organisms usually occurs as a pellicle at the liquid surface (Whittenbury *et al.*, 1970). As soon as active growth is observed the culture is transferred to fresh medium. Failure to do this may result in a poor enrichment since exhaustion of the primary carbon source may lead to the development of other species growing on secondary substrates, such as metabolites excreted or lysis products, other than the defined primary substrate (Johnson, 1972; Wilkinson *et al.*, 1974). Once the enrichment has been through a series of transfers the major components of the mixed culture may be isolated and the culture reconstituted to give a defined mixed culture (Linton and Buckee, 1977). In the isolation of the component heterotrophic species it is difficult to determine what substrate these organisms utilize *in situ*. Linton and Buckee (1977) incorporated lysate of the major organism into agar medium to provide a medium for the isolation of the heterotrophic species present.

2.1.2 Enrichment in Continuous-flow Culture

Although the theory of enrichment in continuous-flow culture was described over 30 years ago (Novick and Szilard, 1950a,b), the technique has not been applied very frequently. Moreover, where it has been used the methodology has not been described in sufficient detail (Sheehan and Johnson, 1971; Linton and Buckee, 1977; Ballerini *et al.*, 1977). Indeed, in a number of cases the logic of the enrichment procedures adopted is difficult to understand. For example, Cremieux *et al.* (1977) described an enrichment procedure for the isolation of a mixed culture suitable for SCP production from methanol. The initial enrichment was carried out in batch culture at 30 to 33 °C in a medium containing yeast extract. This was followed by a further enrichment step in continuous-flow culture at 34 °C with a dilution rate of $0.15\ h^{-1}$ Once the culture had established itself under these conditions the yeast extract was omitted from the medium and the culture gradually adapted to growth at 40 °C at a higher growth rate ($D = 0.30\ h^{-1}$). The resulting mixed culture performed most efficiently in terms of yield from methanol and oxygen at the initial enrichment temperature of 30°C, and these yield values were considerably lower at the higher temperature of 40°C. These results were not surprising. Sukatch and Johnson (1972) demonstrated that five different mixed cultures capable of utilizing hexadecane as their sole source of carbon and energy could be obtained by carrying out batch culture enrichments at 25, 35, 45, 55, and 65°C (Table 2). Mixed cultures enriched at a low temperature (25°C), although capable of growth at temperatures above those at which they were enriched, performed poorly in terms of yield and growth rate at the higher temperatures. However, mixed cultures isolated at high temperatures (65°C) performed better at temperatures lower than those at which they were enriched. For example, one mixed culture enriched at 65°C performed considerably better at 45°C than at

Table 2. The effect of the initial enrichment temperature on the yield and growth rate of various mixed cultures growing on hexadecane as their sole source of carbon and energy (from Sukatch and Johnson, 1972).

Mixed Culture	Temperature of Enrichment (°C)	Maximum Growth Rate (h^{-1})	Yield [g dry weight (g hexadecane)$^{-1}$]
S_1	25	0.62	1.12
S_2	35	0.69	1.10
S_3	45	0.57	1.0
S_4	55	0.27	0.49
S_5	65	0.175	0.26

65°C, although there was a significant alteration in the ratio of the three organisms present (Table 3).

Enrichments in batch culture, followed by a second selection in continuous-flow culture under carbon limitation at low dilution rates on a single carbon substrate, usually result in the selection of one dominant organism capable of utilizing the sole carbon and energy source. Unless the continuous enrichment is continuously inoculated, any subsequent alteration of the environment, such as an increase in the temperature or growth rate, leads to the selection of mutants of the mixture that was initially enriched. Although this may result in the adaptation of the culture for growth under the altered environmental conditions, the resultant culture is unlikely to perform as well as one that was specifically enriched for under these conditions.

The growth rate of the initial enrichment has a profound effect on the type of organism selected. This was admirably demonstrated by Veldkamp and his colleagues (Veldkamp and Kuenen, 1973; Kuenen and Veldkamp, 1973; Harder *et al.*, 1977), who used one reservoir of medium to feed two chemostats under identical environmental conditions. Both chemostats were inoculated with the same source of water but one operated at a low growth rate ($D = 0.03\ h^{-1}$) and the other at a higher growth rate ($D = 0.30\ h^{-1}$). Two different types of organism developed in the two enrichment systems. At low growth rates organisms with high affinity (i.e. low saturation constant values) for the growth-limiting substrate and low maximum specific growth rate (μ_{max}) were enriched. At high growth rates organisms with a relatively poor affinity (high K_s) for the growth-limiting substrate but a higher μ_{max} were selected. In competition experiments, organisms with low μ_{max} and

Table 3. The effect of temperature on the yield and growth rate of a mixed culture isolated at 65°C (from Sukatch and Johnson, 1972). ND = Not detected.

Growth Temperature (°C)	Yield [g dry weight (g hexadecane)$^{-1}$]	Maximum Specific Growth Rate (h^{-1})	Population Ratio		
			A	B	C
25	1.02	0.69	30	70	ND
35	1.02	0.69	50	50	ND
45	0.93	0.57	35	60	5
55	0.43	0.23	40	40	20
65	0.15	0.12	30	40	30

high K_s values predominated at low growth rates and those with relatively poor K_s and high μ_{max} values predominated at high growth rates.

Clearly, enrichments performed under one set of environmenal conditions followed by a gradual change to a radically different set of conditions make little sense. The enrichment procedures adopted should be carefully designed so that they closely simulate those likely to be endured in the final process in which the new organisms are to be employed.

A number of workers have reported that enrichment methods employing mineral salts media and methane, methanol, or higher alkanes as the sole carbon and energy sources resulted in the selection of mixed populations of microorganisms from which single species capable of growth on the elective carbon source could not be isolated (Miller and Johnson, 1966; Vary and Johnson, 1967; Wodzinski and Johnson, 1968; Wilkinson *et al.*, 1974; Snedecor and Cooney, 1974). At the time these reports were published very little was known about the nutritional requirements and physiology of methylotrophic microorganisms and this probably contributed to the difficulty experienced in obtaining pure cultures of these organisms.

These findings stimulated research into the nature of the microbial interaction that occurred in these mixed culture systems. Although these studies have revealed a number of interesting mixed culture interactions, they have generally perpetuated the view that it is still difficult to isolate methane- and methanol-utilizing microorganisms in pure culture and that, in examples where these organisms have been isolated, they invariably perform poorly in terms of growth yield and μ_{max} when compared with the parent mixed culture from which they were isolated (Harrison *et al.*, 1975; Harrison and Wren, 1976; Harrison, 1976, 1978). On the other hand, a number of workers have isolated methane- and methanol-utilizing organisms from mixed cultures with relatively little difficulty (Leadbetter and Foster, 1958; Brown *et al.*, 1964; Whittenbury *et al.*, 1970; Chen *et al.*, 1977). Certainly, in our experience, continuous-flow enrichment under either carbon limitation or in batch culture enrichments with mineral salts medium and methane, ethane, propane, or methanol as the sole carbon and energy source resulted in the development of mixed cultures (Linton and Buckee, 1977; Linton *et al.*, 1980; J. W. Drozd, unpublished observations). However, we have experienced little difficulty in obtaining these hydrocarbon utilizers in pure culture using routine microbiological procedures, namely, plating onto mineral salts medium in the presence of a dissolved or gaseous hydrocarbon as the sole source of carbon and energy.

Why do these enrichments result in the development of mixed cultures of microorganisms? Microorganisms when grown as pure cultures in carbon limited continuous culture, excrete various organic compounds into the culture supernatant. In those cases where full carbon balances have been carried out (Dostalek and Molin, 1975; Linton *et al.*, 1975; Linton and Buckee, 1977) it is seen that, depending on growth rate, supernatant carbon usually accounts for 2 to 45% of the steady-state concentration of bacterial carbon. In the case of the methane-utilizing bacterium *Methylococcus* sp. NCIB 11083 it has been shown that the supernatant carbon probably originated from a high specific death rate of the culture (Linton and Buckee, 1977; Drozd *et al.*, 1978b). The specific death rate increased slightly with dilution rate and, consequently, the amount of organic carbon present in the culture supernatant was inversely proportional to growth rate. A similar finding was reported for the growth of *Methylomonas methanolica* during methanol limited growth in a chemostat (Dostalek and Molin, 1975). It is the presence of the supernatant carbon that is the reason for enrichments developing mixed cultures containing one dominant organism, accounting for 90 to 95% by weight of the culture, and a range of secondary species accounting for ≤1 to 5% by weight of the culture. The dominant species utilized the given carbon source and the secondary species utilize the organic lysis compounds released during, or, more probably, after the death of the dominant species (Linton and Buckee, 1977; Wren and Harrison, 1976). This phenomenon will be discussed in more detail later in this chapter.

3. SURVEY OF MIXED CULTURE SYSTEMS OF INDUSTRIAL INTEREST

3.1 The Symba Process for Single-cell Protein Production

The Symba Process (Jarl, 1969, 1971; Skogman, 1969) was designed by the Swedish Sugar Company primarily as an effluent treatment system to deal with the potato waste produced by one of its subsidiary companies. The process differed from most other effluent treatment systems since, in addition to decreasing the biological oxygen demand of the effluent, the waste was converted into a valuable product, namely, single-cell protein. The process employed a two component mixed culture. Potato starch was hydrolysed to dextrins and low molecular weight sugars by a yeast, *Endomycopsis fibuliger* that grew to high cell densities on potato waste but had a low growth rate. The second species, *Candida utilis* was not capable of hydrolysing starch but rapidly assimilated the soluble products released by the action of *E. fibuliger*. In the mixed culture, *C. utilis* soon became the dominant organism and accounted for approximately 96% of the mixed population. The predominance of *C. utilis* over *E. fibuliger* is desired and deliberately maintained as the former organism has a higher nutritional value and is accepted as a food yeast.

The name Symba Process was given to the system to denote the symbiotic relationship between the two yeast species. This is not strictly correct as the organisms do not have an obligate requirement for each other. *Endomycopsis fibuliger* was capable of growth on starch in the absence of *C. utilis* but the latter organism was totally reliant on the production of soluble sugar by the former organism. Although it has been suggested (Jarl, 1971) that *C. utilis* excreted vitamin B which stimulated the growth of *E. fibuliger*, no experimental evidence was given to substantiate this statement.

3.1.1 Process Conditions

Waste starch and potato peelings account for up to 20% (by wt) of the potatoes processed. After supplementation with a source of nitrogen and phosphorus the waste was inoculated with a mixed culture of the two organisms. The mixed culture was not stable in batch or continuous-flow culture systems unless the saccharification activity was maintained at the required level, a function that was achieved by the continuous reintroduction of *Endomycopsis fibuliger*. This was effected by growing a pure culture of *E. fibuliger* in a separate reactor and using this continuously to feed *E. fibuliger* into the large reactor containing the mixed culture. The saccharification step was the overall rate-limiting reaction, although improved strains of *E. fibuliger* possessing higher amylase activity have been selected.

3.1.2 Product Recovery

After passage through coarse screens to retain large particulate material the yeast was concentrated to approximately 14% (w/v) by passage through separators before it was freeze-dried. The product contained 40 to 50% protein, 36% carbohydrate, 4 to 5% lipid; was rich in lysine and the vitamin B's, and has been marketed as feed for pigs and chickens.

3.1.3 Yield and Productivity

The yield of the process was approximately 0.6 g dry weight (g available starch)$^{-1}$ (Jarl, 1969, 1971) and the productivity was 1 to 2 kg dry weight (m^3 culture)$^{-1}$ h^{-1}. Unfortunately, the yield and productivity of a pure culture of *Endomycopsis fibuliger* were not measured, so it is not possible to compare the performance of the mixed culture with that of a pure culture of this organism.

3.1.4 Commercial Considerations

The Symba Process has been discontinued because it was uneconomical for three main reasons.

(1) The production plant was not large enough. Jarl (1969, 1971) calculated that to be economical the plant had to produce 2000 to 4000 tons y^{-1}. In a plant processing 50 000 tons potato waste y^{-1} the waste treatment cost could be totally offset by the cost of the single-cell protein produced.

(2) The waste must contain more than 2% utilizable material, but as the processing plant became more efficient this value decreased and caused a marked decrease in SCP productivity.

(3) The plant did not operate throughout the year but remained idle because of the short growing and processing season for potatoes.

As single-cell protein is a low value, high volume product, it is essential to run a high productivity (g dry weight l^{-1} h^{-1}) process. However, this was not met in the Symba Process because of the low concentration of carbon substrates in the waste and it is believed that the waste is currently being converted into high value-added compounds, e.g. amino acids, where the productivity of the process is not as critical as in SCP production.

The main advantages of the Symba Process were that both saccharification of starch and the production of SCP occurred in a single vessel, thereby reducing capital costs. Although saccharification was the overall rate-limiting step and considerably faster rates of hydrolysis could be achieved by acid treatment, it is claimed that the products of acid hydrolysis are not completely utilized and a corresponding decrease in yield is observed. Moreover, acid hydrolysis requires the use of expensive corrosion resistant equipment. The Symba Proces is an example of the type of processes where one organism is used to hydrolyse a complex insoluble carbohydrate and a second species is used to convert the soluble sugars produced into a useful product.

3.2 Conversion of Renewable Resources into Useful Products

Oil is a finite resource and the recent rapid price increase has stimulated considerable interest in the development of alternative feedstocks for the chemical and liquid fuel industries. Biomass is a renewable resource and its use as an alternative feedstock is receiving a great deal of attention (Wang *et al.*, 1977; Tsao *et al.*, 1978; Flickinger and Tsao, 1978; Linderman and Rocchiccioli, 1979; Tonge, 1979; Bu'Lock, 1979; Keenan, 1979). The term "biomass" is generally taken to mean herbaceous material of any kind from wood and cellulose waste materials to crops grown specifically for their high contents of particular organic compounds, such as sugars or oils.

Cellulosic material is by far the most abundant form of biomass and it is composed of three main constituents: cellulose, hemicellulose, and lignin. The relative proportion of these three constituents varies e.g. on a dry weight basis most soft woods contain approximately 45 to 50% cellulose, 20 to 30% hemicellulose, and 25 to 35% lignin, whereas hard woods contain approximately 40 to 55% cellulose, 20 to 40% hemicellulose, and 18 to 25% lignin (Cowling and Kirk, 1976). As most microorganisms utilize soluble sugars much more rapidly than insoluble carbohydrates, the cellulosic material must be hydrolysed before it can be fermented to the desired end-product. The resistance of cellulosic material to hydrolysis is generally attributed to its highly crystalline structure and to the protective covering of lignin which prevents access to hydrolysing enzymes or chemicals. Thus, to allow maximum contact between enzyme and cellulose, the material has to be subjected to some form of pretreatment. All the processes using mixed cultures to transform biomass directly into single-cell protein or ethanol have three clearly discernable steps:

(1) pre-treatment, where the cellulosic material is sterilized and comminuted or treated chemically to give the largest possible surface area on which enzymatic attack can occur;
(2) simultaneous saccharification and fermentation in the same reactor;
(3) product recovery.

3.2.1 Production of Ethanol from Biomass

A number of industrial organizations have shown considerable interest in the use of mixed cultures for the production of ethanol directly from cellulose wastes or biomass. Unfortunately, most of the information regarding this interesting application of mixed cultures is confined to the patent literature where the scientific aspects of these processes are poorly described. A brief outline of some of the various processes is given below.

The General Electric Company, Schenectady, New York has described three processes for the production of ethanol directly from cellulosic materials. For convenience these processes are referred to as GEC Process I (Bellamy, 1978); Process II (Brooks *et al.*, 1979); and Process III (Pye *et al.*, 1979).

Process I utilized native cellulose or cellulosic waste, whereas Process II and III used wood chips from poplar trees. In Process I the cellulosic material was comminuted to particles $\leqslant 100\mu$ and materials containing substantial amounts of lignin were treated with 5 to 50% alkali at 100°C for varying periods depending on the lignin content. Process II employed low pressure steam in the presence of sulphur dioxide, followed by rapid decompression which resulted in the fracture of the material. The material was neutralized with ammonia and directly fermented. In Process III the lignin was extracted with hot aqueous butanol which was used as a bunker C type fuel. The cellulose and hemicellulose were extracted into the aqueous phase. In Process I a thermophilic *Sporocytophaga* sp. capable of hydrolysing cellulose to soluble sugars was combined with *Bacillus stearothermophilus* which fermented the sugars to ethanol at 55 to 60°C, but this mixed culture did not utilize the hemicellulose fraction. The mixed culture employed in Process II utilized both cellulose and hemicellulose and also involved two organisms: *Clostridium thermocellum*, which hydrolysed cellulose and converted cellobiose into ethanol but did not utilize hemicellulose, and *Clostridium thermosaccharolyticum* which converted the latter into ethanol. In Process III the fermentation was operated in two stages. In the first stage cellulose was hydrolysed with a cellulase obtained from *Thermoactinomyces* sp. and simultaneously fermented to ethanol by *Clostridium thermocellum*. The residual hemicellulose from the first stage (pentoses) was fermented to butanol and acetone by *Clostridium acetobutylicum* in the second stage.

Unfortunately, it has proved possible only to obtain values for the product yield and productivity for Process I. Although yields ranging between 40 to 80% of the theoretical maximum values were reported when the fermentation was carried out under a partial vacuum (100 to 400 mm Hg in order to remove the ethanol produced), the productivity of the process was exceedingly low, being less than 0.5 g ethanol l^{-1} h^{-1}. Moreover, in the example quoted in the Patent (Bellamy, 1978), the yield of ethanol was four times greater than the theoretical value.

3.2.2 Bio-industries Process

In this process (Hoge, 1977) the cellulosic material was steam-sterilized in the presence of acid which causes partial breakdown of the cellulose structure, making it more susceptible to hydrolysis by cellulases obtained from *Trichoderma viride*. The soluble sugars produced were fermented directly to ethanol which was recovered by applying intermittent vacuum distillation. The consequent cooling effect of the distillation was used to control the temperature of the reactor and it is claimed that this system reduced the cooling water requirements from between 50 and 80 l (l ethanol)$^{-1}$ to almost zero. The yield of ethanol was 38% of the theoretical value but no values for the productivity of this process were given.

3.2.3 The Gulf Oil Process

This process was described by Gauss and Suzuki (1976) and Huff and Yata (1975). Cellulosic waste from agricultural, industrial, or municipal sources was pulped and a portion was sterilized and inoculated with *Trichoderma reesei*. After a suitable incubation period to allow for the growth of the organism, the inoculated sample was mixed with the remaining pulp, fed into a reactor with added mineral nutrients, and a yeast species which fermented the soluble sugars to ethanol was then added. It is claimed that the yield of the mixed culture system was 40% higher and the conversion efficiency 90% higher than that of a monoculture, but no experimental data supporting this statement were given.

3.2.4 Advantages of Mixed Culture Processes for Ethanol Production from Biomass

A number of advantages have been claimed for these mixed culture systems.

(1) The hydrolysis of cellulose occurs under relatively mild conditions, namely 40 to 70°C and pH 4.0 to 5.0, and the yield can approach 100%. Although acid hydrolysis is undoubtedly faster, significant decomposition of the liberated sugar occurs and the final yield is only approximately 50% of the theoretical value (Saeman, 1945; Ladisch, 1979). Moreover, acid hydrolysis necessitates the use of expensive acid resistant materials for the processing equipment.

(2) The saccharification and fermentation steps occur simultaneously in a single reactor. Saccharification proceeds at a rapid rate and the products of hydrolysis are immediately utilized by the other component of the mixed culture. Thus, the steady-state concentration of the hydrolysis products is kept low and the problem of end-product inhibition of saccharification by soluble sugars is avoided.

(3) As saccharification and fermentation occur in the same vessel there is a considerable reduction in capital costs.

(4) Cellulose and hemicellulose are both utilized. This is of the utmost importance as all the major constituents of cellulosic material must be utilized to ensure favourable process economics.

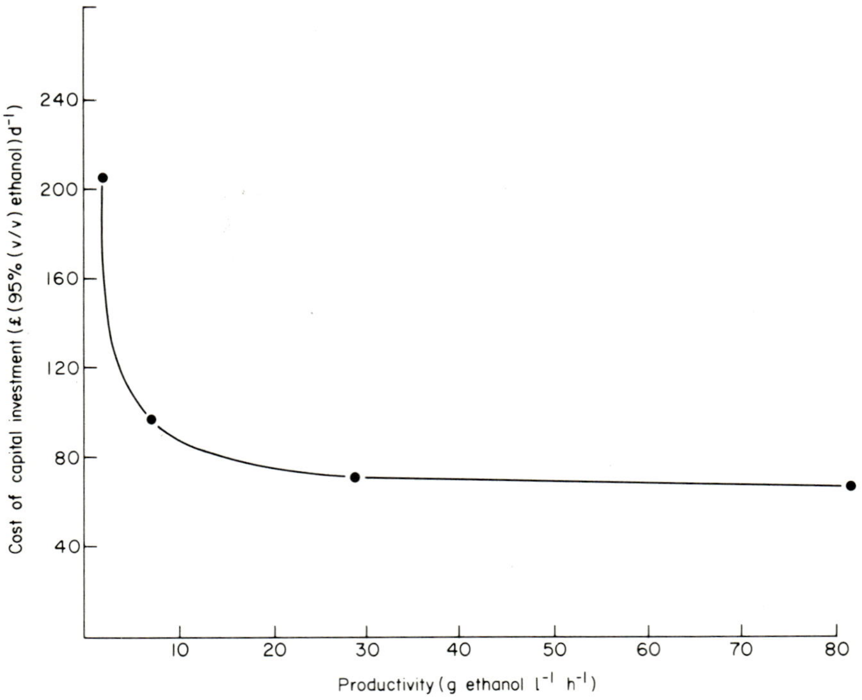

Fig. 1. The relationship between fixed capital investment and productivity of ethanol production, based on the production of 95 000 gal d^{-1} 95% ethanol (after Cysewki and Wilke, 1978).

The major disadvantage of mixed culture systems is that the saccharification step is the overall rate-limiting step and, consequently, the productivities of these mixed culture systems tend to be very poor ($\leqslant 0.5$ gl^{-1} h^{-1}). The relationship between capital costs and productivity for ethanol production is shown in Fig. 1 and illustrates clearly that there is considerable scope for improvement of these processes. Wilke and his co-workers (Cysweski and Wilke, 1976, 1977, 1978; Margaritis and Wilke, 1978a,b) have shown that it is possible to develop a high productivity process for ethanol production from soluble sugar. Using a simple single stage continuous-flow culture system the productivity of ethanol production can be increased from a maximum of 2.2 $gl^{-1} h^{-1}$ in the conventional batch culture system to 7 $gl^{-1} h^{-1}$. By employing a cell recycle system the productivity can be increased to 28 $gl^{-1} h^{-1}$ but ethanol toxicity limits further increase in productivity. This problem can be overcome by operating the fermentation under a partial vacuum (32 mm Hg at 30°C) to facilitate the continuous removal of ethanol as it is formed. Under these conditions the productivity can be increased to 82 $gl^{-1} h^{-1}$. Clearly, the hydrolysis step must be considerably improved before these mixed culture processes for ethanol production from cellulosic material can successfully compete with fermentation processes using soluble cane sugar. As the cost of fermentable sugar accounts for up to 70% of the overall costs of ethanol production (Cysewski and Wilke, 1978), an economic and rapid method of cellulose hydrolysis could markedly influence the overall economics of ethanol production. Recent reports suggest that a combination of acid and enzymatic hydrolysis could lead to significantly faster rates of hydrolysis without concomitant degradation of the soluble sugars produced. The cellulosic material is treated with 35% HCl at 20°C for 30 min, resulting in the selective hydrolysis of hemicellulose with no loss through degradation. This is followed by treatment with 41% HCl which results in the solubilization and cleavage of oligosaccharides. These water soluble polysaccharides are then quantitatively hydrolysed enzymatically. This type of pretreatment followed by fermentation using a mixed culture capable of converting all the constituents of biomass into ethanol, could result in substantial increase in productivity.

3.2.5 Production of Single-cell Protein from Renewable Resources

The arguments that apply to the utilization of mixed microbial cultures for single-cell protein (SCP) production from hydrocarbons and their derivatives and to the conversion of biomass into chemicals, also apply to the production of SCP from a variety of sources. These include: plant-derived cellulosic materials (Bellamy, 1974); carbon dioxide via microbial photosynthesis, usually involving algae (Clement, 1975; Benemann *et al.*, 1977; Pirt and Pirt, 1977); effluent streams from various processing plants, e.g. whey from cheese manufacture (Bernstein *et al.*, 1977) and the Symba Process described earlier (p. 364 *et seq.*); and animal feedlot wastes (Reddy and Erdman, 1977). Some of these processes are designed to treat an effluent as well as to produce biomass which may be used as animal feed, fertilizer or as a feedstock for methane production (Benemann *et al.*, 1977).

The reader is referred to Gaden and Humphrey (1977) for a full discussion of the production of SCP from renewable and non-renewable resources. Dunlop (1975) used a two component mixed culture comprising a *Cellulomonas* sp. and *Alcaligenes faecalis* or an unidentified yeast for batch and continuous-flow culture SCP production from chemically pretreated bagasse. The role of the second organism was to scavenge soluble sugar fractions produced from cellulose by the *Cellulomonas* sp. Productivity up to 1.0 g dry biomass l^{-1} h^{-1} was obtained in continuous-flow culture, although the performance of the pure and mixed cultures at high productivities was not studied in more detail. The most promising system for SCP production from cellulosic wastes may, however, be a pure culture of *Thermoactinomyces* (Bellamy, 1974), but there are no reports that show growth at high productivities in continuous culture with recycling of the water stream. Such studies would be of interest because of the complex enzymic interactions (repression of enzymes, etc.) that occur during growth on cellulosic substrates.

SCP production from ethanol has attracted some interest. Petrochemically derived ethanol would be the substrate used in these processes because there would be little point in converting plant carbohydrate, first into ethanol and then into protein when the carbohydrate could be converted directly into protein. Bacteria grow efficiently on ethanol but in the presence of excess ethanol growth inhibitory concentrations of acetate and acetaldehyde are produced (Laskin, 1977). For example, acetaldehyde concentrations as low as 0.01% (w/v) decreased the growth rate of *Acinetobacter calcoaceticus* by 80% (Laskin, 1977). However, continuous operation under ethanol limitation minimizes acetate and acetaldehyde accumulation but, nevertheless, in a large, imperfectly stirred reactor these compounds are likely to be produced. Thus, a mixed culture with acetate- and acetaldehyde-utilizing bacteria might give a more stable culture. This system is similar to that proposed for the production of SCP from natural gas (described later, p. 380) in which acetate is produced from the co-oxidation of ethane.

Another mixed culture system of interest is one exploiting algae for biomass production or for combined biomass production and effluent treatment. The algae produced are used as a source of SCP, as fertilizer or a source of carbon and nutrients for methane generation (Benemann *et al.*, 1977). In the effluent-treatment process in shallow oxidation ponds the algae generate enough oxygen by photosynthesis to allow bacterial oxidation of organic wastes. The algae behave as "aerators" for the system and the carbon dioxide and other nutrients released by the bacteria are assimilated by the algae. For the development of shallow, high-rate oxidation ponds it is necessary to remove the algae continuously and this is a considerable problem, since the normal algae involved in these processes are unicellular. Only the much less commonly found filamentous cyanobacterium *Oscillatoria* sp. and the spined green algae *Micrachnium* sp. can be easily removed by microstrainers (Benemann *et al.*, 1977). Size-selection by a recycle method can be used to help alleviate this problem. A fraction of the larger, more easily strained algae are continuously returned to the pond and gradually replace the smaller, less easily harvested algae. This system allows some control over the main microbial components of the undefined mixed culture in that the system selects for those algae that are retained by the microstrainer.

Generally, it is likely that mixed cultures will be used for the production of SCP from complex natural substrates not readily attacked by one microbial species, but which may be rapidly broken down by several enzymes produced from a variety of microbial species. A recent report (Ban and Glanser-Soljan, 1979) indicated that a mixed culture containing two yeast species and four bacterial species degraded calcium lignosulphonate at a much higher rate than was observed with various pure cultures. This is another good example where the mixed microbial population has great advantages over the pure culture.

3.3 Single-cell Protein from Methanol, Methane and Natural Gas

3.3.1 Single-cell Protein from Methanol

A continuous process using a mixed culture for the production of single cell protein from methanol was described by Harrison *et al.*, (1974). The mixed culture was composed of a single methanol-utilizing organism *Pseudomonas* sp. strain EN, and four heterotrophic bacteria that were unable to utilize methanol as their sole source of carbon and energy (*Pseudomonas* sp. NCIB 11019; *Pseudomonas* sp. NCIB 11022; *Acinetobacter* sp. NCIB 11020 and *Curtobacterium* sp. NCIB 11021). In a series of experiments conducted in a "U"-shaped vessel, the arms of which were separated by a dialysis membrane, it was demonstrated that *Pseudomonas* sp. EN grew readily in mineral salts medium containing methanol but no added heterotrophic organism. However, when the heterotrophic organisms were inoculated into the other arm of the culture vessel the growth of the methanol utilizer was slightly enhanced, although the two cultures were separated by a dialysis membrane (Harrison *et al.*, 1974). These results suggested that either the methanol utilizer produced a dialysable growth-inhibitory product which was removed by the heterotrophic component of the mixed culture or the heterotrophs produced a dialysable product that promoted the growth of the methanol-utilizing organism. Using [^{14}C]-methanol

(Harrison *et al.*, 1974) it was shown that approximately 20% of the methanol carbon was excreted as products by the methylotroph *Pseudomonas* sp. EN. These excretion products (mainly nucleic acid and protein) were the growth substrates of the heterotrophic components of the mixed culture (Wren and Harrison, 1976).

Earlier reports (Harrison *et al.*, 1974; Wren *et al.*, 1974) suggested that the mixed culture was more stable and performed better in terms of growth yield and maximum specific growth rate (μ_{max}) than a pure culture of the methanol-utilizer *Pseudomonas* sp. EN (Table 4). This work was carried out in media containing citrate (0.125 gl^{-1}) as a chelating

Table 4. A comparison of the maximum growth rate and growth yield of a pure culture of the methanol-utilizing bacterium *Pseudomonas* sp. strain EN with those of the mixed culture from which it was isolated.

	Yield [g dry weight (g methanol)$^{-1}$]		
Dilution Rate (h^{-1})	Pseudomonas sp. EN	Complete Mixed Culture	Comments
0.06	—	0.31	
0.10	0.29	—	
0.16	0.30	0.40	Citrate-containing medium. Harrison *et al.* (1974)
0.19	Washed out	—	
0.64	—	0.54	
0.66	—	Wash out	
0.5	0.35	—	Citrate absent from medium. Wren and Harrison (1976)
0.64	—	0.52	

agent and it was subsequently shown (Wren and Harrison, 1976) that citrate inhibited the growth of a pure culture of *Pseudomonas* sp. EN. In the absence of citrate in the medium the μ_{max} of the pure culture increased from 0.16 h^{-1} to 0.50 h^{-1} which was close to that of the mixed culture (Table 4). However, the maximum growth yield of the pure culture was lower (0.35 $g.g^{-1}$) than that of the mixed culture (0.54 $g.g^{-1}$). This result was rather surprising as the growth yield from methanol was 0.30 $g.g^{-1}$ at a dilution rate of 0.16 h^{-1} and only increased marginally when the organism was grown under methanol limitation at a dilution rate of 0.50 h^{-1}. Further optimization of the growth environment would undoubtedly result in the pure culture performing as well as the mixed culture in terms of growth yield and μ_{max} and, indeed, in a recent review (Harrison, 1978) it was stated that the yield of the pure culture was 0.46 $g.g^{-1}$. For example Goldberg (1977) reported that once the environmental conditions were optimized for the growth of a pure culture, addition of heterotrophic organisms had no effect on growth yield or μ_{max}. Rokem *et al.*, (1980) isolated two non-methanol-utilizing organisms, a *Moraxella* sp. and a *Bacillus* sp., that had grown as contaminants in a chemostat culture of *Pseudomonas* sp. strain C growing on methanol. These organisms were reinoculated into a pure culture of *Pseudomonas* sp. strain C growing in a chemostat and, although they became established in the culture, their numbers were very low and their presence had no effect on growth yield. Indeed, a mixed culture of *Pseudomonas* sp. C and *Bacillus* sp. showed a decrease in growth yield (Table 5). In a similar experiment, a culture of *Pseudomonas* sp. C was inoculated with a sample of soil and again a mixed culture was established, but there was no difference in growth yield between the pure and the mixed cultures (Table 5). The non-methanol-utilizing organism accounted for approximately 2% of the population in the reactor bulk

Table 5. The effect of added heterotrophic bacteria on the growth yield of a chemostat culture of the methanol-utilizing bacteria *Pseudomonas* sp. strain C grown at various dilution rates (after Rokem *et al.*, 1980).

Composition of Culture	Dilution Rate (h^{-1})	Growth Yield [g dry weight (g methanol)$^{-1}$]
Pseudomonas sp. C	0.1	0.43
Pseudomonas sp. C + *Moraxella* sp.	0.1	0.43
Pseudomonas sp. C + *Bacillus* sp.	0.1	0.35
Pseudomonas sp. C + *Escherichia coli*	0.1	0.35
Pseudomonas C	0.3	0.50
Pseudomonas C + soil bacteria	0.3	0.50

liquid phase but examination of growth on the fermenter walls revealed that these organisms accounted for approximately 75% of the attached population. It was shown that potentially pathogenic bacteria, *Escherichia coli*, *Staphylococcus aureus*, and *Salmonella typhimurium*, readily attached themselves to the walls of the fermenter when they were introduced into a pure culture of *Pseudomonas* sp. strain C. The pathogens accounted for 0.1 to 5% of the population in the fermentation liquid phase but were present on the vessel walls in numbers which were five to eight times higher than that of the methanol utilizer. However, in mixed cultures of *Pseudomonas* sp. strain C and soil bacteria, pathogens did not establish themselves in the mixed culture and their numbers on the wall of the vessel were many orders of magnitude lower than that observed with the pure culture. It is probable that most of the available attachment sites on the fermenter wall were occupied by soil bacteria thereby preventing the attachment of other added bacteria.

An interesting mixed culture system containing one obligate methanol utilizer, *Methylomonas methylovora*, and three faculative methanol utilizers, *Xanthomonas* sp., *Flavobacter* sp. and *Pseudomonas pseudoalcaligenes*, was described by Cremieux *et al.* (1977). The mixed culture was stable and its composition remained unaltered after 2 months continuous growth at a dilution rate of 0.30 h^{-1}. The maximum growth yield was 0.44 g dry weight (g methanol)$^{-1}$ and the μ_{max} of the culture was 0.49 h^{-1}. The culture had a high productivity, 9.47 g dry weight l^{-1} h^{-1}, but unfortunately the nature of the interactions involved in the mixture was not examined. It is unlikely that all four organisms utilized methanol as their sole source of carbon and energy, but that the obligate methanol utilizer excreted lysis products during growth on methanol and the faculative methyltrophs grew at their expense.

3.3.2 Single-cell Protein from Methane

Vary and Johnson (1967) using batch enrichment methods isolated a number of mixed cultures that were capable of utilizing methane as the sole source of carbon and energy. One mixed culture, designated HR, contained two Gram-negative rods that grew optimally at pH 6.5 and 30°C with ammonia as the sole source of nitrogen. The culture had a doubling time of 3 h and the growth yield from methane was 0.65 to 0.70 g dry weight (g methane)$^{-1}$. These studies were in batch culture, full carbon balances were given but attempts to isolate a pure culture of the methane-utilizing organism failed. Although single colonies were obtained on mineral salts agar plates incubated in the presence of methane these isolates did not grow in liquid medium. When growth did occur two organisms could be detected, suggesting that, for unexplained reasons, interactions between members of the mixed culture were essential.

One of the first reports of a stable mixed culture growing on methane in continuous-flow culture was by Sheehan and Johnson (1971). The mixed culture, M45, was isolated using continuous-flow enrichment at 45°C and was composed of two Gram-negative non-spore-forming rods. This mixed culture was stable over a period of 3 years, during which time the culture was grown under non-aseptic conditions. It was suggested that cultures grown with ammonia as the nitrogen source were subjected to contamination by nitrifying bacteria as nitrite and nitrate were detected in these cultures. However, no details of the contaminants were given and it now seems likely that the oxidation of ammonia was carried out by the methane-utilizing organism (Drozd *et al.*, 1978a). With nitrate as the source of nitrogen, the growth yield from methane was 0.62 to 0.65 dry weight (g methane)$^{-1}$. The lower growth yield during growth with nitrate instead of ammonia was expected since a significant amount of reducing equivalents are needed for the reduction of nitrate to ammonia prior to assimilation. The interaction between the organisms was not examined.

Wilkinson and Hamer (1972) noted that *Hyphomicrobium* species were frequently found in mixed cultures growing on methane and suggested that the methanol excreted by the methane-oxidizing bacteria supported the growth of the *Hyphomicrobium* sp. Later Wilkinson *et al.* (1974) noted that the presence of the *Hyphomicrobium* sp. in a mixed culture which grew on methane, rendered the culture less susceptible to inhibition by the pulse addition of methanol to the culture. Growth of the methane utilizer, a *Pseudomonas* sp., was inhibited by low concentrations of added methanol. However, when the *Hyphomicrobium* sp. was present, it rapidly utilized the added methanol and quickly prevented the inhibition of *Pseudomonas* sp. growth. A criticism of this work is that it was not shown that the methane utilizer produced methanol: this was an untested assumption. The mixed culture also contained two heterotrophic bacteria, a *Flavobacterium* sp. and an *Acinetobacter* sp., which were thought to be growing on growth and lysis products of the methane utilizer.

When a methane-utilizing bacterium *Methylococcus* sp. NCIB 11083 was isolated from a mixed culture and grown as a monoculture on methane under ammonia limitation, it excreted organic compounds into the culture supernatant (Linton and Buckee, 1977). At low growth rates ($D = 0.03\ h^{-1}$) the supernatant carbon amounted to approximately 39% of the bacterial carbon but at higher growth rates ($D = 0.21\ h^{-1}$) the value decreased to approximately 15% of the steady-state bacterial carbon value (Fig. 2). Oxidation products of methane, such as methanol, formaldehyde and formate, did not constitute a significant part of the supenatant organic carbon which could be accounted for mainly as protein and nucleic acid. The ratio of protein to nucleic acid in the supernatant closely resembled that found in the whole bacteria at various growth rates and the protein profiles of the culture supernatant and whole cells were very similar, indicating that proteins originated from the lysis of the methane-utilizing organism. As the specific lysis rate of the culture increased only slightly with

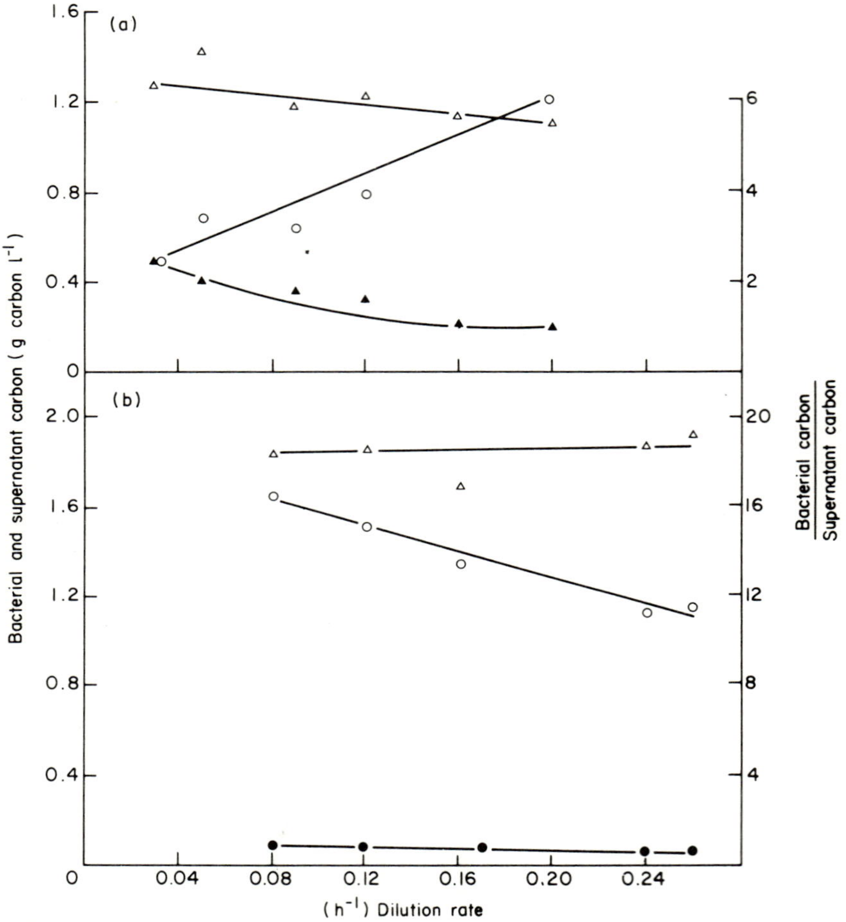

Fig. 2. Steady state values for bacterial carbon, culture supernatant carbon, and ratio of bacterial carbon to culture supernatant carbon as a function of dilution rate of an ammonia limited chemostat culture of: (a) a pure culture of *Methylococcus* sp. NCIB 11083. △, bacterial carbon; ▲, culture supernatant carbon; and ○, the ratio of bacterial carbon to culture supernatant carbon (after Linton and Buckee, 1977). (b) a mixed culture containing *Methylococcus* sp. NCIB 11083. ○, bacterial carbon; ●, culture supernatant carbon; and △, the ratio of bacterial carbon to culture supernatant carbon (J. D. Linton, unpublished observations)

growth rate (Drozd *et al.*, 1978b) a higher concentration of lysis products was present at low growth rates compared with higher growth rates. In a reconstituted mixed culture containing the methane utilizer, *Methylococcus* sp. NCIB 11083, and four heterotrophic bacteria (the heterotrophic bacteria were unable to oxidise methane), the amount of the culture supernatant carbon was low and independent of culture growth rate (Fig. 2). The heterotrophic bacteria were shown to possess extracellular proteases, lipases and nucleases and it was suggested that these enzymes were intimately involved in the metabolism of the excreted polymers, providing carbon sources used by the heterotrophic bacteria. Although the supernatant carbon was reduced markedly compared to that found in a pure culture of *Methylococcus* sp., it was not completely removed. A possible explanation for the residual carbon is that it was composed of compounds which were individually present at concentrations too low to be utilized by the heterotrophic organisms present.

The maximum growth rate and growth yield from methane and oxygen of a pure culture of *Methylococcus* sp. NCIB 11083 was not significantly different from that of the mixed culture (Table 6). Nevertheless, the mixed culture had a number of advantages compared with the pure culture. The relative merits of pure and mixed culture systems for SCP production are discussed later (pp. 391–398).

Table 6. A comparison of the performance of a pure culture of *Methylococcus* NCIB 11083 with that of a mixed culture containing NCIB 11083, both growing on methane (Khosrovi and Downs, unpublished). ND = Not detected.

Growth Parameters	Mixed Culture Containing Methylococcus sp. NCIB 11083	Pure Culture of Methylococcus sp. NCIB 11083
μ_{max} (h^{-1})	0.34	0.31
Y_{CH_4} g dry weight (g methane)$^{-1}$	0.71	0.72
YO_2 g dry weight (g oxygen)$^{-1}$	0.26	ND
Maximum productivity (g dry weight $l^{-1}.h^{-1}$)	4.83	3.2
Supernatant organic matter at above productivity (gl^{-1} dry organic matter), assuming 46% carbon	0.32	3.6

3.3.3. Single-cell Protein from Natural Gas

A great deal of process research has been carried out on SCP production from relatively pure feedstocks, such as methane and industrial methanol. In this section the type of microbial technology required for the production of SCP from the mixed carbon feedstock, natural gas, is considered.

Natural gas is found in pockets by itself or in association with oil. The main component of natural gas is methane but it may contain considerable quantities of higher alkanes (especially ethane and propane) as well as nitrogen, carbon dioxide, and hydrogen sulphide. The relative amounts of these compounds vary from source to source (Table 7). The higher alkanes are often removed by differential liquefaction. However, in many parts of the world very large amounts of associated natural gas are flared or used as an energy source without prior treatment. Such natural gas could be a feedstock for SCP production.

Several Russian investigators (Malashenko *et al.*, 1973, 1975; Romanovskaya *et al.*, 1976) have noted that mixed cultures are essential for the production of SCP from natural gases containing an appreciable content of higher alkanes. Malashenko *et al* (1973) drew attention to the large number of conflicting reports concerning the ability of a single species of microorganism to simultaneously utilize gaseous C_1 to C_4 hydrocarbons. An examination of 147 isolates capable of growing on natural gas showed that some were capable of growth on either C_1 or C_2 to C_4 hydrocarbons but none was capable of simultaneously utilizing C_1 and C_2 to C_4 gaseous hydrocarbons. Earlier reports of organisms

Table 7. Alkane contents of some natural gases (mol %). Other major components may be CO_2, N_2 and H_2S e.g. Qatar tail gas contains (mol %) 8.1 CO_2, 2.5 N_2, 1.7 H_2S. Traces of helium, benzene and toluene may also be present in some gases. It is important to realize that the gas composition varies from source to source, and varies with time as the gas is removed from one source.

Source	Methane	Ethane	Propane	Butane and Pentane
Leman Bank, North Sea	95.9	3.1	0.6	0.4
Groningen, Holland	81.1	2.7	0.4	0.18
Kuwait	58.0	20.0	13.0	6.1
Qatar Tail Gas	76.8	9.3	1.5	0.1
Algeria	83.0	7.1	2.3	1.5
Egypt, Gulf of Suez	83.9	8.4	4.6	3.0

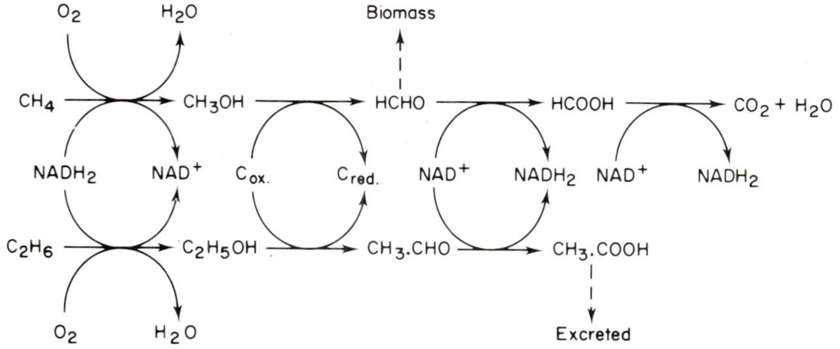

Fig. 3. Schematic outline of methane oxidation and ethane co-oxidation reaction by the methane-utilizing bacterium *Methylococcus* sp.

capable of simultaneously utilizing C_1 to C_4 hydrocarbon were equivocal as the utilization of individual gases was not monitored and in certain cases no data on the composition of the gases were given (Bogdanova, 1966; Wolnak *et al.*, 1967).

All bacteria isolated on methane as sole carbon and energy source can also co-oxidize or co-metabolize higher alkanes to more oxidized products which are assimilated to a very limited extent (Fig. 3) (Leadbetter and Foster, 1958; Whittenbury *et al.*, 1970). In batch culture experiments with pure cultures of several obligate methane-utilizing bacteria, Romanovskaya *et al.* (1976) found that ethane and propane were competitive inhibitors of methane oxidation. Ethane was oxidized to acetaldehyde and ethanol which accumulated in the culture and completely inhibited growth of the methane utilizer at a final concentration of 0.07% (w/v). As predicted from the competitive kinetics, batch culture studies in the presence of excess methane showed that there was very little ethane oxidation. It was stated that obtaining high cell densities of pure cultures of methylotrophic organisms on a natural gas containing higher alkanes would be very difficult because of the accumulation of toxic higher alkane co-oxidation products. However, it was predicted that heterotrophic bacteria in mixed cultures would utilize the co-oxidation products as their carbon and energy sources thereby preventing growth inhibition of the methane-utilizing organisms.

Although this phenomenon has been known for many years and a detailed kinetic model of the competitive inhibition of methane oxidation by ethane or propane based solely on the concentration of these gases in the aqueous phase was developed (Romanovskaya *et al.*, 1976) there have until recently (Drozd *et al.*, 1980; Drozd and McCarthy, 1981) been no experimental or theoretical studies, related to the effects of the higher alkanes on a continuous, high productivity process for the production of SCP from natural gas.

To make predictions on the performance of a mixed culture grown on natural gas a mathematical model was developed (Drozd *et al.*, 1980; Drozd and McCarthy, 1981). The values for the various parameters used were based on experimental data obtained with the obligate methane utilizer, *Methylococcus* sp. NCIB 11083. The mathematical development of the model and the assumptions made are described in detail elsewhere (Drozd and McCarthy, 1981). In the initial development of the model (Drozd *et al.*, 1980) it was assumed that the rate of ethane co-oxidation was so low that the concentration of ethane would equal its saturation concentration. Under these conditions there would of course be a negligible biomass produced from ethane co-oxidation. An error was thereby introduced into the final equation of Drozd *et al.* (1980) which related to the effect of increased total gas pressure on the ratio of biomass derived from methane to that derived from ethane. For this calculation the dissolved ethane concentration was taken to be the saturation concentration for the particular conditions defined, this of course implied that ethane was not co-oxidized and, hence, there would have been no biomass produced from ethane. This point was rectified in a further development of the model (Drozd and McCarthy, 1981) which assumed that the rate of ethane co-oxidation could be as rapid as that of methane oxidation, this assumption was more in keeping with experimental findings. The dissolved ethane and methane concentrations were interdependent and sensible values for their steady-state concentrations were generated by a computer iteration programme. With this programme it was predicted that an increase in total gas pressure would not influence the ratio of biomass from methane to that from ethane. Methane limitation was studied and modelled because under these conditions the rate of methane gas transfer and nitrogen content of the cells (Linton and Cripps, 1978) will be maximal. Under other growth limitations "co-oxidation" will still occur but any problems due to an increased K_s for methane will be minimized. For example, the critical dilution rate will not be decreased because the presence of higher alkanes should not alter the K_s of the *Methylococcus* for nutrients other than methane. In this respect the use of methane limitation represents the greatest challenge.

The solutions to some of the predictive equations for methane-limitation were described by Drozd and McCarthy (1981). If the concentration of ethane in the fermenter gas was increased at a fixed concentration of methane, then, as expected, there was an increase in the steady-state acetate concentration. This acetate could be used as a growth substrate by the heterotrophic bacteria if present in the mixed culture. It was predicted that as the total gas pressure was increased there would be no change in the ratio of *Methylococcus* sp. to heterotrophic bacteria growing on co-oxidation products. This is important because high gas pressures are likely to be used in an industrial fermenter not only because of the considerable hydrostatic pressure of the liquid which will give a pressure gradient up the fermenter but also high pressures would be essential for a high productivity process with a 60% to 80% methane gas conversion. Natural gas with a low ethane content, e.g. Groningen natural gas is approximately 80% v/v methane and 3% v/v ethane, with the given process conditions, gives a predicted steady state acetate concentration of approximately 70mM at a dilution rate of 0.2 h^{-1}. This is the calculated concentration in the absence of any recycle of the process water. In practice water recycle would probably be used and this would increase the steady-state acetate concentration. With a maximum water recycle ratio of eight (Topiwala and Khosrovi, 1978) the steady-state acetate concentration would be 675 mM and obviously it would be impossible to operate a pure culture system under these conditions as 100 mM acetate causes complete inhibition of growth of a pure culture of *Methylococcus* sp. The theoretical predictions indicate that for SCP production from even a natural gas with a low higher alkane content it is essential to use a mixed culture.

The broad outline of the predictions has been borne out in practice. Mixed bacterial cultures have been grown at low and high productivities with simulated and actual natural gases. In a series of oscillations observed during growth of a low productivity mixed culture containing *Methylococcus* sp. NCIB 11083 on Groningen natural gas, acetate accumulated to a maximum concentration of 2.6 mM (Godley, Linton, and Drozd, unpublished observations). The culture was ammonium-limited, operated at a dilution rate $D = 0.1\ h^{-1}$ and had a biomass concentration of 2 to 3 g dry weight l^{-1}. Surprisingly, methanol at 2.5 mM was also detected in the culture supernatant. When wash-out occurred it was often accompanied by the production of even higher concentrations of methanol (up to 10 mM) in the culture supernatant. The production of methanol is possibly due to intracellular ethanol, generated from ethane oxidation competing with methanol at the alcohol dehydrogenase. *In situ*, some or all of the heterotrophic bacterial species (*Flavobacterium* sp. NCIB 11282; *Pseudomonas* sp. NCIB 11309 and NCIB 11310; *Moraxella* sp. NCIB 11308 and *Nocardia* NCIB 11307) oxidized and assimilated acetate, so it was not clear why low concentrations of acetate were detected in the culture. The culture was stable under methane limitation even with high concentrations of ethane, up to an ethane content of 55% (v/v) of the total effluent gas. In steady-state methane limited cultures, acetate, acetaldehyde and methanol concentrations were normally below the limits of detection. However, when there was a sudden increase in the ethane concentration there was a transient appearance of acetate, acetaldehyde, ethanol and methanol (Fig. 4). When the ethane flow rate was returned to its original value the culture recovered. However, culture stability to such transients may not be shown at high productivities when the rates of co-oxidation product formation per unit liquid volume will be much higher. Stepwise changes in ethane concentration are unlikely to occur in an industrial process but there may be significant variations in parameters such as dissolved gas concentrations or pH in a large industrial reactor

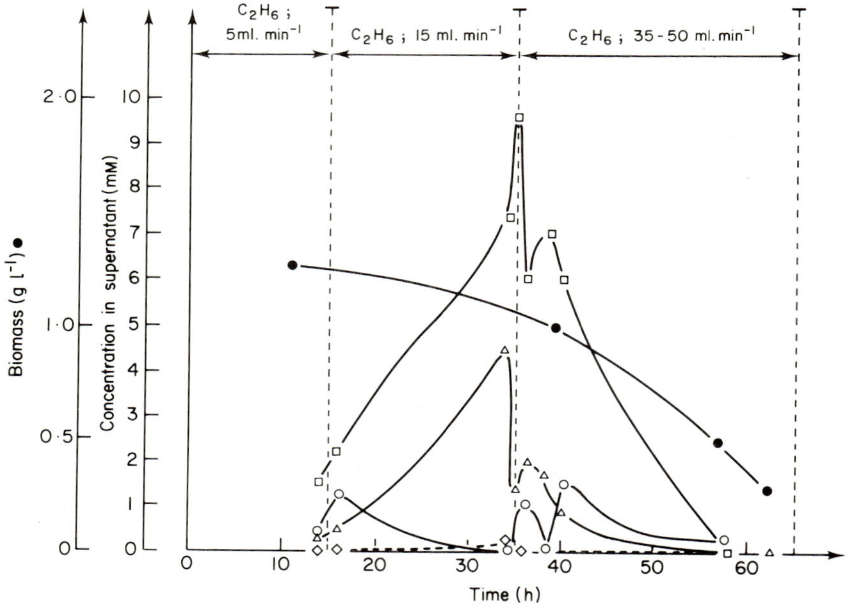

Fig. 4. Transients in a methane limited *Methylococcus* mixed culture; 45°C, pH 6.6, D = 0.12 h^{-1}, CH_4 at 100 ml min^{-1}, air at 300 ml min^{-1}. □, acetate; ●, biomass; △, acetaldehyde; ○, ethanol; ◇, methanol.

operated at high productivities. This could lead to transient accumulations of co-oxidation products which could cause culture wash-out if the growth of the *Methylococcus* sp. was inhibited. An additional problem is that the *Methylococcus* sp. can oxidize ammonia to nitrite and nitrate (Drozd *et al.*, 1978a) and nitrite is a non-competitive growth inhibitor. Any alteration of culture conditions which leads to the accumulation of ammonia at concentrations greater than a few millimolar results in nitrite and nitrate accumulation and culture wash-out. The situation is of a "runaway" type since once the growth rate of the *Methylococcus* sp. is decreased there is a consequential increase in the ammonia concentration, a further decrease in the growth rate and so on. To avoid this situation the ammonia concentration should be controlled to less than 5 to 10 mM.

The performance of high productivity systems (3 to 6 g dry weight l^{-1} h^{-1}) has been described by Drozd *et al.* (1980); Drozd and McCarthy (1981), and Drozd and Linton (1981). High productivity pure cultures ($D = 0.2\ h^{-1}$ and dry weight to 20 g l^{-1}) could not be established with even a low (2 to 5% v/v) concentration of ethane in the methane. Indeed, such cultures were unstable and washed out even when the ethane was added transiently for a few minutes.

Mixed cultures were grown to productivities of up to 5.5 g dry weight l^{-1} h^{-1} with a simulated natural gas containing various concentrations of ethane (up to 15% (v/v) in relation to methane) and propane (up to 10% (v/v) in relation to methane). The performance of such cultures were described by Drozd *et al.* (1980) and Drozd and McCarthy (1981). Table 8 shows typical culture growth yields on methane, ethane and oxygen, for an ammonia-limited culture. From the ethane uptake coefficients it was calculated that the free acetate concentration would theoretically have been greater than 0.1 M if a pure culture could have been used. This concentration had previously been shown to be growth inhibitory. When propane was added to the inlet gas to give a final concentration of 7% (v/v) in relation to the methane (ethane was 10% (v/v) relative to the methane), acetone accumulated to a steady-state concentration of 8mM at a biomass productivity of 4.0 g dry weight l^{-1} h^{-1} under ammonia-limitation. The implication was that none of the heterotrophic bacteria present could oxidize acetone and this was confirmed by oxygen electrode studies on washed culture samples which showed that the culture possessed high affinity acetate, propionate and propionaldehyde oxidation systems, but none for acetone. Pure cultures of *Methylococcus* sp. NCIB 11083 showed no oxidation of acetate, propionaldehyde or propionate, so the oxidation observed in the mixed culture must have been due to the heterotrophic bacteria. With recycle

Table 8. Steady-state data for a mixed bacterial culture based on *Methylococcus* NCIB 11083 grown at high productivities on methane/ethane gas mixtures, ammonia-limited in continuous culture, 45°C, pH 6.5.

Dilution rate (h^{-1})	Steady-state Biomass (gl^{-1})	Gas Outlet* CH_4(%) (% v/v)	Gas Outlet* C_2H_6(%) (% v/v)	Y_{CH_4} g Cells $(g\,CH_4)^{-1}$	$Y_{CH_4+C_2H_6}$ g Cells $(g\,CH_4+C_2H_6)^{-1}$	Y_{O_2} g Cells $(g\,O_2)^{-1}$	Cell Carbon (gl^{-1})	% Biomass Carbon from Ethane	Acetate Concentration in Pure[(1)] Culture (mM)	Supernatant Carbon (gl^{-1})
0.17	20.3	19.7	3.7	0.92	0.80	0.27	9.35	14.7	119	0.39
0.17	27.2	17.2	2.8	0.93	0.79	0.27	12.50	16.6	183	0.72
0.17	21.5	19.2	3.4	0.93	0.81	0.27	9.90	14.0	114	0.39
0.18	14.7	23.8	4.0	0.95	0.83	0.29	6.75	12.4	76	0.25
0.24	17.4	16.7	2.8	—	—	—	8.00	—	—	0.33

* Equivalent to mol. fractions of gas in the fermenter. (1) Theoretical value if a pure culture of *Methylococcus* was present. Values for % of the inlet gas used were in the range: 42 to 52 for CH_4, 18 to 42 for C_2H_6 and 47 to 58% (v/v) for O_2. The freeze-dried biomass had a total nitrogen content of 10.4 to 11.3% (wt/wt).

of the liquid stream the acetone concentration would increase unless bacteria that can oxidize and assimilate acetone are added to the mixed culture.

We have also isolated ethane-utilizing bacteria (Linton *et al.*, 1980) and introduced them into the natural gas mixed cultures expecting that the *Methylococcus* sp. would oxidize and grow on methane, and the ethane-utilizers would oxidize and grow on ethane and also oxidize propane. Unfortunately, none of the ethane utilizers isolated could be established in any of the mixed cultures which suggested that the *Methylococcus* sp. competed more effectively for ethane under these growth conditions. A major process factor which mitigates against this approach is that none of the ethane-oxidizing bacteria isolated at 45°C had a maximum specific growth rate greater than 0.13 h^{-1} under the conditions used for the growth of *Methylococcus* sp. The maximum specific growth rate of the ethane utilizers was less than that required for a high productivity process and meant that these organisms could not establish themselves in the reactor at the dilution rates (0.15 to 0.3 h^{-1}) proposed for SCP production from natural gas.

In summary, it is essential to utilize a mixed microbial culture for SCP production from natural gas, because the methane utilizers can oxidize higher alkanes (Fig. 5) to products that cannot be assimilated by the methane utilizers and which accumulate in the culture and inhibit growth. In a mixed culture these co-oxidation products are effectively removed by the heterotrophic bacteria. Even with a very low concentration of higher alkanes in the natural gas it can be predicted that a mixed culture will still be essential to avoid the build-up of higher alkane "co-oxidation" products and lysis products when recycle of some of the liquid stream is instigated.

3.4 Miscellaneous Mixed Cultures

3.4.1 Effluent Treatment

Mixed microbial populations are an integral part of any biological effluent treatment process. Jones and his associates (Jones and Carrington, 1972; Jones *et al.*, 1973) attempted to unravel some of the complex interactions that occur in the biological treatment of carbonization waste liquors containing phenol and thiocyanate. Although two bacterial strains were isolated which could grow on phenol and one strain that grew on thiocyanate, the growth kinetics in mixed cultures were markedly different from the growth kinetics of the pure culture (Jones and Carrington, 1972). Jones *et al.*, (1973) showed that growth of a pure bacterial culture on phenol in a two-stage continuous-flow culture system was dominated by the inhibitory effect of phenol, even at

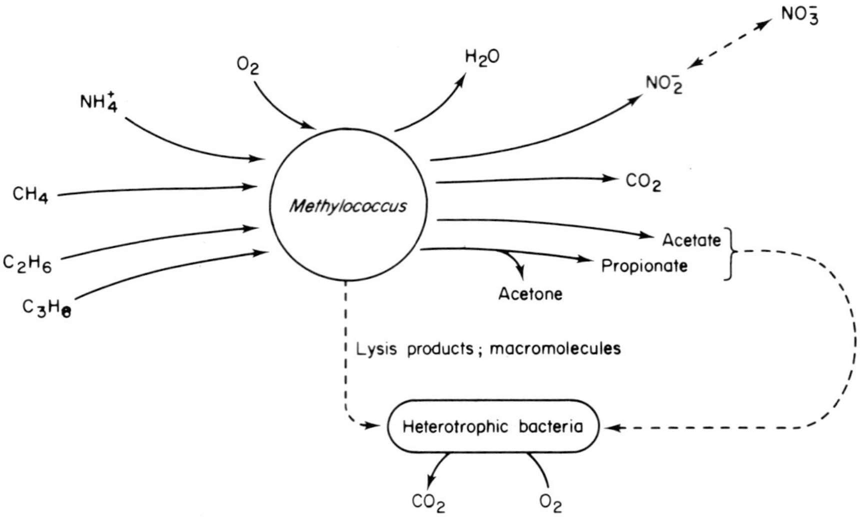

Fig. 5. Schematic representation of the interactions which may occur in a mixed culture grown on natural gas. At low dissolved oxygen concentrations the heterotropic bacteria may utilize NO_2^- or NO_3^- as the terminal electron acceptor(s) for respiration. The dotted lines emphasize those products which may be removed by the heterotrophic bacteria.

low phenol concentrations. Interestingly, it was noted that the rate of phenol consumption for maintenance purposes was very high, which could be due to a high specific death rate produced by the growth-inhibitory phenol. In the mixed culture the growth kinetics were very complex and the results suggested a large number of interactions, including stimulation and inhibition of growth of one species by other species. In another study Jones *et al.* (1975) noted the production of extracellular acetate during aerobic growth of an *Aeromonas* sp. at high dilution rates in a glucose-limited two-stage continuous-flow culture. The acetate produced inhibited the growth of the *Aeromonas* sp. and increased its K_s for glucose. When a second organism was added that metabolized acetate but not glucose then the growth rate of the culture in the second stage was greatly increased compared with that of the pure culture, acetate did not accumulate and the potential rate of glucose oxidation was increased threefold. This study has many similarities to those concerned with the inhibitory action of acetate (produced from ethane) on the growth of *Methylococcus* sp. on natural gas, and the alleviation of this inhibition by added heterotrophic bacteria that can grow on the acetate (see pp. 380–387).

In another study with pure and mixed cultures, Wilkinson and Hamer (1979) examined the biological oxidation of a mixture of methanol, phenol, isopropanol, and acetone. Their conclusion was that, although considerable interactions between the various members in the mixed culture might have been expected, the experimental results showed that the system behaved as if it consisted of three non-interacting systems, with each species oxidizing a separate substrate (methanol, phenol, and acetone) and isopropanol was co-oxidized by the acetone-oxidizing bacteria.

This description indicates some of the interactions that may occur in mixed bacterial cultures used in the biotreatment of pure or mixed component effluents. The natural systems that are undefined mixed cultures grown in open systems are difficult to analyse as the experimental techniques available are inadequate. Most of the experimental systems involve the isolation of the major bacterial species from the mixed culture, and the growth of pure and reconstituted mixed cultures on the compounds in question. There must always be some uncertainty as to whether some important component of the natural population cannot be isolated because it is growing on a co-oxidation product or an overflow metabolite not present in the isolation medium. However, it is important to realise that the same principles of microbial physiology that are used to describe simple mixed cultures can be applied to more complex systems. It is probable that co-oxidation reactions, especially in the metabolism of alkanes and aromatic compounds, and the utilization of co-oxidation products produced by one species as a substrate for another species is quite a common phenomenon. It is also probable that the production of overflow metabolites, e.g. acetate or polysaccharides by one species, (Neijssel and Tempest, 1979) and their utilization by other species is a common phenomenon in many biological effluent treatment processes. The important point is that to date all such processes rely on the development of a natural population of microorganisms. There have been no attempts to use defined mixed cultures. Perhaps a future development will be to add specific microorganisms if a sudden shock-loading with a novel substrate is expected, thereby increasing the capacity of the system to degrade the substrate in question. For a more comprehensive outline of this area the reader is referred to other chapters in this series.

3.4.2 Metal leaching and Bioaccumulation

Metal recovery is discussed in detail by Norris and Kelly (p. 443) but a few interesting examples are discussed here. As the natural resources of the world are becoming exhausted, secondary metal recovery via metal leaching of low-grade ores will assume greater importance. Processes for the recovery of copper and uranium are already in operation (Kelly *et al.*, 1979) and the bioaccumulation of platinum, palladium, mercury, gold, and silver could have industrial application in the near future (Chakrabarty, 1976; Charley and Bull, 1979).

The potential of mixed cultures in metal recovery has been demonstrated by Trevidi and Tsuchiya (1975). A mixed culture of *Thiobacillus ferroxidans* and *Beijerinkia lacticogenes*, an acid-tolerant nitrogen-fixing bacterium, was shown to have considerably higher leaching rates for copper and nickel that a pure culture of *T. ferroxidans*. *T. ferroxidans* supplied fixed carbon to *B. lacticogenes* which in turn supplied fixed nitrogen to the former organism. In addition, polysaccharide produced by *B. lacticogenes* was thought to make the surface of the ore more accessible to attack.

Charley and Bull (1979) described a mixed culture containing *Pseudomonas maltophilia, Staphylococcus aureus*, and a coryneform bacterium that tolerated up to 100mM silver ions. *Pseudomonas maltophilia* was mainly responsible for the resistance to silver, and the other two organisms, in particular *Staphylococcus aureus*, were very sensitive to silver ions. But, surprisingly, the mixed culture was more resistant to silver than a pure culture of *P. maltophilia*. Moreover, the rate of bioaccumulation of silver by the mixed culture was greater than that observed with *Pseudomonas maltophilia* with a maximum accumulation of silver of 316 mg silver (g dry weight)$^{-1}$ for the mixed culture compared with 182 mg silver (g dry weight)$^{-1}$ for the pure culture after 15 h incubation. The interactions in the mixed culture were not examined in detail but it was suggested that cross-feeding probably occurred since a pure culture of *P. maltophilia* has a growth requirement for L-methionine. In the mixed culture grown in the presence of silver, the coryneform bacterium was the dominant organism and probably supplied the pseudomonas with the required growth factor.

3.4.3 Steroid Biotransformation

A number of reports indicate that various steroid transformations that require a series of discrete microbiological steps involving the action of different microorganisms, can be performed more efficiently in a single stage using a mixture of suitable microorganisms (Lee *et al.*, 1969,

1970). By combining two bacteria, *Arthrobacter simplex* (1-dehydrogenator) and *Streptomyces roseochromogenes* (16-α-hydroxylator), the multi-step transformation of 9-α-fluorohydrocortisone to 1-dehydro-16-α-hydroxy-9-α-fluorohydrocortisone was effected in a single fermentation. This transformation using a mixed culture is particularly interesting as it demonstrates that enzyme induction and repression occur as a result of the interaction of two microorganisms. When a pure culture of *A. simplex* was used to carry out the transformation, the enzyme 20-ketoreductase was induced and caused considerable by-product formation. In the mixed culture, this enzyme was repressed and the 1-dehydrogenase and 16-α-hydroxylase were selectively induced, and high yields of the required end-product 1-dehydro-16-α-hydroxy-9-α-fluorohydrocortisone were obtained.

The mixed culture system has two main advantages:

(1) In a transformation requiring the sequential activity of a number of microorganisms the product of the first reactions is the substrate for the next reaction in the sequence. In normal processes, the product of each reaction is isolated before the next step is carried out in a separate reactor. This procedure causes substantial losses of product which is avoided by carrying out transformation in a single reactor with a mixed culture.

(2) In the example quoted above the mixed culture provided a better control of the transformation as an undesirable side reaction, normally expressed by the monoculture, was repressed. The conventional monoculture has to be carefully monitored so that the reaction can be terminated when the product concentration is at its maximum and by-product formation is negligible.

4. MERITS OF MIXED CULTURES IN INDUSTRY

4.1 Yield

It has been calculated (Harrison, 1978) that for SCP production a small, but significant increase (11%) in yield occurs with mixed cultures because the heterotrophic members utilize the growth and lysis products excreted by the primary substrate utilizer. However, the excretion of products into the culture during growth of a pure culture on methane or methanol depresses the growth yield of the C_1 utilizer. Although the heterotrophic components can assimilate the lysis products, these growth substrates are much more oxidized than the starting substrate and therefore support a lower total biomass concentration

Table 9. A comparison of yield coefficients and maximum growth rates reported for various pure and mixed cultures growing on methanol as the sole source of carbon and energy.

Organisms	Growth Conditions	μ_{max} (h^{-1})	Y_{CH_3OH} g dry weight (g methanol)$^{-1}$	Y_{O_2} g dry weight (g oxygen)$^{-1}$	Pathway of C_1 fixation	Productivity (g dry weight l^{-1} h^{-1})	References
Pseudomonas C	32°C	0.49	0.54	—	RM*	low	Goldberg *et al.* (1976) Goldberg (1977) Streglitz and Mateles (1973)
Methylomonas clara	39°C, pH 6.8	0.50	0.50	—	RM	high	Faust *et al.* (1977)
Methylomonas methanolica	30°C, pH 6.8	0.53	0.48	0.53	RM	low	Dostalek and Molin (1975)
Methylomonas methanolica	35°C		0.39	0.57	RM	low	Dostalek *et al.* (1972)
Pseudomonas EN	42°C, pH 6.8	0.51	0.46	—	RM	low	Wren and Harrison (1976) Harrison (1978)
Torulopsis galbrata	—	0.08	0.45	—	—		Asthana *et al.* (1977)
Pichia methanotherm	37 to 40°C, pH 4 to 6		0.39			high	Minami *et al.* (1978)
Mixed Culture	56°C	0.32	0.42			low	Snedecor and Cooney (1974)
MSI (mixed culture)	34°C	0.49	0.44	0.61	RM	high	Cremieux *et al.* (1977) Ballerini *et al.* (1977)
Mixed Culture containing *Pseudomonas* EN	42°C, pH 6.8	0.64	0.52 to 0.54	—	RM	high	Harrison *et al.* (1974)
TM 20 (mixed culture)	31°C, pH 6 to 6.3 batch	0.21	0.41	—	—	—	Haggstrom (1969)
Mixed Culture HR	30°C, pH 6.48 to 6.60 batch	0.32	0.33	0.31		low	Vary and Johnson (1967)

* RM = ribulose monophosphate pathway of C_1 fixation.

(Linton and Stephenson, 1978) than would be obtained if the C_1 compound has been assimilated by a pure culture with no product excretion. Evidence showing that mixed cultures growing on methane or methanol are inherently better than pure cultures in terms of yield and maximum growth rate is poor. Although mixed cultures may exhibit higher growth yield values than the component methane- or methanol-utilizer grown in pure culture (Harrison *et al.*, 1974), in our experience there is no significant difference between a pure culture of *Methylococcus* sp. and a mixed culture containing this organism in terms of growth yield and μ_{max}. A comparison of growth rate and yields of various pure and mixed cultures growing on methane and methanol supports this view (Tables 9 and 10).

4.2 The Problem of Foaming

The complete utilization and elimination of organic compounds from the culture supernatant by a monoculture has not been reported (Goldberg, 1977). Certainly in all pure cultures examined so far, the excreted carbon compounds amount to 2 to 45% of the steady-state bacterial biomass carbon, depending on the growth conditions (Dostalek and Molin, 1975; Linton *et al.*, 1975; Linton and Buckee, 1977). In practice, it is the removal of this organic carbon that gives the mixed culture a number of advantages over the pure culture system. Pure cultures of C_1 utilizers grown to high cell densities (and other pure cultures as well), foam a great deal owing to the release of surface-active proteins and nucleic acids as a result of cell lysis. Although antifoam agents could be used to control this problem it is not an acceptable solution as these agents become associated with the product. Foaming may adversely affect the rate of oxygen mass transfer by altering bubble coalescence and break-up, and the physical characteristics of the liquid-gas interface may be altered. Foaming also causes considerable difficulty in liquid volume control and, consequently, problems with the control of dilution rate. It can also cause blockage of filters and gas lines and this may increase the chance of contamination when these components are replaced.

4.3 Process Effluents

Foaming is a symptom of a much more serious problem, namely, the presence of significant amounts of organic carbon in the supernatant. In a large industrial process, say, for SCP production, employing recycled process water, this supernatant carbon rapidly accumulates to

Table 10. A comparison of maximum yields and growth rates of pure and mixed cultures growing on methane.

	Organism(s)	Growth Conditions	μ_{max} (h^{-1})	Y_{CH_4} g dry weight (g methane)$^{-1}$	Y_{O_2} g dry weight (g oxygen)$^{-1}$	Pathway of C_1 fixation	Remarks	References
Pure cultures	*Methylococcus*	30°C, pH 7.0	≥0.24	1.0	0.6	RM*	No carbon balance, cell density ≤10 mg l^{-1} continuous culture	Nagai *et al.* (1973)
	Methylococcus capsulatus	37°C, methane-limited oxygen-limited		1.01 0.31	0.29 0.51	RM	Carbon balance, low productivity continuous culture	Harwood and Pirt (1972)
	Methylococcus capsulatus	45°C, Ammonia-limited partly Nitrogen fixing		0.63	0.21	RM	No carbon balance, productivity 2.46 $gl^{-1} h^{-1}$. Continuous, BP Patent	Gould *et al.* (1976)
	Methylococcus NCIB 11083	40°C, Ammonia-limited pH 6.5	0.31	0.72	0.70	RM	Carbon balance, low productivity, continuous	Khosrovi, unpublished

Mixed cultures	Mixed culture containing *Methylococcus* NCIB 11083	40°C, Ammonia-limited pH 6.4	0.34	0.71	0.70	RM	Carbon balance, low productivity, continuous	Khosrovi, unpublished
	As above	45°C, Methane-limited pH 6.5, NH_4-grown	0.34	0.81	—	—	Carbon balance, high productivity 4.8 $gl^{-1} h^{-1}$, continuous	Downs *et al.* (1978)
	Methylosinus trichosporium OB 3B	30°C, NH_4-limited	—	0.63	0.41	S	Carbon balance, high productivity	Downs *et al.* (1978)
	Mixed culture TM 10	31°C, NH_4-grown		0.9	0.52	RM	No carbon balance, batch	Bewersdorff and Dostalek (1971)
	Mixed culture containing *Pseudomonas* sp.	32°C, NH_4-grown		0.8	0.24	?	No carbon balance, low productivity	Wilkinson *et al.* (1974)
	Mixed culture	33°C		1.0	0.76	?	No carbon balance	Grigorian and Lalov (1976)
	Mixed culture M45	45°C, NO_3-grown	0.303	0.62	0.22	?	Carbon balance, 2.4 $gl^{-1} h^{-1}$, high productivity, continuous	Sheehan and Johnson (1971)
	Mixed culture HR	30°C, NH_4, pH 6.7	0.14	0.70	0.23	?	Carbon balance, batch	Vary and Johnson (1967)

* RM = ribulose monophosphate pathway of C_1 fixation; S = serine pathway of C_1 fixation.

unacceptably high levels and, in addition to an effluent problem, may cause inhibition of growth of the C_1-utilizing organism (Eroshin *et al.*, 1968; Eccleston and Kelly, 1972). We have observed that when a pure culture of *Methylococcus* sp. NCIB 11083 was grown at a dilution rate of 0.22 h^{-1}, cell density 14.9 g dry weight l^{-1}, and a productivity of 3.2 g dry weight l^{-1} h^{-1} the supernatant organic matter (assuming a 46% carbon content) was approximately 2.8 g dry weight l^{-1}. If a 90% water recycle is used, this organic carbon would be concentrated to 28 g dry weight l^{-1}. By comparison, a mixed culture grown at a productivity of 4.8 g dry weight l^{-1} h^{-1} ($D = 0.22\ h^{-1}$ dry weight, 21.8 gl^{-1}), contained ten times less supernatant carbon, 0.32 g dry weight l^{-1}, even with a 90% water recycle, this would increase to 3.2 g dry weight l^{-1}.

In pure cultures grown to high productivities, the high content of organic matter in the culture supernatant provides a nutrient-rich environment for the growth of contaminants.

4.4 Sterilization and the Risk of Contamination

Can a pure culture be maintained in an industrial process for SCP production?

In most industrial fermentations, all contaminating microorganisms must be excluded from the system by sterilization prior to inoculation. The rigorousness of sterilization depends on economic and safety factors related to the product and the process. For example, rigorous product sterilization is necessary in the canned food industry, while in antibiotic production, the conditions are such that some probability of contamination is allowable.

The predominant method of sterilization in industrial fermenters is the use of heat. The kinetics of thermal death of viable populations is a first-order reaction (Aiba *et al.*, 1973; Bailey and Ollis, 1977): the probability that a batch fermentation may become contaminated can be calculated. For example, in the antibiotic industry a sterilization regime necessary to ensure that only 1 in 100 fermentations become contaminated is used. In many batch fermentations there is also the probability that if contamination occurs in low numbers the fermentation could be completed before the contaminant establishes itself and significantly affects the process. However, with the introduction of very large reactors for time-independent continuous production of single cell protein the situation is rather different:

(1) Only a few, usually one, very large reactor (s) will be operated, thus to ensure sterile conditions prior to inoculation the probability of contamination should be reduced to 1 in 10^{12} fermentations. This value

is used in the canning industry (Bailey and Ollis, 1977) where there must be complete elimination of *Clostridium botulinum* spores.

(2) The very large size of the reactors, e.g. approximately 2000 m^3 liquid volume for a 100 000 ton per annum reactor operated at a maximum productivity of 10 kg m^3 h^{-1} will mean that the total initial organism/spore count will be much higher than that found in smaller conventional reactors which are often more than a hundred times smaller. The large value for total initial organism/spore concentrations will mean that long sterilization times are required to reduce the probability of the fermenter being contaminated to 1 in 10^{12} sterilizations.

(3) The operation of a high productivity continuous process in a time independent manner means that any contaminants that can grow on suitable substrates in the reactor, e.g. on lysis products of the major organism, will reach a final steady-state concentration.

From the above factors it can be concluded that there are several reasons that are against the probable attainment of sterility prior to inoculation and, hence, subsequent pure culture operations in a large continuous SCP reactor. If strict adhesion to monoculture was used then not only would initial sterilization be extremely difficult, but it would mean that the reactor would have to be shut down every time contamination was detected and the liquid contents would have to be disposed of. The latter would pose a considerable effluent treatment problem. The continual shut-down and start-up of the reactor would seriously decrease the productivity of the process.

These arguments suggest that it would be technically difficult and uneconomic to ensure absolute monoculture operation in a large continuous industrial reactor for biomass production. It must also be emphasized that in practice a large amount of the process water will be recycled back to the reactor after removal of the biomass by a centrifugation/flocculation step (Topiwala and Khosrovi, 1978); it is difficult to ensure the complete absence of contamination in these downstream operations, especially during continuous centrifugation. This means that under operational conditions contamination of a monoculture is very likely to occur. To avoid this problem the British Petroleum n-paraffin process was run under non-aseptic conditions (Laine, 1974; Levi, 1976) such that a yeast species was the dominant organism in the mixed culture. The non-aseptic operation allowed substantial savings to be made as the sterilization step was omitted.

The use of mixed cultures for SCP production should be viewed against the downstream processing of the product, the storage of the product and its subsequent use. It is difficult to carry out many downstream processing steps without contamination occurring (Levi, 1976) and it is no good rigorously maintaining a mono-culture in the reactor when considerable contamination of the product may occur downstream from the reactor.

4.5 Resistance to Perturbations in Culture Environment

Mixed cultures suitable for the production of SCP from methane and methanol appeared to be more resistant to perturbations in the culture environment than pure cultures. For example, those caused by failure of culture pH, temperature and the transient build-up of ammonia.

Although it has been suggested that mixed cultures are more resistant to contamination than pure cultures (Harrison *et al.*, 1975) there has only been one serious attempt to check this (Rokem *et al.*, 1980). Rokem *et al.* (1980) demonstrated that potential pathogenic bacteria could not get established when introduced into a mixed culture containing *Pseudomonas* C and a number of heterotrophic soil bacteria, although in a pure culture these organisms easily establish themselves.

5. CONCLUSIONS

The application of mixed cultures to SCP production relies on overcoming two major obstacles. The first is concerned with the IUPAC guidelines, which state that cultures used for SCP production must be defined. In theory this does not pose a problem as defined mixed cultures can be used. However, in practice the heterotrophic bacteria commonly found in mixed cultures growing on methane and methanol rarely fit into the identification schemes generally used (Cowan and Steel, 1974; Buchanan and Gibbons, 1974). Consequently, it is difficult to be absolutely certain that one of the heterotrophic components of the mixed culture has not been replaced by a similar organism. Because there is a strong selection pressure for a particular metabolic type, contaminants are likely to resemble metabolically those organisms they replace. On the other hand contamination of a monoculture of a C_1 utilizer by heterotrophic bacteria is easily detected and, consequently, product quality is easier to control. The second major problem lies in patenting mixed cultures. As mentioned earlier some of the heterotrophic microorganisms are difficult to identify. In certain cases the primary organism (Wilkinson *et al.*, 1974) cannot be isolated in pure culture because it has a symbiotic relationship with another organism. This may prove a considerable problem in patenting these mixed culture systems.

The use of mixed cultures in traditional fermentation processes such as cheese manufacturers has been developed as an empirical "art" rather than a quantitative science. However, the development of industrial mixed culture processed for the production of chemicals from biomass and of SCP from hydrocarbons has greatly increased our scientific understanding of some of the complex interactions that occur in such systems. We can expect to see further developments in this area and especially of structured mixed cultures that are far more suitable for certain industrial processes than monocultures. For example, in the production of chemicals from renewable resources. Such mixed cultures offer a good process alternative to the use of genetically engineered microorganisms for carrying out a wide range of reactions.

REFERENCES

Aiba, S., Humphrey, A. E., and Millis, N. F. (1973). "Biochemical Engineering." Academic Press, New York and London.

Asthana, H., Humphrey, A. E., and Moritz, V. (1977) Growth of yeasts on methanol as the sole source of carbon substrate. *Biotechnology and Bioengineering* **13**, 923–929.

Bailey, J. E. and Ollis, D. F. (1977). "Biochemical Engineering Fundamentals". McGraw-Hill Kogakusha, Tokyo.

Ballerini, D., Parlovar, D., Lepeyonnie, M., and Sri, K. (1977). Mixed culture of bacteria utilizing methanol for growth. *European Journal of Applied Microbiology* **4**, 11–19.

Ban, S. and Glanser-Soljan, M. (1979). Rapid biodegradation of calcium lignosulphonate by means of a mixed culture of microorganisms. *Biotechnology and Bioengineering* **21**, 1917–1928.

Bellamy, W. D. (1974). Single cell proteins from cellulosic wastes. *Biotechnology and Bioengineering* **16**, 869–890.

Bellamy, W. D. (1978). Production of ethanol from cellulose using a thermophilic mixed culture. *United States Patent 4,094,742.*

Benemann, J. R., Weissmann, J. E., Koopman, B. L., and Oswald, W. J. (1977). Energy production by microbial photosynthesis. *Nature, London* **268**, 19–23.

Bewersdorff, M. and Dostalek, M. (1971). The use of methane for the production of bacterial protein. *Biotechnology and Bioengineering* **13**, 49–62.

Bernstein, S., Tzeng, C. H. and Sisson, D. (1977). The commercial fermentation of cheese whey for the production of protein and for alcohol. *In* "Single Cell Protein from Renewable and Non-renewable Resources". *Biotechnology and Bioengineering Symposium* **7**, pp. 1–11. (Gaden, E. L. and Humphrey, A. E., eds.) John Wiley, New York.

Bogdanova, V. M. (1966). Physiological properties of microorganisms isolated in the atmosphere of methane and propane. *Mikrobiologica* **35**, 234–241.

Brown, L. R., Strawinski, R. J. and McCleskey, C. S. (1964). The isolation and characterization of *Methanomonas methano-oxidans* Brown and Strawinski. *Canadian Journal of Microbiology* **10**, 791–799.

Brooks, R. E., Su, T. M., Bellamy, W. D. and Brennan, M. J. (1979). Bioconversion of plant biomass to ethanol. Summary of experimentation carried out by above personnel. Bio-energy Directory: category B. Ethanol Information, 2. Fermentation. March, pp. 306.

Buchanan, R. E. and Gibbons, N. E. (eds) (1974). Bergey's Manual of Determinative Bacteriology, 8th Edn., Williams and Wilkins, Baltimore.

Bu'Lock, J. D. (1979). Industrial alcohol. *In* "Microbial Technology, Current State, Future Prospects", pp. 309–325. (Bull, A. T., Ellwood, D. C., and Ratledge, C., eds.) Cambridge University Press, London.

Bungay, H. R. and Bungay, M. L. (1968). Microbial interactions in continuous culture. *Advances in Applied Microbiology* **10**, 269–290.

Chakrabarty, A. M. (1976). Plasmids in *Pseudomonas*. *Annual Reviews of Genetics* **10**, 7–30.

Charley, R. C. and Bull, A. T. (1979). Bioaccumulation of silver by a multispecies community of bacteria. *Archives for Microbiology* **123**, 239–244.

Chen, B. J., Hirt, W., Lim, H. C., and Tsao, J. T. (1977). Growth characterisation of a new Methylomonad. *Applied and Environmental Microbiology* **33**, 269–274.

Clement, G. (1975). Producing *Spirulina* with carbon dioxide. *In* "Single Cell Protein", vol. II, pp. 467–474. (Tannenbaum, S. R. and Wang, D. I. C., eds.) MIT Press, Cambridge, Mass.

Cowan, S. T. and Steel, K. J. (1974). "Manual for the Identification of Medical Bacteria". 2nd Edn. Cambridge University Press, London.

Cowling, E. B. and Kirk, T. K. (1976). Properties of cellulose and lignocellulosic materials as substrates for enzymatics conversion processes. *Biotechnology and Bioengineering Symposium* **6**, 95–123.

Cremieux, A., Chevalier, J., Combet, M., Dumenil, G., Parlovar, D. and Ballerini, D. (1977). Mixed culture of bacteria utilizing methanol for growth. I. Isolation and Identification. *European Journal of Applied Microbiology* **4**, 1–9.

Cysewski, G. R. and Wilke, C. R. (1976). Utilization of cellulosic materials through enzymatic hydrolyses. 1. Fermentation of hydrolysate to ethanol and single cell protein. *Biotechnology and Bioengineering* **18**, 1297–1313.

Cysewski, G. R. and Wilke, C. R. (1977). Rapid ethanol fermentation using vacuum and cell recycle. *Biotechnology and Bioengineering* **19**, 1725–1138.

Cyewski, G. R. and Wilke, C. R. (1978). Process design and economic studies of alternative fermentation methods for the production of ethanol. *Biotechnology and Bioengineering* **20**, 1421–1444.

Dostalek, M., Haggstrom, L., and Molin, N. (1972). Optimization of biomass production from methanol. *In* Proceedings of IVth International Fermentation Symposium "Fermentation Technology Today". pp. 497. Kyoto, Japan. (Terui, G., ed.) Society of Fermentation Technology, Japan.

Dostalek, M. and Molin, N. (1975). Studies of biomass production of methanol oxidizing bacteria. *In* "Single Cell Protein", vol. 2, pp. 395–401. (Tannenbaum, S. R. and Wang, D. I. C., eds.) MIT Press, Cambridge, Mass.

Downs, J., Drozd, J. W., Khosrovi, B., Linton, J. D., and Barnes, L. J. (1978). An analysis of growth energetics in Methylococcus NCIB 11083. *Proceedings of the Society for General Microbiology* **5**, 45.

Drozd, J. W., Godley, A., and Bailey, M. L. (1978a). Ammonia oxidation by methane-oxidizing bacteria. *Proceedings of the Society for General Microbiology* **5**, 66.

Drozd, J. W., Linton, J. D., Downs, J., and Stephenson, R. J. (1978b). An *In situ* assessment of the specific lysis rate in continuous cultures of *Methylococcus* sp. (NCIB 11083) Grown on methane. *FEMS Microbiology Letters* (1978), **4**, 311–314.

Drozd, J. W., Khosrovi, B., Downs, J., Bailey, M. L., Barnes, L. J., and Linton, J. D. (1980). Biomass production from natural gas. *In* Proceedings of the 7th International Continuous Culture Sympoium, Prague, 1978, pp. 505–519. (Sikyta, B. S., Fencl, Z., and Polacek, V., eds.) Czechoslovak Academy of Science, Prague.

Drozd, J. W. and McCarthy, P. W. (1981). Mathematical model of microbial hydrocarbon oxidation. *In* Proceedings of the 3rd International Symposium on microbial growth on C_1-compounds, Sheffield, 1980, pp. 360–369. (Dalton, H. D., ed.) Heyden & Sons, London.

Drozd, J. W. and Linton, J. D. (1981). Single cell protein production from methane and methanol in continuous culture. *In* "Continuous Culture of Cells", vol. 1, pp. 113–141. (Calcott, P. H., ed.) CRC Press Florida, USA.

Dunlop, C. E. (1975). Production of single cell protein from insoluble agricultural wastes by mesophiles. *In* "Single Cell Protein", vol II, pp. 244–272. (Tannenbaum, S. R. and Wang, D. I. C., eds.) MIT Press, Cambridge, Mass.

Eccleston, M. and Kelly, D. P. (1972). Assimilation and toxicity of exogenous amino acids in the methane-oxidizing bacterium *Methylococcus capsulatus*. *Journal of General Microbiology* **71**, 541–554.

Eroshin, U. K., Harwood, J. H., and Pirt, S. J. (1968). Influence of amino-acids, carboxylic acids and sugars on the growth of *Methylococcus capsulatus* on methane. *Journal of Applied Bacteriology* **31**, 560–567.

Faust, U., Prave, P., and Sukatsch, D. A. (1977). Continuous biomass production from methanol by *Methylomonas clara*. *Journal of Fermentation Technology* (Japan) **55**, 609.

Flickinger, M. C. and Tsao, G. T. (1978). Fermentation substrates from cellulosic materials. Fermentation products from cellulosic materials. *In* "Annual Reports on Fermentation Processes, vol. 2", 23–42. (Perlman, D., ed.) Academic Press, London and New York.

Gaden, E. L. and Humphrey, A. E. (1977). Single cell protein from renewable and non-renewable resources. Biotechnology and Bioengineering Symposium 7. John Wiley, New York.

Gauss, W. F. and Suzuki, S. (1976). Manufacture of alcohol from cellulosic materials using plural ferments. *United States Patent No.*3,990,944.

Goldberg, I., Rock, J. S., Ben-Bassat, A., and Mateles, R. I. (1976). Bacterial Yields on methanol, methylamine, formaldehyde and formate. *Biotechnology and Bioengineering* **18**, 1657.

Goldberg, I. (1977). Production of SCP from methanol-yield factors. *Process Biochemistry* **12**, 12–18.

Gould, P., Moran, F., and Myers, P. A. (1976). Fermentation process for converting methane and elemental nitrogen into proteinaceous material. *British Patent No.* 1,421,135.

Grigorian, A. N. and Lalov, V. V. (1976). Studies in cultivation of microorganisms on natural gas. *In* Natural Gas Proceedings and Utilization Conference. Dublin, *Inst. Chem. Eng. Symp. Ser.* **44**.

Haggstrom, L. (1969). Studies on methanol oxidizing bacteria. *Biotechnology and Bioengineering* **11**, 1043–1054.

Harder, W., Kuenen, J. G., and Martin, A. (1977). A review: Microbial selection in continuous culture. *Journal of Applied Bacteriology* **43**, 1–24.

Harrison, D. E. F. (1976). Making protein from methane. *Chemical Technology* **6**, 570–574.

Harrison, D. E. F. (1978). Mixed cultures in industrial fermentation processes. *Advances in Applied Microbiology* **24**, 129–162.

Harrison, D. E. F., Harwood, H., and Wren, S. J. (1974). Process for the simultaneous production of methanol-utilizing and non-methanol utilizing microorganisms. *British Patent No.* 1,450,412.

Harrison, D. E. F., Wilkinson, T. G., Wren, S. J., and Harwood, J. H. (1975). Mixed bacterial cultures as a basis for continuous production of single cell protein from C_1 compounds. *In* "Continuous Culture 6. Application and New Fields", pp. 122–134. (Dean, A. C. R., Ellwood, D. C., Evans, C. T. G., and Melling, J., eds.) Ellis Horwood, Chichester.

Harrison, D. E. F. and Wren, S. J. (1976). Mixed microbial cultures as a basis for future fermentation process. *Process Biochemistry* **10**, 30–32.

Harwood, J. H. and Pirt, S. J. (1972). Quantitative aspects of growth of the methane oxidizing bacterium *Methylococcus capsulatum* on methane in shake flasks and continuous culture. *Journal of Applied Bacteriology* **35**, 597–607.

Hoge, W. H. (1977). Process for making alcohol from cellulosic material using plural ferments. *United States Patent No.* 4,009,075.

Huff, G. F. and Yata, N. (1975). Enzymatic hydrolysis of cellulose. *United States Patent No.* 3,990,945.

Jannasch, H. W. and Mateles, R. I. (1974). Experimental bacterial ecology studies in continuous culture. *In* "Advances in Microbial Physiology", pp. 165–212. (Rose, A. H. and Tempest, D. W., eds.) Academic Press, London and New York.

Jarl, K. (1969). Production of microbial food from low-cost starch materials and purification of industry's waste starch effluents through the Symba yeast process. *Food Technology* **23**, 23–26.

Jarl, K. (1971). Utilization of waste material by fermentation. Communications from the Swedish Sugar Corporation. *Socker Handlinger* **25**, 4–11.

Johnson, M. (1972). Techniques for selection and evaluation of cultures for biomass production. *In* "Proceedings of IV International Fermentation Symposium", pp. 473, Kyoto, Japan. (Terui, G. ed.) Society of Fermentation Technology, Japan.

Jones, G. L. and Carrington, E. G. (1972). Growth of pure and mixed cultures of microorganisms concerned in the treatment of carbonization waste liquors. *Journal of Applied Bacteriology* **35**, 395–404.

Jones, G. L., Jansen, F., and McKay, A. J. (1973). Substrate inhibition of the growth of bacterium NCIB 8250 by phenol. *Journal of General Microbiology* **74**, 139–148.

Jones, G. L., Loveless, J. E., and Novak, A. J. (1975). Kinetics of the utilization of glucose by *Aeromonas* in the presence and absence of Bacterium NCIB 8250. *Technical memorandum TM 116.* Water Research Centre, Stevenage, Herts, England, pp. 1–13.

Keenan, J. D. (1979). Review of biomass to fuels. *Process Biochemistry* **5**, 9–15.

Kelly, D. P., Norris, P. R., and Brierley, C. L. (1979). Microbiological methods for the extraction and recovery of metals. *In* "Microbial Technology: Current State, Future Prospects", pp. 263–308. (Bull, A. T., Ellwood, D. C., and Ratledge, C., eds.) Cambridge University Press, London.

Kuenen, J. G. and Veldkamp, H. (1973). De Chemostat als lulpmiddel bij de stodie can de oecologie van bacterien. *Vakblad voor Biologen* **6**, 100.

Ladisch, M. R. (1979). Fermentable sugars from cellulosic residues. *Process Biochemistry* **1**, 21–25.

Laine, B. M. (1974). What proteins cost from oil. *Hydrocarbon Processing*, November, 139–142.

Laskin, A. I. (1977). Ethanol as a substrate for single cell protein production. *In* "Single Cell Protein from Renewable and Non-renewable Resources". Biotechnology and Bioengineering Symposium 7. pp. 91–103. (Humphrey, A. E. and Gaden, E. L., eds.) John Wiley, New York.

Leadbetter, E. F. and Foster, J. W. (1958). Bacterial oxidation of gaseous alkanes. *Archives for Mikrobiologie* **35**, 92–104.

Lee, B. K., Brown, W. E., Ryll, D. Y., Jacobson, H., and Thoma, R. W. (1970). Influence of mode of steroid substrate addition on conversion of steroid and growth characteristics in a mixed culture fermentation. *Journal of General Microbiology* **61**, 97–105.

Lee, B. K., Ryu, D. Y., Thoma, R. W., and Brown, W. E. (1969). Induction and repression of steroid hydroxylases and dehydrogenases in mixed culture fermentations. *Journal of General Microbiology* **55**, 145–153.

Levi, J. D. (1976). Bacteriological aspects of the production of "Toprina" dried yeasts from n-paraffins. Abstract of 5th International Fermentation Symposium Berlin. Edited by H. Dellweg, P. 205. Verlag Versuchs und Lehranstalt fur spiritusfabrikation und Fermentation Technologie in Institut fur Garungsgewerke und Biotechnologie Berlin.

Lindeman, L. R. and Rocchiccioli, C. (1979). Ethanol in Brazil: brief summary of the state of Industry in 1977. *Biotechnology and Bioengineering* **21**, 1107–1119.

Linton, J. D. and Buckee, J. C. (1977). Interactions in a methane-utilizing mixed culture in a chemostat. *Journal of General Microbiology* **101**, 219–225.

Linton, J. D. and Cripps, R. E. (1978). The occurrence and identification of intracellular polyglucose storage granules in *Methylococcus* NCIB 11083 grown in chemostat culture on methane. *Archives of Microbiology* **117**, 41–48.

Linton, J. D., Godley, A. R., Bailey, M. L., and Barnes, L. J. (1980). Growth of an ethane-utilizing mixed culture in a chemostat. *Journal of Applied Bacteriology* **48**, 341–347.

Linton, J. D., Harrison, D. E. F., and Bull, A. T. (1975). Molar growth yields, respiration and cytochrome patterns of *Beneckea natriegens* when grown at different medium dissolved oxygen tensions. *Journal of General Microbiology* **90**, 237–246.

Linton, J. D. and Stephenson, R. J. (1978). A preliminary study on growth yields in relation to the carbon and energy content of various organic growth substrates. *FEMS Microbiology Letters* **3**, 95–98.

Malashenko, Y. R., Romanovskaya, V. A., Bogachenko, V. N., Khotyan, L. V., and Voloshin, N. V. (1973). Characterisation of carbon nutrition of microorganisms that grow on natural gas. *Mikrobiologiya* **42**, 403–408.

Malashenko, Y. R., Romanovskaya, V. A., Bogachenko, V. N., Voloshin, L. V., and Kryshtab, T. P. (1975). Peculiarities of assimilation of carbon components of natural gas by one species cultures and their artificial associations. *Transactions USSR Academy of Biological Sciences*. Series I, 45–51.

Margaritis, A. and Wilke, C. R. (1978a). The rotor fermenter. I. Description of the apparatus, power requirements, and mass transfer characteristics. *Biotechnology and Bioengineering* **20**, 709–826.

Margaritis, A. and Wilke, C. R. (1978b). The rotor fermenter, II. Application to ethanol fermentation. *Biotechnology and Bioengineering* **20**, 727–753.

Meers, J. L. (1973). Growth of bacteria in mixed cultures. *CRC Critical Reviews in Microbiology* **2**, 139–184.

Miller, J. L. and Johnson, M. J. (1966). Utilization of normal alkanes by yeasts. *Biotechnology and Bioengineering* **8**, 549–565.

Minami, K., Yamamura, M., Shimizu, S., Ogawa, K. and Sekine, N. (1978). A new ethanol-assimilating, high productivity thermophilic yeast. *Journal of Fermentation Technology* **56**, 1–7.

Nagai, S., Mori, T., and Aiba, S. (1973). Investigation of the energetics of methane-utilizing bacteria in methane and oxygen limited chemostat culture. *Journal of Applied Chemistry and Biotechnology* **23**, 549–562.

Neijssel, O. M. and Tempest, D. W. (1979). The physiology of metabolite over production. *In* "Microbial Technology: Current State, Future Prospects, pp. 53–82. (Bull, A. J., Ellwood, D. C., and Ratledge, C., eds.) Cambridge University Press, London

Novick, A. and Szilard, L. (1950a). Description of the chemostat. *Science* **112**, 715.

Novick, A. and Szilard, L. (1950b). Experiments with the chemstat on spontaneous mutation of bacteria. *Proceedings of the National Academy of Science* (Washington), **36**, 708–719.

Pye, E. K., Humphrey, A. E., Alexander, J., Forror, J. R., Mateles, R. I., Nolan, E., and Rolz, C. (1979). The biological production of liquid fuels from cellulosic biomass. Summary of experimentation carried out by above personnel. *In* "Bio-energy Directory, Category B. Ethanol formation. 2. Fermentation", January, p. 309.

Pirt, M. W. and Pirt, S. J. (1977). Photosynthetic production of biomass and starch by Chlorella in chemostat culture. *Journal of Applied Chemistry and Biotechnology* **27**, 643–650.

Reddy, C. A. and Erdman, M. D. (1977). Production of a ruminant protein supplement by anerobic fermentation of feedlot waste filtrate. *In* "Single Cell Protein from Renewable and Non-renewable Resources". *Biotechnology and Bioengineering Symposium* **7**, pp. 11–22. (Gaden, E. L. and Humphrey, A. E.,eds.) John Wiley, New York.

Rokem, J. S., Goldberg, I., and Mateles, R. I. (1980). Growth of mixed cultures of bacteria on methanol. *Journal of General Microbiology* **116**, 225–232.

Romanovskaya, V. A., Malashenko, Y., R., Sokolov, I. G., and Kryshtab, T. P. (1976). The competitive inhibition of the microbial oxidation of methane by ethane. *In* "Microbial Production and Utilization of Gases". pp. 345–353. (Schlegel, H. G., Goltschalk, G., and Pfennig, N., eds.) Gottingen, E., Goltze, KG.

Saeman, J. F. (1945). Kinetics of wood saccharification. Hydrolysis of cellulose and decomposition of sugars in dilute acid at high temperature. *Industrial and Engineering Chemistry* **37**, 43–52.

Sheehan, B. T. and Johnson, M. J. (1971). Production of bacterial cells from methane. *Applied Microbiology* **21**, 511–515.

Skogman, H. (1969). Production of Symba-yeast from potato wastes.

Slater, J. M. and Bull, A. T. (1978). Interactions between microbial populations. *In* "Companion to Microbiology", pp. 181–206. (Bull, A. T. and Meadow, P. M., eds.) Longman, London.

Snedecor, B. and Cooney, C. L. (1974). Thermophilic mixed culture of bacteria utilizing methanol for growth. *Applied Microbiology* **27**, 1112–1117.

Streglitz, B. and Mateles, R. I. (1973). Methanol metabolism in *Pseudomonas* C. *Journal of Bacteriology* **114**, 390–398.

Sukatch, D. A. and Johnson, M. V. (1972). Bacterial cell production from hexadecane at high temperatures. *Applied Microbiology* **23**, 543–546.

Tonge, G. E. (1979). Industrial chemicals from fermentation. *Enzyme Microbial Technology* **1**, 173–179.

Topiwala, H. H. and Khosrovi, B. (1978). Water recycle in biomass production processes. *Biotechnology and Bioengineering* **20**, 73–85.

Trevidi, N. C. and Tsuchiya, H. M. (1975). Microbial mutation in leaching of Cu-Ni sulphide concentrates. *International Journal of Mineral Processing* **2**, 1–14.

Tsao, G. T., Ladisch, M., Ladisch, C., An Hsu, T., Dale, B., and Chou, T.

(1978). Fermentable substrates from cellulosic materials: Production of fermentable sugars from cellulosic materials. *In* "Annual Reports on Fermentation Processes, vol. 2", pp. 1–23. (Perlman, D., ed.) Academic Press, London and New York.

Vary, P. and Johnson, M. J. (1967). Cell yields of bacteria grown on methane. *Applied Microbiology* **15**, 1473–1478.

Veldkamp, H. and Jannasch, H. W. (1972). Mixed culture studies with the chemostat. *Journal of Applied Chemistry and Biotechnology* **22**, 105–123.

Veldkamp, H. and Kuenen, J. G. (1973). The chemostat as a model system for ecological studies. *Bulletins from the Ecological Research Committee (Stockholm)* **17**, 347.

Wang, D. E. C., Cooney, C. L., Demain, A. L., Gomez, R. F., and Sinskey, A. J. (1977). Degradation of cellulosic biomass and its subsequent utilization for the production of chemical feedstocks. *US Department of Energy, Progress Report* TID–27977.

Whittenbury, R., Phillips, K. C., and Wilkinson, J. F. (1970). Enrichment isolation and some properties of methane utilizing bacteria. *Journal of General Microbiology* **61**, 205–218.

Wilkinson, T. G. and Hamer, G. (1972). Some growth characteristics of a *Hyphomicrobium* sp. in batch culture. *Journal of Applied Bacteriology* **35**, 577–588.

Wilkinson, T. G. and Hamer, G. (1979). The microbial oxidation of mixtures of methanol, phenol, acetone and isopropanol with reference to effluent purification. *Journal of Chemical Technology and Biotechnology* **29**, 56–67.

Wilkinson, J. G., Topiwala, H. H. and Hamer, G. (1974). Interactions in a mixed continuous culture. *Biotechnology and Bioengineering* **16**, 41–59.

Wodzinski, R. S. and Johnson, M. J. (1968). Yields of bacterial cells from hydrocarbon. *Applied Microbiology* **16**, 1886–1891.

Wolnak, B. V., Andreen, B. N., Chisholm, J., and Saaden, M. (1967). Fermentation of methane. *Biotechnology and Bioengineering* **9**, 57–76.

Wren, S. J. and Harrison, D. E. F. (1976). The role of heterotrophic bacteria in a mixed culture growing on methanol. *Proceedings Society for General Microbiology* **4**, 29.

Wren, S. J., Harwood, J. H., and Harrison, D. E. F. (1974). Growth characterisation of a methanol-utilizing mixed culture. *Proceedings Society for General Microbiology* **2**, 14.

Young, F. M. and Wood, J. B. (1974). Microbiology and biochemistry of soy sauce fermentation. *Advances in Applied Microbiology* **17**, 157–188.

11

Fermented Foods and Beverages: the Role of Mixed Cultures

K. H. Steinkraus

1. INTRODUCTION

Microbial communities with their combined physiology and interactions and their enzymatic activities are responsible for the major biochemical and nutritional changes that occur in the substrates of most fermented foods and beverages. There are some fermented foods and beverages in which the enzymes in the raw plant or animal substrates also play important roles. This is particularly true in foods and beverages in which germination or malting is part of the process.

In some fermented foods the major or essential microorganisms and their metabolic sequences have been elucidated. In other cases not even the essential microorganisms have been identified. In some cases, the biochemical and nutritional changes occurring in the substrates have been determined. Some of the interactions among the essential microorganisms have been studied. However, when it comes to the molecular level, the interactions of microorganisms in most of the fermented foods have been inadequately studied.

The ecology of microorganisms in a fermented food environment is very important to the human race since fermentation is a major means of preserving food and also for transforming it to forms and flavours acceptable to the human consumer. Generally fermentation processes have expanded the variety of food flavours and textures available to man and the digestibility is often improved. In some cases, cooking time is shortened. Nutritional value is often modified and in some cases improved.

This chapter describes a number of primarily indigenous fermented foods and what is known of the interrelationships of their component microorganisms.

2. DEVELOPMENT OF MEAT-LIKE FLAVOURS THROUGH FERMENTATION

2.1 Soy Sauce (Japanese Shoyu) and Related Fermentations

Soy sauce is a light-brown to black liquid with a meat-like, salty flavour manufactured by hydrolysing soy beans, with or without the addition of wheat, using enzymes produced by *Aspergillus oryzae* in a strong salt brine (approximately 18%, w/v). An aerobic, solid state fungal fermentation is followed by an anaerobic mixed lactic acid bacteria and yeast submerged fermentation (Yokutsuka, 1977).

The meat-like flavourings of Chinese soy sauce, which are rich in amino acids and peptides, Japanese shoyu and miso are traditional fermentations in the Orient. The discovery of the soy sauce fermentation was a great and ancient contribution to mankind showing for the first time how to produce meat-like flavours from vegetable proteins and lipids. It eventually led to the development of the large modern enzyme, monosodium glutamate and nucleotide flavour-enhancing industries. Since most humans appear to enjoy meat-flavoured foods, the meat-like flavours of Chinese soy sauce, Japanese shoyu, and miso produced by fermentation may become of even greater importance than they are today as the world population increases and per capita availability of animal protein decreases.

Good quality Chinese soy sauce is dark in colour, has a high specific gravity and viscosity, and a high nitrogen content. Japanese shoyu is lower in viscosity, lower in nitrogen, has a lighter red colour and, generally, higher glutamic acid and ethanol contents than the Chinese type. Shoyu also has a distinctive flavour because of its higher wheat content and specific lactic acid and yeast fermentations (Yokutsuka, 1972).

The major substrates for Chinese soy sauce and Japanese shoyu are soy beans and wheat. However, various pulses and corn have been used as substrates in earlier times (Yokutsuka, 1972).

2.1.1 The Koji Principle

Koji is a Japanese term for an enzyme concentrate that is produced by growing a microorganism, generally a mould, or a mixture of organisms on a solid substrate consisting of rice, soy beans, wheat or wheat bran, barley or other cereals. The only requirement is that the microorganism(s) selected produces the desired enzyme(s) and is able to grow on the substrate selected. For shoyu, the traditional substrate is soaked (hydrated), cooked soy beans coated with ground roasted wheat. The mixture is overgrown with selected strains of *Aspergillus oryzae* which elaborates amylases, proteases, and other enzymes. The substrate overgrown with the mould is itself a koji containing a concentration of the necessary enzymes. For Japanese miso, the substrate for the koji is rice or barley which is hydrated, steamed, cooled and overgrown with selected strains of *Aspergillus oryzae.* This rice or barley koji becomes a concentrate of enzymes that is ground and thoroughly mixed with the soy bean substrate which is to be hydrolysed.

Kojis for high protein substrates, such as soy bean, must be rich in proteases, while rice kojis used for Japanese saké production must be rich in amylases which hydrolyse the starch in rice to maltose and glucose that can be fermented to ethanol by yeasts. For further information, the reader should consult Church (1923), Hesseltine (1965), Shurtleff and Aoyagi (1976), and Wang and Hesseltine (1979).

2.1.2 The Essential Microorganisms

The essential microorganisms in soy sauce manufacture are moulds belonging to the *Aspergillus oryzae* (=*Apergillus soyae*) group, salt tolerant homo fermentative lactic acid bacteria, principally *Pediococcus cerevisiae* (=*Pediococcus soyae*), or *Lactobacillus delbreuckii* and salt tolerant yeasts, primarily *Saccharomyces rouxii*, *Zygosaccharomyces soyae*, and *Zygosaccharomyces major* (Yokutsuka, 1977).

Initially a koji is prepared in which hydrated, cooked soy beans, coated with ground roasted wheat, are overgrown in a solid-state fermentation by *Aspergillus oryzae*. In this stage amylases, proteases, and lipases are produced which continue to hydrolyse their substrates in a subsequent anaerobic fermentation in a high concentration salt brine (18% w/v).

Bacillus subtilis and other aerobic spore-forming bacteria grow in the koji reaching populations of 1×10^6 to 1×10^8 organisms (g koji)$^{-1}$ and add their enzymes to the koji (Yokutsuka, 1977), and also contribute to the desirable shoyu flavour (Sakaguchi, 1959).

During the submerged brine fermentation (Japanese moromi), *Pediococcus cerevisiae* starting at 1×10^2 to 1×10^3 organisms ml^{-1} multiply and reach populations of 1×10^8 to 1×10^9 organisms ml^{-1} in about 4 months at room temperature. The pH falls to about 4.9 in about 10 d (Sakaguchi, 1959). *P. cerevisiae* produces acids and other compounds important in soy sauce flavour and aroma.

Saccharomyces rouxii is unable to grow in the 18% (w/v) salt brine until *P. cerevisiae* has lowered the pH to between 4.0 and 5.0. *S. rouxii* and other salt tolerant yeasts ferment glucose and maltose derived by amylolytic hydrolysis of the wheat starch to ethanol and other flavour and aroma compounds. *S. rouxii* reaches populations of 1×10^6 to 1×10^7 organisms ml^{-1}. In the presence of the high salt concentration, it converts as much as 50% of the glucose to glycerol, an important flavour compound in soy sauce (Yong and Wood, 1974).

Torulopsis yeasts which are generally present, contribute to the flavour of soy sauce through the production of 4-ethylguaicol, 4-ethyl phenol and 2-phenylethanol (Yokutsuka *et al.*, 1967a,b). Yeasts also produce furfural, another important flavouring compound (Morimoto and Matsutani, 1969).

In laboratory experiments, Yong and Wood (1976) found that *Lactobacillus delbrueckii* inoculated at a level of 6.8×10^6 organisms (g dry weight of mash)$^{-1}$ in the moromi increased to 2.5×10^7 organisms g^{-1} in 2 d, remained constant for 3 d and then decreased to 1.0×10^4 organisms g^{-1} after 31 d. *Saccharomyces rouxii* decreased from an initial population (inoculated) of 4.6×10^7 organisms g^{-1} to 1.96×10^5 organisms g^{-1} in 10 d then increased to 7.6×10^7 organisms g^{-1} in 14 d and remained at that level.

2.1.3 Biochemical Changes

Initial pH of the moromi is 6.5 to 7.0. The lactic acid bacteria lower the pH to 4.8 to 5.0, allowing the yeast fermentation to begin. In traditional soy sauce fermentations, approximately 65% of the protein in the substrate is hydrolysed whereas under controlled, modern processing conditions 80 to 90% of the total protein is hydrolysed. Good quality Japanese shoyu contains 1.5 to 1.8% (w/w) total nitrogen of which 40 to 50% is lower peptides and peptones, and 40 to 50% is amino acids of which approximately 20% is glutamic acid. Shoyu also contains 2 to 5% (w/w) reducing sugar (60% of which is glucose), 1 to 2% (v/v) ethanol, 1 to 2% (w/w) organic acids (60 to 80% of which is lactic acid), 18% (w/v) NaCl and a pH of 4.6 to 4.9 (Yokutsuka, 1960, 1972, 1977).

Total nitrogen in the substrate remains unchanged but the total carbohydrate decreases by 25% as mould growth proceeds. Reducing sugars increase rapidly during koji preparation and the total soluble nitrogen increases steadily up to 70 h during initial mould growth (Yong and Wood, 1977).

2.1.4 Enzymes Produced by the Mould

Alkaline protease is formed in the largest amounts but neutral protease is most important for shoyu production (Yokutsuka, 1977). Yamamoto (1957) and Yamamoto *et al.* (1972) identified a neutral protease, an acid protease (optimum pH 3.0 to 4.0) and an alkaline protease (optimum pH 7.0 to 10.0) in koji cultured with *Aspergillus soyae*. They reported that the lower the incubation temperature in the range 20 to 35°C, the higher the protease activity but at 30 and 35°C, maximum protease production was markedly lower. Maximum protease activity was reached in about 50 h at 25 to 28°C or 5 d at 20°C.

Invertase production precedes alpha-amylase activity by *Aspergillus oryzae* during koji production (Yong and Wood, 1977; Yamamoto *et al.*, 1972; Yokutsuka, 1972).

2.1.5 Nutritional Considerations

Soy sauce is very easily digested and absorbed. It contributes amino acids, peptides and peptones to the diet and stimulates the appetite. Small amounts of amino acids are destroyed by deamination and amino-carbonyl reactions as the colour darkens during processing and storage, representing a loss of nutritive value.

The strains of mould used in commercial production of soy sauce (shoyu) appear to be non-aflatoxin producers (Hesseltine *et al.*, 1966; Yokutsuka, 1967; Yokutsuka *et al.*, 1967c; Manabe *et al.*, 1967, 1968).

2.2 Japanese Miso

Miso is a protein-rich, salty, soy bean or soy bean and rice or soy bean and barley paste with a meaty flavour which has a smooth or chunky texture produced by a fermentation using enzymes of *Aspergillus oryzae* or *Aspergillus soyae* produced on a rice, barley, or soy bean koji. Miso varies in colour from light yellow to beige to tan to amber to chocolate brown to black, and is produced in a wide variety of flavours, textures, and aromas. It is used mainly as a seasoning but it is also used as a major ingredient for miso soup in Japan. Some types of miso can serve as a replacement for bouillion and other meat-flavoured cubes (Shibasaki and Hesseltine, 1961a,b, 1962; Hesseltine, 1965, 1967; Shurtleff and Aoyagi, 1976; Wang and Hesseltine, 1979).

Miso koji is produced using hydrated, steamed rice, barley, or soy beans as a substrate for growth and enzyme production by *Aspergillus oryzae* or *Aspergillus soyae*. After the koji substrate is completely overgrown by the mould and before sporulation occurs, the koji is ground and mixed with hydrated, steamed soy beans along with dry salt and, generally, an inoculum of previously fermented miso. The miso fermentation is entirely solid state. If soy beans are used for the koji, it may be ground and mixed with additional cooked soy beans (Shibaski and Hesseltine, 1962).

Salt, added at the time of grinding and mixing the koji with the soy beans, varies from 5.5 to 13% (w/w) in fresh miso and from 7 to 24% (w/w) on the basis of dry cereal and soy bean solids weight. The protein content varies from 8 to 21% and carbohydrate varies from 12 to 36% in fresh miso (Ebine, 1977; Shurtleff and Aoyagi, 1976; Wang and Hesseltine, 1979).

During incubation the temperature of the koji gradually rises to 36°C as fungal growth and enzyme synthesis proceeds. The temperature of the rice should not rise above 43°C during growth of the mould and every kernel of rice should be covered with the mycelium. There should be no sporulation (Shibasaki and Hesseltine, 1961b).

2.2.1 The Essential Microorganisms

The essential microorganisms for miso are *Aspergillus oryzae* (Ahlburg) Cohn (= *Aspergillus soyae*) and the yeast *Saccharomyces rouxii* Boutroux identified by Wickerham and Burton (1960). Hesseltine and Shibasaki (1961) demonstrated that these were the only microorganisms essential for the manufacture of miso. However, the Northern Regional Research

Laboratories (Peoria, Illinois) recommends that three different strains of *A. oryzae* ought to be combined for enzyme production in the koji. In a traditional miso fermentation other microorganisms, such as *Pediococcus halophilus*, *Torulopsis versatilis* and *Streptococcus faecalis*, may be present but they are not essential even though they may contribute to flavour of the miso (Wang and Hesseltine, 1979).

2.2.2 Control of the Process

Miso is a fermented food with an almost infinite variety of colour and flavour. The variation is the result of differences in the cereal grains used, namely rice or barley, or no cereal grains, the proportion of cereal to soy bean, the salt concentration used, the total time taken to cook the soy bean, and the fermentation time and temperatures. The amount of starch available for the hydrolysis and the resulting sweetness of the miso, depend directly on the proportion of rice or barley used. The amino acid and peptide nitrogen content of the miso is proportional to the soy bean content. Colour intensity is related to the length of time the soy beans are cooked, the amount of reducing sugars present, and the conditions of storage following fermentation. As the salt content decreases, fermentation time and keeping time also decrease. The finer the koji and soy beans are ground at the time of mixing with salt, the shorter the fermentation time (Ebine, 1977; Shurtleff and Aoyagi, 1976; Hesseltine and Shibasaki, 1961).

2.2.3 Biochemical Changes

The moulds growing in the koji produce amylases that hydrolyse starch to dextrins, maltose, and glucose. Proteases also are produced which hydrolyse proteins to peptones, peptides, and amino acids. Lipases hydrolyse the lipids to free fatty acids. Soluble nitrogen increases from 0.67 to 1.49% (w/w); amino nitrogen increases from 0.13 to 0.50% (w/w). Total acid (mainly lactic acid) increases from 0.57 to 1.05% (w/w) and reducing sugars (glucose) decrease from 11.05 to 9.17% (w/w) (Hesseltine and Shibasaki, 1961).

The protein content of various types of miso range from 8 to 21% (w/w). 60% of the total nitrogen in miso is water soluble and easily digested. Riboflavin (vitamin B_2) and cyanocobalamine (vitamin B_{12}) are produced (Shurtleff and Aoyagi, 1976; Ebine, 1977).

3. DEVELOPMENT OF MEAT-LIKE TEXTURES BY FERMENTATION

While meat-like flavours are highly appreciated by human consumers, meat-like textures also are desirable in protein-rich foods that substitute for meat in the diet. Large Western food companies have invested millions of pounds developing processes by which soy bean proteins are extracted and concentrated to purities above 90% and then spun, by extrusion through fine platinum dies and by chemical modification, into fibrous protein strands which can be oriented to provide meat-like textures for formulation of synthetic meats. With added meat flavours and fats, the products are called meat analogues. Several of these products are already on the market (Odell, 1966; Wanderstock, 1968; Horan, 1974). Similarly, large meat packers have developed processes by which soy beans are flaked, tempered, formulated, and processed in extruders, where the products are subjected to high pressure and temperature for a short time, and emerge from the extruder as chewy, protein-rich, meat-like nuggets which add the flavour, texture, and nutritive value of meats to the Western diet (Smith, 1978). In England, a process has been developed in which a mould mycelium is grown on low-cost carbohydrate to yield a meat-like texture which can be formulated, with added meat flavours and fats, to produce another type of meat analogue (Spicer, 1971a,b). It is obvious that the Western food industry expects these products to play an important role in feeding people in the future.

3.1 Indonesian Tempe

Centuries ago the Indonesians, without modern chemistry and microbiology, developed a meat analogue by a fermentation process in which soy beans are soaked, dehulled, partially cooked, and inoculated with moulds belonging to genus *Rhizopus*. Incubated in a warm place (30 to 33°C), the soy bean cotyledons are knitted into a compact cake by the fibrous mould mycelium in 1 to 3 d. The product, called "tempe kedele", can be sliced thinly and deep-fat fried or cut into chunks and used in place of meat in soup and other dishes. Tempe contains over 40% (w/w) protein and a flavour and texture which appeals to the consumers. It is a good substitute for meat and it has begun to serve the same role in diets of American vegetarians (Steinkraus *et al.*, 1960; Hesseltine, 1961; Saotto *et al.*, 1977a; Shurtleff and Aoyagi, 1979, 1980; Wang and Hesseltine, 1979).

It should be noted that the tempe cake is essentially a koji containing pectinases, proteases, and lipases which exert their activities on the soybean substrate during the fermentation.

3.1.1 The Essential Microorganisms

The essential microorganism for the production of the typical tempe bean cake is a mould belonging to genus *Rhizopus*. A number of species can complete the essential step of knitting the soy bean cotyledons into a firm cake. These include *Rhizopus oligosporus*, *Rhizopus stolonifer*, *Rhizopus oryzae*, and *Rhizopus arrhizus* (Hesseltine *et al.*, 1963; Dwidjosepurto and Wolf, 1970). The best of the moulds so far discovered is *Rhizopus oligosporus*, Northern Regional Research Laboratories (NRRL) strain 2710. This strain was originally isolated by Keith H. Steinkraus at Cornell University from dried powdered tempe brought to the United States by Miss Yap Bwee Hwa in 1958 and identified by Dr C. W. Hesseltine in the Northern Regional Research Laboratories, Peoria, Illinois. The unique characteristics of *R. oligosporus* strain NRRL 2710 include the following:

(1) ability to grow very rapidly at temperature from 30 to 42°C;
(2) inability to ferment sucrose;
(3) a high proteolytic activity;
(4) a high lipolytic activity;
(5) an ability to form a typical tempe cake with characteristic aroma and flavour;
(6) an ability to ferment wheat or other starchy cereals without producing noticeable amounts of organic acids that would sour the product.

In the tropics, the fermentation with *Rhizopus* species is preceded by a bacterial acid fermentation during soaking of the soy beans (Stahel, 1946). This increases the acidity of the soy beans lowering the pH to 4.5 to 5.0 which is favourable for subsequent growth of the mould but inhibits most bacteria that would spoil the tempe if they were allowed to develop. This is important, if not essential, to the tempe fermentation. The bacteria involved in acidification of the beans during soaking have not been identified. They are killed during the subsequent cooking of the soy bean cotyledons prior to inoculation with the mould.

Steinkraus *et al.* (1960) reported that the only essential microorganism in the tempe fermentation was the mould. Subsequently, however, it was found that tempe contained vitamin B_{12} (Steinkraus *et al.*, 1961). When a search was made for the source of the vitamin B_{12}, it was found that all commercial tempes studied contained a Gram-negative rod along with the mould. This rod produced the vitamin B_{12} (Liem *et al.*, 1977). Tempe made with the pure mould contains no vitamin B_{12} activity. The Gram-negative rod was subsequently identified as a non-pathogenic strain of *Klebsiella pneumoniae* (Curtis *et al.*, 1977). Since vitamin B_{12} is so important in the diets of vegetarians, from the nutritional view-point, it must be considered that *Klebsiella pneumoniae* is essential in the tempe fermentation. Interestingly enough, the bacterium does not interfere with growth of the mould nor does the mould interfere with growth of the rod. Both grow over a very wide range of pH (3.5 to 7.0) and both have a wide range of growth temperatures (25 to 42°C).

3.1.2 Biochemical Changes

During the tempe fermentation, total soluble solids in soy beans increases from 13 to 28% (w/w); soluble nitrogen increases from 0.5 to 2.5% (w/w) whilst total nitrogen remains relatively constant. The pH gradually rises to a neutral value or higher if the fermentation is prolonged (Steinkraus *et al.*, 1960).

The mould possesses strong lipolytic activity and hydrolyses over a third of the neutral fat liberating the free fatty acids, in spite of which the pH gradually rises as a result of active proteolysis and deamination of amino acids (Wagenkneckt *et al.*, 1961). The principal sugars in soy bean are sucrose, stachyose and raffinose and there is a slow hydrolysis of stachyose (Shallenberger *et al.*, 1967). Lysine, methionine, and phenylalanine decrease during the tempe fermentation (Steinkraus *et al.*, 1960; 1961; Murata *et al.*, 1967).

The concentration of riboflavin doubles, niacin increases seven times and vitamin B_{12} activity increases to nutritionally significant levels in commercial tempe. The concentration of thiamine decreases whilst pantothenate remains approximately constant (Steinkraus *et al.*, 1961; Murata *et al.*, 1967, 1968). Biotin and total folate compounds increase appreciably (Murata *et al.*, 1970).

3.2 Indonesian Oncom (Ontjom) and Tempe Bongkrek

Indonesian oncom (ontjom) is a tempe-like product made by fermentation of peanut press-cake. In fact, it is possible to utilize the tempe mould *Rhizopus oligosporus* in the fermentation. However, much oncom is made using *Neurospora sitophila* or *Neurospora intermedia* as the fermenting mould. Peanut press-cake is hydrated, broken into small particles, often by hand, steamed in flat cakes, cooled, inoculated with the desired mould, incubated at about 30°C until the particles are knitted into a firm cake that can be thinly sliced and deep-fat fried or used as chunks in soups (Steinkraus *et al.*, 1956; van Veen *et al.*, 1968; Saona *et al.*, 1977b).

Tempe bongkrek is a similar product made by the fermentation of coconut press-cake or coconut grits which are the by-products after the oil has been extracted from the coconut. Again the fermenting mould can be either *Rhizopus oligosporus* or *Neurospora intermedia*. Oncom is also made by fermentation of tapioca waste and the residue left from soy bean curd manufacture.

3.2.1 Biochemical Changes

The fermentation and the changes that occur in oncom are similar to those reported for tempe kedele. The initial pH of the substrate is 4.5 which gradually rises as fermentation continues. Concentrations of approximately 5.3% (w/v) sucrose, 0.1% (w/v) raffinose and 0.5% (w/v) stachyose are reduced to trace amounts in peanuts fermented with *Neurospora sitophila*, indicating that the organism has an active α-galactosidase system (Worthington and Beuchat, 1974). Thiamine and riboflavin increase during peanut fermentation with either *N. sitophila* or *R. oligosporus* (van Veen *et al.*, 1968; Quinn *et al.*, 1975). The niacin content increases during fermentation with *N. sitophila* (Quinn *et al.*, 1975) and vitamin B_{12} is also produced (Liem *et al.*, 1977).

4. FOODS AND BEVERAGES INVOLVING AN ALCOHOLIC FERMENTATION

4.1 Japanese Saké

Japanese saké is a clear, pale-yellow rice wine with an alcoholic content of 16% (v/v) or higher, a characteristic aroma, little acid, and slight sweetness (Murakami, 1972). Although saké is a highly sophisticated industry today, its roots are in the indigenous fermentations. In the very early days, the rice was saccharified by chewing it (Kodama and Yoshizawa, 1977).

Japanese saké, rice wine, is the Oriental counterpart to Western beer as both are made from cereal grains. In the West, barley is malted (germinated) under controlled conditions to provide the amylases necessary to hydrolyse starch to maltose and glucose, thus making it fermentable by yeasts which produce the ethanol. Japanese saké, on the other hand, uses a koji in which rice is hydrated, steamed, cooled, and inoculated with strains of *Aspergillus oryzae* that are high in amylolytic activity. The koji consisting of rice grains overgrown with the mould mycelium, is then mixed thoroughly with additional hydrated, steamed, cooled rice and the mixture is inoculated with *Saccaromyces cervisiae* (= *Saccharomyces saké*) (Steinkraus, 1979).

In beer brewing, saccharification of the starch is a separate step utilizing amylases and proteases in malted barley, followed by filtration and fermentation of the wort. In the saké fermentation, saccharification and fermentation proceed simultaneously in the unfiltered, dense mash. It is a fermentation proceeding at low temperatures ranging from 8 to 18°C. The conditions lead to very high populations of yeast cells and the ethanol content can reach 20% (v/v) or higher (Kodama and Yoshizawa, 1977).

4.1.1 The Essential Microorganisms

Saké manufacture is a complex microbial interaction. The mould *Aspergillus oryzae* produces both amylases and proteases that hydrolyse the starch to fermentable sugars and also degrade a portion of the protein, releasing amino acids and peptides for the growth of the other microorganisms.

The first organisms to develop in the koji and rice mash are nitrate-reducing bacteria, such as species of *Pseudomonas*, *Achromobacter*, *Flavobacterium*, or *Micrococcus*, which all produce nitrite from nitrate in the water (Murakami, 1972). These organisms are followed, or possibly accompanied, by *Leuconostoc mesenteroides* var. *saké* and *Lactobacillus saké* (Kodama and Yoshizawa, 1977). The second set of microorganisms reach levels of 1×10^7 to 1×10^8 organisms (g mash)$^{-1}$ acidify the mash and then disappear. Yeasts take over the fermentation although wild yeasts cannot survive in the presence of the nitrite. *Hansenula anomala* type yeasts survive the nitrite but cannot survive the anaerobic conditions of the mash (Kodama and Yoshizawa, 1977). *Saccharomyces cerevisiae* (= *Saccharomyces saké*), inoculated into the mash at levels of 1×10^5 to 1×10^6 organisms (g mash)$^{-1}$ multiply reaching levels of 3 to 4×10^8 organisms (g mash)$^{-1}$, finally constituting 95 to 98% of the microbial flora (Murakami, 1972).

4.1.2 Biochemical Changes

The principal product is ethanol which reaches levels as high as 20% (v/v) in the mash (Kodama and Yoshizawa, 1977). Starches in the rice are hydrolysed to dextrin, maltose and glucose by amylases in the koji. Rice proteins are hydrolysed to peptides and amino acids by koji proteases produced by *Aspergillus oryzae*. Lactobacilli produce lactic, succinic and other organic acids, whilst the yeast produces ethanol and carbon dioxide. An average saké contains total sugars (as glucose) 4.2% (w/v); acidity 1.52 meq (100 ml)$^{-1}$; total organic acids 115.22 mg (100 ml)$^{-1}$; glutamic acid 20.23 mg (100 ml)$^{-1}$; total nitrogen 0.0726% (w/v); formal nitrogen 0.0288% (w/v) and ethanol 15.0% (v/v) (Kodama and Yoshizawa, 1977).

4.2 Indonesian Tapé Ketan

Indonesian tapé ketan is a sweet-and-sour alcoholic paste in which an amylolytic mould of the *Amylomyces rouxii* type and at least one yeast of the *Endomycopsis burtonii* type, hydrolyse steamed rice starch to maltose and glucose and then ferment the sugars to ethanol and organic acids (Ko, 1972; Cronk *et al.*, 1977). If yeasts of the *Hansenula* type are present, acids and ethanol are esterified producing highly aromatic esters and the fermentation time is generally about 3 d at ambient temperature. Peeled, steamed cassava tubers are also used as a substrate and the fermented product is then called tapé ketella. A product closely related to tapé is Chinese lao chao in which the essential microorganisms are *Rhizopus chinensis* and *Endomycopsis* sp. (Wang and Hesseltine, 1970).

If the fermentation is allowed to continue beyond 3 or 4 d at 30°C nearly all the rice starch is converted to fermentable sugars and the rice is largely liquified. In this form, with or without the addition of water during the fermentation, the product is essentially an indigenous primitive rice wine as consumed in the villages of Thailand and other South-East Asian countries (Steinkraus, 1979).

4.2.1 The Essential Microorganisms

The essential microorganisms are a mould of the *Amylomyces rouxii* type and at least one yeast of the *Endomycopsis burtonii* or *Endomycopsis fibuliger* type. Yeasts belonging to genus *Hansenula* are also frequently present and are essential for ester formation (Dwidjoseputro and Wolf, 1970; Ko, 1972, 1977; Ellis *et al.*, 1976; Cronk *et al.*, 1977).

To provide inocula, village markets sell ragi cakes 2.5 to 3.0 cm in diameter which contain the essential moulds and yeasts grown and dried on rice flour (Ko, 1977).

4.2.2 *Biochemical Changes*

Indonesian tapé fermentation depends upon amylolytic moulds and yeasts growing in the rice substrate, essentially a type of koji, to hydrolyse the rice starch and also to ferment the sugars to ethanol.

The changes involved are the hydrolysis of the starch to maltose and glucose to provide fermentable sugars and sweetness and the fermentation of part of the sugars to ethanol and organic acids. These changes require approximately 48 to 96 h at 30°C (Cronk *et al.*, 1977). Malaysian tapé (tapei) contains 23% (w/v) reducing sugar, 27% (w/v) total sugar, 5% (v/v) ethanol and has a pH of 3.9 (Merican and Yeoh, 1977). Ethanol ranges from 3% (v/v) (Tanuwidjaja, 1972) to as high as 8.5% (v/v) (Cronk *et al.*, 1977). The pH falls from approximately 6.0 to as low as 3.5 but generally levels off at pH 4.0 (Ko, 1972).

In laboratory studies *Amylomyces rouxii* in combination with *Endomycopsis burtonii* reduced total solids of rice by 50% in 192 h at 30°C raising the crude protein content of the rice to 16.5% (w/w) (Cronk *et al.*, 1977). *A. rouxii*, by itself, was able to utilize 30% of the total solids raising the protein content to 12% (w/w). Soluble solids rose to 67% in 36 h and fell to 58% in 96 h. The mould and yeast combination reduced the starch content from 78% to 15% in 48 h and to 8% in 192 h at 30°C. *A. rouxii* by itself was able to reduce the total starch content to 2% in 144 h. The higher starch content (8%) remaining after 192 h with the combined microorganisms probably reflects synthesis of glycogen by the yeast which would be measured as starch in the analysis. The mould and yeast combination produced an ethanol concentration of 8% (v/v) at 144 h. The mould by itself, was able to produce an alcohol content of 5.6% at 96 h. With the combination of organisms, the pH fell from 6.3 to 4.1 in 114 h while *A. rouxii* by itself reduced the pH to 4.0 in 48 h.

Lysine, an amino acid present in the feedstock in limiting quantities, is selectively synthesized by the microorganisms and increases by 15%. Thiamine which is low in polished rice increases from 0.04 mg (100 g^{-1} to 0.12 mg (100 g)$^{-1}$, i.e., a 300% increase which has a considerable nutritional importance (Cronk *et al.*, 1977).

Substitution of *Hansenula subpelliculosa* or *Hansenula anomala* for *Endomycopsis burtonii* resulted in the formation of ethyl acetate (859 to 905 mg l^{-1}) at 96 h at 30°C (Cronk *et al.*, 1979).

4.3 Mexican Pulque

Mexican pulque is a white, viscous, acidic, alcoholic beverage made by fermentation of juice of the century plants, the *Agave*, mainly *Agave atrovirens* or *Agave americana*. It has been a national Mexican drink since the time of the Aztecs (Goncalves de Lima, 1975). Pulque plays an important role in the nutrition of the low income adults and children in the semi-arid regions of Mexico.

4.3.1 The Essential Microorganisms

Sanchez-Marroquin (1953, 1957, 1967, 1970, 1977) and Sanchez-Marroquin and Hope (1953) studied the microbial flora of the pulque fermentation which, generally, is natural. The essential microorganisms are *Saccharomyces cerevisiae*, *Lactobacillus plantarum*, *Leuconostoc* sp., and *Zymomonas mobilis*.

The hetero-fermentative lactobacillus is closely related to *Leuconostoc mesenteroides* and *Leuconostoc dextranicum*. These bacteria play an essential role producing bacterial polysaccharides, dextrans, that contribute a characteristic viscosity to pulque (Sanchez-Marroquin and Hope, 1953). They also increase the acidity of the *Agave* juice rapidly, inhibiting growth of other less desirable bacteria. *Lactobacillus plantarum* also produces lactic acid increasing the final acidity of pulque. *Lactobacillus brevis* is also generally present. *Saccharomyces cerevisiae* appears to be the major producer of ethanol but Swings and DeLey (1977) consider *Zymomonas mobilis* to be the most important ethanol producer in pulque. Under anaerobic conditions, *Zymomonas* sp. uses the Entner-Douderoff pathway transforming 45% of the glucose to ethanol and 45% to carbon dioxide. It also produces some acetic acid and acetylmethylcarbinol and slimy gums which may contribute to the viscous nature of traditional pulque.

4.3.2 Biochemical Changes

Soluble solids in the fresh *Agave* juice decrease from 25 to 30% (w/w) to 6.0% (w/w) in pulque. The pH falls from 7.4 to 3.5–4.0. Total acid (mainly lactic acid) increases from 0.029 to 0.4–0.7% (v/v). Sucrose decreases from 18.6% to less than 1% whilst ethanol increases from 0 to 4–6% (v/v) (Sanchez-Marroquin, 1977).

Bacterial dextrans are produced providing a typical viscosity in traditional pulque. The vitamin B's are present in nutritionally important quantities with amounts ranging from (μg (100 ml)$^{-1}$): thiamine 5.2 to 29.0; niacin 54 to 515; riboflavin 18 to 33; pantothenic acid 60 to 355; *p*-aminobenzoic acid 12 to 20; pyridoxine 14 to 23; and biotin 9 to 32 (Sanchez-Marroquin, 1977).

5. FOODS INVOLVING AN ACID FERMENTATION

Acid fermentations have some distinct advantages since foods become resistant to microbial spoilage and the development of toxins. Furthermore, acid foods are less likely to transfer pathogenic microorganisms and result in modifications of the flavour of the original ingredients and often improve nutritive value.

Since canned or frozen foods are unavailable or too expensive for hundreds of millions of the world's economically deprived and hungry, acid fermentation, combined with salting, remains one of the most practical methods of preserving and often enhancing organoleptic and nutritional quality of fresh vegetables, cereal gruels, and milk-cereal mixtures.

5.1 Sauerkraut

Lactic acid fermentation of cabbage and other vegetables is a common way of preserving fresh vegetables in the Western world, in China, and in Korea where kimchi is a staple in the diet. It also is a simple way of preserving food as the raw vegetable needs only to be sliced or shredded and approximately 2.25% (w/w) salt added. The salt extracts liquid from the vegetable which serves as a substrate for growth of the lactic acid bacteria. Conditions should be maintained as anaerobic as possible to prevent growth of microorganisms that might spoil the sauerkraut.

5.1.1 The Essential Microorganisms

Pederson (1930a,b) determined the sequence of microorganisms that develop in a typical sauerkraut fermentation. *Leuconostoc mesenteroides* initiates growth in the shredded cabbage over a wide range of temperatures and salt concentrations. It produces carbon dioxide and lactic and acetic acids which lower the pH thereby inhibiting development of undesirable microorganisms that might destroy the crispness of the cabbage. The carbon dioxide produced replaces the air and generates the anaerobic conditions required for the fermentation. The fermentation is completed in sequence by *Lactobacillus brevis* and *Lactobacillus plantarum*. *L. plantarum* is responsible for the high acidity. If the fermentation temperature or the salt concentration is higher than usual *Pediococcus cerevisiae* develops and contributes to acid production (Pederson and Albury, 1969).

Stamer *et al.* (1971) and Stamer (1975) stressed that sauerkraut is an excellent model for the study of mixed cultures. Using filter-sterilized cabbage juice containing 2.25% (w/w) salt and pH 6.2, *Leuconostoc mesenteroides* displayed the shortest lag and generation time of the lactic acid bacteria involved in sauerkraut fermentation. It also exhibited the most rapid death rate. *L. mesenteroides*, *Lactobacillus plantarum* and *Pediococcus cerevisiae* reached populations of 3.0×10^8 organisms ml^{-1} in 24 h while *Lactobacillus brevis* required 48 h to reach the same population size. Within 4 d, the viable population of *L. mesenteroides* had fallen by 1×10^3 whilst the viable populations of the other organisms remained unchanged.

Following 10 d incubation, *L. plantarum* and *P. cerevisiae*, both homolactic fermenters, produced 1.4 and 0.9% titratable acidity respectively. Each produced less than 0.1% volatile acidity (as acetic acid). The hetero-fermentative species, *L. mesenteroides* and *L. brevis*, yielded ratios of 0.68% (w/v) lactic: 0.25% (w/v) acetic and 0.40% (w/v) lactic: 0.45% (w/v) acetic respectively. The ratio of non-volatile to volatile acid was 2.7 : 1 for *L. mesenteroides* and 0.88 : 1 for *L. brevis* and had a marked effect on the flavour. *L. brevis* produced a harsh vinegar-like flavour while *L. mesenteroides* produced a mild, pleasantly aromatic flavour. The homo-fermenters generated products unacceptable in flavour.

Increasing the salt concentration to 3.5% (w/w) results in a 90% inhibition of the growth and acid production of both *L. mesenteroides* and *L. brevis* (Stamer *et al.*, 1971).

Under ideal conditions using shredded cabbage containing 2.25% (w/w) salt, *L. mesenteroides* reach maximum population, about 1×10^8 organisms ml^{-1}, in 12 to 14 h and *L. plantarum* and *L. brevis* reach their maximum population about 10 to 22 h later (Stamer *et al.*, 1971).

5.1.2 Biochemical Changes

The principal change occurring during the sauerkraut fermentation is the production of lactic and acetic acids and carbon dioxide. As expected, the fermentation is very slow at 7.5°C. *Leuconostoc mesenteroides* grows slowly producing an acidity of 0.4% (w/v) in about 10 d and an acidity of 0.8 to 0.9% (w/v) in a month. Lactobacilli and pediococci cannot grow well at this temperature and the fermentation may not be complete for 6 months. At 18°C, a total acidity (as lactic acid) of 1.7 to 2.3% (w/v) is reached, with an acetic to lactic acid ratio of 1 : 4, in about 20 d. At 32°C, similar acidity is reached in 8 to 10 d with most of the acid being lactic produced by the homo-fermentative bacteria, *L. plantarum* and *P. cerevisiae* (Pederson and Albury, 1969).

5.2 Indian Idli and Dosa

Indian idli is a small, white, acid, leavened, and steam-cooked cake made by bacterial fermentation of a thick batter made from carefully washed, soaked, coarse ground rice and soaked, finely ground dehulled black gram dhal, a pulse. The cakes are soft, moist and spongy and have a desirable sour flavour. A closely related product is dosa made from the same ingredients which are both finely ground. The batter is generally thinner and dosa is fried like a pancake.

The importance of the idli fermentation is that it is a process by which leavened bread-like products can be made from cereals other than wheat or rye and without yeast. The initial step in the fermentation is to wash carefully both the rice and black gram dhal. They are then soaked during the day, generally for 5 to 10 h. The ingredients are drained and ground separately in a mortar or other grinder. The rice and black gram are then combined with water and 1% (w/w) salt to make a thick batter. The batter is fermented in a warm place (30 to 32°C) overnight during which time acidification and leavening occur. The batter is placed in small cups and steamed or fried as a pancake.

The proportions of rice to black gram vary from 4:1 to 1:2 depending upon the relative cost on the market (Steinkraus *et al.*, 1967; Ramakrishnan, 1977; Purushathaman *et al.*, 1977).

5.2.1 The Essential Microorganisms

Idli and dosa production is a natural fermentation. *Leuconostoc mesenteroides* and *Streptococcus faecalis* develop concomitantly during soaking and continue to multiply following grinding. Each eventually reach more than 1×10^9 organisms g^{-1} 11 to 13 h after forming the batter. These two species predominate until 23 h following batter formation by when, normally all batters have been steamed. If the batter is incubated further, the lactobacilli and streptococci decrease in numbers and *Pediococcus cerevisiae* develops (Mukerjee *et al.*, 1965). *Leuconostoc mesenteroides* is the essential microorganism for leavening of the batter and is also responsible, along with *Streptococcus faecalis*, for acid production; both functions are essential for producing a satisfactory idli (Mukerjee *et al.*, 1965).

5.2.2 Biochemical Changes

In idli made with a 1:1 proportion of rice to black gram, the batter volume increases by about 47%, 12 to 15 h after incubation at 30°C. The pH falls to 4.5 and total acidity rises to 2.8% (w/v) (as lactic acid).

Using a 1:2 ratio of black gram to rice, batter volume increases 113% and acidity rises to 2.2% (w/v) in 20 h at 29°C. Reducing sugars (as glucose) show a steady decrease from 3.3 mg (g dry ingredients)$^{-1}$ to 0.8 mg (g dry ingredients)$^{-1}$ in 20 h reflecting their utilization for acid and gas production (Desikachar *et al.*, 1960). Soluble solids increase while soluble nitrogen decreases (Steinkraus *et al.*, 1967). Flatulence-causing oligosaccharides, such as stachyose and raffinose, are completely hydrolysed (Agarwal, 1976).

Rao (1961) reported an 18% (w/v) increase in methionine but this has not been substantiated by other investigators (Khandwala *et al.*, 1962; Ananthachar and Desikachar, 1962; van Veen *et al.*, 1967).

Rajalakshmi and Vanaja 1967) reported that thiamine and riboflavin increase during fermentation and phytate phosphorus decreases. Ramakrishanan and Rao (1977) reported the isolation of a strain of *L. mesenteroides* that hydrolyses haemagglutinin.

5.3 Sour-dough Breads and Related Fermentations

Wood *et al.* (1975) and Wood (1977) emphasized the close relationships between yeasts and lactic acid bacteria in sour-dough-type breads, soy sauce, miso, and kefir. Sour-dough leaven contains both yeasts and lactobacilli. Wheat, rye or other cereal grain flour is mixed with water and incubated for a few days in a warm place. Initially a wide range of microorganisms develop, but eventually the lactic acid bacteria predominate because of their acid production. Yeasts can also survive as they tolerate the acid conditions. Additional flour is added to make a dough. Then the dough is sub-divided and used to make a batch of bread while the rest of the dough is kept for future bread making. Wherever sour-dough leavens have been studied, the organisms found have been similar (Wood *et al.*, 1975; Cardenas, 1972).

5.3.1 The Essential Microorganisms

The essential microorganisms in sour-dough are a yeast, *Torulopsis holmii* (= *Saccharomyces exiguus*), and a *Lactobacillus* sp. *Saccharomyces inusitatus* has also been isolated and identified in sour-dough leaven (Sugihara *et al.*, 1971a,b; Kline and Sugihara, 1971). Kline and Sugihara (1971) suggested the name *Lactobacillus san francisco* for the lactobacillus which is a hetero-fermentative bacterium producing lactic and acetic acids and carbon dioxide. *Torulopsis holmii* grows on glucose but not on maltose. The lactobacillus has a preference for maltose and uses the maltose phosphorylase pathway to metabolize the sugar (Wood and Rainbow, 1961).

Wood (1977) has emphasized the commensal nature of the interaction between *T. holmii* which utilizes only glucose and the *Lactobacillus* sp. that preferentially utilizes maltose releasing glucose so that both develop in a dough where the amylases reduce starch to maltose. Wood (1977) refers to a private communication from Dr Magdalena Wtodarczyk of the Politechnika in Poland who discovered that the yeast also produces a vitamin essential for the growth of the lactobacillus.

5.3.2 Biochemical Changes

The basic biochemical changes that occur in the sour-dough bread fermentation are the acidification of the dough with lactic and acetic acids produced by the lactobacilli, and the leavening of the dough with carbon dioxide produced by the yeast and the lactobacilli. Typical flavour and aroma development can be traced to biochemical activities of both lactobacilli and yeasts and some of the chewiness, characteristic of sour-dough bread, may be due to production of bacterial polysaccharides by the lactobacilli.

5.4 Nigerian Ogi

Nigerian ogi is a smooth-textured, sour porridge with a flavour resembling yogurt, made by fermentation of corn, sorghum, or millet. It has a solids content of about 8% (w/w). The cooked porridge known as "pap" is gel-like. The first step in the fermentation is steeping of the cleaned grain for 1 to 3 d. During this time, the desirable microorganisms develop and are selected (Akinrele, 1966). It would be expected that some pre-germination reactions also occur in the soaking grains which are then ground with water and filtered to remove coarse particles. The pH following steeping should be 4.3 with a final optimum pH for oji of 3.6 to 3.7. If the pH falls below 3.5, the product is less acceptable (Akinrele, 1970; Akinrele *et al.*, 1970; Onyekwere and Akinrele, 1977a).

5.4.1 The Essential Microorganisms

Oji is a natural fermentation and a wide variety of moulds, yeasts, and bacteria are initially present. *Lactobacillus plantarum* appears to be the essential microorganism in the fermentation (Banigo and Muller, 1972a). It is able to utilize dextrins from the corn following depletion of the fermentable sugars. *Aerobacter cloacae* was shown to be capable of increasing the riboflavin and niacin content of the mash. *Corynebacterium* sp. is reported to hydrolyse the starch and produce organic acids while *Saccharomyces cerevisiae* and *Candida mycoderma* contribute to a desirable flavour (Akinrele, 1970). Banigo *et al.* (1974) developed a mixed culture for ogi using *Lactobacillus plantarum*, *Streptococcus lactis* and *Saccharomyces rouxii*.

5.4.2 Biochemical Changes

The basic biochemical change is the production of acid lowering the pH of the substrate to about pH 3.6 to 3.7. The concentration of lactic acid may reach 0.65% (w/v) and acetic acid 0.11% (w/v) during the fermentation. The ratio of volatile to non-volatile acid is generally about 1:0.13. Lactic and acetic acids are the predominate acids produced but traces of formic and butyric acids are present and contribute to the flavour and aroma (Banigo and Muller, 1972a,b). It is likely that certain pre-germination changes occur in the substrate during steeping. Fermentation markedly increases the swelling and thickening characteristics of maize starch (Banigo *et al.*, 1974).

5.5 Nigerian Gari

Cassava is a major source of food for millions of the world's poor, but it contains low levels of protein which are insufficient to provide the protein needs of the consumer.

Nigerian gari is a granular starchy food made from cassava (*Manihot utilissima* or *Manihot esculenta*) by acid fermentation of the grated pulp, followed by a dry-heat treatment to gelatinize and semi-dextrinize the starch and drying. Cassava tubers are washed, peeled and grated. An inoculum of three-day-old cassava juice or fermented mash liquor is added. The pulp is placed in a cloth bag, excess water is squeezed out and the pulp then undergoes an anaerobic, acid fermentation for 12 to 96 h. When the pH of the mash reaches pH 4.0 with about 0.85% (w/v) total acid (as lactic acid), the gari has the desired sour flavour and a characteristic aroma. Further moisture may be removed and the pulp is toasted (semi-dextrinized) in shallow iron pots and dried to less than 20% (w/w) moisture in village processes.

The process proceeds best at 35°C. Sunlight and frequent mixing of the pulp accelerates the fermentation when *Geotrichum candidum* is the functional organism (Akinrele, 1963). Water is released during the fermentation, so no water is added to the original pulp. For consumption the gari is added to boiling water in which it increases in volume by

300% to yield a semi-solid plastic dough. The stiff porridge is rolled into a ball (10 to 30 g wet weight) and dipped into stew for consumption during the meal (Akinrele *et al.*, 1975; Onyekwere and Akinrele, 1977b; Okafor, 1977a,b; Ogunsua, 1977).

5.5.1 The Essential Microorganisms

Collard and Levi (1959) reported that *Corynebacterium* sp. and *Geotrichum candidum* were the important microorganisms in the gari fermentation. However, they presented no data on changes in numbers during the fermentation. Okafor (1977b) isolated and enumerated the numbers of five different genera present in the gari fermentation. These included species *Leuconostoc*, *Alcaligenes*, *Corynebacterium*, *Lactobacillus*, and *Candida*. Only species of *Leuconostoc*, *Alcaligenes*, and *Candida* were present in significant numbers and *Alcaligenes* species die out after 2 d. *Leuconostoc* species reaches populations of 1×10^8 organisms g^{-1} in 24 h. Yeasts reach 1×10^5 to 1×10^6 organisms g^{-1} but *Corynebacterium* species was isolated only once. Species of *Leuconostoc* and *Candida* appear to be the essential microorganisms in the gari fermentation.

5.5.2 Biochemical Changes

Total acidity of gari should be between 0.58 and 1.2% (w/v) (as lactic acid). Organic acids produced during the fermentation are responsible for the sour flavour and the aldehydes and esters provide the characteristic aroma.

The water content is about 71.50% (w/w) in the freshly peeled cassava tuber with a carbohydrate content of 26.82% (w/w) (wet weight). The protein content of the freshly peeled tuber is 0.74% (w/w) (2.6% (w/w) dry basis) (Akinrele *et al.*, 1965).

The moisture content in village processed gari is about 16.7% (w/w) with a carbohydrate content of 81.8% (w/w) and a crude protein content of 0.9% (w/w). Gari processed in a pilot plant by modern techniques contains less moisture (9.1% w/w) and has therefore better keeping quality than the village processed gari. The starches have been modified by fermentation so that when moistened, the volume increases more than three times.

Akinrele *et al.* (1975) have identified principally acetic, lactic, propionic, succinic and pyruvic acids in gari. Other changes that occur include gelatinization and partial dextrinization of the starch. Cassava contains a considerable quantity of the cyanogenic glucoside linamarin (manihotoxine) (Jansz *et al.*, 1974) which, on crushing or grinding the roots, is broken down by the enzyme linamarese (linase) to glucose, acetone, and hydrogen cyanide. Cassava has a rather high level of ascorbic acid ranging from 122 to 165 mg $(100\ g)^{-1}$ in the fresh tubers. Approximately 4 to 6% of the initial ascorbic acid is retained in the gari (Ogunsua, 1977).

The pH of fresh cassava tuber is about 6.2. The pH falls to 4.0 in 3 d fermentation (Ogunsua, 1977). Titratable acid is initially at an equivalent of 1.2 mg NaOH g^{-1} in the fresh tuber which increases to 3.4 mg NaOH g^{-1} in the gari after 3 d fermentation. The composition of cassava and, therefore, the resulting gari, is subject to variation in climate, soil, time of harvest and cassava variety (Coursey, 1973).

5.6 Kenkey-fermented Maize Dough Balls of Ghana

Kenkey-fermented maize dough balls are an important food in Ghana. Maize kernels are washed, soaked in water 12 to 48 h, ground finely, moistened, packed tightly, and covered in wooden tubs or vats to maintain anaerobiosis and fermented for 1 to 3 d. Then a portion of the fermenting dough is precooked and mixed with the remaining dough with or without added salt, shaped into balls (about 9 cm in diameter) or cylinders and wrapped in dry plantain leaves or maize husks and then cooked by boiling (Christian, 1966; Nyako, 1977).

5.6.1 The Essential Microorganisms

The kenkey fermentation is completely uncontrolled. The stored maize generally contains a variety of moulds including species *Aspergillus*, *Rhizopus*, and *Penicillium* and these, along with a Gram-negative, catalase-positive coccus, dominate the initial fermentation. The fungi die out but may reappear later. The Gram-negative coccus also decreases in numbers as soon as the fermentation starts and disappears after 2 days. Gram-positive, catalase-negative, acid producing cocci, grouped in pairs (probably *Leuconostoc* sp.) appear after about 9 h and reach peak population in 24 to 36 h. Then they decrease steadily if the fermentation is continued. *Lactobacillus brevis* and other *Lactobacillus* sp., *Clostridium* sp. and *Acetobacter* sp. are present in the fermenting dough.

Yeasts, including *Saccharomyces* sp., can be detected by 9 h and appear on the dough surface after the first day of fermentation. The yeasts are wild types ranging from ovoid to mycelial forms. By the fourth days there is a thick, slimy layer of yeasts on the surface of the dough.

It is in this surface layer that the characteristic flavour and aroma resembling a mixture of diacetyl and acetic acid is concentrated. A butyric acid aroma also may be present and is considered a sign of good kenkey. Yeasts also produce some ethanol and esters from the organic acids and alcohol (Christian, 1966; Nyako, 1977).

5.7 Mexican Pozol

Mexican pozol is a fermented maize dough ball consumed by the Indian and mestizo populations of South-eastern Mexico. Balls of pozol, recently made or fermented to varying degrees, are diluted with water to obtain a whitish porridge which is drunk raw as a basic food in the daily diet (Ulloa *et al.*, 1977). White maize kernels are boiled in water containing lime powder (10% (w/v) calcium hydroxide). The kernels swell, their pericarps peel off, the kernels are cooled, rinsed, drained, and ground to obtain a coarse dough which is shaped into balls of various sizes. The balls are wrapped in banana leaves and fermented 1 to 14 d.

5.7.1 The Essential Microorganisms

Most of the microorganisms are killed in the initial boiling (Ulloa, 1974). A variety of organisms are introduced by handling at the dough ball stage including *Geotrichum candidum, Trichosporon cutaneum*, and other *Candida* species. Other moulds, such as *Cladosporium cladosporioides, Cladosporium herbarum, Monilia sitophila, Mucor rouxianus*, and *Mucor racemosus*, are commonly found.

During the first stages of fermentation, the bacteria outnumber the yeasts and moulds and are probably responsible for most of the acid production for the first 24 h. During this time, the pH drops from 7.5 to 5.0. A very wide variety of microorganisms have been found in pozol (Ulloa, 1974; Herrera and Ulloa, 1970, 1971, 1972, 1975; Ulloa and Herrera, 1970, 1971, 1972, 1973; Fuentes *et al.*, 1974; Salinas and Herrera, 1974; Taboada *et al.*, 1971, 1973; Ulloa and Kurtzman, 1975). In most cases, the various species have not been enumerated and their importance in the fermentation is unknown.

The nitrogen-fixing types are important as they increase the total nitrogen and protein content (Taboada *et al.*, 1973). *Agrobacterium azotophilum* was isolated and described by Ulloa and Herrera (1972). It fixes atmospheric nitrogen both aerobically and anaerobically in maize dough (Taboada *et al.*, 1971, 1973). *Klebsiella pneumoniae* (*Aerobacter aerogenes*) is also present in pozol and fixes nitrogen (Salinas and Herrera, 1974). *Agrobacterium azotophilum* also produces a substance antagonistic toward a wide variety of other microorganisms that might grow in pozol. Thus, it likely exerts a selective effect on the microbial flora that develops (Herrera and Ulloa, 1975). The substance appears to be anti-fungal as well as anti-bacterial.

5.7.2 Biochemical Changes

Two essential changes during the pozol fermentation are the development of an acid flavour and a characteristic aroma. The pH of corn kernels (pH 5.6) is raised to pH 7.5 by boiling in lime water. Maize dough with an initial pH of 6.8 decreases to pH 3.9 after 8 d fermentation. Moisture content remains constant at about 30% (Ulloa, 1974). Pozol has a higher content of protein, niacin, riboflavin, lysine, and tryptophan than the starting maize. The protein following fermentation also has a higher nutritional value measured both by amino acid pattern and actual feeding tests with weanling rats.

5.8 Russian Kefir

Russian kefir is an acidic, mildly alcoholic, distinctly effervescent fermented cow, sheep, or goat milk.

5.8.1 Essential Microorganisms

Kefir grains, moist gelatinous, white to yellowish, irregular, cauliflower-shaped granules, ranging in size from wheat grains to walnuts (0.5 to 20 ml), are the essential fermenting agents in kefir (la Riviere, 1963; la Riviere *et al.*, 1967; Hartles *et al.*, 1977). The grains start out small and gradually grow in size doubling their weight in 7 to 10 d at ambient temperatures (17 to 23°C optimum). The grains are slimy but extremely resilient with the microorganisms firmly imbedded in a gelatinous matrix. The kefir grains are stable conglomerates of lactic acid bacteria and yeasts held together by a polysaccharide gum called kefiran produced by the bacterial species and thereby creating a natural immobilized cell system. Disintegration of the kefir grains in a homogenizer or blender produces a viscous suspension of microorganisms.

The kefir grains cannot be dehydrated and survive but they do survive for considerable periods when stored in milk at about 4°C.

The predominate yeasts include the lactose negative *Torulopsis holmii* and *Saccharomyces delbrueckii* in a ratio of 10:1. Total number of lactose-negative yeasts is 1.4 to 3.3×10^8 (g wet weight kefir grain)$^{-1}$ (la Riviere, 1969; la Riviere *et al.*, 1967; Hartles *et al.*, 1977).

The predominate bacterium is a Gram-positive, catalase negative, non-motile, non-sporeforming, microaerophilic to anaerobic, lactose-fermenting rod, that produces the gum kefiran which embeds the microorganisms in the grain matrix. It is unable to grow in milk unless it is supplemented with yeast extract factors likely to be supplied by the yeasts within the kefir grains. The organism closely resembles *Lactobacillus brevis*. The bacterium rapidly loses its ability to produce kefiran when removed from the kefir grain and grown on a typical lactobacillus medium such as MRS (Difco). The total number of rods by direct microscopic count is 8.0×10^9 rods (g wet weight kefir grains)$^{-1}$ (la Riviere, 1963, 1969; la Riviere *et al.*, 1967).

The kefir granules behave as a single organism and efforts to produce new granules by growing the two component organisms separately and then combining them have not been successful (la Riviere, 1963; la Riviere *et al.*, 1967; la Riviere, 1969).

5.8.2 Biochemical Changes

The kefir fermentation is a hetero-fermentative lactic acid fermentation combined with an alcoholic yeast fermentation. The pH may drop to 3.0 or below with total acid (as lactic acid) reaching 0.85 to 1.0% (w/v). Kefir is effervescent due to the carbon dioxide produced by both the lactic acid bacterium and the yeast. Generally ethanol is less than 1.0% (v/v) and acetoin and diacetyl are produced.

The gum, kefiran amounts to about 25% (w/w) of the dry weight of the grains and is synthesized along with new cells. It is a polymer consisting of galactose and glucose in the ratio 1:1 and has been characterized chemically (la Riviere *et al.*, 1967).

5.9 Vinegar (Acetic Acid) Fermentation

The vinegar (acetic acid) fermentation naturally follows the alcoholic fermentation and both are very ancient. The vinegar fermentation begins as ethanol and carbon dioxide production subsides and the fermentation surface becomes aerobic. The vinegar fermentation has been extensively reviewed by Conner & Allgeier (1976) and Nickol (1979). The vinegar fermentation like the ethanol (wine) fermentation occurs naturally. Vinegar can be made from any substrate which can first undergo an alcholic fermentation which includes most fruit juices and also starchy substrates hydrolysed to maltose and glucose. Wines and fermented apple cider are common substrates in the Western World. Optimum temperature is 24 to 27°C. Ethanol content in the substrate should be 10 to 13% (v/v).

5.9.1 The Essential Microorganisms

The essential microorganisms for vinegar production are a yeast *Saccharomyces cerevisiae* producing ethanol from sugar substrates and *Acetobacter aceti* producing acetic acid from ethanol. A bacterium, such as *Zymomonas mobilis*, which produces ethanol from glucose could substitute wholly or in part for *S. cerevisiae* and it is probably involved in some vinegar fermentations.

5.9.2 Biochemical Changes

The initial ethanol fermentation is catalyzed by *Saccharomyces cerevisiae* anaerobically. The basic overall reaction is very simple:

$$C_6H_{12}O_6 \xrightarrow[\textit{S. cerevisiae}]{\text{anaerobic}} 2CH_3CH_2OH + 2CO_2$$

The intermediates in the glucose to ethanol fermentation have been completely elucidated and are reviewed by Prescott and Dunn (1959). The second overall reaction is transformation of ethanol to acetic acid aerobically by *Acetobacter aceti* via acetaldehyde as an intermediate:

$$CH_3CH_2OH \xrightarrow[\textit{Acetobacter}\text{ sp.}]{O_2} CH_3CHO + H_2O \xrightarrow[\textit{Acetobacter}\text{ sp.}]{O_2} CH_3COOH + H_2O$$

The ethanol and the acetic acid fermentations are fairly self-controlled as the substrates for ethanol production are generally acid and the production of ethanol by *Saccharomyces cerevisiae* under anaerobic conditions is accompanied by production of substantial quantities of carbon dioxide which further ensure anaerobiosis and make the growth of either molds or aerobic bacteria unlikely. The ethanol produced also inhibits a wide range of microorganisms.

The bacterium best adapted to development on an acid and alcoholic substrate is *Acetobacter aceti* if conditions are aerobic; and, as carbon dioxide production subsides and the surface of the fermenting substrate becomes aerobic, *Acetobacter aceti* is able to dominate the fermentation because of the acid/alcoholic substrate which becomes even more inhibitory toward other organisms as acetic acid is produced. The ethanol/acetic acid fermentation is a beautiful sequence of microorganisms leading to a food/condiment/preservative for man.

6. SUMMARY

Microbial communities in fermented foods are among the most complex in the microbial world. They are also among the most important to man since they influence the variety of foods, flavours, textures and aromas available to the consumer. The microbial communities in fermented foods are responsible for the preservation of vast quantities of food for man. Unfortunately, our knowledge of the interrelationships of the microbes involved in specific food fermentations in many cases is nil or extremely limited. It is hoped that this article will stimulate further research at the molecular level on the interrelationships of the microbial communities in various fermented foods.

REFERENCES

Agarwal, A. K. (1976). Pattern of various carbohydrates in soybean in various processings. MSc Thesis, University of Baroda, Baroda, India.

Akinrele, I. A. (1963). Further studies on the fermentation of cassava. Research Report No. 20 of the Federal Institute of Industrial Research. Oshodi, Nigeria.

Akinrele, I. A. (1966). A biochemical study of the traditional method of preparation of ogi and its effects on the nutritive value of corn. PhD Thesis. University of Ibadan, Nigeria.

Akinrele, I. A. (1970). Fermentation studies on maize during the preparation of a traditional African starch-cake food. *Journal of Science Food Agriculture* **21**, 619–625.

Akinrele, I. A., Cook, A. S., and Holgate, R. A. (1965). The manufacture of gari from cassava in Nigeria. *In* "First International Congress of Food Science and Technology", vol. 4, pp. 633–644. (Leitch, J. M., ed.) Gordon & Breach, New York.

Akinrele, I. A., Adeyinka, O., Edwards, C. C. A., Olatunji, F. O., Dina, J. A., and Ioleoso, O. A. (1970). The development and production of Soy-Ogi. FIIR Research Report No. 42. Federal Ministry of Industries, Lagos, Nigeria.

Akinrele, I. A., Ohankhai, S. I., and Koleoso, O. A. (1975). Chemical constituents of gari flavor (Conference Proceedings) IV Int'l. *Congress of Food Science Technology* **19**, 64–66.

Ananthachar, T. K. and Desikachar, H. S. R. (1962). Effect of fermentation on the nutritive value of *Idli*. *Journal of Science Industrial Research* **21**, 191–192.

Banigo, E. O. I. and Muller, H. G. (1972a). Manufacture of Ogi (a Nigerian fermented cereal porridge): comparative evaluation of corn, sorghum and millet. *Journal of the Canadian Institute of Food Science and Technology* **5**, 217–221.

Banigo, E. O. I. and Muller, H. G. (1972b). Carboxylic acid patterns in ogi fermentation. *Journal of Science Food Agriculture* **23**, 101–111.

Banigo, E. O. I., deMan, J. M., and Duitschaever, C. L. (1974). Utilization of high-lysine corn for the manufacture of ogi using a new, improved processing system. *Cereal Chemistry* **51**, 559–572.

Cardenas, O. A. (1972). Sour dough microorganisms; behaviour and effect on the sour dough process. MSc Thesis, University of Strathclyde, Glasgow, Scotland.

Christian, W. F. M. (1966). Fermented Foods of Ghana. Presented at the Symposium on Food Science & Technol. Accra, Ghana. (May 1966).

Church, M. B. (1923). Soy and related fermentations. U.S. Dept. Agric. Bull. No. 1152.

Collard, P. and Levi, S. (1959). A two stage fermentation of cassava. *Nature* **183**, 620–621.

Conner, H. A. and Allgeier, R. J. (1976). Vinegar: its history and development. *Advances in Applied Microbiology* **20**, 81–133.

Coursey, D. G. (1973). Cassava as food; toxicity and technology. *In* "Chronic Cassava Toxicity", pp. 27–36. (Nastel, B. and MacIntyre, R., eds.) International Development Center, Ottawa, Canada.

Cronk, T. C., Steinkraus, K. H., Hackler, L. R., and Mattick, L. R. (1977). Indonesian tape ketan fermentation. *Applied and Environmental Microbiology* **33**, 1067–1073.

Cronk, T. C., Mattick, L. R., Steinkraus, K. H., and Hackler, L. R. (1979). Production of higher alcohols during the Indonesian tapé ketan fermentation. *Applied and Environmental Microbiology* **37**, 892–896.

Curtis, P. R., Cullen, R. E., and Steinkraus, K. H. (1977). Identity of a bacterium producing vitamin B-12 activity in tempe. Symposium on Indigenous Fermented Foods (SIFF). Bangkok, Thailand, Nov. 21–27.

Desikachar, H. S. R., Radhadrishnamurty, R., Rama Rao, G., Kadkol, S. B., Srinivasan, N., and Subrahmanyan, V. (1960). Studies on *Idli* fermentation I. Some accompanying changes in the batter. *Journal of Science Industrial Research* **19**, 168–172.

Dwidjoseputro, D. and Wolf, F. T. (1970). Microbiological studies of Indonesian fermented foodstuffs. *Mycopathologia et Mycologia Applicata* **41**, 211–222.

Ebine, H. (1977). Japanese Miso. Symposium on Indigenous Fermented Foods. Bangkok, Thailand, November 21–27.

Ellis, J. J., Rhodes, L. J., and Hesseltine, C. W. (1976). The genus *Amylomyces*. *Mycologia* **68**, 131–142.

Fuentes, I., Herrera, T. and Ulloa, M. (1974). Descripcion de una especie nueva de *Pseudomonas, P. mexicana*, y determinacion de *Escherichia coli* var. *neapolitana* aislados del pozol. *Rev. lat-amer. Microbiol.* **16**, 99–103.

Goncalves de Lima, O. (1975). Pulque, Balche E. Pajuaru. Universidade Federal de Pernambuco. Recife, Brasil.

Hartles, P., van Hooidonk, J., and la Riviere, J. W. M. (1977). Kefir. Symposium on Indigenous Fermented Foods (SIFF). Bangkok, Thailand, November 21–27.

Herrera, T. and Ulloa, M. (1970). Aspectos generales sobre la microbiologia del pozol. *Rev. lat-amer. Microbiol.* **12**, 103–108.

Herrera, T. and Ulloa, M. (1971). Estudia de *Candida krusei* y *Trichosporum cutaneum* aislados del pozol. *Rev. lat-amer. Microbiol.* **13**, 255–261.

Herrera, T. and Ulloa, M. (1972). Estudio de *Hansenula fabianii* aislado del pozol. An Inst. Biol. Univ. Nal. Auton. Mexico, 43, *Ser. Bot.* **1**, 1–8.

Herrera, T. and Ulloa, M. (1975). Antagonismo del pezol y de *Agrobacterium azotophilum* sobre diversas especies de bacterias y hongos, algunas patogenas del hombre. *Rev. lat-amer. Microbiol.* **17**, 143–147.

Hesseltine, C. W. (1961). Research at Northern Regional Research Laboratory on fermented foods. *In* Proceedings of Conference on Soybean Products for Protein in Human Foods, pp. 67–74. U.S. Department of Agriculture, Peoria, Illinois.

Hesseltine, C. W. (1965). A millenium of fungi, food, and fermentation. *Mycologia* **57**, 149–197.

Hesseltine, C. W. (1967). Fermented Products, Miso, sufu, and tempeh. *In* Proceedings of the International Conference on Soybean Protein Foods, pp. 170–179. U.S. Department of Agriculture, Peoria, Illinois.

Hesseltine, C. W. and Shibasaki, K. (1961). Pure culture fermentation with *Saccharomyces rouxii*. *Applied Microbiology* **9**, 515–518.

Hesseltine, C. W., Smith, M., Bradle, B., and Djien, K. S. (1963). Investigation of tempeh, an Indonesian food. *Developments in Industrial Microbiology* **4**, 275–287.

Hesseltine, C. W., Shotwell, O. L., Ellis, J. J., and Stubblefield, R. D. (1966). Aflatoxin formation by *Aspergillus flavus*. *Bacteriological Reviews* **30**, 795–805.

Horan, F. E. (1974). Meat analogues. *In* "New Protein Foods", vol. 1a, pp. 366–413. (Altschul, A. M., ed.) Academic Press, New York and London.

Jansz, E. R., Pieris, N., Jeya Raj, E. E., and Abeyratne, D. J. (1974). Cyanogenic glucoside content of Manioc. II. Detoxification of manioc chips and flour. *J. Nat. Sci. Coun. Sri Lanka* **2**, 129–134.

Khandwala, P. K., Ambegaokar, S. D., Patel, S. M. and Radhakrishna Rao, M. V. (1962). Studies on fermented foods. Part I—Nutritive value of idli. *J. Sci. Ind. Res.* **21**, 275–278.

Kline, L. and Sugihara, T. F. (1971). Microorganisms of the San Francisco sour dough bread process. II. Isolation and characterization of undescribed species responsible for the souring activity. *Applied Microbiology* **21**, 459–465.

Ko, S. D. (1972). Tape fermentation. *Applied Microbiology* **23**, 976–978.

Ko, S. D. (1977). Indonesian ragi. Symposium on Indigenous Fermented Foods (SIFF). Bangkok, Thailand, November 21–27.

Kodama, K. and Yoshizawa, K. (1977). Sake. *In* "Economic Microbiology", vol. 1, pp. 432–475. (Rose, A. H., ed.) Academic Press, New York and London.

la Riviere, J. W. M. (1963). Studies on the kefir grain. *Journal of General Microbiology* **31**, v.

la Riviere, J. W. M. (1969). Ecology of yeasts in the kefir grain. *Antonie van Leeuwenhoek* **35**; (Supplement: Yeast Symposium) pp. 15–16.

la Riviere, J. W. M., Kooiman, P., and Schmidt, K. (1967). Kefiran, a novel polysaccharide produced in the kefir grain by *Lactobacillus brevis*. *Archive für Mikrobiology* **59**, 269–278.

Liem, I. T. H., Steinkraus, K. H., and Cronk, T. C. (1977). Production of vitamin B_{12} in tempeh, a fermented soybean food. *Applied and Environmental Microbiology* **34**, 773–776.

Manabe, M., Matsura, S., and Nakano, M. (1967). Isolation and quantitative analysis of four aflatoxins (B_1, B_2, G_1, G_2) by thin layer chromatography. *Nippon Nogei Kagaku Kaishi* **45**, 592–598.

Manabe, M., Matsura, S., and Nakano, M. (1968). Studies on the fluorescent compounds in fermented foods. I. Chloroform-soluble fluorescent compounds produced by Koji-molds. *Journal of Food Science and Technology of Japan* **15**, 341–346.

Merican, Z. and Yeoh, Q. L. (1977). Malaysian Fermented Food. Symposium on Indigenous Fermented Foods (SIFF). Bangkok, Thailand, Nov. 21–27.

Morimoto, S. and Matsutani, N. (1969). Flavor components of soy sauce. Isolation of furfuryl alcohol and the formation of furfuryl alcohol from furfural by yeasts and molds. *Journal of Fermentation Technology* **47**, 518–525.

Mukherjee, S. K., Albury, M. N., Pederson, C. S., van Veen, A. G., and Steinkraus, K. H. (1965). Role of *Leuconostoc mesenteroides* in leavening the batter of idli, a fermented food of India. *Applied Microbiology* **13**, 227–231.

Murakami, H. (1972). Some problems in sake brewing. *In* "Fermentation Technology", pp. 639–643. (Terui, G., ed.) Society of Fermentation Technology (Japan), Kyoto, Japan.

Murata, K., Ikehata, H., and Miyamoto, T. (1967). Studies on the nutritional value of tempeh. *Journal of Food Science* **32**, 580–586.

Murata, K., Miyamoto, T., and Taguchi, F. (1968). Biosynthesis of B vitamins with *Rhizopus oligosporus*. *Journal of Vitaminology* **14**, 191–197.

Murata, K., Miyamoto, T., Kokufu, E., and Sanke, Y. (1970). Studies on the nutritional value of tempeh. III. Changes in biotin and folic acid contents during tempeh fermentation. *Journal of Vitaminology* **16**, 281–284.

Nickol, G. B. (1979). Vinegar. *In* "Microbial Technology", pp. 155–172. (Peppler, H. J. and Perlman, D., eds.) Academic Press, New York and London.

Nyako, K. O. (1977). Kenkey—a fermented staple in Ghana. Symposium on Indigenous Fermented Foods (SIFF). Bangkok, Thailand. Nov. 21–27.

Odell, A. D. (1966). Meat analogues—a new food concept. *Cornell Hotel and Restaurant Administration Quarterly* **7**, 20–24.

Ogunsua, A. O. (1977). Fermentation of cassava tubers. Symposium on Indigenous Fermented Foods (SIFF). Bangkok, Thailand. Nov. 21–27.

Okafor, N. (1977a). Nigerian gari. Symposium on Indigenous Fermented Foods (SIFF). Bangkok, Thailand. Nov. 21–27.

Okafor, N. (1977b). Microorganisms associated with cassava fermentation for gari production. *Journal of Applied Bacteriology* **42**, 279–284.

Onyekwere, O. O. and Akinrele, I. A. (1977a). Ogi, a Nigerian fermented maize beverage. Symposium on Indigenous Fermented Foods (SIFF). Bangkok, Thailand. Nov. 21–27.

Onyekwere, O. O. and Akinrele, I. A. (1977b). Nigerian gari. Symposium on

Indigenous Fermented Foods (SIFF). Bangkok, Thailand. Nov. 21–27.

Pederson, C. S. (1930a). Floral changes in the fermentation of sauerkraut. N.Y.S. Agr. Exp. Sta. Tech. Bull. 168. Geneva, NY.

Pederson, C. S. (1930b). The effect of pure culture inoculation on the quality and chemical composition of sauerkraut. N.Y.S. Agr. Exp. Sta. Tech. Bull. 169. Geneva, NY.

Pederson, C. S. and Albury, M. N. (1969). The sauerkraut fermentation. N.Y.S. Agr. Exp. Sta. Bull. 824. 84 pages. Geneva, NY.

Prescott, S. C. and Dunn, C. G. (1959). "Industrial Microbiology". McGraw-Hill Publishers, New York.

Purushothaman, D., Dhanapal, N., and Rangaswami, G. (1977). Microbiology of idli fermentation. Symposium on Indigenous Fermented Foods (SIFF). Bangkok, Thailand. Nov. 21–27.

Quinn, M. R., Beuchat, L. R., Miller, J., Young, C. T., and Worthington, R. E. (1975). Fungal fermentation of peanut flour: Effects on chemical composition and nutritive value. *Journal of Food Science* **40**, 470–474.

Rajalakshmi, R. and Vanaja, K. (1967). Chemical and biological evaluation of the effects of fermentation on the nutritive value of foods prepared from rice and gram. *British Journal of Nutrition* **21**, 467–473.

Ramakrishnan, C. V. (1977). The use of fermented foods in India. Symposium on Indigenous Fermented Foods (SIFF). Bangkok, Thailand. Nov. 21–27.

Ramakrishnan, C. V. and Rao, G. S. (1977). Isolation, identification and characterization of a microorganism (*Leuconostoc mesenteroides*) HA in fermented soy-idli batter capable of hydrolyzing soybean hemagglutinins. *Baroda Journal of Nutrition* **4**, 9–21.

Rao, N. V. R. (1961). Some observations on fermented foods. *In* "Meeting Protein Needs of Infants and Children", pp. 291–293. National Academy of Sciences. National Research Council Publication 843.

Sakaguchi, K. (1959). Studies on the activities of bacteria in soy sauce brewing. V. The effects of *Aspergillus sojae*, *Pediococcus soyae*, *Bacillus subtilis* and *Saccharomyces rouxii* in pure cultured soy sauce brewing. *Bulletin Agricultural Chemistry Society of Japan* **23**, 100–106.

Salinas, C. and Herrera, T. (1974). Aislamiento de *Aerobacter aerogenes* del pozol del estado de Campeche. *Rev. lat-amer. Microbiol.* **16**, 95–98.

Sanchez-Marroquin, A. (1953). The biochemical activity of some microorganisms of pulque. Mem. Congr. Cient. Mex. IV Centenario Univ. Mex. **2**, 471–484.

Sanchez-Marroquin, A. (1955). Estudios sobre la Microbiologia de Pulque. XIV. Propagacion de *C. parapsilosis* y otras levaduras de este substrato. *Ciencia* **15**, 167–174.

Sanchez-Marroquin, A. (1957). Microbiology of pulque. XVIII. Chemical data on the fermentation of agave juice with pure microbiol cultures. *Rev. Soc. Quim. Mex.* **1**, 167–174.

Sanchez-Marroquin, A. (1967). Estudios sobre la Microbiologia de Pulque. XX. Proceso industrial para la ealboracion tenica de la bebida. *Rev. lat-amer. Microbiol. Parsitol.* **9**, 87–90.

Sanchez-Marroquin, A. (1970). Investigaciones realizadas en la Facultad de Quimca de la Universidad Nacional Autonoma de Mexico, tendientes a la industrializacion del agave. *Rev. Soc. Quim. Mex.* **14**, 184–188.

Sanchez-Marroquin, A. (1977). Mexican pulque—a fermented drink from *Agave* juice. Symposium on Indigenous Fermented Foods (SIFF). Bangkok, Thailand. Nov. 21–27.

Sanchez-Marroquin, A. and Hope, P. H. (1953). Agave juice fermentation and chemical composition studies of some species. *Agricultural and Food Chemistry* **1**, 246–249.

Saono, S., Brotonegoro, S., Basuki, T., Sastraatmadja, D. D., Jutono, Badjre, I. G. P., and Gandjar, I. (1977a). Tempe. Symposium on Indigenous Fermented Foods (SIFF). Bangkok, Thailand. Nov. 21–27.

Saono, S., Busuki, T., and Sastraatmadja, D. D. (1977b). Oncom. Symposium on Indigenous Fermented Foods (SIFF). Bangkok, Thailand. Nov. 21–27.

Shallenberger, R. S., Hand, D. B., and Steinkraus, K. H. (1967). Changes in sucrose, raffinose, and stachyose during tempeh fermentation. Report of Eighth Dry Bean Research Conference, pp. 67–71. Bellaire: United States Department of Agriculture.

Shibasaki, K. and Hesseltine, C. W. (1961a). Miso. I. Preparation of soybeans for fermentation. *Journal of Biochemical and Microbiological Technology and Engineering* **3**, 161–174.

Shibasaki, K. and Hesseltine, C. W. (1961b). Miso. II. *Fermentation Development in Industrial Microbiology* **2**, 205–214.

Shibasaki, K. and Hesseltine, C. W. (1962). Miso fermentation. *Economic Botany* **16**, 180–195.

Shurtleff, W. and Aoyagi, A. (1976). "The Book of Miso". Autumn Press, Kanagawa-ken, Japan.

Shurtleff, W. and Aoyagi, A. (1979). "The Book of Tempeh". Harper and Row, New York.

Shurtleff, W. and Aoyagi, A. (1980). "Tempeh Fermentation". New Age Foods, Lafayette.

Smith, O. B. (1978). Extrusion—cooked foods. *In* "Encyclopedia of Food Science", pp. 245–250. (Peterson, M. S. and Johnson, A. H., eds.) Avi Publishing, Westport.

Spicer, A. (1971a). Synthetic proteins for human and animal consumption. *The Veterinary Record* **89**, 482–487.

Spicer, A. (1971b). Protein production by microfungi. *Tropical Science* **13**, 239–250.

Stahel, G. (1946). Foods from fermented soybeans as prepared in the Netherlands Indies. II. Tempe, a tropical stable. *Journal of the New York Botanical Garden* **47**, 285–296.

Stamer, J. R. (1975). Recent developments in the fermentation of sauerkraut. *In* "Lactic Acid Bacteria in Beverages and Foods", pp. 267–280. (Carr, J. G., Cutting, C. V., and Whiting, G. S., eds.) Academic Press, New York and London.

Stamer, J. R., Stoyla, B. O., and Dunckel, B. A. (1971). Growth rates and

fermentation patterns of lactic acid bacteria associated with the sauerkraut fermentation. *Journal of Milk Food Technology* **34**, 521–525.
Steinkraus, K. H. (1979). Nutritionally significant indigenous foods involving an alcoholic fermentation. *In* "Fermented Foods Beverages in Nutrition", pp. 35–59. (Gastineau, C. F., Darby, W. J., and Turner, T. B., eds.) Academic Press, New York and London.
Steinkraus, K. H., Lee, C. Y., and Buck, P. A. (1956). Soybean fermentation by the ontjom mold *Neurospora*. *Food Technology* **19**, 119–120.
Steinkraus, K. H., Hwa, Y. B., Van Buren, J. P., Provvidenti, M. I., and Hand, D. B. (1960). Studies on tempeh—an Indonesian fermented soybean food. *Food Research* **25**, 777–778.
Steinkraus, K. H.. Hand, D. B., Van Buren, J. P., and Hackler, L. R. (1961). Pilot plant studies on tempe. *In* "Proceedings of Conference on Soybean Products for Protein in Human Foods, pp. 75–84. U.S. Department of Agriculture, Peoria, Illinois.
Steinkraus, K. H., van Veen, A. G., and Thiebeau, D. B. (1967). Studies on Idhi—an Indian fermented black gram food. *Food Technology* **21**, 110–113.
Sugihara, T. F., Kline, L., and Miller, M. W. (1971a). Microorganisms of the San Francisco sour dough bread process. I. Yeasts responsible for leavening action. *Applied Microbiology* **21**, *Applied Micro 456–458.*
Sugihara, T. F., Kline, L., and Miller, M. W. (1971b). Microorganisms of the San Francisco sour dough bread process. II. Isolation and characterization of undescribed species responsible for the souring activity. *Applied Microbiology* **21**, 459–465.
Swings, J. and DeLey, J. (1977). The biology of *Zymomonas*. *Bacteriological Reviews* **41**, 1–46.
Taboada, J., Herrera, T., and Ulloa, M. (1971). Prueba de la reduccion del acetileno para la determinacion de microorganismos fijadores de nitrogeno aislados del pozol. *Rev. lat-amer. Quim.* **2**, 188–191.
Taboada, J., Ulloa, M., and Herrera, T. (1973). Fijacion de nitrogeno *in vitro* por *Agrobacterium azotophilum* en diversos substrates, principalmente tierra y derivados de la industria azacarera. *Rev. lat-amer. Microbiol.* **15**, 143–146.
Tanuwidjaja, L. (1972). Sticky rice fermentation (tape ketan fermentation). Bull. L.K.N. K-12. Indonesian Institute Sciences. National Institute Chem. Bandung.
Ulloa, M. (1974). Mycofloral succession in pozol from Tabasco, Mexico. Bol. Soc. Mex. Mic. **8**, 17–48.
Ulloa, M. and Herrera, T. (1970). Persistencia de las aflatoxinas durante la fermentacion del pozol. *Rev. lat-amer. Microbiol.* **12**, 19–25.
Ulloa, M. and Herrera, T. (1971). Mohos aislados del pozol en medios con dificiencia o carencia de nitrogeno. Bol. Soc. Mex. Mic. **5**, 13–21.
Ulloa, M. and Herrera, T. (1972). Descripcion de dos especies neuvas de bacterias aisladas del pozol: *Agrobacterium azotophilum* y *Achromobacter pozolis*. *Rev. lat-amer. Microbiol.* **14**, 15–24.
Ulloa, M. and Herrera, T. (1973. *Phialophora richardsiae*, un hongo causante de

feosporotricosis en el hombre, aislado del pozol. *Rev. lat-amer. Microbiol.* **15**, 199–202.

Ulloa, M. and Kurtzman, C. P. (1975). Occurrence of *Candida parapsilosis, C. tropicalis*, and *Saccharomyces cerevisiae* in pozol from Tabasco, Mexico. Bol. Soc. Mex. Mic. **9**, 7–12.

Ulloa, M., Herrara, T., and Taboada, J. (1977). Pozol, a fermented maize dough consumed in southeastern Mexico. Symposium on Indigenous Fermented Foods (SIFF). Bangkok, Thailand. Nov. 21–27.

van Veen, A. G., Hackler, L. R., Steinkraus, K. H., and Mukherjee, S. K. (1967). Nutritive quality of idli, a fermented food of India. *Journal of Food Science* **32**, 339–341.

van Veen, A. G., Graham, D. C. W., and Steinkraus, K. H. (1968). Fermented peanut presscake. *Cereal Science Today* **13**, 96–99.

Wagenknecht, A. C., Mattick, L. R., Lewin, L. M., Hand, D. B., and Steinkraus, K. H. (1961). Changes in soybean lipids during temperh fermentation. *Journal of Food Science* **26**, 373–376.

Wanderstock, J. J. (1968). Food analogues. *The Cornell Hotel and Restaurant Administration Quarterly* , 29–33.

Wang, H. L. and Hesseltine, C. W. (1970). Sufu and Lao-Chao. *Journal of Agricultural and Food Chemistry* **18**, 572–575.

Wang, H. L. and Hesseltine, C. W. (1979). Mold-modified foods. *In* "Microbial Technology",vol.2, pp. 95–129. (Peppler, H. H. and Perlman, D., eds.) Academic Press.

Wickerman, L. J. and Burton, K. A. (1960). Heterothallism in *Saccharomyces rouxii. Journal of Bacteriology* **80**, 492–495.

Wood, B. J. B. (1977). Lactobacillus/yeast interactions; their contribution to fermented foodstuffs, especially traditional bread. Symposium on Indigenous Fermented Foods (SIFF). Bangkok, Thailand. Nov. 21–27.

Wood, B. J. B. and Rainbow, C. R. (1961). The maltophosphorylase of beer *Lactobacilli. Biochemical Journal* **78**, 204–209.

Wood, B. J. B., Cardenas, O. S., Yong, F. M., and McNulty, D. W. (1975). *Lactobacilli* in production of soy sauce, sourdough bread and Parisian barm. *In* "Lactic Acid Bacteria in Beverages and Food", pp. 325–335. (Carr, J. G., Cutting, C. B., and Whiting, G. C., eds.) Academic Press, London and New York.

Worthington, R. W. and Beuchat, L. R. (1974). α-Galactosidase activity of fungi on intestinal gas-forming peanut oligosaccharides. *Agricultural and Food Chemistry* **22**, 1063–1066.

Yamamoto, K. (1957). Studies on koji. II. Effect of some conditions of medium on the production of mold protease. *Bulletin of the Agricultural Chemistry Society of Japan* **21**, 313–318.

Yamamoto, Y., Yangida, F., and Suminoe, K. (1972). Changes of enzyme activities during preparation of the koju for raw soy sauce. *Nippon Shokuhin Kogyo Gakkai-Shi* **19**, 29–30.

Yokotsuka, T. (1960). Aroma and flavor of Japanese soy sauce. *Advances in Food Research* **10**, 75–135.

Yokotsuka, T. (1967). Toxic substances produced by molds and non-productivity of aflatoxin by Japanese koji-molds. *Seasoning Science* **14**, 23–37.

Yokotsuka, T. (1972). Some recent technological problems related to the quality of Japanese shoya. *In* "Fermentation Technology Today", pp. 659–662. (Terui, G., ed.) Society of Fermentation Technology (Japan).

Yokotsuka, T. (1977). Japanese shoyu. Symposium on Indigenous Fermented Foods (SIFF). Bangkok, Thailand. Nov. 21–27.

Yokotsuka, T., Sakasai, T., and Asao, Y. (1967a). Studies on flavorous substances in shoyu. Part 25. Flavorous compounds produced by yeast fermentation. *Journal of Agricultural Chemistry Society of Japan* **41**, 428–433.

Yokotsuka, T., Asao, Y., and Sakasi, T. (1967b). Studies on flavorous substances in shoyu. Part 27. The production of 4-ethyl-guaicol during shoyu fermentation and its role for shoyu flavor. *Journal of Agricultural Chemistry Society of Japan* **41**, 442–447.

Yokotsuka, T., Sasaki, M., Kikuchi, T., Asao, Y., and Nobuhara, A. (1967c). Compounds producted by molds. II. Fluorescent compounds produced by Japanese industrial molds. *Journal of Agricultural Chemistry Society of Japan* **41**, 32–38.

Yong, F. M. and Wood, B. J. B. (1974). Microbiology and biochemistry of soy sauce fermentation. *Advances in Applied Microbiology* **17**, 157–230.

Yong, F. M. and Wood, B. J. B. (1976). Microbial succession in experimental soy sauce fermentation. *Journal of Food Technology* **11**, 525–536.

Yong, F. M. and Wood, B. J. B. (1977). Biochemical changes in experimental soy sauce koji. *Journal of Food Technology* **12**, 163–175.

12

The Use of Mixed Microbial Cultures in Metal Recovery

P. R. Norris and D. P. Kelly

1. INTRODUCTION

Of the various interactions of microorganisms with metals (Summers and Silver, 1978; Kelly *et al*., 1979), the current and potential uses of organisms in metal recovery mainly concern metal accumulation, metal precipitation, and the oxidation of minerals.

Typical studies of the metal recovery potential, of the intracellular accumulation of metals and of the binding of metals at organism surfaces, have used pure cultures (Friedmann and Dugan, 1967; Norris and Kelly, 1979; Shumate *et al*., 1978), although silver accumulation has been studied using a community of three bacteria (Charley and Bull, 1979). The mechanisms and sites of silver accumulation were not described, beyond the observation that *Pseudomonas maltophilia* was the major accumulator of metal, but the community was able to tolerate and accumulate considerably more silver than *P. maltophilia* in pure culture. Microbial removal of pollutant metals from waters, which could lead to later metal recovery, can occur through metal accumulation and by trapping of colloidal or particulate metals in algal growth below mine tailing ponds in stream systems (Jennett *et al*., 1975) and in settling ponds (Jackson, 1978). Inevitably, mixed populations are involved in these processes but the microflora composition, dynamics and interactions in the communities are so far little studied. The precipitation of metals as sulphides when such metal-rich algal blooms decompose (Jackson, 1978) and when metal-rich industrial effluents become depleted of oxygen (Ilyaletdinov *et al.*, 1977), arises through the activity of sulphate-reducing bacteria, activity which has been implicated in the formation of biogenic, sedimentary, sulphide minerals (Trudinger *et al.*, 1972; Trudinger, 1976; Krouse and McCready, 1979; Kelly, 1980). As with metal accumulation studies, there has been little reported work involving mixed cultures in the development of metal recovery processes based on sulphide generation and, where mixed cultures have been used (Tomizuka and Yagisawa, 1978), it is not evident how the metal recovery might have compared with that by a pure culture.

The oxidation of minerals by bacteria contributes to the economically important leaching of copper and uranium and also contributes to acid and metal pollution of drainage waters emanating from pyrite-rich mines or coal spoil dumps. The principal reactions of bacterial leaching are the direct oxidation of sulphide minerals and the indirect dissolution of metal sulphides and uranium oxides by ferric iron, which can be formed from bacterial oxidation of ferrous iron or iron sulphide minerals. The chemistry, basic microbiology, and biochemistry of leaching processes are reviewed extensively elsewhere (Dutrizac and MacDonald, 1974; Tuovinen and Kelly, 1972, 1974; Kelly, 1976; Karavaiko *et al.*, 1977; Brierley, 1978; Kelly *et al.*, 1979; Lundgren and Silver, 1980). Most bacterial leaching studies have concentrated almost exclusively on *Thiobacillus ferrooxidans* as the organism responsible for mineral oxidation but the more recent of the above reviews have indicated an increasing number of isolations of other mineral-oxidizing organisms and the existence of positive and negative interactions between microorganisms in leaching systems.

In contrast to microbial accumulation and precipitation of metals, the bacterially-assisted leaching of metals is currently of considerable industrial significance and has been reported to be influenced by the use of mixed cultures. It is with these reports and with the increasing industrial importance of bacterial leaching in mind that we devote this chapter to the microbial interactions and the use of mixed cultures in the microbially-assisted, acid leaching of metal-bearing materials.

In considering microbial interactions in leaching, we are only concerned with that part of the metal recovery process in which organisms are active, i.e. the extraction of metals into solution rather than their ultimate recovery which involves chemical and physical hydrometallurgical processes. Microbially-assisted leaching relies on bacterial oxidation of iron and sulphur and therefore is principally involved with the extraction of metals found as or with sulphide minerals. These ores may be treated in dump, heap, underground *in situ* and vat leaching operations (Brierley, 1978). The variation in scale, complexity and degree of control over the chemical and physical conditions which can be maintained in these operations at least potentially affects the composition of the microbial population and the range and extent of microbial interactions. Thus with underground *in situ*, heap, and particularly with dump leaching on a scale which can involve up to 4 billion tons of material (Brierley, 1978), we are concerned with a "semi-natural" microflora and possibly with different populations or communities in different parts of these operations, such as the ore in contact with leaching solutions, pregnant liquor in collecting ponds, and in ferrous iron-rich barren solutions in which ferric iron regeneration is required after copper or uranium removal. In contrast, vat or fermenter leaching of metal concentrates, which has been studied on a semi-industrial scale as a potentially economically and environmentally attractive alternative to conventional concentrate treatment (Torma, 1978; McElroy and Bruynesteyn, 1978) or of metal-bearing waste materials which do not have an indigenous microflora, offers more scope for selection and control of the leaching organisms.

2. OBSERVATIONS ON THE MICROFLORA AND MICROBIAL INTERACTIONS IN ACID LEACHING OF METAL SULPHIDES

2.1 Observations Related to Ore Dump and Mine Drainage Studies

The diversity of organisms which can be expected in the semi-natural environment of dump, heap and leaching operations *in situ* can be seen from observations on the acid drainage from a North American copper mine (Ehrlich, 1963) and on the microflora of a low-grade copper ore dump in Bulgaria at which commercial leaching began in 1972 (Groudev *et al.*, 1978). Both studies recorded protozoa, particularly *Amoeba* sp. feeding on yeasts and bacteria, and species of *Euglena* and *Eutrepia*. In some parts of the Bulgarian ore dump *Amoeba* sp. was found at numbers up to 500 ml^{-1} and, in laboratory experiments, decreased the rate and extent of leaching and reduced the viable, iron-oxidizing bacteria from 1×10^8 to 5×10^3 ml^{-1} (Groudev *et al.*, 1978).

Yeasts and filamentous fungi were found in the studies described above, particularly species of *Rhodotorula, Trichosporon, Cladosporium*, and *Penicillium*. Utilization of organic compounds by fungi in leaching environments could influence the activity of organisms and of ions important in metal extraction. Inhibition by pyruvate, one of many organic compounds which is toxic to iron-oxidizing bacteria (Tuttle and Dugan, 1976), of iron oxidation and uranium leaching by *Ferrobacillus ferrooxidans* was alleviated in laboratory experiments by *Rhodotorula* sp. isolated from mine water (Nerkar *et al.*, 1977). An active fungal population (including species of *Penicillium* and *Paecilomyces*) was found in laboratory tests of heap leaching of uranium from organic-rich, alum shales and appeared to be involved in the utilization of organic matter, consequently establishing reducing conditions and causing a reduction of ferric iron (Napier and Wakerley, 1969).

Of the algae found in leaching environments, the most noticeable is probably *Hormidium fluitans* which grows profusely in some dump effluents in several countries (Groudev *et al.*, 1978; Madgwick and Ralph, 1977). An extracellular heat-stable substance produced by the alga was reported to increase the bacterial oxidation of ferrous iron and the leaching of chalcopyrite by some strains of *Thiobacillus ferrooxidans* (Madgwick and Ralph, 1977).

Sulphur bacteria, iron bacteria, such as *Gallionella* spp. and *Leptothrix* spp., sulphate-reducing and other heterotrophic bacteria (including species of *Bacillus* and *Pseudomonas*) were recorded at the Bulgarian ore dump (Groudev *et al.*, 1978) and various heterotrophic bacteria have been found in acid (pH 2.8) coal mine drainage (Dugan *et al.*, 1970a). The activity of heterotrophic bacteria in such environments is uncertain. *Desulfovibrio* sp. was found particularly in the deeper levels of the Bulgarian ore dump but, although some activity may be possible in acid conditions (Satake, 1977), sulphate-reducing bacteria are not generally associated with acidic environments. Similarly, although a *Bacillus* sp. appeared mainly responsible for the formation of polysaccharide, acid streamers in mine drainage, the optimum pH for the organism was near neutrality (Dugan *et al.*, 1970b). Heterotrophic organisms could interfere with metal leaching by physically blocking sulphide surfaces and obstructing leach solution flow (Ehrlich and Fox, 1967). In contrast, certain autotrophic bacteria have long been associated with the promotion of leaching. The presence and activities of various types of thiobacilli found in metal leaching and other environments where sulphide mineral oxidation occurs, have been reviewed (Tuovinen and Kelly, 1974; Pol'kin *et al.*, 1975; Brierley,

1978; Groudev *et al.*, 1978; Kelly *et al.*, 1979; Ralph, 1979). Thiobacilli, such as *Thiobacillus thioparus* and *Thiobacillus denitrificans*, with pH optima nearer to neutrality than the values prevailing in metal sulphide leaching environments, are nevertheless found in such environments and at least *Thiobacillus neapolitanus* (Goroll, 1976) can apparently reach numbers as high as the acidophilic *Thiobacillus ferrooxidans* and *Thiobacillus thiooxidans*. pH-dependent successions involving non-acidophilic sulphur-oxidizing thiobacilli and acid-tolerant iron-oxidizers, such as *Metallogenium* sp., have been proposed to account for the progressive development of acidity and iron oxidation in freshly-exposed pyritic material which culminates in the dominance of acidophilic bacteria (Walsh and Mitchell, 1972; Tuovinen and Kelly, 1974; Harrison, 1978; Ralph, 1979; Le Roux *et al.*, 1980). Such successions could occur in some parts of dump leaching operations but application of acid leaching solutions, some chemical oxidation of pyrite, and activity of *Thiobacillus ferrooxidans*, initially in microhabitats or at higher pH values than those usually associated with the organism (as suggested by Kleinmann and Crerar (1979) to account for pyritic coal acidification), could all contribute to reducing the significance of microbial successions. Therefore, it is likely that in metal leaching operations in which it is desirable and possible to maintain a low pH in order to keep iron in solution, and desirable to attain high dissolved concentrations of metals toxic to most organisms, *T. ferrooxidans* and possibly *T. thiooxidans* will be the most active organisms. It is also likely that most of the other organisms and microbial interactions mentioned so far will be relatively unimportant. The study of microbial populations of various parts of the Bulgarian copper leaching operation described above and laboratory experiments have shown that, although the biomass of organisms other than *T. ferrooxidans* reached 30% of the total biomass at pH 3.5 to 4.0 with copper concentrations below 1 g l^{-1}, the iron oxidizers increased from 76 to 93% of the population with an increase of acidity from pH 3.2 to 2.4 and reached 97% of the population at pH 2.1. In the dump leaching operation, lowering the pH from 4.0 to 2.5 also increased the numbers of *T. ferrooxidans* and increased the copper extraction (Groudev *et al.*, 1978).

2.2 Observations Related to Laboratory Studies of Metal Concentrate leaching

Most workers in bacterial leaching use enrichment cultures from local sites of metal leaching, thus making comparative assessments of the usefulness of various cultures difficult, especially as the ores or concentrates used in different laboratories frequently have diverse chemical composition and physical characteristics. However, the use of mixed enrichment cultures from various copper and coal mine sites in India has shown copper extraction from low grade ores (mostly chalcopyrites) to vary with different cultures (Rangachari *et al.*, 1978). Our comparison of pyrite leaching by mixed enrichment cultures, including some from sites of significance in commercial metal-recovery operations, has revealed almost complete, rapid dissolution of pyrite compared with poorer leaching by a pure culture of a previously characterized strain of *Thiobacillus ferrooxidans* (Norris and Kelly, 1978) (Fig. 1). The experimental conditions for the pyrite leaching shown in Fig. 1 comprised flask cultures of 100 ml culture medium shaken at 150 rev/min and 30°C with 200 mesh (45 μ) pyrite (10 g l^{-1}) in a medium adjusted initially to pH 2.0 with sulphuric acid and containing (g l^{-1}): $(NH_4)_2SO_4$, 3.0; $MgSO_4.7H_2O$, 0.5; K_2HPO_4, 0.5; and KCl, 0.1. The mixed enrichment cultures are named after the mines or areas of their origins. Thus, the Chino culture was derived from a sample of a copper leach dump effluent at the Kennecott Chino Mine, New Mexico; the Agnew Lake culture from pregnant liquor of the Agnew Lake uranium leach dump in Ontario, Canada; the Alvecote culture from a coal waste spoil tip near Alvecote, Warwickshire; and the Los Amoles culture from mine shaft water of the pyrite-rich uraniferous Los Amoles Mine, Sonora, Mexico. It should be stressed that the illustrated activity of the enrichment cultures, maintained in the laboratory by serial transfer on pyrite, did not reflect activity of the organisms *in situ* at the sites from which they were obtained. Similar activity has been shown also by samples from other coal and metal mine waste sites in Great Britain and is probably typical of enrichment cultures from such habitats but is possibly specific for the given mineral (pyrite) and experimental conditions with, for example, a different leaching pattern shown with chalcopyrite by the Chino mixed culture relative to the pure strain of *T. ferrooxidans* (see Section 3.1.3). Mixed enrichment cultures have also been reported to remove pyritic sulphur from coal slurries under conditions in which a pure culture of *T. ferrooxidans* was ineffective (Dugan and Apel, 1978).

For the optimization of the microbial involvement in potential coal desulphurization and in metal leaching operations, it would be useful to know whether more efficient pyrite-oxidizing organisms than the characterized strains of *Thiobacillus ferrooxidans* are present among the many types of bacteria in mixed, pyrite-enrichment cultures or whether there are microbial interactions which influence pyrite dissolution. Microbial interactions have been observed in controlled laboratory studies which have involved levels of acidity and rates of mineral leaching comparable to those desired in potential concentrate or waste material leaching processes. Leaching of copper sulphides by ferrobacilli in laboratory experiments was reported to be decreased when other organisms were added in quantities and ratios characteristic of the microflora of a copper leach dump (Groudev *et al.*, 1978), but the potential improvement of leaching through microbial interactions has received more study than such adverse effects. For example, mutualism between *T. ferrooxidans* and *Beijerinckia lacticogenes* has been described resulting in an improved rate of and extent of leaching of a copper-nickel concentrate. This was achieved apparently as a result of an increase in available nitrogen for the mineral oxidizer through nitrogen fixation by the heterotrophic organism (Tsuchiya *et al.*, 1974; Tsuchiya, 1978). Whether other factors were also involved is uncertain since *T. ferrooxidans* can also fix nitrogen (MacIntosh, 1978) and other work with different ores has failed to show any benefit through the presence of *B. lacticogenes*(Rangachari *et al.*, 1978). A second example of a potentially beneficial influence on the *T. ferrooxidans* population in a mixed culture at low pH could be removal of organic material inhibitory to *T. ferrooxidans* by mixotrophic, acidophilic bacteria, such as *Thiobacillus acidophilus* which has been found in close association with *T. ferrooxidans* in cultures previously considered to be pure (Guay and Silver, 1975) and with recent *T. ferrooxidans* isolates from acid sulphate soils (Arkesteyn, 1980). Inhibition of iron-oxidizing *T. ferrooxidans* by some organic compounds (e.g. 1 mM-glucose, ethanol, lactate, succinate, serine and aspartate) was relieved by *T. acidophilus*, whereas inhibition by compounds more toxic to the iron oxidizer (e.g., 1 mM-pyruvate and α-ketoglutarate) occurred even in the mixed culture (Arkesteyn, 1980). Inhibition by lower concentrations of pyruvate was not tested but it might have been alleviated as was inhibition by 0.1 mM pyruvate by *Rhodotorula* sp. in the interaction described previously (Nerkar *et al.*, 1977). Pyruvate did not support growth of *T. acidophilus* (Guay and Silver, 1975; Arkesteyn, 1980) but neither did lactate or succinate and yet inhibition of iron oxidation by these compounds was alleviated in mixed cultures. Further work is required in this area with various concentrations of organic compounds which can act as both substrates and inhibitors for some thiobacilli.

Thiobacillus acidophilus, and the similar *Thiobacillus organoparus* (Markosyan, 1973), might also have to be considered in interactions which could directly influence leaching rates through cooperative action in mineral degradation with iron-oxidizing organisms, particularly such as *Leptospirillum ferrooxidans* (see Section 3.2). Whether *T.*

acidophilus oxidizes sulphide as well as elemental sulphur is in doubt (Guay and Silver, 1975; Arkesteyn, 1980) but its presence in mixed culture has been reported not to influence the rate of pyrite leaching by *T. ferrooxidans* (P. R. Norris, unpublished observations; Arkesteyn, 1979). Mixed chemostat cultures of *L. ferrooxidans* or *T. ferrooxidans* with *T. acidophilus* on ferrous iron contained about 1% *T. acidophilus* at pH 2.0 to 2.5 but only 0.02 to 0.05% at pH 1.7 (A. L. Smith and D. P. Kelly, unpublished observations) which is consistent with the lower limit for growth of *T. acidophilus* on sulphur or glucose at about pH 1.6. Arkesteyn (1980) found that *T. acidophilus* comprised about 30% of a mixed culture of *T. ferrooxidans* and *T. acidophilus* grown on ferrous iron at pH 3.0. The low tolerance to high acidity shown by sulphur-oxidizing *T. acidophilus* suggests that such a role for the mixotroph in mixed, mineral-oxidizing cultures would be less significant in metal recovery processes than the activity of sulphur-oxidizing organisms better adapted to pH values below 2.0.

Acidophilic, sulphur-oxidizing *Thiobacillus thiooxidans* is probably unable to degrade directly most minerals of commercial importance but it can increase the dissolution of some metal sulphides, particularly in the presence of excess sulphur (Kelly *et al.*, 1979), and could be active in the removal of sulphur layers which might form on the surfaces of minerals being leached by ferric iron or directly oxidized by iron-oxidizing bacteria. Such sulphur oxidation by *T. thiooxidans* could involve a bacterially mediated reduction of ferric iron (Brock and Gustafson, 1976). Interactions between *T. thiooxidans* and *T. ferrooxidans* have been described, but general agreement on the influence of *T. thiooxidans* is lacking, mainly as a result of the diverse experimental conditions under which the mixed cultures have been studied. *T. thiooxidans* has been reported to have no influence on the rate of pyrite leaching by *T. ferrooxidans* (Norris and Kelly, 1978); to have slightly increased the yield of nickel from low grade pentlandite-bearing ore when added with *T. ferrooxidans* (Bosecker, 1977); to have increased the yield of copper from high carbonate ores by 30% compared to the leaching by *T. ferrooxidans* alone and to have reduced the acid consumption of the leaching (Bosecker, 1976; Bosecker *et al.*, 1978); and to have slightly increased the rate of sulphur removal from coal when mixed with one strain of *T. ferrooxidans* but to have reduced the rate when in mixed culture with a different strain of *T. ferrooxidans* (Groudev and Genchev, 1979). Thus, strain differences of *T. ferrooxidans* could determine, at least with certain substrates, whether other organisms with similar but potentially more active aspects of *T. ferrooxidans* metabolism (e.g. sulphur oxidation), could influence metal leaching processes which were, nevertheless, essentially dependent on the iron-oxidizing bacteria. Furthermore, leaching of a sulphidic dust concentrate from a copper acid-leaching process was improved with a mixed culture of *T. ferrooxidans* strains compared to using either strain separately (Ebner, 1978). In order to clarify the basis of such reports of improved leaching by mixed cultures involving *T. ferrooxidans*, variation in this organism's characteristics which are relevant to metal leaching are considered in the next section. Other iron-oxidizing organisms potentially able to replace or interact with *T. ferrooxidans* are considered in subsequent sections (pp. 458–464).

3. IRON-OXIDIZING BACTERIA: SOME CHARACTERISTICS RELEVANT TO THE ASSESSMENT OF MICROBIAL INTERACTIONS

3.1 Thiobacillus ferrooxidans

3.1.1 Strain Variation

Thiobacillus ferrooxidans is generally considered to be the most important organism in bacterial leaching and has been the subject of extensive study and reviews (Tuovinen and Kelly, 1972; Torma, 1977; Schwartz, 1977; Murr *et al.*, 1978). Different isolates of the organism were initially described in accordance with apparently different capacities to oxidize sulphur. Thus, *Thiobacillus ferrooxidans* (Temple and Colmer, 1951), *Ferrobacillus ferrooxidans* (Leathen *et al.*, 1956) and *Ferrobacillus sulfo-oxidans* (Kinsel, 1960) were proposed before further nutritional and taxonomic studies led to the proposal that these were all strains of the same organism and that *T. ferrooxidans* be accepted as the only species (Kelly and Tuovinen, 1972). The current position has been summarized by Ralph (1979) as "a substantial consensus that most of the numerous isolations of acidophilic, chemoautotrophic and facultatively heterotrophic, iron- and sulphur-oxidizing bacteria from acidic, sulfidic mineral habitats are all biochemical and physiological variants of the species *T. ferrooxidans*." There is variation in colony types of different strains plated on solid media (Manning, 1975; A. L. Smith and P. R. Norris, unpublished observations) and there are different capacities of adapted strains to oxidize particular substrates (Silver and Torma, 1974) as well as different DNA base compositions of the adapted strains (Guay *et al.*, 1976). *T. ferrooxidans* grown on ferrous iron, chalcopyrite, and a lead sulphide concentrate exhibited %GC values of 56.0, 60.1, and 54.4, respectively. These variations could indicate significant population heterogeneity, the different substrates allowing selective growth of different members of a mixed population. The "ferrobacilli" of a commercial leach dump have been described as a very mixed population (Groudev *et al.*, 1978) with some organisms incapable of sulphur oxidation despite attempts to induce such activity in the laboratory: thus, these could be organisms which fit the earlier designation of *Ferrobacillus ferrooxidans*. Pol'kin *et al.* (1975) found that the differences among 30 strains of *T. ferrooxidans* were primarily in their resistance to metals and acidity and in the rates of oxidation of various sulphide minerals and sulphur; all the differences were reported to be an adaptive characteristic. With respect to metal extraction, variants of these iron-oxidizing organisms can be compared most usefully by their relative capacities for ferrous iron oxidation and mineral oxidation.

3.1.2 Ferrous Iron Oxidation

With strain purification and adaptation procedures often poorly documented, assessment of variations in the reported iron-oxidation rates is difficult. *Thiobacillus ferrooxidans* isolates obtained from two coal mining areas (Montana and Wyoming) showed a range of generation times from 7.7 to 15.9 h for iron oxidation at room temperature (Olson *et al.*, 1979). Apart from those organisms incapable of sulphur oxidation, Groudev *et al.* (1978) found that the ferrobacilli of the Vlaikov vrah mine, Bulgaria, could be grouped into two typical types:

(1) isolates with pronounced iron-oxidizing activity and negligible (arising only after induction) sulphur-oxidizing activity;
(2) isolates with moderate iron-oxidizing activity but pronounced sulphur-oxidizing activity.

Further screening of 120 "different strains" of *Thiobacillus ferrooxidans* isolates from acid mine drainage water and ore lumps from various mines and dumps in Bulgaria (Groudev, 1979) did not reveal a clear relationship between iron- and pyrite-oxidizing activity among the various strains, but the most active iron-oxidizing strains did possess the greatest pyrite-degrading activity. Iron oxidation activity by some organisms isolated from our pyrite-degrading mixed cultures (Fig. 1) has been examined in an initial phase of a study of the various organisms in the cultures, in an attempt to explain their efficient dissolution of pyrite and particularly to isolate organisms more efficient in pure culture than the previously characterized strain of *T. ferrooxidans*. Pure cultures of iron-oxidizing organisms were obtained from pyrite-maintained stock cultures via iron-enrichment cultures and two, successive, single colony isolations on ferrous iron agar plates. The number of serial subcultures in ferrous iron medium before single colony isolation varied with the time the culture had been in the laboratory with 10 subcultures for the Alvecote sample to 50 serial subcultures for the Chino sample. In each case, pure strains selected from the Los Amoles, Chino, Alvecote, and Agnew Lake samples were morphologically similar to the characterized *T. ferrooxidans* strain and all the strains obtained showed similar rates of iron-oxidation with doubling times for the oxidation of between 6.7 and 7.5 h at 30°C. With the Chino and Alvecote samples, three single colonies of each were selected from plates as there was some slight colony morphology variation, but all the cultures obtained showed similar iron oxidation rates. Some notable differences in the lag phases between the strains were apparent, however, particularly with the Agnew Lake strain which showed longer lag phases than the others. Differences in metal resistance were also apparent with, for example, the Los Amoles strain being more resistant to uranium than the Chino isolate or the characterized *T. ferrooxidans* strain. The iron enrichment serial subcultures and subsequent plating probably selected for organisms which were most able to oxidize iron at high rates under the given conditions. Pure strains inoculated into pyrite medium and subsequently transferred into ferrous iron medium after pyrite oxidation, retained their similar iron oxidation rates. In contrast, uniform iron oxidation rates were not observed when samples of the pyrite-maintained mixed cultures from various sites (Fig. 1) were transferred to a ferrous iron medium. The initial rates of iron oxidation on this first transfer from pyrite ranged from 8 h iron oxidation doubling time for the Agnew Lake sample, to 13 h for the Alvecote sample, and to a range of rates from 13 to 20 h for the Los Amoles sample with transfer from pyrite at different phases of the bacterial growth-leaching cycle. Microscopic examination of the iron-oxidizing organisms deriving from the Los Amoles culture in this way, revealed that the dominant organisms was motile, spiral bacteria resembling *Leptospirillum ferrooxidans* (see Section 3.2).

3.1.3 Mineral Oxidation

Pyrite oxidation by pure strains of *Thiobacillus ferrooxidans* isolated, as described above, from our mixed, enrichment cultures has been measured (Fig. 2). Pyrite oxidation by another strain isolated by enrichment culture and single colony isolation from iron-stained snowfields in the Hardangervidda, Norway (and subsequently referred to as

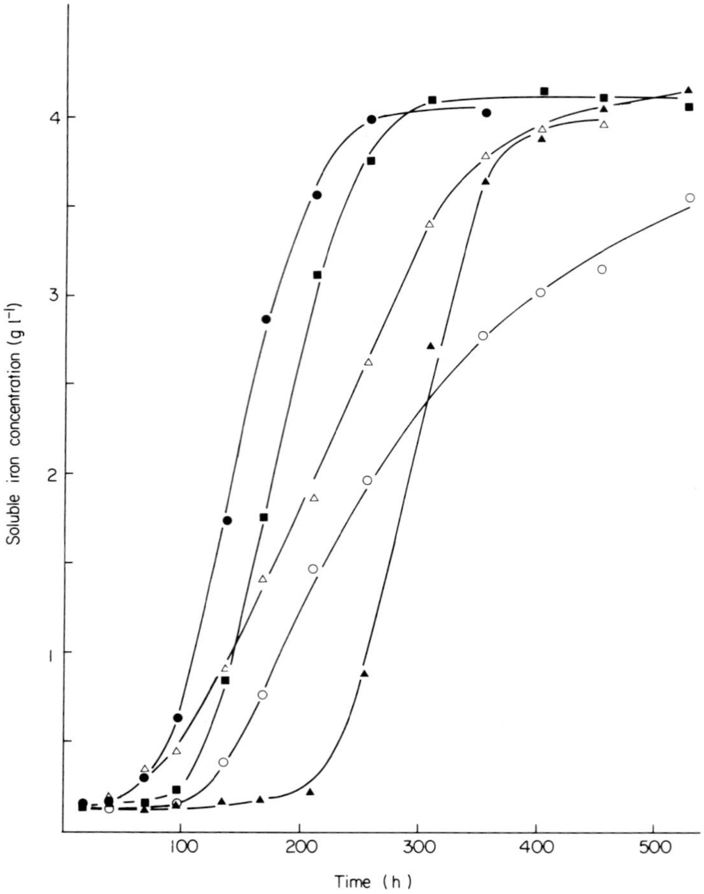

Fig. 1. Pyrite leaching by *Thiobacillus ferrooxidans* (○) and by mixed cultures from Chino (●), Los Amoles (△), Alvecote (■), and Agnew Lake (▲).

the Hardanger strain) is included in Fig. 2 and discussed in Section 3.3. Pyrite oxidation by cultures derived from each of the three single colony isolations from the Chino and Alvecote cultures (p. 16) was identical and the soluble iron yield was similar to that by the characterized strain of *T. ferrooxidans*. More efficient pyrite leaching, in terms of the soluble iron yield rather than the initial rate of iron release, was shown by the Agnew Lake and Los Amoles strains. The Agnew Lake strain showed a 350 h lag phase prior to the iron release curve illustrated in Fig. 2. The different efficiencies of pyrite leaching by the strains were reproducible and have been re-examined at higher levels of

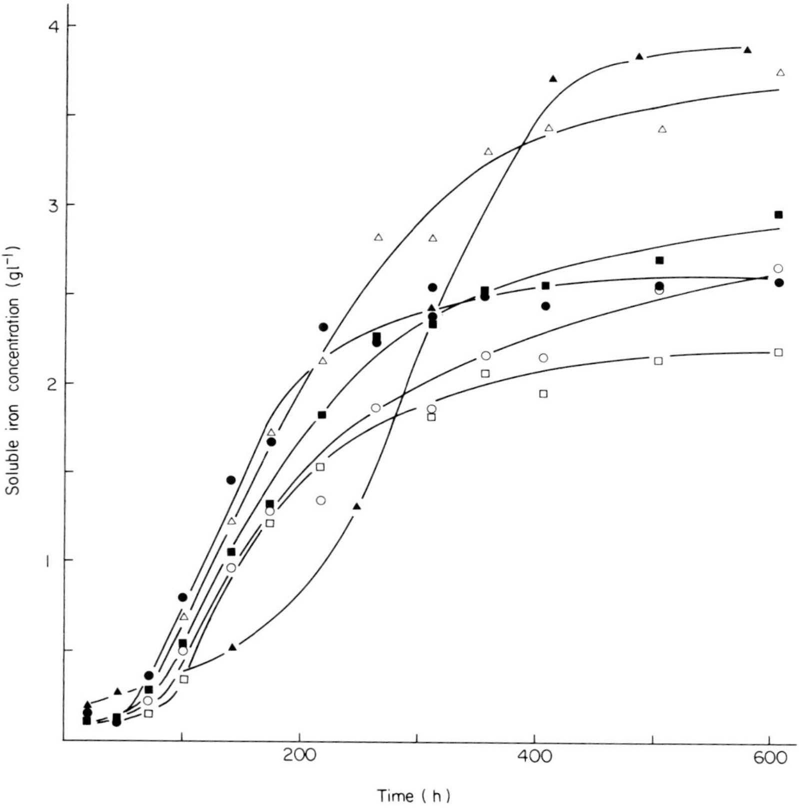

Fig. 2. Pyrite leaching by strains of *Thiobacillus ferrooxidans* previously characterized (○); isolated from mixed cultures designated as the Chino (●); Los Amoles (△); Alvecote (■); Agnew Lake (▲) and Hardanger (□).

pyrite (50 g l^{-1}) to show more clearly the different strain characteristics. Typical results with the Chino (final pH 1.4) and Los Amoles (final pH 1.3) strains are compared (Fig. 3) with the Chino mixed, enrichment culture (final pH 1.2). The rate of iron release from pyrite (50 g l^{-1} at 30°C) by the Chino mixed culture was 50 mg iron l^{-1} h^{-1} which compared favourably with other reported rates of 30 mg iron l^{-1} h^{-1} (50 g l^{-1}—350 mesh (75 μ) pyrite at 35°C—Atkins, 1978) and 35 mg iron l^{-1} h^{-1} (50 g l^{-1}—325 (70 μ) mesh pyrite at 30°C—Groudev, 1979). Groudev (1979) found two distinguishable types of rate curves for iron release from pyrite by different strains of *T. ferrooxidans* with most strains showing biphasic iron-release with a step in the iron-release curves as opposed to the gradual transition from the maximum

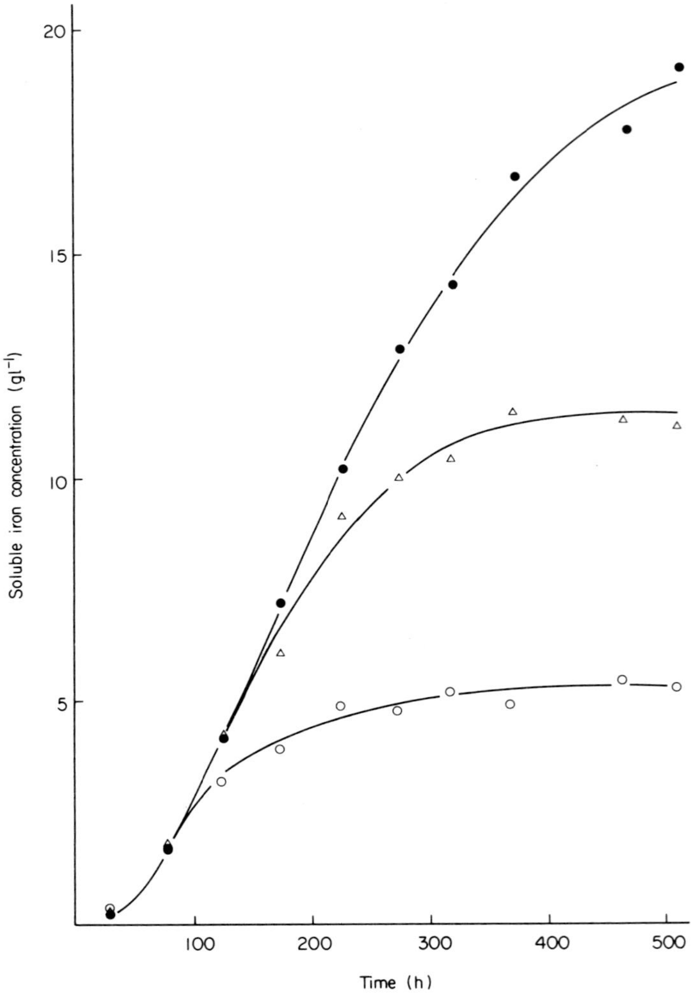

Fig. 3. Pyrite leaching by the Chino mixed culture (●) and by strains of *Thiobacillus ferrooxidans* obtained from the Chino (○) and Los Amoles (△) cultures.

iron release rate to culture stationary phase shown by the other strains and by all the strains in our experiments. A smoother transition to the maximum dissolved iron yield was obtained after preliminary adaptation to pyrite and to low pH (Groudev, 1979).

Adaptation of organisms to mineral substrates must be considered in comparison of leaching by different cultures. Groudev and Genchev

(1978) found that their 120 *Thiobacillus ferrooxidans* isolates from mineral environments showed a wide range of chalcopyrite leaching rates and could be placed in three groups defined by their propensity for adaptation to oxidation of the mineral. The rates of chalcopyrite leaching appeared to be positively correlated with the sulphur-oxidizing activity of the strains whereas a clear relationship with ferrous iron-oxidizing activity was not found. Some strains showed a rate of chalcopyrite dissolution which was independent of whether previous growth was on ferrous iron, elemental sulphur or chalcopyrite, whereas other strains showed improved leaching of the mineral after previous growth on sulphur or, particularly, on chalcopyrite. Other strains showed greatest leaching rates after growth on ferrous iron (Groudev and Genchev, 1978). Earlier work showing increased sulphur oxidation by sulphur-grown, rather than by iron-grown, *Ferrobacillus ferrooxidans* raised an awareness to the possibility that heterogeneous cultures, rather than enzyme induction, accounted for such observations (Margalith *et al.*, 1966). Even allowing for the possibility of heterogeneous cultures, adaptation to mineral substrates remains a complex phenomenon which, with chalcopyrite, for example (Groudev and Genchev, 1978), may result from alterations in toxic metal (copper) tolerance and in the capacity of the organism to attach to the solid substrate as well as from induction of enzymes. Adaptive mechanisms of some strains were reported to be sufficiently selective to distinguish between different samples of chalcopyrite (Groudev and Genchev, 1978). The leaching rate and soluble iron yield through the action of the characterized strain of *T. ferrooxidans* in our work (Fig. 1) were not significantly increased after many subcultures of the organism on pyrite but work with minerals at different pH values has indicated that the pH during leaching is probably an important factor in the observed greater efficiency of the mixed cultures compared with this characterized strain of *T. ferrooxidans*. Oxidation of chalcopyrite, a mixed pyrite-chalcopyrite concentrate and a nickel-bearing (violarite) concentrate by *T. ferrooxidans* and by some of the mixed cultures in our laboratory (data not shown) proceeded without the progressive accumulation of acid which occurred during pyrite leaching. With these mineral concentrates, final pH values at stationary phases of culture growth were between 1.8 and 2.2 (initial medium at pH 2.0) and the rates of leaching and yields of soluble metals by the characterized *T. ferrooxidans* strain were similar to the mixed cultures, with the exception of the improved chalcopyrite leaching by a mixed culture maintained on the mineral for two years prior to the leaching experiments. However, when leaching of chalcopyrite by the Chino mixed culture and by the characterized strain of

T. ferrooxidans proceeded at a lower, controlled pH (1.5), the maximum rate of copper leaching by the mixed culture was much greater than by the pure culture, the maximum rate by the latter being less than in culture without pH control.

3.2 Leptospirillum ferrooxidans

Although most mesophilic, acidophilic iron-oxidizing bacteria isolated from leaching environments are regarded as strains of *Thiobacillus ferrooxidans*, it is likely that some other iron-oxidizing organisms are active in nature but are overlooked through being less competitive in the usual isolation or enrichment media. *Leptospirillum ferrooxidans* was isolated from Armenian Copper deposits in 1972 (Balashova *et al.*, 1974) but has not so far been recognized as of widespread distribution. *L. ferrooxidans* is vibrioid, capable of forming spirals of joined cells and is thus distinct from *T. ferrooxidans* in morphology, in DNA base composition (% GC content of 54.0 (D. P. Kelly, unpublished observations) compared with 58.0 for *T. ferrooxidans* (Tuovinen *et al.*, 1978); and in rates of iron oxidation. We have obtained colony formation of *L. ferrooxidans* on iron agar plates only with difficulty in comparison with *T. ferrooxidans* and have found that the organism requires more frequent subculturing than *T. ferrooxidans* if maintenance in liquid ferrous iron medium is required. These observations, coupled with a lower maximum observed doubling time for iron oxidation (11 h at 30°C) than for *T. ferrooxidans*, could explain the lack of isolations of *L. ferrooxidans* from iron enrichment cultures, the usual procedure for isolation of acidophilic iron-oxidizing bacteria. We have noted spiral organisms in natural mixed cultures on pyrite (Norris and Kelly, 1978) but only recently have such organisms achieved dominance in our pyrite-maintained culture from the Los Amoles mine. This fortuitous change in composition of the pyrite-oxidizing population during growth and serial transfers in the laboratory has enabled isolation (through serial dilution and plating techniques) of an organism which resembles *L. ferrooxidans* but which has a slower growth rate (22 h iron oxidation doubling time at 30°C) under comparable conditions. As this doubling time is slower than that observed following the pyrite to ferrous iron medium transfers described earlier, it is possible that it does not represent the maximum growth rate of the organism. Initial studies of total protein electrophoresis patterns have indicated some differences between the new isolate and *L. ferrooxidans* but the patterns of the two organisms are relatively similar in comparison with the different, characteristic patterns we have obtained with *T. ferrooxidans* strains (A. L. Smith and P. R. Norris, unpublished observations).

Earlier studies of *Leptospirillum ferrooxidans* indicated that it could not oxidize sulphur compounds and thus could only degrade minerals, such as pyrite and chalcopyrite in mixed culture with a sulphur-oxidizing acidophile such as *Thiobacillus organoparus* (Balashova *et al.*, 1974), *T. ferrooxidans* or *T. acidophilus* (Norris and Kelly, 1978). Recently we have shown rapid dissolution of pyrite at similar rates by pure cultures of *L. ferrooxidans* and by the similar organism isolated from the Los Amoles mixed culture. Subculture from a pyrite-oxidizing culture of *L. ferrooxidans* into medium containing elemental sulphur or copper sulphide did not result in further growth of the organism whereas transfer (under the same conditions) of, for example, the pure *T. ferrooxidans* strain from the Chino mixed culture resulted in rapid growth and extensive sulphur or copper sulphide oxidation. Thus, the previously noted essential lack of sulphur oxidation by *L. ferrooxidans* is apparently confirmed. Serial transfer of pyrite-oxidizing *L. ferrooxidans* cultures has been readily achieved but successful initial transfer of ferrous iron-oxidizing cultures onto the solid mineral substrate has been less reproducible. Further study of mineral oxidation by the organism is required as these observations raise important questions concerning the mechanisms of microbial pyrite dissolution which, to proceed rapidly, has been considered to require bacterial iron- and sulphur-oxidation. The rates of iron release from pyrite by *L. ferrooxidans* and by the similar isolate from the Los Amoles culture were only slightly faster in the presence of sulphur-oxidizing *T. thiooxidans* (Fig. 4). However, in the absence of the sulphur-oxidizing organism, an accumulation of some elemental sulphur and less acid production occurred during pyrite dissolution by the iron-oxidizing organisms. The final pH of the culture medium after pyrite dissolution (Fig. 4) by *L. ferrooxidans* was 1.7 compared with 1.5 in the mixed culture of *L. ferrooxidans* and *T. thiooxidans*, suggesting that the latter organism oxidized sulphur as it became available through ferric iron leaching of the pyrite. The rate of release and the yield of iron from 50 g pyrite l^{-1} were similar in cultures of *L. ferrooxidans* and of the Chino mixed, enrichment culture (Fig. 5). It remains to be resolved whether, as in the earlier studies (Balashova *et al.*, 1974; Norris and Kelly, 1978), the sulphur-oxidizing organisms were participating in cooperative attack on the minerals or were oxidizing sulphur as it became available. It is also possible that they were involved in aiding the establishment of a ferric iron leaching cycle catalyzed by *L. ferrooxidans* through their oxidation and removal of

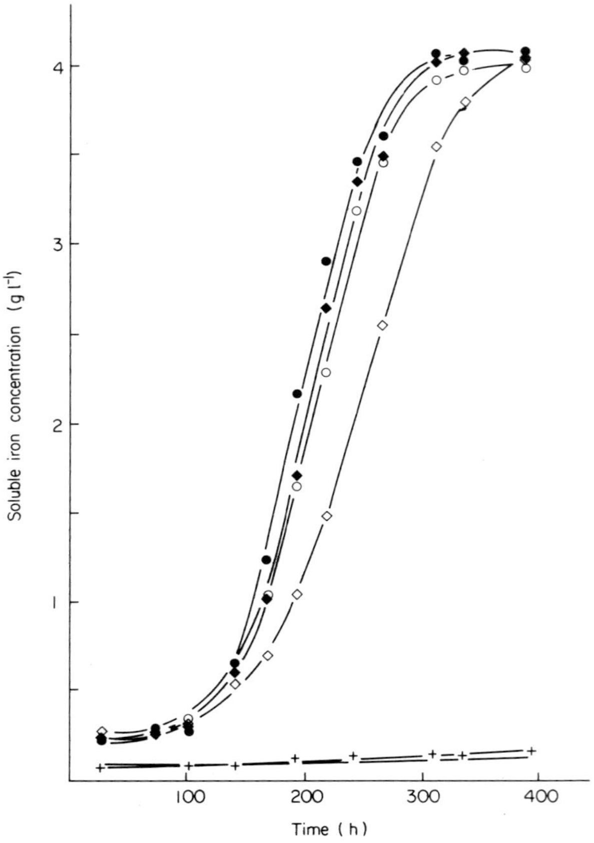

Fig. 4. Pyrite leaching by *Leptospirillum ferrooxidans* with (◆) and without (◇) *Thiobacillus thiooxidans*; by the *Leptospirillum*-like organism (from the Los Amoles culture) with (●) and without (○) *T. thiooxidans*; and soluble iron release in the presence of *T. thiooxidans* alone (+) and in a sterile control (•).

an initial passivation layer on the solid substrates (Golding *et al.*, 1977) or through provision of extracellular surface active agents *T. thiooxidans* is known to produce (Jones and Benson, 1965).

3.3 Iron-oxidizing Acidophiles Active at Low and High Temperatures

Iron-oxidizing organisms with temperature optima outside the range most frequently employed in isolations and enrichments from metal

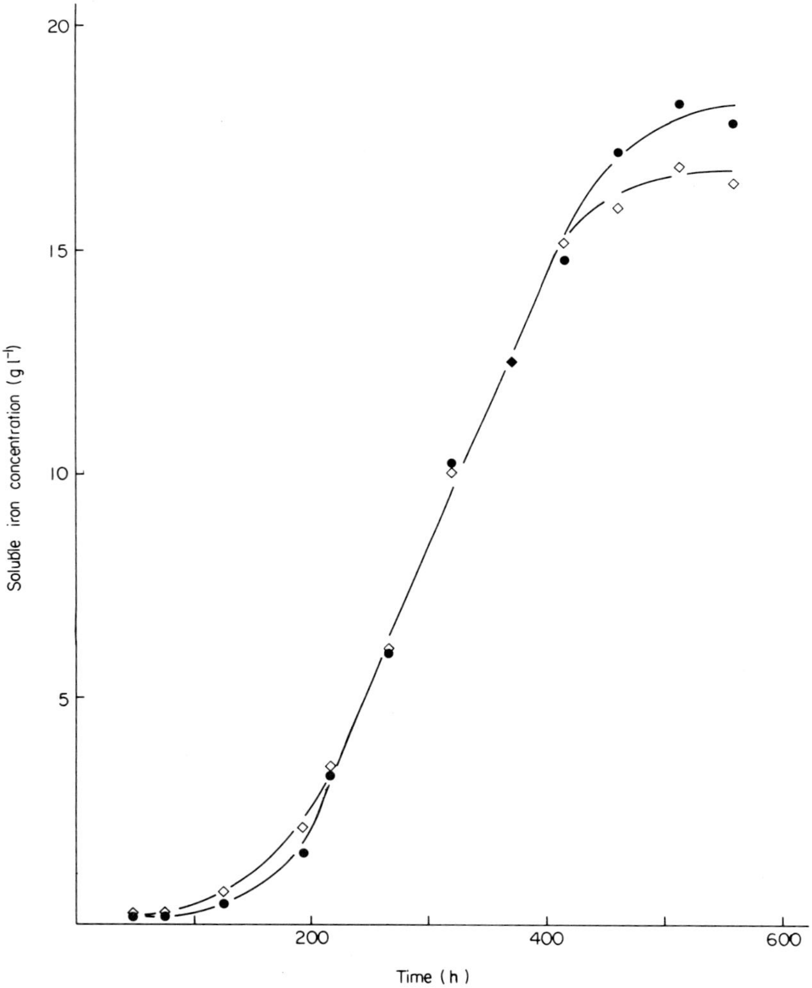

Fig. 5. Pyrite leaching by *Leptospirillum ferrooxidans* (◇) and by the Chino mixed culture (●).

leaching environments (room temperature to 30°C) have received little attention until relatively recently, yet these organisms could be useful in metal extraction operations.

Development of rotating-disc biological treatment of acid mine drainage (Olem and Unz, 1980) has shown that iron oxidation occurs at lower temperatures than expected from laboratory studies with *Thiobacillus ferrooxidans*, possibly suggesting the activity of different strains in the natural mixed cultures of the mine waters. The Hardanger isolate (p. 453) appears to be the first *T. ferrooxidans* strain described showing a reduced sensitivity to a decrease in temperature. Initial studies have given a doubling time for iron oxidation of 14.5 h at 30°C compared with 6.5 h for the characterized *T. ferrooxidans* strain and of 16.5 h at 15°C compared with 25 h for the previously characterized strain (P. R. Norris and D. P. Kelly, unpublished observations). At 30°C on iron or pyrite, the Hardanger strain grew as chains of elongated cells instead of the typical short rod morphology characteristic of *T. ferrooxidans* which it exhibited at lower temperatures. The experiment shown in Fig. 2 illustrated a poor yield of soluble iron from pyrite but the rate of iron release was enhanced by the Hardanger strain relative to that of the characterized *T. ferrooxidans* at 10°C (Fig. 6a). The iron release rate appeared to increase during incubation of the Chino mixed culture at 10°C, possibly indicating adaptation of the mixed culture or adaptation or selection of an uncharacterized organism at the lower temperature. Transfer of inocula to pyrite medium at 30°C after 1300 h at 10°C (Fig. 6b) confirmed the leaching pattern previously shown by the cultures at the higher temperature.

Isolations and subsequent studies of thermophilic, acidophilic, iron-, sulphur-, and mineral-oxidizing bacteria have been recently reviewed (Brierley, 1978; Brock, 1978; Brierley *et al.*, 1980). The potential importance of thermophiles in metal recovery could include a maintenance of biogenic leaching at the high temperatures reached in some leach dumps (Beck, 1967); a capacity to catalyse more effective leaching of some copper concentrates than occurs at lower temperatures (Brierley and Brierley, 1978; Norris *et al.*, 1980); and a capacity to leach molybdenum which is toxic to mesophilic, mineral-oxidizing organisms from molybdenite (Brierley, 1974).

Two major groups of organisms have become apparent from studies over the last 10 years: moderately thermophilic, rod-shaped bacteria with an optimum growth temperature about 50°C and extremely thermophilic, generally spherical, *Sulfolobus*-type bacteria with an optimum growth temperature about 70°C. A correlation between temperature and the composition and activity of the mineral-oxidizing microflora has been demonstrated in column leaching tests sufficiently large (containing about 1.7×10^5 kg waste rock) to allow above ambient temperatures to be maintained in the ore body following exothermic oxidation of pyrite (Murr and Brierley, 1978). The numbers of *T. ferrooxidans* fell from 1×10^7 organisms (g low grade copper-bearing ore waste)$^{-1}$ to less than 1×10^2 organisms g^{-1} in areas where the temperature reached about 55°C. Moderately thermophilic bac-

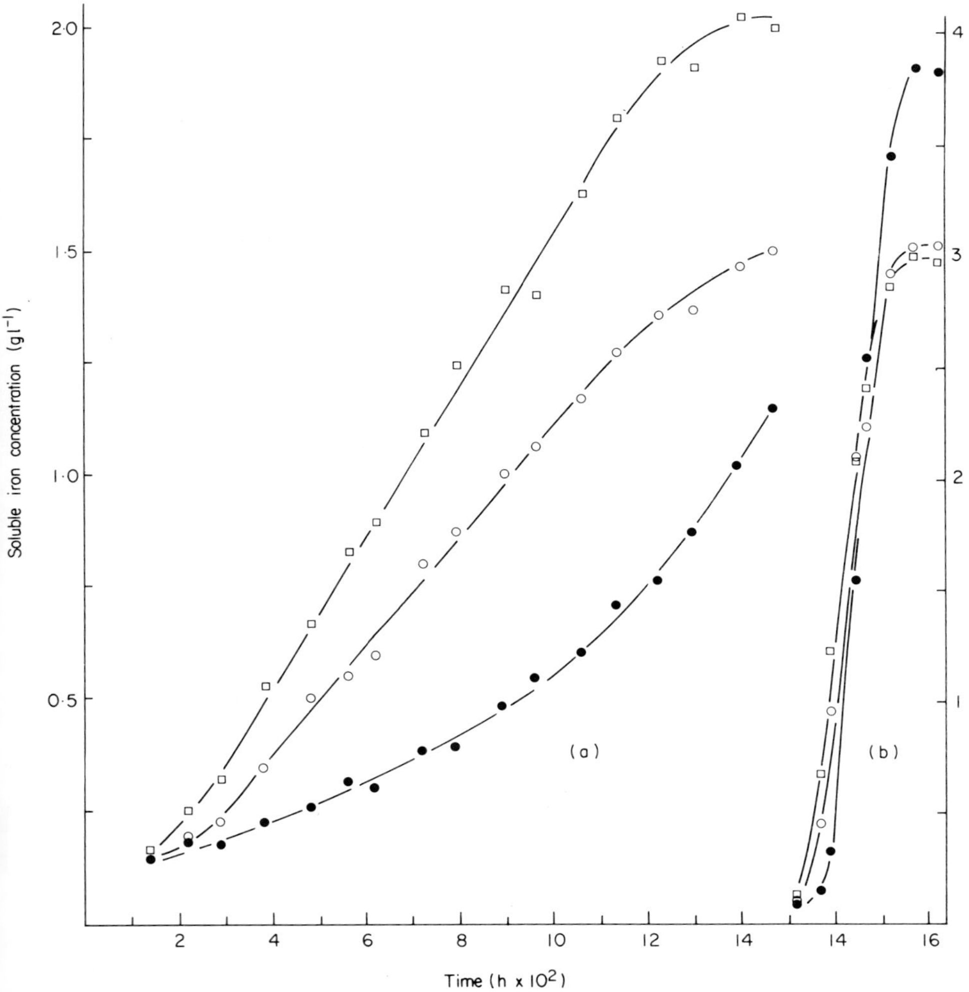

Fig. 6. Pyrite leaching (*a*) 10°C and (*b*) 30°C by the characterized strain of *Thiobacillus ferrooxidans* (○), the Hardanger strain (□), and the Chino mixed culture (●).

teria were found in the warmer areas (Brierley and Lockwood, 1977) and their activity *in situ* was suggested to account for the continuing oxygen consumption in the ore despite the decline in the numbers of *T. ferrooxidans* (Murr and Brierley, 1978). A similar correlation of temperature and the constituent microflora has been observed in hot, acid soils with organisms resembling *T. thiooxidans* present up to 55°C and *Sulfolobus acidocaldarius* active between 55 and 85°C (Fliermans and Brock,

1972). Thus, successions of different organisms or their activity might be expected with transitions to higher temperatures in sulphur-rich or mineral sulphide-rich environments or metal recovery systems, but the wide range of temperatures over which some thermophiles are active could allow interactions between moderate and extreme thermophiles, and between moderate thermophiles and mesophiles. At least two of the moderate thermophiles were almost as active at 30°C as *T. ferrooxidans* in chalcopyrite leaching in laboratory tests (Norris *et al.*, 1980). Bacteria-free filtrates from cultures of *T. ferrooxidans* have supported growth of the moderate thermophiles (Norris *et al.*, 1980) so mixed cultures of the mesophile and the thermophiles could be maintained in metal leaching systems even in the absence of autochthonous or applied organic nutrients which are required by the chemolithotrophic, heterotrophic thermophiles.

4. CONCLUSION

Thus far we hope to have demonstrated that the degradation of sulphide minerals in natural and engineered systems is more complex microbiologically than earlier studies were able to indicate. Much of our review has necessarily dealt with the properties and capacities of pure cultures of organisms rather than interactions in mixed cultures. This is inevitable in this field, in which fundamental information about the individuals in such environments is a prerequisite to understanding the actual or potential interactions in leaching. Observations on the microflora of acid mine drainage and ore dumps indicate that, although these environments tend to be extreme (with respect to acidity, metal content, and sometimes temperature) and have a limited species diversity, a sufficient variety of organisms and conditions can exist for diverse microbial interactions to occur. Thus, there is some scope for predator–prey interactions and for microbial successions with transient competition between organisms which become dominant in the different pH and temperature regimes developing within mineral- or coal-spoil dumps.

A variety of interactions related to the level and nature of organic matter in ore dumps may be envisaged. These may be dependent on whether the ecosystem is primarily "inorganic" (with, as suggested by Ehrlich (1963), primary production of organic carbon from chemolithoautotrophic carbon dioxide fixation rather than photosynthetically as in most environments) or whether it is a relatively organic-rich ecosystem, such as some uranium-bearing deposits, in which organic matter could be toxic to the mineral-oxidizing bacteria (Brierley, 1978). We have reviewed a potentially mutualistic association involving the exchange of organic material and fixed nitrogen (*Thiobacillus ferrooxidans* with *Beijerinckia lacticogenes*) and examples of organic compound toxicity alleviation (*T. acidophilus* and *Rhodotorula* sp. with *T. ferrooxidans*). Although Groudev *et al.* (1978) found no correlation between the numbers of ferrobacilli (40% of which could utilize organic compounds) and the quantity of organic matter in effluents of a copper ore leach dump, it is conceivable that further studies could reveal examples of: excreted organic matter inhibiting iron-oxidizing bacteria; cross-feeding of heterotrophs (whether these be chemolitho- or chemoorganotrophic) by chemolithotrophic bacteria releasing organic matter, possibly resulting in improved growth of the latter; and competition for essential organic nutrients between heterotrophs and mineral-degrading, chemolithotrophic-heterotrophic bacteria such as the moderate thermophiles.

Interactions between different bacterial types during sulphide mineral breakdown may also be envisaged which result in inorganic substrate recycling between the bacteria catalysing the reactions. Such a process cited earlier (Brock and Gustafson, 1976) (p. 450) could be of widespread significance in leach dumps which become oxygen-limited in their interiors. Thus, ferric iron produced by iron-oxidizing bacteria could be reduced to ferrous iron by *Thiobacillus ferrooxidans* or *Thiobacillus thiooxidans* by means of anaerobic sulphur oxidation or by the latter organism even in the presence of oxygen. A recycle system for iron oxidation could thus be established with a possible benefit from such a reaction sequence being the anaerobic removal of sulphur coatings from the surfaces of minerals attacked chemically by ferric iron in the anoxic zones of large leach dumps. Direct inorganic substrate (sulphur) provision for one group of organisms (sulphur oxidizers) may also occur simply through sulphide mineral dissolution by iron oxidizers, such as *Leptospirillum ferrooxidans*.

Study of ore dump microflora *in situ*, of which there has been very little, would be useful in assessing whether any of these potential microbial interactions involving organic or inorganic materials may influence metal recovery on the industrial scale. With the microbial activity in ore dumps and leaching operations *in situ* so far little understood, it is too early to say how stable the mixed cultures of leaching environments might be and whether cooperative microbial communities with strongly interdependent component populations occur in such environments in the manner demonstrated for some other microbial communities (Slater, 1978). Laboratory leaching data with natural and contrived mixed cultures are also so far inadequate to indicate whether such cultures are stable or significant, but interactions do occur despite the relatively more extreme and homogeneous conditions in laboratory concentrate leaching experiments in comparison with *in situ* or dump leaching environments. From the data we have reviewed and introduced, a preliminary assessment of the potential uses of mixed cultures in metal recovery can be made and the reasons for maintaining a cautious attitude in relation to such evaluations can be stressed.

The nature of the *Thiobacillus ferrooxidans* strain and of the substrate composition should be taken into account when considering the influence of *T. thiooxidans* on mineral or waste material leaching by these acidophilic bacteria. In the work we cited earlier, Ebner (1978) reported best leaching of lead-rich fly ash and slag wastes and of a jarosite from a zinc electrolysis plant by *T. thiooxidans*. Wastes of this type would presumably be leached more effectively by strains of *T. ferrooxidans* with high rather than low sulphur-oxidizing activity, with leaching by the latter strains perhaps being amenable to improvement by mixed culture with *T. thiooxidans*. The choice of *T. ferrooxidans* strain, and whether a mixed culture with *T. thiooxidans* might be beneficial, could also be significant in leaching minerals, such as chalcopyrite, for which the sulphur-oxidizing activity of iron oxidizers appears related to leaching rates (Groudev and Genchev, 1978). Interactions between these bacteria and such waste materials and minerals could be viewed as:

(1) competition between *T. ferrooxidans* and *T. thiooxidans* for sulphur readily available for either organism or for sulphur released from a mineral by initial action of the iron oxidizer or;
(2) a cooperative attack on a mineral if the *T. ferrooxidans* strain had such a low sulphur-oxidizing activity that sulphur, which could otherwise reduce leaching by coating the mineral surface, was oxidized predominantly by *T. thiooxidans*.

It is not yet clear which interactions, if any, result in the more efficient leaching of pyrite by mixed cultures than by any of the *T. ferrooxidans* strains we have isolated. Uncertainty also surrounds the general application of this phenomenon in metal recovery because of the limited range of conditions and minerals so far used and because of the slight industrial value of pyrite concentrate leaching. Although pyrite oxidation might be important for uranium recovery and potentially important for coal desulphurization, development of concentrate leaching primarily concerns more complex and recalcitrant minerals, such as chalcopyrite, which, in contrast to pyrite, usually exhibits incomplete bacterial leaching, unless part-leached material is reground. Mixed culture chalcopyrite leaching has yet to be compared with pure culture leaching under conditions giving maximum copper release and under conditions simulating potential industrial recovery of the metal. So far, therefore, the efficient mixed culture pyrite leaching may serve best as a model system for investigation of potential interactions in leaching and for gaining further understanding of the culture characteristics most important in mineral oxidation.

As the initial rate of pyrite dissolution was similar in pure cultures of *Thiobacillus ferrooxidans* strains and in mixed cultures, the prolonged maintenance of the initial iron release rate in the latter suggests better adaptation of the cultures or of certain component organisms to factors arising during growth on pyrite. Increasing acidity from sulphide and sulphur oxidation during leaching could selectively affect different strains of *T. ferrooxidans*. Strains with greater acid tolerance than those so far isolated could be present in the mixed cultures or strains which have been isolated could show a greater tolerance, perhaps of indirect effects of acidity, when they are growing in mixed culture. As the strains of *T. ferrooxidans* and *Leptospirillum ferrooxidans* show similar tolerance of acidity at least during oxidation of soluble ferrous iron (Norris and Kelly, 1978), an indirect effect of acidity on organism activity, possibly on the extent of organism interaction with the solid substrate, could contribute to the different soluble iron yields from pyrite with different cultures. Successful leaching of pyrite to completion by *T. ferrooxidans* has been described (Atkins, 1978) as requiring pH adjustment by resuspension in fresh medium when the acidity (pH 1.0) inhibited bacterial activity. In the experiments we have described, pH 1.0 was reached only after dissolution of almost 50 g pyrite l^{-1} by the Chino mixed culture. However, we have confirmed the observations of Atkins in that premature cessation of pyrite leaching by our characterized *T. ferrooxidans* strain, after the pH had decreased to 1.5, was reversed by resuspending the partly leached pyrite and the bacteria in fresh medium at pH 2.0 (P. Norris and L. A. Lockhart, unpublished observations). Resuspension of the pyrite and bacteria in fresh medium at pH 1.5, when pH 1.5 was reached during pyrite leaching, did not lead to a further release of iron by the characterized, pure *T. ferrooxidans* in contrast to an immediate continuation of leaching by the Chino mixed culture.

Similar iron release from 50 g pyrite l^{-1} by the Chino mixed culture and by *Leptospirillum ferrooxidans* occurred (Fig. 5) but, although organisms resembling *L. ferrooxidans* have been observed in the mixed cultures (with the exception of the Los Amoles serial subcultures in which they have become numerous), such organisms are usually present in very small numbers compared with the density they attain in comparable pyrite leaching by pure *L. ferrooxidans* cultures, making it unlikely that their activity could explain the efficient leaching effected by all of the mixed cultures. Again, further experiments are required to quantify the relationship of organism numbers with pyrite dissolution. *L. ferrooxidans* and the similar organism we have described, together with *T. ferrooxidans*, offer a good opportunity for competition experiments with mineral-oxidizing bacteria. *L. ferrooxidans* and *T. ferrooxidans* mixed cultures showed pyrite dissolution typical of *L. ferrooxidans* cultures rather than of pure *T. ferrooxidans* cultures (Norris and Kelly, 1978) and the *Leptospirillum*-like organism we have described appears to have been more successful than *T. ferrooxidans* in the Los Amoles pyrite serial subcultures. Whether the apparent relative competitive advantage of these iron oxidizers compared to the iron- and sulphur-oxidizing *T. ferrooxidans* would be maintained with pyrite other than that of fine particle size remains to be examined.

Apart from the possibility of different *Thiobacillus ferrooxidans* strains not yet isolated, the most likely remaining explanation for the efficiency of mixed culture leaching is that microbial interactions of an as yet unknown nature are involved. The nature of possible interactions may be revealed with further isolations of various organisms from the mixed cultures followed by their recombination under controlled conditions.

Finally, problems of the introduction of particular organisms, including natural or contrived mixed cultures to industrial scale processes must be considered. Groudev *et al* (1978) have described the introduction over a period of 2 years of a genetically-marked *Thiobacillus ferrooxidans* strain to a section of a copper leach dump. The introduced strain which possessed high activity against the copper ore in laboratory tests, apparently eventually comprised 55% of the ferrobacilli population in the dump section but there was no increase in the total ferrobacilli count or in the rate of leaching. However, the marked organisms had not retained the high mineral leaching activity of the laboratory strain. This report illustrated that introduction on a large scale is feasible but again emphasizes the lack of current understanding of the activity *in situ* of leach dump microflora. Use of particular cultures in fermenter concentrate leaching would present less of a problem of introduction but maintenance of a pure culture would be unlikely, probably precluding for example, use of an organism such as *Leptospirillum ferrooxidans* which might leach some minerals with less acid production than would occur in the presence of sulphur-oxidizing organisms. Possible greater stability of natural mixed cultures if their activity is based on "tight" microbial communities (Slater, 1978) could be useful in ensuring consistent leaching kinetics in a continuous process open to contamination.

REFERENCES

Arkesteyn, G. J. M. W. (1979). Pyrite oxidation by *Thiobacillus ferrooxidans* with special reference to the sulphur moiety of the mineral. *Antonie van Leeuwenhoek* **45**, 423–435.

Arkesteyn, G. J. M. W. (1980). Contribution of microorganisms to the oxidation of pyrite. Doctoral thesis, University of Wageningen.

Atkins, A. S. (1978). Studies on the oxidation of sulphide minerals (pyrite) in the presence of bacteria. *In* "Metallurgical Applications of Bacterial Leaching and Related Microbiological Phenomena, pp. 403–426. (Murr, L. E., Torma, A. E., and Brierley, J. A., eds.) Academic Press, New York and London.

Balashova, V. V., Vedenina, I. Ya., Markosyan, G. E., and Zavarzin, G. A. (1974). The autotrophic growth of *Leptospirillum ferrooxidans*. *Mikrobiologiya* **43**, 581–585 (English translation pp. 491–494).

Beck, J. V. (1967). The role of bacteria in copper mining operations. *Biotechnology and Bioengineering* **9**, 487–497.

Bosecker, K. (1976). Studies in the bacterial leaching of the German copper shale. Abstract Fifth International Fermentation Symposium (Dellweg, H., ed.), p. 451. Berlin.

Bosecker, K. (1977). Studies in the bacterial leaching of nickel ores. *In* "Conference, Bacterial Leaching", pp. 139–144. (Schwartz, W., ed.) Verlag Chemie, Weinheim.

Bosecker, K., Neuschütz, D., and Scheffler, U. (1978). Microbiological leaching of carbonate-rich German copper shale. *In* "Metallurgical Applications of Bacterial Leaching and Related Microbiological Phenomena", pp. 389–401. (Murr, L. E., Torma, A. E., and Brierley, J. A., eds.) Academic Press, New York and London.

Brierley, C. L. (1974). Molybdenite-leaching: use of a high temperature microbe. *Journal of the Less Common Metals* **36**, 237–247.

Brierley, C. L. (1978). Bacterial leaching. *CRC Critical Reviews in Microbiology* **5**, 207–262.

Brierley, J. A. and Brierley, C. L. (1978). Microbial leaching of copper at ambient and elevated temperatures. *In* "Metallurgical Applications of Bacterial Leaching and Related Microbiological Phenomena", pp. 477–490. (Murr, L. E., Torma, A. E., and Brierley, J. A., eds.) Academic Press, New York and London.

Brierley, J. A. and Lockwood, S. J. (1977). The occurrence of thermophilic iron-oxidizing bacteria in a copper leaching system. *FEMS Microbiology Letters* **2**, 163–165.

Brierley, C. L., Brierley, J. A., Norris, P. R., and Kelly, D. P. (1980). Metal-tolerant microorganisms of hot, acid environments. *In* "Microbial Growth and Survival in Extremes of Environment", pp. 39–51. (Gould, G. W. and Corry, J. E. L., eds.) Academic Press, London and New York.

Brock, T. D. (1978). "Thermophilic Microorganisms and Life at High Temperatures". Springer-Verlag, New York.

Brock, T. D. and Gustafson, J. (1976). Ferric iron reduction by sulfur- and iron-oxidizing bacteria. *Applied and Environmental Microbiology 32*, 567–571.

Charley, R. C. and Bull, A. T. (1979). Bioaccumulation of silver by a multispecies community of bacteria. *Archives of Microbiology* **123**, 239–244.

Dugan, P. R. and Apel, W. A. (1978). Microbial desulfurization of coal. *In* "Metallurgical Applications of Bacterial Leaching and Related Microbiological Phenomena", pp. 223–250. (Murr, L. E., Torma, A. E., and Brierley, J. A., eds.) Academic Press, New York and London.

Dugan, P. R., MacMillan, C. B., and Pfister, R. M. (1970a). Aerobic heterotrophic bacteria indigenous to pH 2.8 acid mine water: microscopic examination of acid streamers. *Journal of Bacteriology* **101**, 973–981.

Dugan, P. R., MacMillan, C. B., and Pfister, R. M. (1970b). Aerobic heterotrophic bacteria indigenous to pH 2.8 acid mine water: predominant slime-producing bacteria in acid streamers. *Journal of Bacteriology* **101**, 982–988.

Dutrizac, J. E. and MacDonald, R. J. C. (1974). Ferric iron as a leaching medium. *Minerals Science Engineering* **6**, 59–100.

Ebner, H. G. (1978). Metal recovery and environment protection by bacterial leaching of inorganic waste materials. *In* "Metallurgical Applications of Bacterial Leaching and Related Microbiological Phenomena", pp. 195–206. (Murr, L. E., Torma, A. E., and Brierley, J. A., eds.) Academic Press, New York and London.

Ehrlich, H. L. (1963). Microorganisms in acid drainage from a copper mine. *Journal of Bacteriology* **86**, 350–352.

Ehrlich, H. L. and Fox, S. I. (1967). Environmental effects on bacterial copper extraction from low-grade copper sulfide ores. *Biotechnology and Bioengineering* **9**, 471–485.

Fliermans, C. B. and Brock, T. D. (1972). Ecology of sulfur-oxidising bacteria in hot acid soils. *Journal of Bacteriology* **111**, 343–350.

Friedmann, B. A. and Dugan, P. R. (1967). Concentration and accumulation of metallic ions by the bacterium, *Zoogloea*. *Developments in Industrial Microbiology* **9**, 381–388.

Golding, R. M., Harris, B., Ralph, B. J., Richard, P. A. D., and Vanselow, D. G. (1977). The nature of the passivation film on covellite exposed to oxygen. *In* "Conference, Bacterial Leaching", pp. 191–200. (Schwartz, W., ed.) Verlag Chemie, Weinheim.

Goroll, D. (1976). Ökologie von *Thiobacillus neapolitanus* und seine mögliche Mitwirkung im Leaching-Prozess. *Zeitschrift für Allgemeine Mikrobiologie* **16**, 3–7.

Groudev, S. (1979). Mechanism of bacterial oxidation of pyrite. *Mikrobiologija* (*Acta Biologica Iugoslavica*) **16**, 75–87.

Groudev, S. N. and Genchev, F. N. (1978). Mechanisms of bacterial oxidation of chalcopyrite. *Mikrobiologija* (*Acta Biologica Iugoslavica*) **15**, 139–152.

Groudev, S. N. and Genchev, F. N. (1979). Microbial coal desulphurization: effect of the cell adaptation and mixed cultures. *Comptes rendus de l'Academie bulgare des Sciences* **32**, 353–355.

Groudev, S. N. and Genchev, F. N., and Gaidarjiev, S. S. (1978). Observations on the microflora in an industrial dump leaching copper operation. *In* "Metallurgical Applications of Bacterial Leaching and Related Microbiological Phenomena", pp. 253–274. (Murr, L. E., Torma, A. E., and Brierley, J. A., eds.) Academic Press, New York and London.

Guay, R. and Silver, M. (1975). *Thiobacillus acidophilus* sp. nov.; isolation and some physiological characteristics. *Canadian Journal of Microbiology* **21**, 281–288.

Guay, R., Silver, M., and Torma, A. E. (1976). Base composition of DNA isolated from *Thiobacillus ferrooxidans* grown on different substrates. *Revue Canadienne de Biologie* **35**, 61–67.

Harrison, A. P. (1978). Microbial succession and mineral leaching in an artificial coal spoil. *Applied and Environmental Microbiology* **36**, 861–869.

Ilyaletdinov, A. N., Enker, P. B., and Loginova, L. V. (1977). Role of sulfate-reducing bacteria in the precipitation of copper. *Mikrobiologiya* **46**, 113–117. (English translation pp. 92–95.)

Jackson, T. A. (1978). The biogeochemistry of heavy metals in polluted lakes and streams at Flin Flon, Canada, and a proposed method for limiting heavy-metal pollution of natural waters. *Environmental Geology* **2**, 173–189.

Jennett, J. C., Bolter, E., Gale, N., Tranter, W., and Hardie, M. (1975). The Viburnum Trend, southeast Missouri: the largest lead-mining district in the world—environmental effects and controls. *In* "International Symposium on Minerals and the Environment", pp. 13–26. Institute of Mining and Metallurgy, London.

Jones, G. E. and Benson, A. A. (1965). Phosphatidyl glycerol in *Thiobacillus thiooxidans*. *Journal of Bacteriology* **89**, 260–261.

Karavaiko, G. I., Kuznetsov, S. I., and Golomzik, A. I. (1977). "The Bacterial Leaching of Metals from Ores". Technicopy, Stonehouse.

Kelly, D. P. (1976). Extraction of metals from ores by bacterial leaching: present status and future prospects. *In* "Microbial Energy Conversion", pp. 329–338. (Schlegel, H. G. and Barnea, J., eds.) E. Goltze, Göttingen.

Kelly, D. P. (1980). The sulphur cycle: definitions, mechanisms and dynamics. *In* "Sulphur in Biology", pp. 3–18. Ciba Foundation Symposium 72, Elsevier, Amsterdam.

Kelly, D. P. and Tuovinen, O. H. (1972). Recommendation that the names *Ferrobacillus ferrooxidans* Leathen and Braley and *Ferrobacillus sulfooxidans* Kinsel be recognized as synonyms of *Thiobacillus ferrooxidans* Temple and Colmer. *International Journal of Systematic Bacteriology* **22**, 170–172.

Kelly, D. P., Norris, P. R., and Brierley, C. L. (1979). Microbiological methods for the extraction and recovery of metals. *In* "Microbial Technology: Current State, Future Prospects", pp. 263–308. (Bull, A. T., Ratledge, C. and Ellwood, D. C., eds.) Cambridge University Press.

Kinsel, N. A. (1960). New sulfur oxidizing iron bacterium: *Ferrobacillus sulfooxidans* sp. n.: *Journal of Bacteriology* **80**, 628–632.

Kleinmann, R. L. P. and Crerar, D. A. (1979). *Thiobacillus ferrooxidans* and the formation of acidity in simulated coal mine environments. *Geomicrobiology Journal* **1**, 373–388.

Krouse, H. R. and McCready, R. G. L. (1979). Reductive reactions in the sulfur cycle. *In* "Biogeochemical Cycling of Mineral-Forming Elements", pp. 315–368. (Trudinger, P. A. and Swaine, D. J., eds.) Elsevier, Amsterdam.

Landesman, J., Duncan, D. W., and Walden, C. C. (1966). Iron oxidation by washed cell suspensions of the chemoautotroph, *Thiobacillus ferrooxidans*. *Canadian Journal of Microbiology* **12**, 25–33.

Le Roux, N. W., Dacey, P. W., and Temple, K. L. (1980). The microbiological role in pyrite oxidation at alkaline pH. *In* "Biogeochemistry of Ancient and Modern Environments", pp. 515–520. (Trudinger, P. A., Walter, M. R., and Ralph, B. J., eds.) Australian Academy of Science, Canberra.

Leathen, W. W., Kinsel, N. A., and Braley, S. A. (1956). *Ferrobacillus ferrooxidans*: a chemosynthetic autotrophic bacterium. *Journal of Bacteriology* **72**, 700–704.

Lundgren, D. G. and Silver, M. (1980). Ore leaching by bacteria. *Annual Review of Microbiology* **34**, 263–283.

Mackintosh, M. E. (1978). Nitrogen fixation by *Thiobacillus ferrooxidans*. *Journal of General Microbiology* **105**, 215–218.

Madgwick, J. C. and Ralph, B. J. (1977). The metal-tolerant alga *Hormidium fluitans* (Gay) Heering from acid mine drainage waters in Northern Australia and Papua-New Guinea. *In* "Conference, Bacterial Leaching", pp. 85–91. (Shwartz, W., ed.) Verlag Chemie, Weinheim.

Manning, H. L. (1975). New medium for isolating iron-oxidizing and heterotrophic acidophilic bacteria from acid mine drainage. *Applied Microbiology* **30**, 1010–1016.

Margalith, P., Silver, M., and Lundgren, D. G. (1966). Sulfur oxidation by the iron bacterium *Ferrobacillus ferrooxidans*. *Journal of Bacteriology* **92**, 1706–1709.

Markosyan, G. E. (1973). A new mixotrophic sulfur bacterium developing in acid media, *Thiobacillus organoparus* sp. n. *Doklady̆ Akademii Nauk SSSR* **211**, 1205–1208. (English translation, pp. 318–320.)

McElroy, R. O. and Bruynesteyn, A. (1978). Continuous biological leaching of chalcopyrite concentrates: demonstration and economic analysis. *In* "Metallurgical Applications of Bacterial Leaching and Related Microbiological Phenomena", pp. 441–462. (Murr, L. E., Torma, A. E., and Brierley, J. A., eds.) Academic Press, New York and London.

Murr, L. E. and Brierley, J. A. (1978). The use of large scale test facilities in studies of the role of microorganisms in commercial leaching operations. *In* "Metallurgical Applications of Bacterial Leaching and Related Microbiological Phenomena", pp. 491–520. (Murr, L. E., Torma, A. E., and Brierley, J. A., eds.) Academic Press, New York and London.

Murr, L. E., Torma, A. E., and Brierley, J. A. (1078). "Metallurgical Applications of Bacterial Leaching and Related Microbiological Phenomena". Academic Press, New York and London.

Napier, E. and Wakerley, D. S. (1969). Fungal growth and iron reduction in Swedish alum shales. *Nature, London* **223**, 289–290.

Nerkar, D. P., Kumta, U. S., and Lewis, N. F. (1977). Pyruvate inhibition of *Ferrobacillus ferrooxidans* reversed by *Rhodotorula* yeast. *Journal of Applied Bacteriology* **43**, 117–121.

Norris, P. R. and Kelly, D. P. (1978). Dissolution of pyrite (FeS_2) by pure and mixed cultures of some acidophilic bacteria. *FEMS Microbiology Letters* **4**, 143–146.

Norris, P. R. and Kelly, D. P. (1979). Accumulation of metals by bacteria and yeasts. *Developments in Industrial Microbiology* **20**, 299–308.

Norris, P. R., Brierley, J. A., and Kelly, D. P. (1980). Physiological characteristics of two facultatively thermophilic mineral-oxidising bacteria. *FEMS Microbiology Letters* **7**, 119–122.

Olem, H. and Unz, R. F. (1980). Rotating-disc biological treatment of acid mine drainage. *Journal Water Pollution Control Federation* **52**, 257–269.

Olson, G. J., Turbak, S. C., and McFeters, G. A. (1979). Impact of western coal mining. II. Microbiological studies. *Water Research* **13**, 1033–1041.

Pol'kin, S. I., Panin, V. V., Adamov, E. V., Karavaiko, G. I., and Chernyak, A. S. (1975). Theory and practice of utilizing microorganisms in processing difficult-to-dress ores and concentrates. 11th International Mineral Processing Congress, Cagliari, Italy, pp. 1–23.

Ralph, B. J. (1979). Oxidative reactions in the sulfur cycle. *In* "Biogeochemical Cycling of Mineral-Forming Elements", pp. 369–400. (Trudinger, P. A. and Swaine, D. J., eds.) Elsevier, Amsterdam.

Rangachari, P. N., Krishnamachar, V. S., Pail, S. G., Sainani, M. N., and Balakrishnan, H. (1978). Bacterial leaching of copper sulfide ores. *In* "Metallurgical Applications of Bacterial Leaching and Related Microbiological Phenomena", pp. 427–439. (Murr, L. E., Torma, A. E., and Brierley, J. A., eds.) Academic Press, New York and London.

Satake, K. (1977). Microbial sulphate reduction in a volcanic acid lake having pH 1.8 to 2.0. *Japanese Journal of Limnology* **1**, 33–35.

Schwartz, W. (1977). "Conference, Bacterial Leaching". Verlag Chemie, Weinheim.

Shumate, S. E., Strandberg, G. W., and Parrot, J. R. (1978). Biological removal of metals from aqueous process streams. *In* "Biotechnology in Energy Production and Conservation", pp. 13–20. (Scott, C. D., ed.) Wiley, New York.

Silver, M. and Torma, A. E. (1974). Oxidation of metal sulfides by *Thiobacillus ferrooxidans* grown on different substrates. *Canadian Journal of Microbiology* **20**, 141–147.

Slater, J. H. (1978). The role of microbial communities in the natural environment. *In* "The Oil Industry and Microbial Ecosystems", pp. 137–154. (Chater, K. W. A. and Somerville, H. J., eds.). Institute of Petroleum, London.

Summers, A. O. and Silver, S. (1978). Microbial transformations of metals. *Annual Review of Microbiology* **32**, 637–672.

Temple, K. L. and Colmer, A. R. (1951). The autotrophic oxidation of iron by a new bacterium: *Thiobacillus ferrooxidans*. *Journal of Bacteriology* **62**, 605–611.

Tomizuka, N. and Yagisawa, M. (1978). Optimum conditions for leaching of uranium and oxidation of lead sulfide with *Thiobacillus ferrooxidans* and recovery of metals from bacterial leaching solution with sulfate-reducing bacteria. *In* "Metallurgical Applications of Bacterial Leaching and Related Microbiological Phenomena", pp. 321–344. (Murr, L. E., Torma, A. E., and Brierley, J. A., eds.) Academic Press, New York and London.

Torma, A. E. (1977). The role of *Thiobacillus ferrooxidans* in hydrometallurgical processes. *In* "Advances in Biochemical Engineering", vol. 6, pp. 1–37. (Ghose, T. K., Fiechter, A., and Blakebrough, N., eds.) Springer-Verlag, Berlin.

Torma, A. E. (1978). Complex lead sulfide concentrate leaching by microorganisms. *In* "Metallurgical Applications of Bacterial Leaching and Related Microbiological Phenomena", pp. 357–387. (Murr, L. E., Torma, A. E., and Brierley, J. A., eds.) Academic Press, New York and London.

Trudinger, P. A. (1976). Microbiological processes in relation to ore genesis. *In* "Handbook of Strata-bound and Stratiform Ore Deposits", pp. 135–190. (Wolfe, K. H., ed.) Elsevier, Amsterdam.

Trudinger, P. A., Lambert, I. B., and Skyring, G. W. (1972). Biogenic sulfide ores: a feasibility study. *Economic Geology* **67**, 1114–1127.

Tsuchiya, H. M. (1978). Microbial leaching of Cu–Ni sulfide concentrate. *In* "Metallurgical Applications of Bacterial Leaching and Related Microbiological Phenomena", pp. 365–373. (Murr, L. E., Torma, A. E., and Brierley, J. A., eds.) Academic Press, New York and London.

Tsuchiya, H. M., Trivedi, N. C., and Schuler, M. L. (1974). Microbial mutualism in ore leaching. *Biotechnology and Bioengineering* **16**, 991–995.

Tuovinen, O. H. and Kelly, D. P. (1972). Biology of *Thiobacillus ferrooxidans* in relation to the microbiological leaching of sulphide ores. *Zeitschrift für Allgemeine Mikrobiologie* **12**, 311–346.

Tuovinen, O. H. and Kelly, D. P. (1974). Use of microorganisms for the recovery of metals. *International Metallurgical Reviews* **19**, 21–31.

Tuovinen, O. H., Kelly, D. P., Dow, C. S., and Eccleston, M. (1978). Metabolic transitions in cultures of acidophilic thiobacilli. *In* "Metallurgical Applications of Bacterial Leaching and Related Microbiological Phenomena", pp. 61–81. (Murr, L. E., Torma, A. E., and Brierley, J. A., eds.) Academic Press, New York and London.

Tuttle, J. H. and Dugan, P. R. (1976). Inhibition of growth, iron and sulfur oxidation in *Thiobacillus ferrooxidans* by simple organic compounds. *Canadian Journal of Microbiology* **22**, 719–730.

Walsh, F. and Mitchell, R. (1972). A pH-dependent succession of iron bacteria. *Environmental Science and Technology* **6**, 809–812.

13

Carbon Mineralization by Mixed Cultures

R. G. Burns

1. INTRODUCTION

Carbon mineralization is defined strictly as the decomposition of carbon-containing organic matter to produce carbon dioxide or methane. Thus, it parallels biogeochemical processes in the nitrogen and sulphur cycles where ammonia and hydrogen sulphide are evolved, respectively. This chapter adopts a rather narrower view of mineralization and concentrates upon the predominantly extracellular events which occur during the saprophytic degradation of carbohydrate polymers and associated carbon-containing structures. Thus, the absorption and intracellular catabolism of relatively low molecular weight carbon compounds will not be considered in any detail nor will the enzymic activities of pathogenic microorganisms and the autolytic reactions concerned with microbial morphogenesis.

Nevertheless, any self-imposed constraint based upon the location of enzyme-substrate interactions or the physiological role of the enzyme must be somewhat flexible for a number of reasons. For example, microorganisms have variable capabilities in terms of their absorption of substrates either due to cell wall porosity (e.g. the cell walls of Gram-positive bacteria have much larger pores than those of Gram-negative bacteria) or to the presence of specific active transport mechanisms. As a consequence, a potential substrate which is absorbed and metabolized by one species may require additional extracellular transformations before it can be utilized by another microorganism. An appropriate example is that some bacteria may find cellobiose or even glucose oligomers suitable as intracellular substrates; others may require monomeric glucose.

The molecular size and solubility of a substrate are not the only factors determining its ease of absorption. Stereochemical and in particular ionogenic properties are also important and may fluctuate according to the immediate environment. In soils and sediments, the clay and humic components have an extensive unit surface area (up to 800 m^2 (g soil colloid)$^{-1}$) and are predominantly negative charged. As a result organic moieties tend to accumulate at the solid–liquid interface where a host of adsorptive phenomena will influence their availability as substrates and, therefore, their rate of mineralization (Burns, 1979, 1980). As a consequence, the rates and routes of mineralization *in vivo* may bear little relationship to that recorded *in vitro*. This is especially true for ionic substrates—proteins, amino acids and a host of xenobiotic compounds—but is also valid for carbohydrates.

Another factor to consider when defining the boundaries of extracellular mineralization is that many intracellular hydrolytic enzymes become externalized following the death and lysis of their parent cell. For example, urease, a key carbon and nitrogen cycle enzyme, is cytoplasmic in terms of its original functional location, yet will survive and retain its activity for long periods in cell debris or if complexed with soil clay and humic colloids. Thus, urea mineralization (and, therefore, the final stages of purine and pyrimidine breakdown) may occur extracellularly even though the substrate has a low molecular weight and is highly soluble, and despite the presence of large numbers of ureolytic microorganisms.

Finally, one needs to define extracellular, a term which means different things to different microbiologists. Some (Glenn, 1976) have suggested that all catalyses outside the cytoplasm should be regarded as extracellular, including events within the periplasmic space of Gram-negative bacteria and those within the outer wall. Others regard enzymes attached to and projecting from the cell wall as extracellular whilst still others would prefer to restrict the adjective to enzymes which are spatially removed from their cellular origins: that is, have diffused away from their parent cells during normal growth and reproduction (Pollock, 1962). These various interpretations are further confused by such factors as the changing location of many enzymes with time: periplasmic enzymes may leak during growth, cytoplasmic enzymes may escape from lysed cells and at any particular moment many truly extracellular enzymes may be *en route* to the outer membrane and thence to the external environment. In addition, the same enzyme may not have the same location in different species: extracellular enzymes (e.g. penicillinase) in Gram-positive bacteria may be periplasmic in Gram-negative organisms. Indeed free extracellular enzymes are uncommon in Gram-negative species (Lory and Colliers, 1978; Pavlovskis and Wretlind, 1978) and most of the secreted enzymes are found in the periplasmic space or the outer membrane.

For the purposes of this chapter extracellular enzymes are regarded as those which catalyse extracellular events and, therefore, have at least some physical contact with the ambient medium. I shall, however, be less rigid when it is uncertain exactly where a reaction has occured or when an intracellular event is an essential step within a predominantly extracellular sequence.

2. CARBON-CONTAINING SUBSTRATES: ORIGINS AND COMPOSITION

2.1 Introduction

Enormous quantities of organic debris are deposited each year into terrestrial and aquatic environments: plant litter alone accounts for between 1.0 and 15.3 tonnes of organic matter hectare $^{-1}$ annum^{-1} (Williams and Gray, 1974) and the root systems of green plants produce cell debris and soluble organic materials during growth. Estimates of root biomass in the top 30 cm of soil are of the order 440 to 1575 g m^{-2} (Dickinson, 1974). In addition, animals and their excreta and dead microbial cells provide a significant residue that, like the plant fractions, is degraded by the microflora in order to maintain the steady-state condition between carbon assimilation and mineralization. Estimates of soil microbial biomass range from 50 to 3710 μg biomass carbon (g soil)$^{-1}$ (Jenkinson and Ladd, 1981). One estimate of fungal biomass associated with decomposing pine needles is 4 to 5 mg (g plant material)$^{-1}$, a figure equivalent to 3000 to 4000 m mycelium g^{-1} (Berg and Soderstrom, 1979).

A large proportion of the annual input of organic matter is carbon (*ca.* 3.0 × 10^{10} tonnes—Norkrans, 1967)—often as simple sugars (hexoses and pentoses), their derivatives (e.g. amino sugars, uronic acids, sugar alcohols, methylated sugars, deoxysugars) or as their appropriate polymers. Carbon also occurs as a component of lignin, urea, alcohols, fatty acids, purine, and pyrimidine bases and lipids together with a host of xenobiotic compounds. Somewhere between 50% and 75% of the dry weight of plant tissues is carbohydrate, while microbial cells contain up to 60% organic carbon (Lieth, 1975). The precise amount of carbon in, as well as its proportional contribution to, a cell, tissue, or entire organism is determined by species and age. For instance, mature plants have a higher percentage of cellulose, hemicellulose, and lignin even though water-soluble carbohydrates may predominate in young plants.

The principal substrates for carbon cycle microorganisms are listed in Table 1 and their origins and chemical structure outlined below.

2.2 Plant Carbon

2.2.1 Cellulose

Cellulose is the most abundant polymeric constituent of plant material. Mature wood contains 40 to 50% cellulose, leaves 10% and cotton 98% and it has been estimated that the annual global production is somewhere in the region of 1×10^{11} tonnes. A significant proportion of this cellulose (*ca.* 5 to 10%) occurs as municipal, industrial, and agricultural waste and it is not surprising that great efforts have been made to understand the physical, chemical, and microbiological factors which influence its decay. Increasingly cellulosic waste is seen as a renewable resource (Gaden *et al.*, 1976; Tsao, 1978; Ghose and Ghosh, 1979) and recent studies have emphasized the potential of the polysaccharide not only as a source of methane and fermentable sugars (Ladisch, 1979), but also as a substrate for microbial biomass or single cell protein production (Flickinger and Tsao, 1978). As a result, polysaccharide microbiology and biochemistry is dominated by cellulose, and it is appropriate that a discussion of carbon mineralization should be centred upon this plentiful substrate.

Cellulose is a linear homopolymer composed of β-(1-4) linked glucose residues. In the case of cotton, as many as 10 000 residues make up the cellulose molecule indicating a molecular weight for native cellulose of $>1.5 \times 10^6$ and a total length in the region of 5 μm. Model cellulose substrates, such as filter paper, have a much lower degree of polymerization (500 to 2000 glucose units) as does the cellulose of mature wood (1500 to 2100 glucose units).

In the plant cell wall individual cellulose molecules associate to form

Table 1. Principal carbon-containing polymers of plant, animal, and microbial organic matter.

Polymer	Predominant Structure (where known) and Component Monomers	Origin
Cellulose	β-(1-4)-D-glucan	Plants, fungi and a few bacteria
Hemicellulose	β-(1-4)-D-xylan xyloglucan arabinogalactan glucomannan arabinoxylan galactoarabinoxylan glucoarabinoxylan	Plants
Pectin	α-(1-4)-D-galacturonan arabinogalactan rhamnogalacturonan xyloglucan	Plants and fungi
Starch (amylose)	α-(1-4)-D-glucan	Plants and fungi
Starch (amylopectin)	α-(1-4)-D-glucan with α-(1-6)-D linked branches	
Inulin	β-(2-1)-D-fructosan	Plants
Fructosan	β-(2-6)-D-fructosan	Plants
Lignin	coniferyl, sinapyl and *p*-coumaryl alcohols	Plants
Glycoprotein	L-arabinose D-galactose D-glucose D-mannose L-rhamnose uronic acids	Animals, plants and microorganisms
Chitin	β-(1-4)-N-acetylglucosamine	Animals (arthropods), fungi and a few protozoa and diatoms
Chitosan	β-(1-4)-glucosamine	Fungi (Zygomycotina)
Glycogen	α-(1-4)-D-glucan α-(1-4)-D-glucan with α-(1-6)-D linked branches	Plants, fungi and animals
Peptidoglycan	N-acetylglycosamine with β-(1-4)-D linked N-acetylmuramic acid	Bacteria

Table 1 (*Continued*)

Polymer	Predominant Structure (where known) and Component Monomers	Origin
Teichoic acid	Glycerol or ribitol phosphate containing sugar, amino sugar and D-alanine substituents	Bacteria (Gram-positive)
Lipopolysaccharides	D-abequose D-glucose D-galactose D-mannose L-rhamnose L-fucose N-acetylglycosamine L-glycero-D-mannoheptose 3-deoxy-D-mannooctulosonate + D-glucosaminyl β-(1-4) or β-(1-6)-D glucosamine	Bacteria (Gram-negative)
Glucans	α-(1-3)-D-glucan α-(1-6)-D-glucan α-(1-3)-D-glucan with α-(1-4) and α-(1-6)-D linked branches	Fungi and bacteria (e.g. *Leuconostoc* spp.)
Mannans	α-(1-6)-D-mannan with α-(1-3)-D and α-(1-2)-D linked branches	Fungi
Dextrans	α-(1-6)-D-glucan	Bacteria (Lactobacteriaceae)
Levans	β-(2-6)-D-fructan	Bacteria (e.g. pseudomonads, *Bacillus* spp.)
Pullulan	α-(1-4)-D and α-(1-6)-D maltotriose	Fungi (*Aureobasidium pullulans*)
Xanthans	β-(1-4)-D-glucan with α-(1-3) linked side chains of mannose and glucuronic acid	Bacteria (*Xanthomonas* spp.)
Alginate	β-(1-4)-D-mannuronic acid and α-(1-4)-L-guluronic acid	Brown algae and a few bacteria

microfibrils which, in turn, aggregate to produce fibrils. The fibrils are stabilized by the formation of hydrogen bonds between the hydroxyl groups of adjacent chains. Some areas of the fibril are tightly bound and are known as ordered or crystalline regions whilst more loosely associated zones are described as amorphous or paracrystalline. This distinction between physically different regions of the cellulose is important when considering the accessibility of the substrate to cellulolytic microorganisms and cellulases. Cellulose structure is discussed in detail by Sihtola and Neimo (1975), Tsao *et al.* (1978), Cowling and Brown (1979), and Fan *et al.* (1980).

2.2.2 Hemicelluloses

Hemicelluloses are a diverse group of alkali soluble polysaccharides structurally associated with cellulose in plant cell walls. Mature wood may contain greater than 30% (w/w) hemicellulose, whilst straw contains about 20% (w/w). Hydrolysis of hemicullulose yields a mixture of hexos sugars (e.g. D-galactose, D-mannose), pentose sugars (D-xylose, L-arabinose) and uronic acids (glucuronic acid, galacturonic acid). Xylose polymers (xylans) may account for 7 to 30% of the plant weight whereas wheat straw hemicellulose is 90% (w/w) xylan. The occurrence and biochemistry of many of the hemicellulose polymers has been reviewed by Dekker and Richards (1976).

2.2.3 Pectin

Pectin substances form a minor component of the cell walls of higher plants rarely contributing more than 5% of the total weight. Pectins are complex polysaccharides composed of galacturonic acid moieties in which the carboxyl groups may be esterified to various degrees with methyl groups. Pectic substances are often divided into pectins, pectinic acids, protopectin, and pectic acids according to their water solubility and the prevalence of methyl ester linkages.

2.2.4 Starch

Starch is a linear or branched glucan serving as a reserve food source for plants and stored in roots (tubers), stems (corms), and swollen leaf structures (bulbs). It is also found in high concentrations in cereal grains (e.g. barley, maize, wheat, oats) and is second only to cellulose as the most common hexose polymer in the plant world. Starch is composed of an essentially linear structure (in which the glucose dimer, maltose, is linked in the α-(1-4) position) as well as of side chains attached through α-(1-6) linkages. The former is known as amylose; the latter as amylopectin.

Agar and alginates

Agar is a complex polysaccharide found in marine red algae and is composed of a neutral agarose fraction and an ionic agaropectin fraction. Both components are linear polymers and thought to consist of alternating units of β-(1-3)-D-galactose and α-(1-4)-3,6-anhydro-L-galactose. The galactoside units of the agaropectin may be methylated or contain pyruvic or sulphuric acid residues (Duckworth and Yaphe, 1971; Izumi, 1972). Porphyran and carrageenan are closely related red algal polysaccharides but which are sulphated to different degrees (Percival and McDowell, 1967).

Alginates are major structural polysaccharides of marine brown seaweeds (e.g. the genera *Laminaria*, *Ascophyllum*, *Macrocystis*) but are also deposited as exopolysaccharides by certain bacteria (e.g. *Azotobacter vinelandii, Pseudomonas aeruginosa*). The function of bacterial alginates may be to protect the cell from dehydration (Carlson and Matthews, 1966), to prevent heavy metals from entering the cell (Den Dooren de Jong, 1971) or as a diffusion barrier to oxygen (Postgate, 1974). Alginate synthesis and structure have been discussed recently by Jarman (1979).

Lignin

Lignin is an important carbon-containing constituent of vascular plants and forms some 15 to 35% by weight of wood. Predictably, lignin composition and quantity vary with the age and species of the plant: mature plants have more lignified tissue than young plants; conifers generally contain more lignin than do hardwood species. Lignins are polymers of coniferyl, sinapyl and *p*-coumaryl alcohols joined by a variety of intermonomer linkages. Ether bonds are the dominant linkages but alkyl-alkyl, alkyl-aryl and aryl-aryl also exist.

2.3 Animal Carbon

Animal tissues generally contain a smaller proportion of carbohydrate than do those of plants and microorganisms. Chitin, however, is a major organic component of arthropod exoskeletons and it has been estimated that a single species of crab produces millions of tonnes of chitin per year. The structure of this abundant carbon (and nitrogen) source is discussed on p. 483.

A second important animal carbohydrate is glycogen. This substance is similar in both structure and function to the starch of plant cells. In other words, it is a polymer composed of α-(1-4) and α-(1-6) linked glucose units and acts as a medium for energy storage.

A large number of animal carbohydrates occur as glycoproteins. Glycoproteins are proteins to which carbohydrates are linked through glycosidic bonds. The carbohydrate moiety varies in size from mono- to polysaccharide. Glycoproteins are common constituents of animals but are also found in some plants and microorganisms. They are also extremely diverse in structure and function, for example, virtually all the proteins of human plasma (excepting albumin) are glycoproteins and glycoproteins have structural (collagen, hyaluronic acid, chondroitin sulphates, heparin), protective (immunoglobulins), hormonal (thyroglobulin), and enzymic (acetylcholinesterase) functions. They also serve as food reserves (casein, ovalbumin).

Sugars which are commonly found in glycoproteins include: galactose, mannose, glucose, L-fucose, N-acetyl-glucosamine, N-acetyl-galactosamine, xylose, and L-arabinose. The proportion of sugar varies from 0.5% in some collagens to greater than 80% in blood-group substances of unknown function. Animal carbohydrate is also contained in nucleic acids and glycosides.

2.4 Microbial Carbon

2.4.1 Chitin and chitosan

It is often stated that chitin is second only to cellulose in abundance and it was once estimated that marine copepods alone produce 1×10^9 tonnes of chitin per year (Tracey, 1957) and that cellulose and chitin together account for 1×10^{10} to 1×10^{12} tonnes of the annual output of new carbohydrate. Within the animal kingdom the exoskeletons of arthropods (insects, spiders, crabs, lobsters) may contain up to 80% chitin (Jeuniaux, 1963). Chitin is also an essential structural component of fungi whose cell walls frequently contain greater than 10% by weight of this polymer. A few protozoa and marine diatoms also contain small amounts of chitin. Chitin is a long chain polymer of N-acetylglucosamine linked by β-(1-4) bonds and is thus chemically related to cellulose. Like cellulose, chitin also forms crystalline structures which are arranged side by side and linked through hydrogen bonds. A detailed description of chitin and its properties can be found in Muzzarelli (1977).

Chitosan occurs in the hyphal walls of zygomycete fungi, such as the genera *Mucor, Mortierella*, and *Rhizopus*, and may contribute as much as 33% of the dry weight of the wall (Bartnicki-Garcia, 1968). Chitosan is a close relative of chitin in two ways: it is physically associated with chitin and is probably derived from chitin by deacetylation.

2.4.2 *Peptidoglycan*

The major structural polymer of virtually all bacterial walls is peptidoglycan (=mucopeptide, murein) which contributes from 5 to 10% (*Escherichia coli*) to 60 to 70% (*Micrococcus luteus*) of the dry weight of the cell wall. Peptidoglycan is similar to chitin in that it is a linear chain made up of β-(1-4) linked acetyl amino sugars but differs in that every other residue is an N-acetylmuramic acid. Short peptides link the parallel chains of amino sugars to form a rigid network. Details of peptidoglycan structure can be found in the review by Rogers (1974).

2.4.3 *Lipopolysaccharides*

Lipopolysaccharides form part of the outer membrane of Gram-negative bacteria and comprise 20 to 30% of the dry weight of the cell wall. As their name suggests they are composed of a polysaccharide and a lipid, yet are extremely varied and complex macromolecules. The lipid fraction consists of β-(1-4) or β-(1-6) linked glucosamine carrying long chain fatty acids (e.g., myristic acid). The lipid is linked, through an eight carbon sugar, to the polysaccharide which contains, in addition to a number of common sugars (glucose, galactose, mannose), L-rhamnose and heptoses.

2.4.4 *Teichoic and teichuronic acids*

Teichoic and teichuronic acids are major cell wall components of Gram-positive bacteria and may constitute 30% to 50% of the dry weight of the wall or about 10% of the total cell. There are several types of teichoic acid but all contain a backbone of polyglycerol phosphate or polyribitol phosphate. In many bacterial species the polyphosphate chain contains sugar and amino sugar substituents. Teichuronic acids are polymers made up of alternating units of uronic acid and a hexose or hexosamine.

The function of these cell wall polyanions is still debated but they may be important in cation binding, regulation of autolytic enzymes, sensitivity to antibiotics, and reception of bacteriophages. Teichoic and teichuronic acids have been reviewed by Duckworth (1977).

Other carbon-containing microbial polymers found in cell walls include mannans (yeasts), lipoproteins, and phospholipids (Gram-negative bacteria), and glycans (filamentous fungi).

2.4.5 *Extracellular polysaccharides*

Several species of bacteria are enveloped in an extracellular layer of polysaccharide. This coat is either of an amorphous nature, in which case it is described as a slime layer, or has a rather more ordered structure, a capsule. The possible functions of this extracellular polysaccharide have been discussed by Dudman (1977) although it is well to remember that the conditions favouring their production *in vitro* (high nutrient and oxygen levels) may be encountered rarely *in vivo*.

Bacterial exopolysaccharides which have been investigated include: cellulose (produced by certain species of *Acetobacter*), dextrans (*Leuconstoc*), levans (*Pseudomonas, Xanthomonas, Bacillus*), xanthan (*Xanthomonas*), curdlan (*Alcaligenes, Agrobacterium*), alginate (*Azotobacter, Pseudomonas*) and a large variety derived from *Klebsiella, Rhizobium* and *Arthrobacter* species. Pullulan, is produced by the fungus *Aureobasidium*. In many instances detailed structures have not been determined but a number of component sugars have been identified: D-glucose, D-galactose, D-mannose, L-rhamnose, and L-fucose. Some polysaccharides resemble the teichoic acids of Gram-positive bacteria and thus contain glycerol and ribitol, others are composed of the N-acetyl derivatives of D-glucosamine and D-galactose whilst many extracellular polymers consist of uronic acids such as D-glucuronic, D-galacturonic, D-mannuronic, and L-guluronic acid. Powell (1979) has recently described many of these exopolysaccharides.

2.4.6 Cytoplasmic contents

The cytoplasm and cytoplasmic membrane of microorganisms become an important source of carbon-substrates upon the death and lysis of the cell. Phospholipids constitute 30 to 40% of the cytoplasmic membrane of all microorganisms and some yeasts and fungi contain polymers of mannose and other sugars. All Gram-positive bacterial membranes contain lipoteichoic acid. The cytoplasmic contents of prokaryotes include a plethora of enzymes, substrates and metabolites too numerous to be mentioned here. However, carbohydrate storage materials, notably starch and glycogen, occur in many fungi, yeasts, protozoa, and algae, as well as in species of *Clostridium* and coliform bacteria.

3. MINERALIZATION OF CELLULOSE

3.1 Introduction

The structural polysaccharides of plants, animals, and microorganisms are not easily mineralized: clearly their function is in support and protection. Microorganisms, however, have developed a variety of polysaccharases and a range of strategies which enable them to overcome the physical and chemical barriers to degradation. Many of these strategies will invoke collections of enzymes often acting in a synergistic manner and either produced by a single species or, more often, by a community of taxonomically diverse microorganisms.

Nonetheless the task faced by microbial polysaccharases is enormous. Even homopolysaccharides, which contain only one type of monomeric unit and are either linear (cellulose) or branched (dextran), present problems to microbes attempting to utilize them as substrates. For instance, they may be poorly soluble, neighbouring chains are stabilized by hydrogen bonds, branch points may retard hydrolysis and non-carbohydrate substituents also effect susceptibility to attack. Heteropolysaccharides, with up to six different monomeric components, may be even more difficult to saccharify. Other constraints on degradation are discussed later in the chapter.

There has been much debate concerned with the early enzymic events in polysaccharide decay. In general, the synergistic activities of two types of extracellular enzymes are implicated: endopolysaccharases and exopolysaccharases. Endopolysaccharases cleave glycosidic linkages within the polymer whilst exopolysaccharases attack linkages to terminal sugar residues. The specificity of these two classes of enzyme is believed to be due to their requirement to bind to other sugar residues or subsites on the polysaccharide during catalysis. For example, exopolysaccharases act more readily as the chain length increases suggesting that a number of sugar residues need to be bound to the enzyme in order to achieve maximum activity. The synergy may be due to the enzymes forming a loose complex with each other and thus optimizing their spatial relationship to the substrate and consequently the rate of hydrolysis. In crystalline structures, such as cellulose, the complex may prevent reformation of ordered zones after endocellulase attack (Wood, 1980). Glycosidases show specificity towards short chains of sugar residues and release monosaccharides.

The mode of action of polysaccharases has been discussed recently by Bacon (1979) as have the methodological problems associated with the enzymology of poorly soluble heterogenous substrates (Lee *et al.*, 1980).

Cellulose is a widespread and abundant source of carbon and thus the possession of enzymes to degrade it is an important attribute of microorganisms. This is particularly true of free-living saprophytic microbes utilizing dead plant tissues, but a cellulolytic capacity may also be essential to pathogenic microorganisms attempting to penetrate plant tissue and to symbiotic microorganisms associated with ruminants and a number of insects.

However, there are a number of physical and chemical constraints on the enzymic degradation of even "pure" cellulose (cotton fibres, filter papers). Some of these have been discussed by Fan *et al.* (1980) and include:

(1) the moisture content of the cellulose fibre—in addition to the obvious requirement for water in hydrolytic events, wet and swollen cellulose presents a greater surface area for enzyme attachment;
(2) the degree of crystallization of the cellulose—the ordered hydrogen-bonded zones of the molecule are far less accessible than the amorphous areas. The degree of crystallinity varies from 50 to 90% according to the type of cellulose;
(3) the conformation and steric rigidity of glucose units;
(4) the degree of polymerization; and
(5) the size and diffusibility of cellulolytic enzymes—in relation to the spaces between the microfibrils and the cellulose molecules. Lower molecular weight endoglucanases (*ca.* 12 500) are suggested as being more effective than their higher molecular weight counterparts (*ca.* 50 000) in the early stages of cellulose depolymerization.

If one now considers cellulose in its native state (plant cell walls), there are a number of additional factors influencing cellulolysis, and in particular the nature of the polymers with which the cellulose is associated. These will be predominantly hemicelluloses, pectin, and lignin in varying proportions depending upon the plant species and its age. Thus, in reality, the cellulolytic microflora must contend with a wide range of heterogeneous substrates with which cellulose forms a chemical and intimate physical relationship. In order to degrade cellulose efficiently the cellulolytic microorganisms must overcome these numerous constraints, a task which could involve one or more of the following strategies:

(i) the ability of an individual species to produce, in addition to its cellulases, a large number of pectinases, hemicellulases and phenolases;
(ii) the capacity to co-operate within a community of microorganisms—some specializing in hemicellulose or pectin decay, others in the depolymerization of lignin;
(iii) the ability to overcome the physical barriers imposed by cell wall constituents by vigorous and sustained growth through regions containing unsuitable substrate; fungi and actinomycetes are said to have this advantage and are viewed by

many as having a key role in the early stages of macroscopic organic matter decay; and

(iv) a capacity to resist the bacteriostatic and fungistatic capacity of phenolics produced during lignin decay and synthesized during humic polymer formation.

The ultimate measure of a truly cellulolytic microorganism is its ability to degrade highly ordered, crystalline cellulose. This is unfortunate from the viewpoint of the enzymologist because crystalline cellulose (as cotton fibres, filter paper, etc.) is insoluble and has a poorly defined and variable structure. Cellulase attack on these substrates is often slow and can only be measured in a semi-quantitative manner by weight loss, change in tensile strength, the release of free fibres, or the solubilization of a dye. Clearly it is impossible to express enzyme activity in terms of standard units or to obtain kinetic data. Frequently, therefore, cellulase activities are assessed using the soluble pseudosubstrates carboxymethylcellulose (CMC) or hydroxyethylcellulose (HEC). However, these two have their limitations in that they also have ill-defined structures and the effect of the substituents upon the enzymes is unknown. As a result CMC-degrading enzymes are often referred to as carboxymethylcellulases (CMCases). It is possible to express activity in absolute units by relating viscosity changes to numbers of broken β-(1-4) bonds (Almin and Eriksson, 1967; Almin *et al.*, 1967). The ability to reduce the viscosity of carboxymethylcellulose is usually regarded as being due to the presence of one or more endoglucanases (EC 3.2.1.4): enzymes which randomly cleave internal 1-4 bonds within the cellulose molecule making it more accessible for subsequent hydrolysis. There are no specific substrates with which to measure exocellulase activity and thus for direct measurements a purified enzyme is necessary. However, the production of reducing sugars from an amorphous cellulose reflects the combined activities of endoglucanase and exoglucanase and the difference between this result and CMCase activity (only due to endoglucanase) gives a measure of exoglucanase. According to Wood and McCrae (1978b) a rapidly acting endoglucanase would be expected to produce a rapid decrease in viscosity in relation to the appearance of reducing sugars (i.e., high ratio of viscometric change to reducing sugars); a less rapidly acting endoglucanase or an exoglucanase would produce a lower ratio.

β-(1-4)-Glucosidases hydrolyse cellobiose and short chain cello-oligosaccharides to glucose. Various simple and reliable assays are available using cellobiose, salicin and the pseudosubstrates *o*-nitrophenyl-β-D-glucanopyranoside (ONPG) and *p*-nitrophenyl-β-D-glucopyranoside (PNPG).

Techniques for measuring cellulases have recently been assessed by Eriksson and Johnsrud (1982).

3.2 Axenic Culture Studies

3.2.1 Fungi

The ubiquitous soil hyphomycete genus *Trichoderma* has been the most widely studied cellulolytic fungus (Mandels, 1975; Pettersson, 1975; Enari and Markkanen, 1977; Wood and McCrae, 1978a). This is due to fact that the two principal species, *T. viride* and *T. koningii*, are capable of synthesizing and secreting the entire cellulase complex. In other words, the distinguishing feature of the *Trichoderma* cellulase system is that it contains extracellular endocellulases, exocellulases and β-glucosidases which work in concert to solubilize highly-ordered cellulose. *Trichoderma* species share this ability with a few other fungi, notably *Fusarium solani* (Wood and Phillips, 1969), *Sporotrichum pulverulentum* (Eriksson and Rzedowski, 1969) and *Penicillium funicolosum* (Wood and McCrae, 1978b) but cell-free culture filtrates from most other cellulolytic fungi and bacteria only effect some of the hydrolytic steps *en route* to glucose. This fact suggests that the successful depolymerization of cellulose in many environments will demand the co-operative efforts of more than one microbial species.

The starting point for all theories of cellulose depolymerization is the C_1-C_x hypothesis of Reese *et al.* (1950). They proposed that the native cellulose was first attacked by an extracellular non-hydrolytic enzyme (C_1) which loosened the fibrils by breaking the hydrogen bonds holding the cellulose chains together. This, in effect, increased the accessibility of the substrate to the subsequent activities of hydrolytic enzymes (C_x). Although this model has proved inadequate, and sometimes misleading, many research workers still attempt to relate their findings to it. Thus the C_1 enzyme is viewed by many as a specific exoglucanase, cellobiohydrolase (Wood and McCrae, 1972; Berghem and Pettersson, 1974; Halliwell and Griffin, 1978) whilst the C_x enzymes are equivalent to endoglucanase. Therefore, if a serial process of saccharification is countenanced it is more likely to be a C_x-C_1 sequence; the reverse of that originally proposed. However, although it appears logical that endoglucanases initiate cellulose degradation, it may be more useful (Wood and McCrae, 1972) to envisage a number of parallel and synergistic enzymic events occurring during the solubilication of crystalline cellulose. The original notion of a non-hydrolytic, but so far unidentified, disaggregating enzyme is still not discounted by some (Leatherwood, 1969; Reese, 1977), and the subject of a preliminary enzymic disaggregation or disturbance of cellulose continues to be warmly debated (Reese, 1977; Wood, 1980; Lee and Fan, 1980).

Berghem *et al.* (1975, 1976) have isolated and purified four cellulases from *Trichoderma viride*. These include two endoglucanases capable of attacking the internal bonds of carboxymethylcellulose. Endoglucanase I has a molecular weight of 12 500 and is more active than endoglucanase II (M.W. 50 000) at releasing free fibres from cotton and filter paper. It is possible that its lower molecular size increases its penetration of the insoluble cellulose molecule. A single exoglucanase (M.W. 42 000) was also isolated and was extremely efficient at removing cellobiose but not glucose units from the non-reducing ends of microcrystalline cellulose (Avicel). This enzyme was also somewhat active against cotton and was product inhibited. Others have reported more than one form of exoglucanase in *T. viride* (Gum and Brown, 1977; Gritzali and Brown, 1979). The fourth enzyme reported by Berghem *et al.* (1975, 1976) was a β-glucosidase (M.W. 47 000) which cleaved cellobiose, various higher oligosaccharides (e.g. cellotetraose), and the pseudosubstrate *p*-nitrophenyl-β-D glucoside. Not surprisingly when the exoglucanase and the β-glucosidase are combined degradation of Avicel was accelerated because of the removal of cellobiose. The β-glucosidase is associated with the cell fraction (Berg and Pettersson, 1977)—a common location in other cellulolytic microorganisms. Incidentally, endoglucanase may exhibit activity towards cellobiose (Ladisch *et al.*, 1980) and is product inhibited.

Using *Trichoderma koningii*, Wood and McCrae (1978a) identified four major endoglucanases with molecular weights ranging from 13 000 to 48 000. By examining various combinations of these endoglucanases with the cellobiohydrolase from the same species they were able to propose that the efficiency of the enzymes was related to their ability to form endoglucanase-cellobiohydrolase complexes on the surface of the cellulose chain. Thus something more than a casual synergy is envisaged in cellulose breakdown. In addition, cellobiose was recorded as being strongly inhibitory to three of the four endoglucanases and also to the cellobiohydrolase. This may be a limiting factor in cellulose hydrolysis for many *Trichoderma* strains as β-glucosidase is produced in low amounts relative to the endo- and exoenzymes (Sternberg, 1976). In general all reactions are product inhibited, usually by competitive inhibition (Halliwell *et al.*, 1972; Howell and Stuck, 1975; Sternberg, 1976), although glucose is a non-competitive inhibitor of cellobiase in *Trichoderma viride* (*T. reesei*) (Ladisch *et al.*, 1980).

Sophorose is the most potent inducer of cellulases in *Trichoderma* spp. (Sternberg and Mandels, 1979) although the natural inducer is probably the somewhat less effective cellobiose (Mandels *et al.*, 1962). Of course, sophorose is itself hydrolysed intracellularly to glucose which then acts as a repressor of cellulase formation (Nisazawa *et al.*, 1972; Loewenberg and Chapman, 1977). Inglin *et al.* (1980) isolated an intracellular β-glucosidase (M.W. 98 000) from *Trichoderma viride* (*T. reesei*) that hydrolysed both sophorose and cellobiose and was distinct from the previously-isolated extracellular β-glucosidases (Berghem and Pettersson, 1974; Gong *et al.*, 1977). They speculated that the intracellular β-glucosidase controls cellulase induction by destroying the inducer and producing a repressor. Gritzali and Brown (1979) have recently shown that sophorose will cause *T. viride* (*T. reesei* strain QM9414) to excrete endoglucanases into the growth medium.

Cellulases of the white-rot fungus *Phanerochaete chrysosporium* (=*Chrysosporium lignorum* = *Sporotrichum pulverulentum*) have been extensively studied by Eriksson and co-workers (Eriksson and Rzedowski, 1969; Almin *et al.*, 1975; Deshpande *et al.*, 1978; Eriksson, 1978). They have reported the presence of five endoglucanases, one exoglucanase (releasing cellobiose or glucose—cf. *Trichoderma* sp.) and two β-glucosidases and have demonstrated a strong synergistic response between the endo- and exoglucanase components. Two other non-hydrolytic enzymes were also described as part of the cellulase complex: a cellobiose oxidase (M.W. 100 000) converting cellobiose to cellobionic acid (Ayers *et al.*, 1978) and cellobiose-quinone oxidoreductase (CBQ), which also oxidizes cellobiose. CBQ is implicated in the degradation of lignin (Westermark and Erikkson, 1975) and will be discussed later in this chapter. *Polyporus versicolor* also produces CBQ enzyme but neither cellobiose oxidase nor cellobiose-quinone oxidoreductase had, until recently, been described for *Trichoderma* sp. However, Vaheri (1980) has implicated oxidative enzymes in cellulose breakdown by *Trichoderma reesei* (formerly *T. viride* strain QM 6a). The regulation of cellulases in *Phanerochaete chrysosporium* has been discussed by Eriksson(1979) who has commented upon the number of differences in this system when compared to *Trichoderma viride*. Smith and Gold (1979) have reported that *Phanerochaete chrysosporium* produces intracellular soluble and particulate β-glucosidases as well as an extracelluar β-glucosidase. The extracellular enzyme (MW 90 000; pH optimum 5.5) is induced by cellulase but repressed by glucose. In contrast, the intracellular enzyme (MW 410 000; pH optimum 7.0) is induced by cellobiose and not affected by glucose.

Penicillium funiculosum when grown on cellulose as the sole carbon and energy source, secretes enzymes that are active against all forms of cellulose (Selby, 1968; Wood and McCrae, 1977). Recently Wood *et al.* (1980) have isolated the sole cellobiohydrolase (M.W. 46 000) from this fungus and found that, although it was capable of hydrolysing phosophoric acid-swollen cellulose, it had little activity against highly ordered celluloses (e.g. cotton). However, when recombined in the original proportions with its endoglucanases (three or more) and β-glucosidases (two or three), 98% of the original activity against this substrate was recovered. Synergistic activity was also observed when the *Penicillium* sp. cellobiohydrolase was mixed with endoglucanases from other fungal sources, notably *Trichoderma koningii* and *Fusarium solani* (Wood *et al.*, 1980), and this discovery is discussed elsewhere in this chapter.

Many other fungi have been investigated for their cellulolytic activities. Germinating microcysts of the cellular slime mould, *Polysphondylium pallidum*, excrete endocellulase (CMCase) and β-glucosidase during the emergence of amoeba (O'Day and Paterno, 1979). However, this is essentially a cyst cell wall digesting system and the cellulases may have little effect once they have passed through the cell wall. *Rhizoctonia lamellifera* exhibits both endocellulase and cellobiase activities enabling it to grow on a range of soluble and insoluble cellulose substrates (Selby and Maitland, 1967; Olutiola and Ayers, 1973; Olutiola, 1976). The white rot fungi are particularly important in cellulose decay *in vivo* because of their involvement in lignin degradation (Lundquist *et al.*, 1977).

The cellulases of the thermophilic fungus *Talaromyces emersonii* have been studied by Folan and Coughlan (1978) and those of *Sclerotium rolfsii* have been investigated by Shewale and Sadana (1978, 1979). The latter Basidiomycete fungus secretes high amounts of cellulases, including cellobiase, which are not inhibited by the presence of cellobiose and glucose. Other Basidiomycetes which have been studied in this context include *Schizophyllum commune* (Wilson and Niederpruem, 1967), *Lenzites trabea* (Herr *et al.*, 1978), *Stereum sanguinolentum* (Bucht and Eriksson, 1969) and *Stachybotrys atra* (Jermyn, 1967).

3.2.2 Bacteria

The cellulose-degrading capacity of the genus *Cytophaga* has been known for many years (Walker and Warren, 1938). More recently Chang and Thayer (1977) have used this flexibacterium to investigate the activities and functional location of the various enzymes of the cellulase complex. They found that in crude extracts cellulases (measured against carboxymethylcellulose) were located either on or within the cytoplasm and periplasm, and that the membrane-bound component reduced CMC viscocity whilst the soluble fraction produced soluble sugars. Cell-free endocellulases were not found. The cytoplasmic endocellulase had a molecular weight of 8650 whereas the periplasmic equivalent had a molecular weight of 6250. In addition to the CMC-hydrolysing endoglucanase, the periplasmic fraction contained an exocellulase, active against microcrystalline cellulose, and most of the bacteria's β-glucosidase activity.

Another flexibacterium, *Sporocytophaga myxococcoides* has been described as one of the most active cellulolytic microorganisms, perhaps because it is able to penetrate fibres to obtain maximum substrate-cell contact. Thus, it degrades insoluble cellulose fibres by direct contact and was originally seen under the electron microscope by the appearance of cavities at contact sites (Berg *et al.*, 1972b). It produces several endoglucanases active against CMC (Osmundsvag and Goksøyr, 1975). Recent studies by Vance *et al.* (1980) have shown that *S. myxococcoides* produces an extracellular CMCase and activity towards Avicel (an exocellulase). The former observation confirms the work of Berg *et al.* (1972b), whereas the latter is a contradiction.

Beguin and Eisen (1978) purified three extracellular endocellulases from cultures of *Cellulomonas flavigena*; one free in solution, the other two bound to the substrate (cellulose powder). The soluble endocellulase had a molecular weight of 118 000 and the two bound ones were around 50 000. The two bound enzymes were glycosylated which is also a feature of cellulases from other microbial sources (Berghem *et al.*, 1976; Erikkson and Pettersson, 1975). The authors state that these three enzymes do not constitute the total cellulolytic complex of *C. flavigena*; indeed at least two other CMC-ases were present in the growth medium. All enzymes were subject to repression by readily assimilatable carbon sources and, in the absence of cellulose, cellobiose and even sophorose proved to be weak inducers (Beguin *et al.*, 1977). Stewart and Leatherwood (1976) have obtained mutants of a *Cellulomonas* sp. which do not exhibit catabolite repression.

The capacity of *Cellvibrio* sp. to degrade various forms of cellulose fibre depends upon the structure and complexity of these fibres (Berg *et al.*, 1972b). However, there is no doubt that members of the genus *Cellvibrio* produce cellulases capable of attacking crystalline cellulose (Berg *et al.*, 1972a). Berg (1975), studying *Cellvibrio vulgaris*, found that the location of cellulases depended upon the carbon source and the age of the culture. When cells were grown on soluble carbon (i.e. cellobiose or glucose) all the CMCase was cell-bound (i.e. associated with the periplasm and on the cell surface). Growth on cellulose resulted in cell-free CMCase. Oberkotter and Rosenberg (1978, 1980) also observed the active secretion of endoglucanases into the culture medium during exponential and maximum population phases of growth.

Themophilic actinomycetes have long been known as dominant mineralizing microorganisms in compost and other high temperature environments which contain organic debris (Hankin *et al.*, 1976). For instance, *Thermomonospora curvata* is just one cellulolytic prevalent in compost (Stutzenberger, 1972).

Su and Paulavicius (1975) using a *Thermoactinomyces* sp. indicated that two β-glucosidases were secreted into the culture medium and that they had different thermal stabilities. In contrast, Hagerdal *et al.* (1978, 1979) using the same genus, provided evidence that the β-glucosidase component was exclusively a soluble intracellular enzyme and that the culture filtrate contained other cellulases (e.g. cellobiohydrolase). This work in effect supports the view of Mandels (1975) that the culture filtrate of *Thermoactinomyces* sp. is incapable of complete cellulose hydrolysis, although she believed that it was the cellobiohydrolase that was absent.

The general model of cellulose breakdown by *Thermoactinomyces* species (Humphrey *et al.*, 1977) is essentially that proposed for the fungi. Namely one or more endoglucanases release oligosaccharides which are hydrolysed by exoglucanases to produce cellobiose or glucose. Several β-glucosidases exist which are able to cleave the cellobiose.

Enzymes of the mesophilic cellulolytic actinomycete, *Streptomyces flavogriseus*, have been described by Ishaque and Kluepfel (1980). This species when grown on Avicel, produced considerable amounts of extracellular endoglucanase whilst β-glucosidase and cellobiase activities were predominantly associated with the mycelial fraction and could only be released by sonication. The actinomycete displayed good overall cellulolytic activity towards filter paper and cotton and the enzymes had an enhanced stability in the presence of the substrate.

Cellulases have been recorded in the thermophilic bacterium *Clostridium thermocellum* (Lee and Blackburn, 1975; Weimer and Zeikus, 1977). Ait *et al.* (1979) isolated an extracellular complex (M.W. 125 000) from this anaerobe which expressed CMCase and cellulase activities and contained carbohydrate residues. The authors suggested that the cellulases bind to cellodextrins produced during the hydrolysis of cellulose. Saddler and Khan (1979) isolated from sewage sludge a member of the Bacteriodaceae (probably *Bacteriodes succinogenes*) which produced a cellulase complex degrading cellulose to cellobiose (an endo-exo system) and showed that the cellobiose was utilized fermentatively producing hydrogen, carbon dioxide, glucose, and acetic acid.

There are a number of anaerobic cellulolytic bacterial genera, many of them isolated from ruminant animals, e.g., the genera *Bacteriodes* and *Ruminococcus* (Halliwell and Bryant, 1963; Hungate, 1966) or from cellulose-digesting insects (Thayer, 1978). The anaerobic digestion of cellulose has recently been discussed by Scharer and Moo-Young (1979). An unidentified *Pseudomonas* species secretes extracellular endoglucanases (as does *P. fluorescens* var. *cellulosae*—Yamane *et al.*, 1970, 1971) and has cell-bound endoglucanase and β-glucosidase activity (Bevers, 1976; Hwang and Suzuki, 1976).

Ramasamy and Verachtert (1980) also studied the cellulases of a *Pseudomonas* species. Two endoglucanases were found in the culture medium during growth on cellulose but not on cellobiose, although cell-bound endoglucanases were present with both substrates. Three cell-bound β-glucosidases were detected, two in the periplasmic space and one in the cytoplasm. The more external of the periplasmic glucosidases increased its concentration when grown on cellobiose. These authors conclude that the location (and, not surprisingly, the concentration of enzyme) is determined by the particular substrate: endoglucanase was more external for growth on cellulose than for growth on cellobiose. As occurs with *Trichoderma* spp., sophorose induces cellulase synthesis in *Pseudomonas* spp. (Suzuki *et al.*, 1969; Yamane *et al.*, 1970).

3.3 Community Studies

We have already seen that the total breakdown of cellulose demands the combined efforts of a number of different enzymes and that the various microbial species involved are not equally effective. Some fungi, such as *Trichoderma* spp., have the capacity to produce the complete compliment of cellulases necessary to convert a highly ordered cellulose (e.g. cotton fibre, filter paper) into glucose. Many bacteria, however, find it difficult or are totally unable to degrade native cellulose, yet are active against pre-treated cellulose or soluble derivatives. A considerable number of microbial species, though not producing endo- or exoglucanases (and therefore not considered to be truly cellulolytic), have β-glucosidase activity and are capable of growing on the soluble, low molecular weight products of cellulolysis. The inability of individual microbial species to degrade cellulose may be due to many factors:

(i) the total absence of certain essential cellulolytic enzymes;
(ii) the absence or incorrect concentration of inducer molecules;
(iii) the presence of enzyme repressor or inhibitor molecules;
(iv) the relationship between substrate adsorption and desorption of the various cellulases such that the optimum ratios of active components are not maintained;
(v) the inability of the cell to locate itself in close proximity to the substrate—crucial for those with cell-bound cellulases and suggesting chemotaxis;
(vi) the absence of enzymes whose function is to release the cellulose from its cell-wall associates (hemicellulose, lignin); and
(vii) the lack of a capacity to overcome the bacteriostatic and fungistatic properties of certain components of the organic debris.

Taking these constraints into account, it is plain to see that if it retains its independence, a high proportion of the cellulolytic microflora will have enormous problems in degrading native cellulose in natural environments. The obvious strategy, therefore, and one which may be frequently adopted, is to form a mutually beneficial association with one or more different microbial species.

One approach to understanding multiple interactions in the breakdown of cellulose is to monitor the cellulose-degrading properties of known permutations of cellulases taken from the growth medium of different microbial species. This type of experiment simplifies the investigation by eliminating inter-microbial interaction along with problems associated with induction and suppression and merely studies the enzymic capacity of the system. It is well known that many *Trichoderma* strains produce insufficient β-glucosidase for the rapid saccharification of cellulose. Sternberg *et al.* (1977) screened a number of fungi for β-glucosidase activity and found that two *Aspergillus* species (*A. niger* and *A. phoenicis*) were superior producers. When *Trichoderma* spp. cellulase preparations were supplemented with *Aspergillus* spp. β-glucosidase, the rate of cellulolysis (of Avicel and purified wood cellulose) increased and glucose was the major product. The co-operative and synergistic effects of cellulolytic enzymes from different sources has also been demonstrated in an extensive study by Wood *et al.* (1980). These workers measured the solubilization of cotton cellulose by mixtures of cellobiohydrolase (contained in culture filtrates of *Penicillium funiculosum*) with endoglucanases from a number of other fungi. Table 2 reveals that synergism was apparent in all combinations but was most evident when cellobiohydrolase was mixed with the endoglucanases of *Trichoderma koningii* or *Fusarium solani*. Significantly it is these two species that normally release a cellobiohydrolase (although

Table 2. Cotton-solubilizing activity of *Pencillium funiculosum* cellobiohydrolase in mixtures of endo-β (1-4) glucanases produced by other fungi. (From Wood *et al.*, 1980).

	Solubilization of Cotton (%)	
Source of Endoglucanase	Endoglucanase alone	Endoglucanase + *P. funiculosum* cellobiohydrolase
Trichoderma koningii	1	51
Fusarium solani	1	45
Myrothecium verrucaria	6	20
Stachybotrys atra	4	11
Memnoniella echinata	5	18

previously removed in this experiment) into the culture medium along with the endoglucanase. The other three fungi only release the endoglucanases. According to Wood (1980) the variation in co-operative action between the two cellulases may be due to the stereochemical requirements of their active sites or the need to form a cellulase complex on the surface of the cellulose crystallite before hydrolysis can take place. Since this experiment was concerned with the combined activities of extracellular cellulases and since it is apparent that cellulose can be degraded in this manner, it is not difficult to envisage combinations of two or more of these fungi co-operating in the natural environment. Certainly one will often isolate species of *Penicillium, Trichoderma*, *Fusarium*, and *Stachybotrys* from the same soil sample. The author, however, is unaware of any detailed studies concerning cellulose degradation by a defined community containing two or more of these fungi. Indeed, there have been few detailed investigations of any microbial communities involved in aerobic cellulose mineralization—other than those with an ecological bias described on p. 506. Recently, however, Lynch *et al.* (1980) have looked at microbial populations concerned with the breakdown of a crude cellulose fraction extracted from straw. Their preliminary studies implicated an eleven-membered community comprised of seven fungi (including a *Trichoderma* sp., *Aspergillus* spp., a *Mucor* sp., and a *Penicillium* sp.) two yeasts and two bacteria (one an actinomycete). The yeasts and bacteria did not exhibit cellulase activity but were probably contributing to cellulose breakdown by depolymerizing the associated xylan and arabinan which represented

11% of the cellulose extract. Possible communal relationships in cellulose decay are depicted in Fig. 1.

Some of the most convincing data implicating microbial communities in cellulose mineralization have come from investigations of anaerobic environments. The anaerobic fermentation of cellulose and other sugar polymers to produce methane and carbon dioxide is an essential component of carbon mineralization in marshes, peat bogs, and aquatic sediments as well as in sewage digesters and the ruminant stomach. There is no doubt that the entire sequence involves a microbial community composed of as many as four physiologically-distinct groups (Wolfe, 1979): one group effecting the breakdown of cellulose and fermenting the products to fatty acids, carbon dioxide, and hydrogen; another concerned with the conversion of fatty acids to acetate and hydrogen; a third oxidizing hydrogen to acetate with the reduction of carbon dioxide; and a fourth group converting acetate, hydrogen and carbon dioxide, methanol, and formate to methane. The number of groups and particular species concerned will be determined by the source of cellulose, pH, temperature, and other factors. In sediments, marshes, bogs, and sewage digesters, all four groups have been implicated. In the rumen the fatty acids are absorbed into the blood stream and are not converted to acetate and hydrogen. In fact in the rumen (and unlike other anoxic environments) methane does not arise from acetate.

Perhaps the best evidence for the involvement of microbial communities in the anaerobic mineralization of carbon polymers has come from studies of the rumen (Bryant, 1977; Hobson and Summers, 1978; Hobson, 1979). Clearly bacteria which produce a

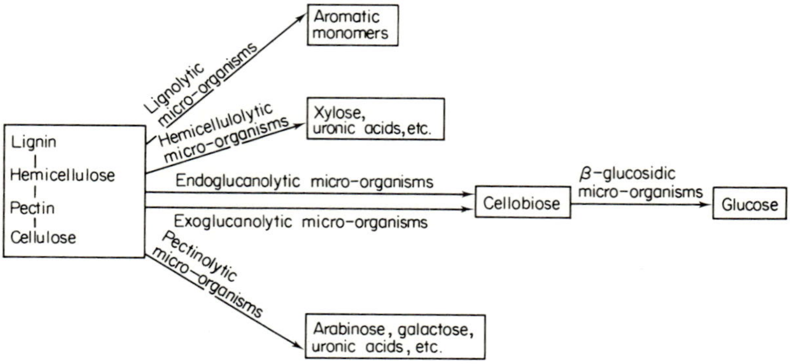

Fig. 1. Possible communal relationships between microorganisms during native cellulose depolymerization.

variety of polysaccharases are well-prepared for the decay of whole plant material as it enters the ruminant stomach. Indeed, there are a number of rumen bacteria which ferment hemicellulose constituents and pectins as well as cellulose (e.g. *Ruminococcus albus, Ruminococcus flavefaciens, Butyrivibrio fibrisolvens*). Others, such as *Bacteriodes succinogenes*, will degrade cellulose but not xylan whilst a few can only utilize starch (e.g. *Selenomonas ruminantium, Bacteriodes amylophilus*). Perhaps surprisingly some bacteria may solubilize xylans and pectins without being able to ferment the resulting oligosaccharides, possibly relying upon associated cellulolytic microorganisms to provide a suitable carbon source. A synergistic relationship between a non-cellulolytic but hemicellulolytic bacterium (*Bacteriodes ruminicola*) was shown to increase cellulose breakdown by a cellulolytic bacterium (Dehority and Scott, 1967). Incidentally, Coen and Dehority (1970) reported that a two-membered community composed of *Bacteriodes ruminocola* and *Ruminococcus flavefaciens* could totally mineralize grass hemicellulose whereas axenic cultures could only perform a portion of the sequence. Scheifinger and Wolin (1973) demonstrated that *Selenomonas ruminantium* and *Bacteriodes succinogenes* grew well in mixed cultures with cellulose as their sole energy source. *S. ruminantium*, however, is non-cellulolytic and thus will not grow on cellulose, although it will utilize glucose—a possible metabolite in the mixed culture. The selenomad certainly produced propionate from the carbohydrates as well as from the decarboxylation of any succinate. Unfortunately, cross-feeding supported by the soluble products of polymer hydrolysis is as poorly understood in the rumen as in other environments, and the success of non-polysaccharase producers may be due to the presence of a range of fermentation products or even cell debris.

Dilworth *et al.* (1980) have recently summarized their investigations of reconstituted anaerobic mesophilic cellulolytic communities. In one experiment, mixed cultures containing cellulase producers (primary bacteria) and cellobiose fermenters (secondary bacteria) were grown at 37°C for 25 d in the presence of cellulose as a sole carbon source. Whereas neither the primary nor the secondary bacteria grew rapidly on cellulose in pure culture, the community utilized 48% of the substrate during the course of the experiment and produced hydrogen, carbon dioxide, ethanol, and acetate. The secondary bacteria were almost certainly accelerating cellulolysis by removing cellobiose which is known to repress cellulase synthesis and activity. A nutritional synergism was not suggested. In addition, the cellobiase producer was the major influence as far as the products of cellulose fermentation were concerned. In another experiment a different secondary bacterium (identified as *Clostridium glycocolum*) was used in the community with the result that acetate was the major product to the exclusion of hydrogen and ethanol. *C. glycocolum* grown axenically on cellobiose yielded large amounts of acetate.

Bacterial methanogenesis represents the final stage of biopolymer mineralization in anaerobic environments. Dilworth *et al.* (1980) have described the conversion of cellulose to methane in mixtures of primary, secondary and hydrogen-utilizing methane bacteria. The combination of cellulolytic species was again required for cellulose degradation but the methanogenic species (*Methanobacterium* sp.), although growing successfully, only marginally stimulated fermentation and should be considered as a commensal component of the community. However, when the *Methanobacterium* species was replaced by *Methanosarcina barkerii*, a more significant increase in the rate of cellulose utilization was seen. Experiments with other combinations of micro-organisms indicated that *M. barkerii* stimulated growth of cellulolytic organisms by removing inhibitory fatty acids, such as acetate or butyrate, and not by utilizing cellobiose or hydrogen.

The resolution of the symbiotic association "*Methanobacterium omelianskii*" (Bryant *et al.*, 1967) has led to a greater understanding of the interactions between methanogens and heterotrophic anaerobes. Interspecies hydrogen transfer, although not demonstrated in the mixtures described above, has been suggested as an important factor in the stimulation of fermentation rates in mixed cultures (Bryant *et al.*, 1977). In this process the hydrogen-evolving bacteria directly support the hydrogen-utilizing methanogenic bacteria and in return the methanogens serve as electron sinks allowing the cellulolytic bacteria to dispose of electrons as hydrogen rather than as other reduced and possibly inhibitory products (e.g. ethanol). Iannotti *et al.* (1973) studied the interaction of a mixed culture of rumen anaerobes in continuous culture and discovered that the growth of *Vibrio succinogenes* depended upon the hydrogen produced from glucose by *Ruminococcus albus*. Weimer and Zeikus (1977) grew the thermophilic anaerobes *Methanobacterium thermoautotrophicum* and *Clostridium thermocellum* together in batch culture with cellulose as their source of carbon. The presence of the methanogenic species in the mixture altered the fermentation pattern of the *Clostridium* species by causing a shift in the conversion of acetyl-CoA from ethanol to acetic acid. The result was that more electrons became available for the production of hydrogen which was then utilized by *M. thermoautotrophicum*—the methanogen acting as an "electron sink".

Other syntrophic communities reported to be involved in the anaerobic mineralization of cellulose, cellobiose and glucose to methane and carbon dioxide include: *Acetobacter woodii* and *Methanosarcina barkerii* (Winter and Wolfe, 1979); *Ruminococcus flavefaciens* and *Methanobacterium ruminatium* (Latham and Wolin, 1977); and *Citrobacter* sp. and *Methanobacterium fromicicum* (Sineriz and Pirt, 1977).

4. MINERALIZATION OF NATIVE POLYSACCHARIDES

4.1 Introduction

Microorganisms must contend with a large number of polysaccharides in addition to cellulose. These will include comparatively simple structures, such as starch, chitin, dextran, and glucans; complex heteropolymers, such as hemicellulose, pectin, and teichoic acids; and ill-defined polysaccharides, such as gums and mucilages.

The same general principles which govern microbial and enzymic co-operation during cellulose mineralization can be applied to all polysaccharides. Thus, as previously described for cellulose, a sequence of events is envisaged during which various microbial species contribute towards the total decay of a particular substrate. The number of species (and enzymes) involved and the rate of the reaction will be determined by structural and chemical features of the substrate, such as linear or branched chains, homo- or heteropolymers, molecular size, types of bonds connecting the monomers, type and rate of substitution, as well as the substrate's relationship with the other cell constituents.

Not surprisingly, attempts to understand the mineralization of other polysaccharides have paralleled those detailed for cellulose. Unfortunately, however, the literature is not extensive and there is little doubt that the same stimuli that have encouraged the study of cellulose have tended to deflect interest away from polymers such as starch, pectin, chitin, and the hemicelluloses. However, there are studies of axenic cultures and the action of isolated enzymes against pure substrates, and some attempts have been made to understand the decay of plant and microbial debris. The former aspect has been reviewed recently (Sturgeon, 1979a,b; Kennedy, 1979; Dekker, 1979; Manners, 1979; Pilnik and Rombouts, 1979) the latter is considered within this section.

There have been a number of studies of the degradation of ill-defined polysaccharide-containing substrates, including plant materials (lignocellulose, straw, leaf litter, wood, lignin), microbial tissues (entire cells, membranes, cytoplasm) and manufactured products (newsprint, cardboard, paper tissue). Many of these investigations have again focused on the cellulose component and have fallen into two broad areas. The first is concerned with the substrate as a renewable resource and concentrates on maximizing fermentation products and biomass production; the second involves the microbial ecology of substrate colonization and succession. There is a tacit agreement in these diverse approaches that the breakdown of organic debris is likely to be a multi-stage, multi-organism process and thus much of this work demands (yet may not receive) the study of dynamic microbial communities.

4.2 Plant Organic Matter

4.2.1 Lignin and lignocellulose

It is clear that the presence of lignin in virtually all mature plant materials is a severe handicap to many polysaccharase-producing microorganisms. Not only does lignin form a mechanical barrier against microbial and enzymic penetration (Kirk, 1975) but some of its aromatic constituents may be bacteriostatic and fungistatic, whilst others may selectively inhibit certain carbohydrate depolymerizing enzymes. For instance, Varadi (1972) demonstrated that a number of phenolic substances (e.g. *p*-hydroxybenzyl alcohol, vanillin, syringaldehyde, *p*-coumaric acid, cinnamic acid) repressed the production of cellulase and xylanase in the fungus *Schizophyllum commune* and recently Vohra *et al.* (1980) reported that ferulic acid reduced β-glucosidase activity to zero whilst partially inhibiting endo- and exocellulase activity. Thus it may be realistic to view lignin degradation as the primary rate-limiting step in polysaccharide mineralization. Certainly pretreatment of organic matter to remove lignin accelerates the subsequent decay of residual polysaccharides (Chahal *et al.*, 1979).

Most studies of the breakdown and enzymology of lignin have involved the white-rot Basidiomycetes (e.g., *Coriolus versicolor, Phanerochaete chrysosporium, Pleurotus ostreatus*) all of which are able to mineralize the substrate totally to carbon dioxide. However, the details are far from conclusive and many questions about lignin decay remain unanswered. To a large extent the very nature of the lignin polymer creates problems in that, in contrast to the polysaccharides, it does not contain repeating units nor are the linking bonds easily hydrolyzed. Indeed, hydrolases are probably of little importance in lignin depolymerization. In addition, lignin is chemically heterogenous and large quantities of a standardized lignin substrate are not easily available. Traditionally heterogeneous substrates, such as milled wood lignin or lignin derived from pulping operations (kraft lignins, lignin sulphonates), have been used but in recent years [^{14}C]-labelled lignins have become available, sensitive and definitive assays have been developed and considerable progress made in describing the events which occur during degradation (Kirk *et al.*, 1977).

The white-rot fungi may attack the lignin polymer in a number of ways:

(i) oxidation of propoid side chains to yield aromatic residues;
(ii) cleavage of β-aryl ether linkages; or
(iii) oxidative cleavage of aromatic rings still attached to the polymer.

The low molecular weight aromatic structures produced by these activities (e.g., vanillin, vanillate, syringaldehyde, guaiacylglycerol, *p*-hydroxybenzoate, coniferaldehyde, *p*-coumarate) are mineralized intracellularly by a large number of fungal and bacterial species (Cain, 1980) or are converted into novel humic polymers (Martin and Haider, 1980). Intriguingly an extracellular aromatic ring cleavage is suggested although how the coenzyme-requiring mono- and di-oxygenases can function has not been revealed.

Lignin breakdown is an obligately aerobic process (Hackett *et al.*, 1977) even though its metabolites are substrates for both aerobic and anaerobic microorganisms. As mentioned previously, the intermonomer linkages of lignin are not attacked by hydrolases but require oxygenases (Dagley, 1978). Clearly oxygen has a mechanistic role in lignin decay but its exact function is uncertain. Two possibilities discussed by Zeikus (1980) and Shimada (1980) are the need to convert ether-containing linkages to esters prior to hydrolysis and the requirement for superoxide radicals. The latter may even contribute to the chemical degradation of lignins.

The enzymes participating in lignin metabolism have yet to be unequivocally identified and described. No doubt part of the problem is the difficulty of obtaining specific lignolytic activities in cell-free preparations and it is probable that lignases are cell wall bound and unstable when released and that direct contact between hyphae and substrate is required for degradation (Rosenberg, 1978). Nevertheless, there is no shortage of speculation concerning lignin enzymology and one group of enzymes, the phenol oxidases, are frequently implicated (Freudenberg and Neish, 1968). Phenol oxidases may have a variety of functions: demethoxylation, α-carbonyl oxidation, side-chain elimination, and polymerization. More specifically, *o*-demethylase catalyses the demethylation of lignin to yield formaldehyde, and the commonly produced laccase stimulates demethoxylation and methanol production. However, some 80% of the lignin phenolic groups are esterified and therefore cannot be substrates for laccase. Incidentally, laccases produced by the white-rot genus *Trametes* have been implicated in the polymerization of coniferyl alcohol to lignin. Peroxidase will also cause substantial demethoxylation in some cases. It has been suggested (Ander and Eriksson, 1976) that phenol oxidases may have a regulatory role in cellulose (and lignin) degradation in that they destroy phenols which are potential enzyme inhibitors. Cellobiose: quinone oxidoreductase (CBQ) is an extracellular enzyme which has been found in culture filtrates of *Polyporus versicolor* and *Phanerochaete chrysoporium* (Westermark and Eriksson, 1974, 1975), and *Sporotrichum* (*Chrysoporium*) *thermophile* (Canevascini and Meier, 1978) and is believed to have a dual role in lignin and cellulose decay. This enzyme which, according to Westermark and Eriksson (1975), has a molecular weight of 58 000, reduces quinnones (produced by the activity of phenol oxidases) to catechols, and simultaneously oxidises cellobiose to cellobiono-σ-lactone (cellobionic acid). The catechols are known to be suitable substrates for the dioxygenases responsible for ring cleavage.

There has been some effort directed towards determining the role of microorganisms other than the white-rot Basidiomycetes in lignin biodegradation. Certainly bacteria are able to degrade a wide range of low molecular weight (<500) lignin-related aromatic monomers and dimers and many of the pathways may be plasmid-encoded. However, studies of bacterial degradation of the lignin macromolecule have not been conclusive and even if it occurs the rate is slow compared to that achieved by the white-rot fungi (Haider and Trojanowski, 1980). Yeasts can grow on di-lignols as can a range of common soil Fungi Imperfecti (*Fusarium, Aspergillus, Penicillium*) yet no degradation of native lignin has been reported. The lignolytic activity of brown-rot fungi is limited to demethylation and some oxidation reactions (Kirk, 1971). Probably the greatest, and as yet unfaced, challenge to those investigating lignin decay is to understand the activities of such a diverse microflora for there is little doubt that the mineralization of lignin in natural environments proceeds slowly and is achieved by the combined and sequential activities of a variety of organisms. Figure 2 illustrates one possible series of events in lignin mineralization.

Due to the intimate relationships between lignin and glucans most naturally-occurring plant-derived substrates are referred to as lignocelluloses. By their very nature lignocelluloses are non-reproducible substrates although some have attempted standardization—at least within a series of experiments. The study of lignocellulose breakdown, like that of lignin itself, has been stimulated by the ability to label the various components preferentially (Crawford and Crawford, 1976; Crawford, 1980) such that ^{14}C-[lignin]-lignocelluloses and ^{14}C-[glucan]-lignocel-

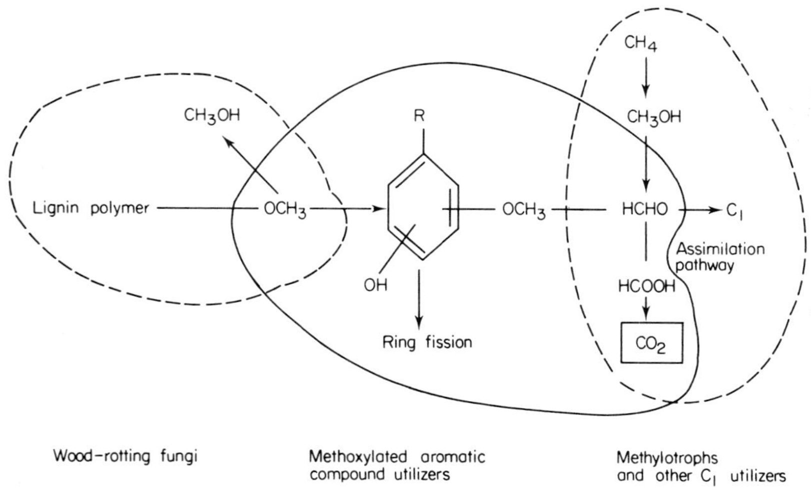

Fig. 2. Microbial communities implicated in lignin mineralization and their probable function (after Kuwahara, 1980).

luloses are now available. Consequently, it is possible to differentiate between those microorganisms merely degrading the glucan fraction, those degrading the lignin, and those degrading both. Using labelled lignocellulose Crawford and Crawford (1976) showed that *Thermonospora fusca* growing at 55°C was primarily utilizing the cellulose and not the lignin, whereas the white-rot fungus, *Polyporus versicolor*, degraded the entire substrate. Crawford and Sutherland (1979) discovered that whilst some *Streptomyces* species would decompose both principal components of lignocellulose and that others would only attack cellulose, none would solely degrade lignin. Phelan *et al.* (1979) examined a large number of actinomycetes and found six which decomposed lignin and cellulose simultaneously, although the former was more completely degraded than the latter. It is likely that some actinomycetes only remove side chains or methoxyl groups from lignin whilst other species are required in order to cleave aromatic rings (Trojanowski *et al.*, 1977; Phelan *et al.*, 1979). However, Crawford *et al.* (1980) have described a single *Nocardia* species which oxidises [^{14}C]-lignin completely to $^{14}CO_2$. Many of the studies of lignocellulose breakdown (Kirk *et al.*, 1976; Ander *et al.*, 1980; Hall *et al.*, 1980) suggest that microbial growth on the lignin component requires the presence of a co-substrate, such as cellulose, glucose, or even starch, so that whilst the presence of lignin is likely to depress cellulose hydrolysis, the presence of cellulose is essential for lignin decay to occur at all.

It appears that cellulases can be induced in the absence of close physical contact between the microorganism and its substrate (Rosenberg, 1979), presumably by a suitable diffusible molecule (Enari and Markkanen, 1977) or possibly through derepression in the absence of a diffusible molecule (Hulme and Stranks, 1971). In contrast, the production of lignin-degrading enzymes may require intimate contact (Drew and Kadam, 1979).

4.2.2 Ecological studies

Fungal species which effect the decay of polymeric wood tissue are broadly classified according to the nature of their substrates and the visual effect they have upon them. Accordingly, white-rot fungi extensively degrade all cell wall components, including lignin and cellulose, rendering the wood fibrous, spongy and bleached in appearance. Brown-rot fungi remove the cellulose and other polysaccharides but not the lignin leaving the wood in a cracked, brittle, and darkened condition. Soft-rots colonize the surface layers, rather than penetrating the wood, and degrade the cellulose, hemicellulose, and pectic components but generally not the lignin. They are particularly prevalent when the wood has a high water content but even then are regarded as having a limited degradative capacity when compared to the white and brown rots, especially when confronted with intact cell walls.

A fourth class of fungi, involved in plant material decay in general, is the sugar fungi (Garrett, 1963). Cells contain a variety of soluble sugars, amino acids, and uronic acids which form a suitable carbon source for the many fungi (and no doubt bacteria) unable to degrade the cell wall polymers. Those fungi utilizing the indigenous soluble carbohydrates present in undecayed plant material are sometimes referred to as primary sugar fungi; those growing on carbohydrates released by the depolymerizing activities of white, brown and soft rot fungi are known as secondary sugar fungi.

Swift (1977) differentiated between wood-inhabiting fungi (white- and brown-rots) and wood-invading fungi (soft-rots). Secondary sugar fungi were renamed secondary saprotrophs and their definition broadened to embrace fungi and bacteria (and arthropods) consuming the products of decay, microbial tissues, or even humus. A large and diverse group of microorganisms are concerned in mineralization processes at the secondary saprophyte stage once the plant material has been comminuted and degraded.

There have been few studies which have combined observations of microbial interactions during the degradation of carbon-containing organic debris with specific enzyme-substrate interactions. However, there have been numerous investigations concerned with microbial colonization and succession in the breakdown of plant materials—some involving frequent sampling from plant debris where it is naturally deposited, some using enclosed and possibly defined substrates buried *in vivo* (e.g. litter bags), others involving decomposer communities *in vitro*. Thus we have a good deal of information as to the microorganisms involved in the mineralization of wood and leaves from a variety of species. A few representative studies are outlined here which, in addition to their ecological value, should be viewed as pointers towards community studies. In other words, likely combinations of bacteria and fungi could be selected for use in experiments designed to reveal the precise relationships of the organisms to each other and to a changing substrate.

Rayner (1978), in a study of fungal interactions in organic matter decay, observed the interactions *in vitro* between 26 different fungal species concerned in the degradation of hardwood. The experiments were conducted by inoculating malt extract agar plates with different pairs of fungi, one species on each side of the plate. Altogether some 200 combinations were tested and a range of responses recorded:

(i) formation of a pigmented zone at the point of contact;
(ii) development of a clear zone between the colonies;
(iii) production of leathery mycelium between the colonies,
(iv) lysis of one fungus by its pair;
(v) replacement of one fungus by its pair; and
(vi) stimulation of fruiting.

Species, such as *Hypholoma fasciculare, Phanerochaete vetulina, Phlebia merismoides* and *Scytalidium album*, were most effective at replacing other species whereas *Chondrostereum purpureum* was non-competitive and easily replaced by others. *Pha. vetulina* and *Phl. merismoides* lysed many of the opposing species before colonization.

With many of the combinations, opposing pairs of fungi were mutually antagonistic due either to the production of antibiotics (*Stereum hirsutum/Heterobasidion annosum, Pseudotrametes gibbosa/Coriolus versicolor*) or to the formation of a dense, leathery mycelial barrier (*Ganoderma adspersum/Daedaleopsis confragosa*). In an attempt to simulate a more natural environment Carruthers and Rayner (1979) paired various fungi on Petri dishes containing oak sawdust. *Phlebia merismoides* and *Hypholoma fasciculare* again proved to be amongst the most aggressive. A third level of study using inoculated wood stumps confirmed many of the results observed using malt extract agar.

Rayner and Todd (1979) summarized these and other studies and tentatively suggested a fungal succession sequence in the decay of hardwood. Initially the virgin wood becomes colonized by a variety of parasitic and sparophytic microorganisms. The composition of this pioneer community will depend upon a number of factors including the part of the plant exposed (branch, bole, root), whether the wood is in contact with the soil, and the residual host resistance of the plant (occurrence of living cells, undamaged tissue). Aerially-exposed material tends to be colonized by soft-rot and stain fungi arising from spores; those portions on or below the ground are likely to be colonized by vegetative mycelium. Parasitic fungi commonly recorded during the early stages of decay include the familar basiodiomycetes *Fomes annosus* and *Armillaria mellea*. Several other species, which are not particularly active in decay and are subject to replacement by more aggressive fungi, are present at this stage: *Botrytis* spp., *Phialophora* spp., *Acremonium* spp., and the basidiomycetes *Chondrostereum purpurem, Shizophyllum commune, Crepidotus variabilis* and *Flammulina velutipes*.

In time, more active degraders expand to dominate the degradative community and these include the white rots *Coriolus versicolor, Stereum hirsutum* and *Hypholoma fasciculare*. Frequently these and other fungi present at this stage are mutually antagonistic and contact between them is restricted. The relatively undecayed interaction zones between antagonists may be colonized by species of *Cladosporium, Rhinocladiella, Endophragmiella* and *Catenularia* which may function as commensal secondary saprophytes dependant upon the products of their fungal neighbours (Rayner, 1976).

Eventually this somewhat stable phase is disrupted by other aggressive colonizers and a second replacement occurs. Fungal species involved in the second phase include *Phlebia merismoides, Phallus impudicus, Phanerochaete vetulina* and again *Hypholoma fasciculare*. As the wood becomes more extensively decayed the importance of these species may decline as the substrate is now suitable and accessible to a wide range of fungi (Mucorales, *Trichoderma* spp., *Scytalidium* spp.) and, of course, bacteria.

Plant litter decomposition was reviewed extensively by Dickinson and Pugh (1974) and recent advances have been discussed by Hayes (1979). For example, the fungal succession of aspen leaf litter in June and October was studied by Visser and Parkinson (1975). Rather than observe succession in progress, they isolated organisms from the litter, fermentation and humus layers of an aspen stand as representative of various stages of decomposition. As is to be expected, fallen leaves carry a microbial population with them which developed when the leaf was on the tree. A phylloplane flora consisting of 2×10^7 bacteria cm^{-2} and numerous fungi and yeasts was recorded by Ruinen (1961). In Visser and Parkinson's (1975) study three principal saprophytes, *Aureobasidium pullulans, Pleurophomella spermatiospora* and *Cladosporium* spp., were present on net-caught leaves and of these *Cladosporium* spp. persisted for some while in the litter layer. Rapid replacement colonization occurred in the litter (L_1) and species of *Phoma, Discula, Mortierella, Phialophora,* and *Penicillium* were recorded. Many of these primary saprophytes are regarded as sugar fungi. Deeper litter layers (L_2), which represented the previous season's leaf fall, contained species of *Altenaria, Trichoderma, Mucor, Penicillium, Sclerotium,* and *Paecilomyces.* The authors suggested that these fungi were concerned either directly with cellulose decay or were associated with secondary saprotrophs (*Mucor, Mortierella*).

Most of the fungal genera in the L_2 layer persisted into the fermentation layer (F_1) but their frequency either increased (*Phoma, Trichoderma*) or decreased (*Discula, Paecilomyces*). Many of the fungi in the F_1 layer were assumed to be cellulolytic and lignolytic genera, whilst others (*Mucor, Absidia, Penicillium*) had a secondary saprotroph role. The same mineralization events were probably occurring in the F_2 layer, although some changes in microbial community structure were observed.

The final stage of succession in aspen leaf degradation was represented by the humus (H) layer. Many of the genera present in the F_2 layer also occurred in numbers here (excepting *Mucor*), although *Mortierella* and *Penicillium* increased in frequency whilst *Cylindrocarpon* appeared for the first time. It was proposed that the active depolymerization of leaf (cellulose, lignin) and fungal (chitin) constituents was taking place in the humus together with numerous secondary saprotrophic activities. Total fungal mycelium increased with depth. For instance in the October sampling the biomass (in g wet wt m^{-2}), in the various litter (L), fermentation (F) and humus (H) horizons, was L_1, 7.55; L_2, 16.99; F_1, 37.90; F_2, 40.15, and H, 188.34.

Frankland (19744, 1976) has reviewed the mycological and biochemical changes occurring during the decomposition of bracken litter (cellulose 37.5%, hemicelluloses 26%, lignin 28%, soluble carbohydrates 1.2%). She recorded approximately 390 species of fungi that were associated with senescent or dead *Pteridium aquilinum* petioles. Over a period of 5 to 6 years the changes in the community were monitored. Weak parasites, such as *Aureobasidium pullulans* and *Rhopographus pteridis,* were important primary colonizers of senescent and dead tissues and no doubt utilized the soluble cell carbohydrates. These species were replaced during the first year by primary saprophytes, such as *Cladosporium herbarum, Cylindrocarpon destructans,* and *Epicoccum nigrum*, and Basidiomycetes (especially *Mycena galopus*) attacking lignin, cellulose, and hemicellulose. During the second and third years other Basidiomycetes accompany *Mycena* spp. in the succession along with a range of cellulolytic Fungi Imperfecti (*Trichoderma, Chloridium, Pestalotiopsis, Penicillium*). By the fourth year the fungal population was declining and was subject to predation by springtails and mites. However the nematode-trapping fungus *Dactylella megalospora* was becoming established and secondary sugar saprophytes (Mucorales, especially *Mucor hiemalis*) utilizing the products of cellulolysis were beginning to dominate the residual fungal community. The role of bacteria was more marked hereafter as the definable leaf litter merges into the soil organic matter.

Boois (1976) investigated fungal development on oak leaf litter and reported that *Aureobasidium pullulans*, as found by other workers using different substrates, was a dominant pioneer fungus in the decay of freshly fallen leaves. *Trichoderma, Penicillium*, and *Mortierella* species were common in the litter from 6 months onwards, *Mucor* species a little later.

Hering (1967) also looked at the fungal decomposition of oak leaves but chose a more classic approach: the incubation of previously sterilized leaves with individual fungal species. Of the ten fungal species tested the most active, in terms of loss of dry weight of oak litter at 3 and 6 months, were *Collybia peronata*, *Mycena galopus*, and *Cryptocline cinerescens*. *M. galopus* was particularly important as a degrader of cellulose, hemicelluloses and lignin. *C. peronata* was active against lignin and hemicelluloses but not cellulose, while *C. cinerescens* was ineffective against the lignin fraction. *Polyscytalum fecundissimum* was an abundant organism which decomposed cellulose and hemicellulose. Interestingly, although *Trichoderma viride* was common on old litter and is well-known as a cellulolytic fungus it had very little effect on the fresh leaves during the six month study. It seems that the cellulose in oak leaves, although theoretically a suitable substrate, is in someway unavailable to *Trichoderma*. Pioneer species, including the ubiquitous cellulose decomposers, *Aureobasidium pullulans* and *Cladosporium herbarum*, were present during the 6-month period but made a rather small contribution to weight loss.

Suberkropp and Klug (1976) investigated the microbial and chemical changes associated with the decay of leaves in an aquatic environment. Leaves of *Quercus alba* (white oak) and *Carya glabra* (pignut hickory) were incubated in a woodland stream and sampled biweekly for up to 32 weeks. Direct counts of bacteria increased exponentially with time although viable counts were generally lower. Bacterial numbers were consistently lower on oak than on hickory but the composition was quite similar on both substrates. No obvious bacterial succession emerged but a range of genera were isolated from the leaves and identified: *Flavobacterium, Flexibacter, Pseudomonas, Acinetobacter, Achromobacter, Chromobacterium, Serratia, Alcaligenes, Bacillus, Cytophaga, Sporocytophaga*, and *Arthrobacter*. *Flavobacterium* and *Flexibacter* isolates were generally able to hydrolyse starch, pectin, casein, gelatin, and inulin, and only *Cytophaga* and *Sporocytophaga* used cellulose and cellobiose as carbon sources. The authors suggested that most of the bacteria may obtain their carbon from fungal metabolites, lysed fungal cells, or the products of proteolysis as they were unable to degrade the structural cell wall carbohydrates. Bacterial biomass increased in the later stages of decomposition and was paralleled by a decline in the fungal population.

Suberkropp and Klugg (1976) further showed that aquatic phycomycetes (e.g. *Alatospora acuminata, Flagellospora curvula, Tetracladium marchalianum*) were the dominant mycoflora on both leaf species although abundance and patterns of succession differed. *F. curvula* was dominant during the first 4 to 6 weeks on both oak and hickory and declined rapidly thereafter. *Lemonniera aquatica* was also an early colonizer but maintained its presence throughout the experimental period. With the decline of *Flagellospora curvula, Alatospora acuminata* become the dominant species on oak and the co-dominant species (with *Tetracladium marchalianum*) on hickory. A large number of soil fungi were also present (e.g. *Alternaria, Penicillium, Fusarium, Aspergillus, Cladosporium*) but in the dormant state, only growing when plated onto rich media. All the aquatic hyphomycetes were capable of producing extracellular pectinase and cellulase.

The differences in microbial colonization between the two leaf species was considered to be due to inherit differences in their chemical composition as well as changes occurring during decay. For instance, soluble polyphenolics and lignins are more abundant in oak than hickory (28% compared with 19%). It is well known that phenolics complex with proteins (Benoit and Starkey, 1968) and therefore these aromatics may be inactivating the exoenzymes and thus reducing fungal growth as well as that of any bacteria which depend upon fungal metabolites. Changes in the dominant microflora during substrate breakdown are no doubt influenced by changes in soluble carbohydrates (largely disappeared during the first two weeks), cellulose (gradual decline after 2 to 4 weeks) and hemicellulose (gradual decline after 12 weeks—oak, and after 2 weeks—hickory).

Goodfellow and Dawson (1978) chose to investigate bacterial populations in the L, F, and H layers of a highly acid spruce forest soil over a period of one year. The highest numbers of bacteria were recorded in the humus layer although actinomycetes and microfungi did not follow this pattern. There was a dramatic drop in bacterial counts in the top layer (A horizon) of the underlying mineral soil. The commonest, positively identified bacteria in the L layer were *Arthrobacter* spp. and *Bacillus firmus*. The occurrence of these species declined sharply in the F layer where other bacilli (*B. lentus*, B. polymyxa, *B. coagulans*, *B. sphaericus*) and *Micrococcus* spp. were dominant. In the H layer yet other bacilli became important (*B. chitinosporus*, *B. cereus*, *B. globisporus*, *B. laterosporus*). The absence of *Achromobacter*, *Pseudomonas*, and *Flavobacterium* species from the litter horizons is somewhat surprising as these are generally considered to be amongst the commonest litter-degrading bacteria (Jensen, 1974), but there is no doubt than Gram-positive organisms were in abundance in this soil. Although the streptomycetes were not described other than by colony characteristics, a similar pattern was seen with certain strains being dominant in one layer yet missing from another.

It is never easy to explain these observations without further information regarding the enzymic capacity of the microflora, the changing morphology and biochemistry of the leaf litter, and the physicochemical properties of the various soil horizons. Appropriately, a recent study adopted an enzymological approach to leaf decay. Spaulding (1977) measured the activities of a number of enzymes extracted from coniferous leaf litter including cellulase, xylanase, peroxidase, amylase, and invertase. This investigation revealed a definite correlation between carbon dioxide production and amylase, cellulase, and xylanase activities although amylase may be involved in microbial cell turnover rather than the mineralization of litter which contains little starch. Enzymic determinations such as these, together with observing substrate destruction *in situ* with the electron microscope should enable us to construct a more accurate picture of leaf carbon mineralization. An example of the latter is provided by Kilbertus and Reissenger (1975) who incubated plant debris both on and within soil and then used transmission electron microscopy to examine thin plant sections for microbial species and associated areas of enzymic erosion. They were able to observe the cellulolysis caused by *Cladosporium herbarum*, *Stachybotrys chartarum*, and a *Chaetomium* sp., and the destruction of starch by *Mortierella ramanniana*. After 60 days there were signs of pectolytic, lipolytic and lignolytic activities. The progress of wood colonization and decay by actinomycetes has also been followed using scanning electron microscopy (Baecker and King, 1980).

A different approach to understanding the enzymic basis of carbohydrate decay is illustrated by a recent study to assess the likely involvement of various *Aspergillus* species in the deterioration of stored grain (Flannigan and Bana, 1980). These workers tested a range of aspergilli for amylase, α- and β-glucosidase, β-(1-3)-glucanase, pectinase, lipase, xylanase, CMCase, and other enzymes. On the basis of their enzyme activities certain species (e.g. *Aspergillus glaucus* group) were regarded as less likely to cause extensive degradation of grain than were other species (*A. flavus, A. terreus*). This work illustrates a separate source of information concerning the contribution of microorganisms to carbon mineralization, namely that derived from biodeterioration research (Oxley *et al.*, 1980).

Just a few of the many studies of microbial succession during the decay of plant materials have been mentioned here. They suffice to indicate the plethora of data which can be used to design experiments which may reveal the interactive nature of microbial species in the mineralization of complex organic substrates.

Whilst this chapter is primarily concerned with the direct action of microorganisms on complex botanical substrates it should not be forgotten that, in general, microfauna (nematodes, protozoa, rotifers), mesofauna (collembola, diptera, larvae, termites) and macrofauna (millipedes, insects, molluscs, earthworms) all have important comminutive and catabolic roles in the degradative community. More specialized contributions are made by wood-boring insects which stimulate breakdown by forming channels in the substrate, thus encouraging aeration and exposing more of the material to microbial colonization. Insects also serve as vectors for microbial inoculum. In addition some insects harbour microorganisms in their gut which are capable of depolymerizing cellulose, other polysaccharides, and possibly even lignin (Seifert and Becker, 1965). Wood digestion by termites has been reviewed recently by Stradling (1977), and Grosovsky and Margulis (1982). Some termite species cultivate (and subsequently graze upon) cellulolytic and lignolytic fungi on the walls of their burrows. For a detailed description of the whole spectrum of decomposer organisms the reader is referred to Swift *et al.* (1979).

4.2.3 Biotechnological studies

Many of the fermentation studies concerned with complex substrates have utilized two or more microbial species. Srinivasan and Han (1969) investigated the breakdown of sugar cane residues (bagasse) by a *Cellulomonas* sp. and discovered that protein production (i.e. yield of microbial biomass) was increased if an *Alcaligenes* species was included in the fermenter. In this two-membered community the *Alcaligenes* consumed the cellobiose which would otherwise have inhibited cellulose production by the *Cellulomonas* species. *Candida guillermondi* and *Trichosporon cutaneum* have been used to remove cellobiose produce by *Cellulomonas* (Srinivasan, 1975). A different relationship was shown by Ghose *et al.* (1976) who used the same form of crude cellulose (46% cellulose, 25% hemicellulose, 21% lignin) but looked at the effect of *Trichoderma viride* and *Aspergillus wentii* culture filtrates on sugar yield. A 1:1 mixture of the enzymes increased the rate of reducing sugar production when compared to. *T. viride* enzymes alone. Apparently, this relationship depends upon the contribution of *Aspergillus wentii* to xylan degradation and presumably the resulting increase in the accessibility of cellulose to the *T. viride* cellulases. The mixed population had no such stimulatory effect upon cellulose alone.

Wheat straw (40% cellulose, 29% hemicellulose, 14% lignin) has been chosen as a realistic cellulosic substrate by a number of workers. The lignin component in straw is bound tightly to the cellulose and hemicellulose fractions, effectively reducing their accessibility to degradative enzymes (Cowling and Brown, 1979; Kirk, 1975). Thus, in a great many experiments, designed to study protein production, physical or chemical pretreatment, is used prior to inoculation. Chahal *et al.* (1979) subjected wheat straw to sodium chlorite to remove the lignin component and found that all microorganisms tested were more effective at degrading the modified holocellulose (cellulose + hemicellulose) substrate. The microorganisms, which included *Aspergillus niger*, *Aspergillus terreus*, *Cochliobolus specifer*, *Myrothecium verrucaria*, *Rhizoctonia solani*, *Spicana fusispora*, *Penicillium* sp., and *Gliocladium* sp., were all originally isolated from decomposing wheat straw. *Cochloibolus specifer* was the most efficient (in terms of protein production) of all species with all substrates: untreated straw, lignin-free straw (helocellulose) and cellulose alone. Hemicelluloses may also slow down the rate of fungal degradation of straw due to the presence of somewhat resistent sugars arabinose and galactose, and methylation (Moo-Young *et al.*, 1978). Other axenic culture studies of wheat straw decay have been carried out by Moo-Young and co-workers (Chahal *et al.*, 1977; Moo-Young *et al.*, 1978, 1979) especially with *Chaetomium cellulolyticum* which they say compares favourably with other fungi in terms of biomass production. No doubt one of the main reasons for this efficiency is that *C. cellulolyticum* rapidly utilizes cellobiose.

Peitersen (1975) used a mixed culture of *Trichoderma viride* and a yeast (*Candida utilis* or *Saccharomyces cerevisiae*) for the breakdown of pretreated straw. The yeast was inoculated some 24 to 32 h after the fungus. In comparison to *T. viride* alone, the time taken for maximum yield of cellulases and cell protein was reduced by several days. This was assumed to be due to the removal of glucose by the yeast which otherwise repressed *Trichoderma* cellulase production. An additional nutritional benefit may arise from lysed yeast cells. Communities of microorganisms involved with other fermentations of industrial importance are fully described elsewhere in this volume (Drozd and Linton).

4.3 Microbial Debris

Despite the accepted importance of microbial residues as substrates for microbial growth, there have been few attempts to describe the communal and successional events concerned in their mineralization. Furthermore, only occasionally are the actual decomposer organisms named, with research efforts having been concentrated on overall mineralization rates. Many recent mixed culture studies have also acknowledged the importance of the products of cell lysis to one or more members of the community but rarely is the nutritional basis of the relationship explained.

Research prior to 1970, particularly as it refers to the mineralization of microbial debris in soil, is reviewed by Webley and Jones (1971). Not surprisingly, chemically-different cellular constituents (lipopolysaccharides, chitin) and therefore cell components (cell wall, cytoplasm) are subject to different rates of mineralization. Generally cytoplasmic fractions are degraded more rapidly than cell wall fractions (Verma and Martin, 1976) whilst according to Nakas and Klein (1980) bacterial cell walls were more susceptible to decomposition than their fungal counterparts, the reverse being true for the cytoplasmic components. It has been known for some while that cell wall pigments, such as melanin, will retard mineralization (Martin *et al.*, 1959; Kuo and Alexander, 1967). Recently Linhares and Martin (1978) measured the $^{14}CO_2$ evolution from the breakdown of labelled fungi incorporated in soil. Over a 12-week period considerably less carbon dioxide was released from melanic fungi (*Hendersonula toruloidea*, *Aspergillus glaucus*, *Eurotium echinulatum*) than from hyaline species (*Penicillium vinaceum*, *Penicillium nigricans*). The rate of carbon mineralization was also reduced by complexing cells and cell components with various phenolics (Verma and Martin, 1976; Nelson *et al.*, 1979), a process that is known to occur naturally as the products of microbial decay become associated with humic matter (Mayaudon and Simonart, 1963; Mayaudon, 1966). Jones and Webley (1968) approached the study of microbial cell degradation somewhat differently. These workers incorporated fungal wall material (from *Fusarium culmorum* and *Mucor ramannianus*) into soil aggregates and monitored colonization and monitored colonization and β-(1-3)-glucanase activity. There was a good relationship between the occurrence of *Streptomyces* spp., the degradation of *Fusarium* and β-(1-3)-glucanase activity.

5. CARBON MINERALIZATION IN THE SOIL ENVIRONMENT

5.1 Introduction

Approximately 10% of the organic matter in soil occurs as carbohydrate and most of that is as polysaccharide. In mineral and cultivated soils carbohydrates account for less than 2% of the soil weight, whereas in organic soils (e.g. peat) values may be as high as 30%. The degradation products of plants, insects, mammals, and microorganisms all contribute to the carbon pool and, in addition, some novel polysaccharides and polyphenolics are synthesized. Soil polysaccharides of microbial origin contain hexoses and deoxyhexoses; those containing arabinose and some of those containing xylose are believed to be derived from plants (Cheshire, 1977).

Microorganisms degrade carbon-containing substrates in soil as in other environments in order to build new cells. In biogeochemistry this process is termed immobilization, as it removes carbon from the immediate attentions of the decomposer community. However, with the death of the first flush of degraders the microbial cells are mineralized and a further proportion of the original carbon is released as carbon dioxide. In other words, each generation of microorganisms is actively involved in immobilization and mineralization yet subsequently serves as substrate for the next generation. Microbial tissues may even be more accessible to decomposers than plant material (Kaszubiak *et al.*, 1976) and the proportion of the original substrate carbon that is mineralized compared to that which is immobilized in new cell material varies with microbial species, substrate, availability of other essential elements (i.e. carbon:nitrogen; carbon:phosphorus; carbon:sulphur ratios) and the physico-chemical nature of the environment (e.g. pH, redox potential, adsorptive properties). The aerobic degradation of substrates in soil may be reasonably efficient. Jenkinson (1968) and Shields *et al.* (1973) estimated that during the primary decomposing cycle, the microflora converts between 40 and 60% of the original carbon to microbial tissues. Anaerobes, by comparison, may assimilate less than 5% of their carbon supply (Wagner, 1975). Fungi and bacteria are believed to make equal contributions to soil respiration (Clarholm and Rosswall, 1980). Assuming that, at least in an aerobic environment, equal proportions of carbon are immobilized and mineralized, then primary decomposition releases 50% of the carbon dioxide, whilst successive waves of secondary decomposers utilizing microbial cell carbon will release 25% of the residual carbon (75% total), 12.5% (87.5%) and so on. Theoretically after six generations (one primary, five secondary) 98.5% of the initial carbon source will be mineralized (i.,e. evolved as carbon dioxide). However, this is an over-simplified view of mineralization because at least one other pool of organic carbon is formed during substrate decay—humic matter.

The humic fraction of soils and sediment contains a variety of chemically and physically associated carbonaceous substrates including amino acids and proteins, incompletely degraded carbohydrates and lignin components, as well as novel polyphenolics synthesized during microbial proliferation. The heterogeneity of humic materials means that its components display different susceptibilities to microbial attack or to abiological release and therefore have different residence times in soil. A frequent estimate is that between 2 and 5% of humus carbon is mineralized each year (Alexander, 1977), although additions of easily degradable substrate, such as green manure, may accelerate carbon dioxide release—a process known as "priming" (Jenkinson, 1966). The most recalcitrant aromatic polymeric components of humic colloids may have half-lives measured in hundreds of years (Campbell *et al.*, 1967; Sorensen, 1975). Thus, in a temporal sense, there is a third phase of mineralization; the gradual conversion of humic carbon to carbon dioxide. Many attempts have been made to assess residence times of various forms of organic carbon in soil (Wood, 1974; Lousier and Parkinson, 1976; Paul and Voroney, 1980) and the influence of clay and organic colloids upon substrate persistence (Cheshire, 1979). Primary, secondary, and tertiary processes in mineralization are depicted in Fig. 3.

5.2 Factors Affecting Persistence of Carbon-containing Substrates in Soil

It appears that certain soil constituents have a protective effect upon polysaccharides (Cheshire, 1977) and it has been shown that purified enzymes have a reduced degradative capacity when presented with polysaccharides from extracted soil (Cheshire and Anderson, 1975).

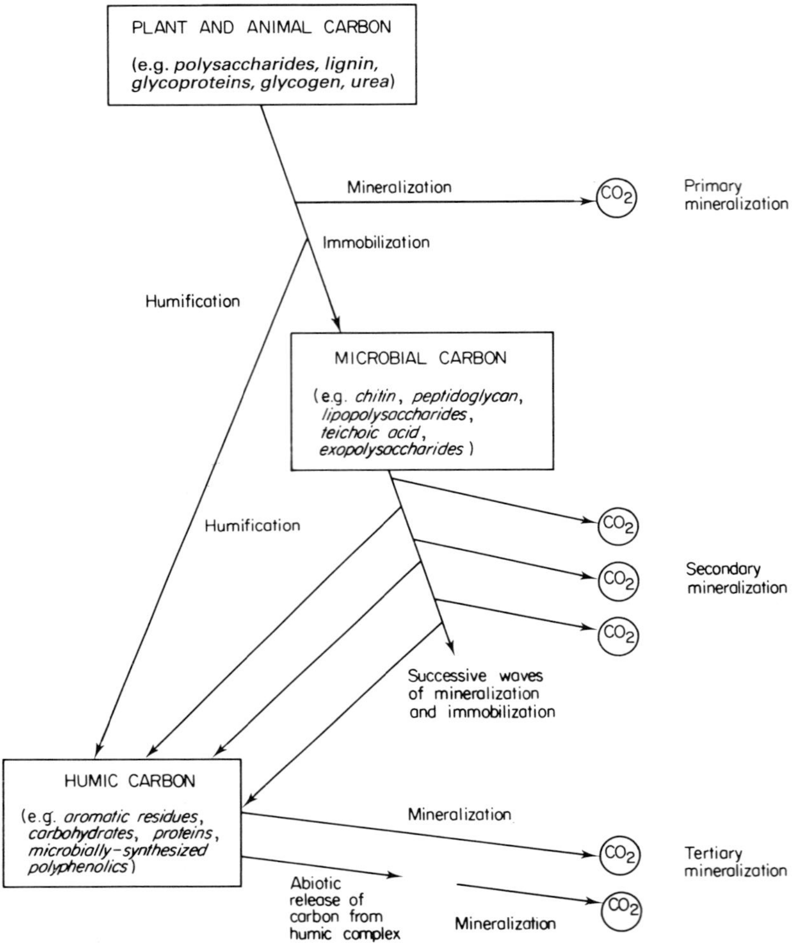

Fig. 3. Levels of carbon mineralization.

Without doubt any differences in stability between polysaccharides observed *in vitro* can be related to intrinsic factors such as sugar composition, the nature of substituents and the presence of other cell components. *In vivo*, however, the soil inorganic and abiotic organic components are believed to contribute significantly in three different ways to the unexpected recalcitrance of polysaccharides in soil.

Firstly, polysaccharides are known to be adsorbed to a certain extent by clays (Parfitt and Greenland, 1970). Polysaccharides can enter the interlamellar spaces of expanding-lattice clays (Greenland, 1956; Finch *et al.*, 1967), a location which will protect the carbohydrate from the direct attention of microorganisms and which may even hinder the approach of exoenzymes. Incidentally, chemical degradation may also be slowed down due to adsorption. The adsorption of a polysaccharide in excess of that predicted by its molecular size and the surface area of the clay adsorbant (Olness and Clapp, 1973, 1975) may occur if only part of the carbohydrate polymer is attached (Hayes and Swift, 1978) or if a multilayer is formed. Various adsorptive mechanisms have been proposed to account for the attachment of polysaccharides to clays and these include: hydrogen bonding (Kohl and Taylor, 1961; Finch *et al.*, 1967); complex formation (Parfitt, 1972); and metal ion bridging (Saini and MacLean, 1966; Guckert *et al.*, 1975). Parfitt and Greenland (1970), however, believe that multiple dispersion forces and entropy changes, arising from the replacement of small molecules, such as water, at the clay surfaces, are sufficient to account for the adsorption of polysaccharides.

Secondly, many polysaccharides form complexes with metal ions, such as calcium, zinc, iron, and aluminium, which will reduce their rate of degradation (Martin *et al.*, 1966, 1972).

Finally, the modifying effect of phenolic compounds on polysaccharide degradation has been recognized for many years (Handley, 1954; Benoit and Starkey, 1968). Phenolics may occur either as a component of plant material which is released during the decay of lignin-containing organic matter or be synthesized *de novo* by the microflora. Either way they may protect polysaccharides in a physical sense (Swincer *et al.*, 1969) or by forming a chemical association (Martin *et al.*, 1978). Phenolics may also protect substrates by specifically inhibiting the microorganisms which would otherwise degrade them (Ivarson, 1977). The persistence of polysaccharides concerned in soil aggregate formation may be due to their subsequent tanning (Griffiths and Burns, 1972).

5.3 Extracellular Carbohydrases in Soil

Soil and sediment-dwelling microbial communities, which depend upon exogenous carbon sources for their nutrition, are confronted with enormous problems. For instance, suitable substrates are discontinuous in both time and space, and there may be prolonged periods during which the microbial cell must rely upon maintenance factors, particularly if it is unable to find refuge in endospore formation. Those microorganisms that utilize polysaccharides have even greater problems in that they need to excrete suitable enzymes whose chance of success is remote. In other words, extracellular polysaccharases are released into an environment in which adsorption, denaturation and degradation take a rapid toll of diffusible enzymes. Any enzyme surviving these obstacles must then encounter a suitable substrate, react with it (if the pH is favourable) and release a lower molecular weight product with increased solubility. The product must subsequently return to the microorganism which secreted the enzyme in the first place, in order that the energetic deficit caused by the production and secretion of extracellular enzymes should be made up and that the microbe should gain a nutritional advantage from its efforts. However, the product is now a suitable substrate for a large number of species in addition to the specialist microorganism which produced the depolymerase, or it may become attracted to and immobilized on a particulate surface (e.g. clay). Whatever, both factors serve to reduce the amount of substrate reaching the enzyme-producing microorganisms.

There is some evidence to support the notion that the soil is a hostile environment for free enzymes. This derives from two sources: the first is that enzymes added to soil (or stimulated within it) are rapidly inactivated (Drozdowicz, 1971; Zantua and Bremner, 1976); the second is that enzymes have great difficulty in diffusing through soil barriers to reach their substrates (R. G. Burns and C. F. A. Hope, unpublished observations).

Because of these constraints it is difficult to envisage a successful strategy for substrate utilization and microbial growth in soil founded upon the production of diffusible exoenzymes. Certainly the continual release of enzymes from microorganisms, whilst profitable in a flask containing abundant substrate, seems doomed to failure in the soil environment. Now we have already seen that in some but not all cellulolytic bacteria the diffusion of extracellular enzymes is curtailed by their retention at the cell wall. This approach will certainly improve the survival and efficiency of the cellulase, even though the enzyme cannot at the same time be involved in scavenging for substrate. Nonetheless, direct contact between cell wall-associated enzyme and substrate must improve the microorganism's chance of success in that the soluble products of any catalysis are available for immediate uptake, and are not subject to attenuation by physical, chemical, and biological forces. This type of intimate interaction does not, however, explain how the microbe detects and locates the substrate in the first instance. A chemotactic response must be initiated by a soluble attractant, and yet we have already stated that it is unlikely that microbial exoenzymes survive in the soil aqueous phase and even if they did the likelihood of a soluble product reaching the producing cell (and stimulating a tactic response) is remote. Therefore, when considering the microbial mineralization of carbon-containing substrates *in vivo*, it may be necessary to search for a more satisfactory strategy—possibly involving "accumulated" hydrolases.

Accumulated enzymes form an extracellular catalytic-component of soils and sediments that is stabilized through its association with the colloidal organic (humic) and inorganic (clay) constituents. These enzymes have a longevity not normally characteristic of free enzymes. The origins of accumulated enzymes and the nature of their association with the soil colloids has been debated exhaustively (Skujins, 1976; Burns, 1977, 1978) and a current view is that these enzymes are derived predominantly from microorganisms and are co-polymerized with humic matter during its formation. The resulting complex affords the enzyme some protection whilst still allowing it to retain a proportion of its activity (Rowell *et al.*, 1973; R. G. Burns and J. P. Martin, unpublished observations). In some instances the clay colloid, as such, may have a protective effect on enzymes (Stotzky, 1972) although it is more common for the clays to protect enzymes indirectly in the sense that they are stabilizing the organic matter. Accumulated hydrolase levels are at a steady-state condition in mature, climax soils. In other words, any leakage and destruction of enzyme is counterbalanced by the continuing process of humification during which enzyme is incorporated into novel polyphenolics. A large number of carbon substrate hydrolases have been reported as existing as colloid-bound moieties in soil (Ladd, 1978—Table 3) and, even allowing for the sometimes less than adequate methods of detection employed, it is apparent that humic matter has a significant enzymic capacity associated with it. By far the most frequently described hydrolase is urease (EC 3.5.1.5) which is suggested as responsible for a high proportion of the rapid urea hydrolysis in soil (Paulson and Kurtz, 1969; Pettit *et al.*, 1976). This ubiquitous accumulated enzyme illustrates an important characteristic of soil enzymes: they are not derived solely from enzymes whose normal functional location is extracellular. Urease, of course, is a cytoplasmic enzyme depending upon a soluble, low molecular weight substrate. Thus it may be inferred that a proportion of enzymes arising from lysed cells and those leaking from live cells can be incorporated into humic polymers as well as extracellular enzymes *sensu stricto*. Other accumulated enzymes involved in carbon mineralization include amylases, a variety of cellulases, β-D-glucosidase, β-1,3-glucanase, xylanase, and pectinase. A comprehensive list is shown in Table 3.

Table 3. Accumulated carbohydrases (Ladd, 1978).

Recommended Name of Enzyme	EC Number
α-amylase	3.2.1.1
β-amylase	3.2.1.2
cellulase	3.2.1.4
endo-1,3(4)-β-D-glucanase	3.2.1.6
inulinase	3.2.1.7
endo-1,4-β-D-xylanase	3.2.1.8
dextranase	3.2.1.11
polygalacturonase	3.2.1.15
α-D-glucosidase	3.2.1.20
β-D-glucosidase	3.2.1.21
α-D-galactosidase	3.2.1.22
β-D-galactosidase	3.2.1.23
levanase	3.2.1.65

Are accumulated enzymes important entities in soil or do they merely make a fortuitous and trivial contribution to microbe–substrate interactions? Unfortunately, there is no unequivocal answer to this question although an ecological role for accumulated enzymes has been proposed (Burns, 1979; 1980). This model suggests that the humic-enzyme complex serves as a stable detector of exogenous substrates which passes on the information (in the form of soluble inducer or derepressor molecules) to adjacent microorganisms. If the substrate concentration is high enough then the signal received will cause the microorganisms to assume the major role in substrate decay. Thus communal activities *in vivo* should perhaps be envisaged as involving both cells and immobilized exoenzymes juxtaposed and interacting at the soil/liquid interface.

6. CONCLUSIONS

It would be naive to view the mineralization of any carbon-containing polymer as the function a single microbial species. Even though a few microbial species in axenic culture are capable of producing the array of enzymes necessary to convert pure forms of cellulose, starch, or xylan to carbon dioxide *in vitro* it is unlikely that these organisms will perform the same function independently when confronted with the substrate *in situ*. This is due to four principal differences: an increase in the physical and chemical complexity of the native substrate (e.g. pure cellulose fibres compared to those imbedded in and associated with lignins, pectins, and hemicelluloses); the intense inter-species rivalry inevitable in a largely oligotrophic habitat; an environment where the temperature, pH, and level of hydration and aeration may be a long way from the optimum conditions provided *in vitro*; and the extensive charged, and therefore reactive, surface areas provided by colloidal organic matter and clays. In order to overcome these constraints it is apparent that microorganisms have surrendered their independance and instead formed mixed species communities which co-operate in the breakdown of natural and synthetic substrates.

It is thus helpful to think of the mineralization of organic matter in terms of microbial communities and indeed a holistic view of soil as a complex multicellular body collectively responding to substrates is not new (Quastel, 1965; McLaren and Peterson, 1967). Furthermore the function of persistent colloid-bound enzymes, proposed on p. 522, implies that the overall microbial community has spatial and temporal dimensions reaching beyond those delineated by proliferating microorganisms. Recently the editors of *Contemporary Microbial Ecology* (Ellwood *et al.*, 1980) wrote that it was useful "to consider the total gene pool of a given mixed microflora located in a particular environment in determining the response of individuals or interacting groups of organisms to . . . whatever influences the growth of microorganisms or communities". The implications of this statement extend even beyond the cellular co-operation involved in mineralization to include the exchange of metabolic capabilities through bacterial conjugation (Slater and Godwin, 1980).

Microbial ecologists acknowledge that the remarkable capacity of terrestrial and aquatic environments for dealing with an enormous volume and range of substrates is due, at least in part, to the co-operative nature of microbial communities. Further to this it is usually accepted that microbial communities have a degree of flexibility and adaptability when placed under stress during natural and man-made perturbations. Some of this resilience may be due to the facultative nature of microbial communities, not in the sense that its members do not require a communal existence for success but rather that a variety of different microbial combinations are capable of the mineralization of a particular substrate and that their relationship to each other is casual in terms of the particular species involved. In other words, if one component of a community is inhibited another will take its place.

Studies of microbial communities and carbon mineralization have tended to concentrate upon the colonization and succession aspects of macro-organic matter decay. Unfortunately, these ecologically-oriented studies have rarely revealed the nutritional relationships between microorganisms and their substrates nor have they emphasized the communal aspects of the various microbial species involved. Nonetheless Rayner and Todd (1979) are quite unambiguous about the importance of interacting microbial communities in organic matter decay, and have warned against considering the process as performed by broad waves of microorganisms influenced solely by physical and nutritional changes in the substrate. Indeed they say that "rather than ask what is succession (in organic matter breakdown) a more pertinent question is what is the fungal community, how is it maintained and what forces change it?"

More recently our understanding of the nature of microbial communities has been advanced by an ever-increasing number of mixed culture studies. These investigations have been carried out largely in continuous culture although the relevance of this environment to a soil, sediment or aquatic situation, where substrates are in low concentrations and are unevenly dispersed (Duursma, 1961; Gray, 1976), is questionable. Nonetheless, the concept of biogeochemical cycles implies the mobility of continuous culture rather than the closed environment of the batch culture; in reality, natural environments may fall somewhere between the two: chemostats subject to irregular pulses of substrate. Major considerations in the design of chemostat studies for monitoring community degradation of polysaccharides include:

(i) the changing chemical and physical nature of the substrate during mineralization;

(ii) the difficulty of including antagonistic microorganisms, such as protozoa, which may be important factors in maintaining a favourable nutritional balance between the primary organisms concerned in polysaccharide decay and therefore the stability of the community (Fenchel and Jørgensen, 1977);

(iii) the feasibility of introducing an insoluble, particulate carbon source;

(iv) the use of growth-limiting concentrations of substrate which may select a different (and more realistic) microbial community with low saturation constants and low maximum specific growth rates (Matin and Veldkamp, 1978); and

(v) whether soil or its components should be included in order to provide extensive anionic surface areas as well as a physical refuge for the less competitive and yet important members of the community.

An isolated attempt to simulate a complex soil/microorganism/macro-organic matter community in a fermenter vessel has been described by Lynch and Gunn (1978). These workers monitored the physico-chemical changes occurring during wheat straw breakdown in soil slurries but did not describe the microflora concerned.

Another approach to understanding carbon mineralization which has attracted some attention involves the use of mathematical models (Bunnell and Scoullar, 1975; Smith, 1979a,b; Swift *et al.*, 1979) to describe the transit of carbon through biological cycles. As with all models these descriptions are only useful if they allow the meaningful interpretation of existing data, stimulate the production of new data, and are flexible enough to embrace unforeseen results.

It is quite obvious from this review that our knowledge of microbial communities involved in naturally-occurring substrate mineralization is paltry. It is not difficult to find reasons for this when one considers the physical and chemical diversity of the substrates and the properties of the terrestrial and aquatic environments where degradation occurs. These factors have conspired with conventional pure culture microbiology to persuade the research worker to study defined and homogeneous substrates in axenic culture. Thus I make no apology for presenting an extensive list of substrates and a detailed description of the mineralization of the commonest carbon polymer, cellulose. This is where the current emphasis lies. However, this information should now form the basis of a gradual return to the study of communities composed of two or more microbial species degrading heterogenous substrates. There are a great many guide-lines deriving from existing investigations of the microbial succession and colonization of organic debris as well as from studies of the rumen and of fermentation processes in general. In addition, the expanding interest of microbial physiologists in mixed cultures and their mathematical description is an important stimulus. McFadyen (1975), when writing about ecology in general, warned of the dangers of a totally descriptive approach, "a bottomless pit of precious time, effort and enthusiasm". Perhaps, as the appearance of this volume implies, we are now ready to pass from the descriptive phase of the role of microbial communities, to an analytical phase, and thence to a synthetic and even predictive phase.

REFERENCES

Ait, N., Creuzet, N., and Forget, P. (1979). Partial purification of cellulase from *Clostridium thermocellum*. *Journal of General Microbiology* **113**, 399–402.

Alexander, M. (1977). "Introduction to Soil Microbiology", 2nd Edition. John Wiley & Sons, New York.

Almin, D. E. and Erikkson, K-E. (1967). Enzymic degradation of polymers. I. Viscometric method for the determination of enzymic activity. *Biochimica et Biophysica Acta* **139**, 238–247.

Almin, K. E., Eriksson, K-E., and Jansson, C. (1967). Enzymic degradation of polymers. II. Viscometric determination of cellulose activity in absolute terms. *Biochimica et Biophysica Acta* **139**, 248–253.

Almin, K. E., Eriksson, K-E., and Pettersson, B. (1975). Extracellular enzyme system utilized by the fungus *Sporotrichum pulverulentum* (*Chrysosporium lignorum*) for the breakdown of cellulose. *European Journal of Biochemistry* **51**, 207–211.

Ander, P. and Eriksson, K-E. (1976). The importance of phenol oxidase activity in lignin degradation by the white-rot fungus *Sporotrichum pulverulentum*. *Archives of Microbiology* **109**, 1–8.

Ander, P., Hatakka, A., and Eriksson, K-E. (1980). Degradation of lignin and lignin-related substances by *Sporotrichum pulverulentum*. *In* "Lignin Biodegradation: Microbiology, Chemistry and Potential Applications", vol. II, pp. 1–15. (Kirk, T. K., Higuchi, T., and Chang, H-M., eds.) CRC Press, Boca Raton, Florida.

Ayers, A. R., Ayers, S. B., and Eriksson, K-E. (1978). Cellobiose oxidase, purification and partial characterization of a hemoprotein from *Sporotrichum pulverulentum*. *European Journal of Biochemistry* **90**, 171–181.

Bacon, J. D. S. (1979). Factors limiting the action of polysaccharide degrading enzymes. *In* "Microbial Polysaccharides and Polysaccharases", pp. 269–284. (Berkeley, R. C. W., Gooday, G. W., and Ellwood, D. C., eds.) Academic Press, London and New York.

Baecker, A. A. W. and King, B. (1980). Decay of wood by Actinomycetales. *In* "Biodeterioration", pp. 53–58. (Oxley, T. A., Becker, G., and Allsopp, D., eds.) Pitman, London.

Bartnicki-Garcia, S. (1968). Cell wall chemistry, morphogenesis and taxonomy of fungi. *Annual Review of Microbiology* **22**, 87–108.

Béguin, P. and Eisen, H. (1978). Purification and partial characterization of three extracellular cellulases from *Cellulomonas* sp. *European Journal of Biochemistry* **87**, 525–531.

Béguin, P., Eisen, H., and Roupas, A. (1977). Free and cellulose-bound cellulases in a *Cellulomonas* species. *Journal of General Microbiology* **101**, 191–196.

Benoit, R. E. and Starkey, R. L. (1968). Inhibition of decomposition of cellulose and some other carbohydrates by tannin. *Soil Science* **105**, 291–296.

Berg, B. (1975). Cellulase location in *Cellvibrio fulvus*. *Canadian Journal of Microbiology* **21**, 51–57.

Berg, B. and Pettersson, G. (1977). Location and formation of cellulase in *Trichoderma viride*. *Journal of Applied Bacteriology* **42**, 65–75.

Berg, B. and Soderstrom, B. (1979). Fungal biomass and nitrogen in decomposing Scots pine needle litter. *Soil Biology and Biochemistry* **11**, 339–341.

Berg, B., v. Hofsten, B., and Pettersson, G. (1972a). Growth and cellulase formation by *Cellvibrio fulvus*. *Journal of Applied Bacteriology* **35**, 201–214.

Berg, B., v. Hofsten, B., and Pettersson, G. (1972b). Electron microscopic observations on the degradation of cellulose fibres by *Cellvibrio fulvus* and *Sporocytophaga myxococcoides*. *Journal of Applied Bacteriology* **35**, 215–219.

Berghem, L. E. R. and Pettersson, L. G. (1974). The mechanism of enzymatic cellulose degradation. Isolation and some properties of a β-glucosidase from *Trichoderma viride*. *European Journal of Biochemistry* **46**, 295–305.

Berghem, L. E. R., Pettersson, L. G., and Axio-Fredriksson, U-B. (1975). The mechanism of enzymatic cellulose degradation. Characterization and enzymatic properties of a β-1,4-glucan cellobiohydrolase from *Trichoderma viride*. *European Journal of Biochemistry* **53**, 55–62.

Berghem, L. E. R., Pettersson, L. G., and Axio-Fredriksson, U-B. (1976). The mechanism of enzymatic cellulose degradation. Purification and some properties of two different 1,4-β-glucan glucanohydrolases from *Trichoderma viride*. *European Journal of Biochemistry* **61**, 621–630.

Bevers, J. (1976). Biochemistry of the breakdown of cellulose by cellulolytic bacteria isolated from waste water treatment units. *Agricultura* **22**, 1–115.

Boois, H. M. de (1976). Fungal development on oak leaf litter and decomposition potential of some fungal species. *Revue d'Ecologie et de Biologie du Sol* **13**, 437–448.

Bryant, M. P. (1977). Microbiology of the rumen. *In* "Duke's Physiology of Domestic Animals", 9th Edition, pp. 28–30. (Stevensen, M. J., ed.) Cornell University Press, Ithaca.

Bryant, M. P., Campbell, L. L., Reddy, C. A., and Crabhill, M. R. (1977). Growth of *Desulfovibrio* in lactate or ethanol media low in sulfate in asociation with H_2-utilizing methanogenic bacteria. *Applied and Environmental Microbiology* **33**, 1162–1169.

Bryant, M. P., Wolin, E. A., Wolin, M. J., and Wolfe, R. S. (1967). *Methanobacillus omelianskii*, a symbiotic association of two species of bacteria. *Archives fur Mikrobiologie* **59**, 20–31.

Bucht, B. and Eriksson, K-E. (1969). Extracellular enzyme system utilised by the rot fungus *Stereum sanguinolentum* for the breakdown of cellulose. IV. Separation of cellobiase and aryl β-glucosidase activity. *Archives of Biochemistry and Biophysics* **129**, 416–420.

Bunnell, F. L. and Scoullar, K. A. (1975). ABISKO II. A computer simulation model of carbon flux in tundra ecosystems. *In* "Structure and Function of Tundra and Ecosystems", pp. 207–226. (Holding, A. J., Heal, O. W., MacLean, S. F., and Flanagan, P. W., eds.) Tundra Biome Steering Committee, Stockholm.

Burns, R. G. (1977). Soil enzymology. *Science Progress* (Oxford) **64**, 275–285.

Burns, R. G. (1978). Enzymes in soil: some theoretical and practical considerations. *In* "Soil Enzymes", pp. 295–339. (Burns, R. G., ed.) Academic Press, London and New York.

Burns, R. G. (1979). Interaction of microorganisms, their substrates and their products with soil surfaces. *In* "Adhesion of Microorganisms to Surfaces", pp. 109–138. (Ellwood, D. C., Melling, J., and Rutter, P., eds.) Academic Press, London and New York.

Burns, R. G. (1980). Microbial adhesion to soil surfaces: consequences for growth and enzyme activities. *In* "Microbial Ahesion to Surfaces", pp. 249–262. (Berkeley, R. C. W., Lynch, J. M., Melling, J., Rutter, P. R., and Vincent, B., eds.) Ellis Horwood, Chichester.

Cain, R. B. (1980). The uptake and catabolism of lignin-related aromatic compounds and their regulation in microorganisms. *In* "Lignin Biodegradation: Microbiology, Chemistry and Potential Applications", vol. I, pp. 21–60. (Kirk, T. K., Higuchi, T., and Chang, H-M., eds.) CRC Press, Boca Raton, Florida.

Campbell, C. A., Paul, E. A., Rennie, D. A., and McCallum, K. J. (1967). Factors affecting the accuracy of the carbon-dating method in soil humus studies. *Soil Science* **104**, 81–85.

Canevascini, G. and Meier, H. (1978). Celluloylic enzymes of *Sporotrichum thermophile*. *Abstract XII International Congress of Microbiology*, Munich, 1978.

Carlson, D. M. and Matthews, L. W. (1966). Polyuronic acids produced by *Pseudomonas aeroginosa*. *Biochemistry* **5**, 2817–2828.

Carruthers, S. M. and Rayner, A. D. M. (1979). Fungal communities in decaying hardwood branches. *Transactions of the British Mycological Society* **72**, 283–289.

Chahal, D. S., Swan, J. E., and Moo-Young, M. (1977). Protein and cellulase production by *Chaetomium cellulolyticum* grown on wheat straw. *Developments in Industrial Microbiology* **18**, 433–442.

Chahal, D. S., Moo-Young, M., and Dhillon, G. S. (1979). Bioconversion of wheat straw components into single-cell protein. *Canadian Journal of Microbiology* **25**, 793–797.

Chang, W. and Thayer, D. W. (1977). The cellulase system of a *Cytophaga* species. *Canadian Journal of Microbiology* **23**, 1285–1292.

Cheshire, M. V. (1977). Origins and stability of soil polysaccharide. *Journal of Soil Science* **28**, 1–10.

Cheshire, M. V. (1979). "Nature and Origin of Carbohydrates in Soils". Academic Press, London and New York.

Cheshire, M. V. and Anderson, G. (1975). Soil polysaccharides and carbohydrate phosphates. *Soil Science* **119**, 356–362.

Clarholm, M. and Rosswall, T. (1980). Biomass and turnover of bacteria in a forest soil and a peat. *Soil Biology and Biochemistry* **12**, 49–57.

Coen, J. A. and Dehority, B. A. (1970). Degradation and utilisation of hemicellulose from intact forages by pure cultures of rumen bacteria. *Applied Microbiology* **20**, 362–368.

Cowling, E. B. and Brown, W. (1979). Structural features of cellulosic materials in relation to enzymatic hydrolyses. *Advances in Chemistry Series* **95**, 152–187.

Crawford, R. L. (1980). Degradation of Douglas fir lignin by *Streptomyces viridostorus*. *In* "Colloque Cellulolyse Microbienne", pp. 191–193. CNRS, Marseilles.

Crawford, D. L. and Crawford, R. L. (1976). Microbial degradation of lignocellulose: the lignin component. *Applied and Environmental Microbiology* **31**, 714–717.

Crawford, D. L. and Sutherland, J. B. (1979). The role of actinomycetes in the decomposition of lignocellulose. *Developments in Industrial Microbiology* **20**, 143–151.

Crawford, R. L., Robinson, L. E., and Cheh, A. M. (1980). ^{14}C-labeled lignins as substrates for the study of lignin biodegradation and transformation. *In* "Lignin Biodegradation: Microbiology, Chemistry and Potential Applications", vol. I, pp. 61–76. (Kirk, T. K., Higuchi, T., and Chang, H-M., eds.) CRC Press, Boca Raton, Florida.

Dagley, S. (1978). Microbial catabolism, the carbon cycle and environmental pollution. *Naturwissenschaften* **65**, 85–95.

Dehority, B. A. and Scott, H. W. (1967). Extent of cellulose and hemicellulose digestion in various forages by pure cultures of rumen bacteria. *Journal of Dairy Science* **50**, 1136–1141.

Dekker, R. F. H. (1979). The hemicellulase group of enzymes. *In* "Polysaccharides in Foods", pp. 93–108. (Blanchard, J. M. V. and Mitchell, J. R., eds.) Butterworths, London.

Dekker, R. F. H. and Richards, G. N. (1976). Hemicellulases: Their occurrence, purification physicochemical properties and mode of action. *Advances in Carbohydrate Chemistry and Biochemistry* **32**, 227–352.

Den Dooren de Jong, L. W. (1971). Tolerance of *Azotobacter* for metallic and non-metallic ions. *Antonie van Leeuwenhoek* **37**, 119–124.

Deshpande, V., Eriksson, K-E., and Pettersson, B. (1978). Production, purification and partial characterization of 1,4-β-glucosidase enzymes from *Sporotrichum pulverulentum*. *European Journal of Biochemistry* **90**, 191–198.

Dickinson, C. H. (1974). Decomposition of litter in soil. *In* "Biology of Plant Litter Decomposition", pp. 633–658. (Dickinson, C. H. and Pugh, G. J. F., eds.) Academic Press, London and New York.

Dickinson, S. and Pugh, G. J. F. (1974). "Biology of Plant Litter Decomposition", vols. 1 and 2. Academic Press, London and New York.

Dillworth, G., Weigel, J., Ljungdahl, L. G., and Peck, H. D. (1980). Reconstruction of mesophilic microbial associations which ferment cellulose to various products. *In* "Colloque Cellulolyse Microbienne", pp. 187–190. CNRS, Marseilles.

Drew, S. W. and Kadam, K. L. (1979). Lignin metabolism by *Aspergillus fumigatus* and white-rot fungi. *Developments in Industrial Microbiology* **20**, 153–161.

Drozdowicz, A. (1971). The behaviour of cellulase in soil. *Revues of Microbiology* **2**, 17–23.

Duckworth, M. (1977). Teichoic acids. *In* "Surface Carbohydrates of the Prokaryotic Cell", pp. 177–208. (Sutherland, I. W., ed.) Academic Press, London and New York.

Duckworth, M. and Yaphe, W. (1972). The structure of agar. II. The use of bacterial agarase to elucidate structural features of the charged polysaccharides in agar. *Carbohydrate Research* **16**, 435–445.

Dudman, W. F. (1977). The role of surface polysaccharides in natural environments. *In* "Surface Carbohydrates of the Prokaryotic Cell", pp. 357–414. (Sutherland, I. W., ed.) Academic Press, London and New York.

Duursma, E. K. (1961). Dissolved organic carbon, nitrogen and phosphorus in the sea. *Netherlands Journal of Sea Research* **1**, 1–148.

Ellwood, D. C., Hedger, J. N., Latham, M. J., Lynch, J. M., and Slater, J. H. (1980). *Contemporary Microbial Ecology*. Academic Press, London and New York.

Enari, P. and Markkanen, A. F. (1977). Production of cellulolytic enzymes by fungi. *Advances in Biochemical Engineering* **5**, 1–24.

Eriksson, K-E. (1978). Enzyme mechanisms involved in cellulose hydrolysis by the white rot fungus *Sporotrichum pulverulentum*. *Biotechnology and Bioengineering Symposium* **10**, 317–332.

Eriksson, K-E. (1979). Biosynthesis of polysaccharases. *In* "Microbial Polysaccharides and Polysaccharases", pp. 285–296. (Berkeley, R. C. W., Gooday, G. W., and Ellwood, D. C., eds.) Academic Press, London and New York.

Eriksson, K-E. and Johnsrud, S. C. (1982). Mineralisation of carbon. *In* "Experimental Microbial Ecology", pp. 134–153. (Burns, R. G. and Slater, J. H., eds.) Blackwell Scientific Publications, Oxford.

Eriksson, K-E. and Pettersson, B. (1975). Extracellular enzyme system utilised by the fungus *Sporotrichum pulverulentum* for the breakdown of cellulose. *European Journal of Biochemistry* **51**, 193–206.

Eriksson, K-E. and Rzedowski, W. (1969). Extracellular enzyme system utilised by the fungus *Chrysosporium lignorum* for the breakdown of cellulose. I. Studies of the enzyme production. *Archives of Biochemistry and Biophysics* **129**, 688–693.

Fan, L. T., Lee, Y-H., and Beardmore, D. H. (1980). Major chemical and physical features of cellulosic materials as substrates for enzymatic hydrolyses. *Advances in Biochemical Engineering* **14**, 101–117.

Fenchel, T. M. and Jørgensen, B. B. (1977). Detritus food chains in aquatic ecosystems: the role of bacteria. *Advances in Microbial Ecology* **1**, 1–58.

Finch, P., Hayes, M. H. B., and Stacey, M. (1967). Studies on soil polysaccharides and their interactions with clay preparations. *International Soil Science Society Transactions*, Commissions II and IV, pp. 19–32.

Flannigan, B. and Baba, M. S. O. (1980). Growth and enzyme production in *Aspergilli* which cause deterioration in stored grain. *In* "Biodeterioration", pp. 229–236. (Oxley, T. A., Allsopp, D., and Becker, G., eds.) Pitman Publishing Ltd and the Biodeterioration Society, London.

Flickinger, M. C. and Tsao, G. T. (1978). Fermentation substrates from cellulosic materials: fermentation products cellulosic materials. *In* "Annual Reports on Fermentation Processes", vol. 2, pp. 23–42. (Perlman, D., ed.) Academic Press, London and New York.

Folan, M. A. and Coughlan, M. P. (1978). The cellulase complex in the culture filtrate of the thermophilic fungus *Talaromyces emersonii*. *International Journal of Biochemistry* **9**, 717–722.

Frankland, J. C. (1974). Decomposition of lower plants. *In* "Biology of Plant Litter Decomposition", Vol. 2, pp. 3–36. (Dickinson, C. H. and Pugh, G. J. F., eds.) Academic Press, London and New York.
Frankland, J. C. (1976) Decomposition of bracken litter. *Botanical Journal of the Linnean Society* **73**, 133–143.
Freudenberg, K. and Neish, A. C. (1968). "Constitution and Biosynthesis of Lignin". Springer-Verlag, Berlin.
Gaden, E. L., Mandels, M. H., Reese, E. T., and Spano, L. A. (1976). "Enzymic Conversion of Cellulosic Materials: Technology and Applications". Wiley, New York.
Garrett, S. D. (1963). "Soil Fungi and Soil Fertility". Pergamon Press, Oxford.
Ghose, T. K. and Ghosh, P. (1979). Cellulase production and cellulose hydrolysis. *Process Biochemistry* (November), 20–24.
Ghose, T. K., Fiechter, A., and Blakebrough, N. (1976). "Advances in Biochemical Engineering", vol. 4. Springer-Verlag, Berlin.
Glenn, A. R. (1976). Production of extracellular proteins by bacteria. *Annual Review of Microbiology* **30**, 41–62.
Gong, C-S., Ladisch, M. R., and Tsao, G. T. (1977). Cellobiase from *Trichoderma viride*: purification, properties, kinetics and mechanism. *Biotechnology and Bioengineering* **19**, 959–981.
Goodfellow, M. and Dawson, D. (1975). Qualitative and quantitative studies of bacteria colonizing *Picea sitchensis* litter. *Soil Biology and Biochemistry* **10**, 303–307.
Gray, T. R. G. (1976). Survival of vegetative microbes in soil. *Symposium of the Society for General Microbiology* **26**, 327–364.
Greenland, D. J. (1956). The adsorption of sugars by montmorillonite. I. X-ray studies. *Journal of Soil Science* **7**, 319–333.
Griffiths, E. and Burns, R. G. (1972). Interaction between phenolic substances and microbial polysaccharides in soil aggregation. *Plant and Soil* **36**, 599–612.
Gritzali, M. and Brown, R. D. (1979). The cellulase system of *Trichoderma*: relationships between extracellular enzymes from induced or cellulose-grown cells. *In* "Hydrolysis of Cellulose: Mechanisms of Enzymatic and Acid Catalysis", pp. 237–260. (Brown, R. D. and Jurasek, L., eds.) American Chemical Society, Washington.
Grosovsky, B. D-D. and Margulis, L. (1982). Termite microbial communities. *In* "Experimental Microbial Ecology", pp. 519–532. (Burns, R. G. and Slater, J. H., eds.) Blackwell Scientific Publications, Oxford.
Guckert, A., Valla, M., and Jacquin, F. (1975). Adsorption of humic acids and soil polysaccharides on montmorillonite. *Pochvovedenye* **2**, 41–47.
Gum, E. K. and Brown, R. D. (1977). Comparison of four purified extracellular 1,4-β-glucan cellobiohydrolase enzymes from *Trichoderma viride*. *Biochimica et Biophysica Acta* **492**, 225–231.
Hackett, W. F., Connors, W. J., Kirk, T. K., and Zeikus, J. G. (1977).

Microbial decomposition of synthetic ^{14}C-labelled lignins in nature: lignin biodegradation in a variety of natural materials. *Applied and Environmental Microbiology* **33**, 43–51.

Hagerdal, B. G. R., Ferchak, J. D., and Pye, E. K. (1978). Cellulolytic enzyme system of *Thermoactinomyces* sp. grown on microcrystalline cellulose. *Applied and Environmental Microbiology* **36**, 606–612

Hagerdal, B., Harris, H., and Pye, E. K. (1979). Association of β-glucosidase with intact cells of *Thermoactinomyces. Biotechnology and Bioengineering* **21**, 345–355.

Haider, K. and Trojanowski, J. (1980). A comparison of the degradation of ^{14}C-labeled DHP and corn stalk lignins by micro- and macrofungi and bacteria. *In* "Lignin Biodegradation: Microbiology, Chemistry and Potential Applications", vol. I, pp. 111–134. (Kirk, T. K., Higuchi, T., and Chang, H-M., eds.) CRC Press, Boca Raton, Florida.

Hall, P. L., Glasser, W. G., and Drew, S. W. (1980). Enzymatic transformations of lignin. *In* "Lignin Biodegradation: Microbiology, Chemistry and Potential Applications", vol. II, pp. 33–49. (Kirk, T. K., Higuchi, T., and Chang, H-M., eds.) CRC Press, Boca Raton, Florida.

Halliwell, G. and Bryant, M. P. (1963). The cellulolytic activity of pure strains of bacteria from the rumen of cattle. *Journal of General Microbiology* **32**, 441–448.

Halliwell, G. and Griffin, M. (1978). Affinity chromatography of the cellulase system of *Trichoderma koningii. Biochemical Journal* **169**, 713–715.

Halliwell, G., Griffin, M., and Vincent, R. (1972). The role of component C_1 in cellulolytic systems. *Biochemical Journal* **127**, 43P.

Handley, W. R. C. (1954). Mull and mor formation in relation to forest soils. *Forestry Commission Bulletin* **23**, H.M.S.O.

Hankin, L., Poincelot, R. P., and Anagnostakis, S. L. (1976). Microorganisms from composting leaves: ability to produce extra-cellular degradative enzymes. *Microbial Ecology* **2**, 296–308.

Hayes, A. J. (1979). The microbiology of plant litter decomposition. *Science Progress (Oxford)* **66**, 25–42.

Hayes, M. H. B. and Swift, R. S. (1978). The chemistry of soil organic colloids. *In* "The Chemistry of Soil Constituents", pp. 179–320. (Greenland, D. J. and Hayes, M. H. B., eds.) John Wiley & Sons, Chichester.

Hering, T. F. (1967). Fungal decomposition of oak leaf litter. *Transactions of the British Mycological Society* **50**, 267–273.

Hobson, P. N. (1979). Polysaccharide degradation in the rumen. *In* "Microbial Polysaccharides and Polysaccharases", pp. 377–397. (Berkeley, R. C. W., Gooday, G. W., and Ellwood, D. C., eds.) Academic Press, London and New York.

Hobson, P. N. and Summers, R. (1978). Anaerobic bacteria in mixed cultures. Ecology of the rumen and sewage digesters. *In* "Techniques for the Study of Mixed Populations", pp. 125–141. (Davies, R. and Lovelock, D. W., eds.) Academic Press, London and New York.

Howell, J. A. and Stuck, J. D. (1975). Kinetics of Solka Floc cellulose hydrolysis by *Trichoderma viride* cellulase. *Biotechnology and Bioengineering* **17**, 873–893.
Hulme, M. A. and Stranks, D. W. (1971). Regulation of cellulase production by *Myrothecium verrucaria* grown on non-cellulosic substrates. *Journal of General Microbiology* **69**, 145–155.
Humphrey, A. E., Moriera, A., Armiger, W., and Zabriskie, D. (1977). Production of single cell protein from cellulose wastes. *Biotechnology and Bioengineering Symposium* **7**, 45–64.
Hungate, R. E. (1966). "The Rumen and its Microbes". Academic Press, London and New York.
Hwang, J. T. and Suzuki, H. (1976). Intracellular distribution and some properties of β-glucosidases of a cellulolytic pseudomonad. *Agricultural and Biological Chemistry* **40**, 2169–2175.
Iannotti, E. L., Kafkewitz, D., Wolin, M. J., and Bryant, M. P. (1973). Glucose fermentation products of *Ruminococcus albus* grown in continuous culture with *Vibrio succinogenes*: changes caused by interspecies transfer of H_2. *Journal of Bacteriology* **114**, 1231–1240.
Inglin, M., Feinberg, B. A., and Loewenberg, J. R. (1980). Partial purification and characterization of a new intracellular β-glucosidase of *Trichoderma reesei*. *Biochemical Journal* **185**, 515–519.
Ishawue, M. and Kluepfel, D. (1980). Cellulase complex of a mesophilic *Streptomyces* strain. *Canadian Journal of Microbiology* **26**, 183–189.
Izumi, K. (1972). Chemical heterogeneity of the agar from *Gracilaria verrucosa*. *Journal of Biochemistry* **72**, 135–140
Ivarson, K. C. (1977). Changes in decomposition rate, microbial population and carbohydrate content of an acid peat bog after liming and reclamation. *Canadian Journal of Soil Science* **57**, 129–137.
Jarman, T. R. (1979). Bacterial alginate synthesis. *In* "Microbial Polysaccharides and Polysaccharases", pp. 35–50. (Berkeley, R. C. W., Gooday, G. W., and Ellwood, D. C., eds.) Academic Press, London and New York.
Jenkinson, D. S. (1966). The priming action. *In* "The Use of Isotopes in Soil Organic Matter Studies". International Atomic Energy Agency, Braunschweig, Germany, 1963 Vienna.
Jenkinson, D. S. (1968). Chemical tests for potentially available nitrogen in soil. *Journal of the Science of Food and Agriculture* **19**, 160–168.
Jenkinson, D. S. and Ladd, J. N. (1981). Microbial biomass in soil—measurement and turnover. *In* "Soil Biochemistry", vol. 5, pp. 415–471. (Paul, E. A. and Ladd, J. N., eds.) Marcel Dekker, New York.
Jensen, V. (1974). Decomposition of angiosperm tree leaf litter. *In* "Biology of Plant Litter Decomposition", vol. I, pp. 69–104. (Dickinson, C. H. and Pugh, G. J. F., eds.) Academic Press, London and New York.
Jermyn, M. A. (1967). Fungal cellulases. XX. Some observations on the induction and inhibition of cellobiase of *Stachybotrys atra*. *Australian Journal of Biological Science* **20**, 193–220.

Jeuniaux, C. (1963). "Chitine et Chitinolyse". Masson et Cie, Paris.

Jones, D. and Webley, D. M. (1968). A new enrichment technique for studying lysis of fungal cell walls in soil. *Plant and Soil* **28**, 147–157.

Kaszubiak, H., Kaczmarek, W., and Durska, G. (1976). Feeding of the soil microbial community on organic matter from its dead cells. *Ekologia Polska* **24**, 391–397.

Kennedy, J. F. (1979). Enzymes. *Carbohydrate Chemistry* **11**, 371–429.

Kilbertus, G. and Reissenger, O. (1975). Decompoition of plant material. The *in vitro* and *in situ* activity of some micro-organisms. *Revue Ecologie et de Biologie du Sol* **12**, 363–374.

Kirk, T. K. (1971). The effects of micro-organisms on lignin. *Annual Review of Phytopathology* **9**, 185–210.

Kirk, T. K. (1975). Chemistry of lignin degradation by wood-destroying fungi. *In* "Biological Transformations of Wood by Microorganisms", pp. 153–164. (Liese, W., ed.) Springer-Verlag, Berlin.

Kirk, T. K., Connors, W. J., and Zeikus, J. G. (1976). Requirement for a growth substrate during lignin decomposition by two wood-rotting fungi. *Applied and Environmental Microbiology* **32**, 192–194.

Kirk, T. K., Connors, W. J., and Zeikus, J. G. (1977). Advances in understanding the microbiological degradation of lignin. *Recent Advances in Phytochemistry* **11**, 369–394.

Kohl, R. A. and Taylor, S. A. (1961). Hydrogen bonding between the carbonyl group and Wyoming bentonite. *Soil Science* **91**, 223–227.

Kuo, M-J. and Alexander, M. (1967). Inhibition of the lysis of fungi by melanins. *Journal of Bacteriology* **94**, 624–629.

Kuwahara, M. (1980). Metabolism of lignin-related compounds by bacteria. *In* "Lignin Biodegradation: Microbiology, Chemistry and Potential Applications", vol. II, pp. 127–146. (Kirk, T. K., Higuchi, T., and Chang, H-M., eds.) CRC Press, Boca Raton, Florida.

Ladd, J. N. (1978). Origin and range of enzymes in soil. *In* "Soil Enzymes", pp. 51–96. (Burns, R. G., ed.) Academic Press, London and New York.

Ladisch, M. R. (1979). Fermentable sugars from cellulosic residues. *Process Biochemistry* **14**, 21–25.

Ladisch, M. R., Gong, C-S., and Tsao, G. T. (1980). Cellobiose hydrolysis by endoglucanase (glucan glucanohydrolase) from *Trichoderma reesei*: kinetics and mechanism. *Biotechnology and Bioengineering* **22**, 1107–1126.

Latham, M. J. and Wolin, M. J. (1977). Fermentation of cellulose by *Ruminococcus flavefaciens* in the presence and absence of *Methanobacterium ruminatium*. *Applied and Environmental Microbiology* **34**, 297–301.

Leatherwood, J. M. (1969). Cellulase complex in *Ruminococcus* and a new mechanism for cellulose degradation. *In* "Cellulases and Their Applications", Advances in Chemistry Series 95, pp. 53–59. (Gould, R. F., ed.) American Chemical Society, Washington.

Lee, B. H. and Blackburn, T. H. (1975). Cellulase production by a thermophilic *Clostridium* species. *Applied Microbiology* **30**, 346–353.

Lee, Y-H. and Fan, L. T. (1980). Properties and mode of action of cellulase. *Advances in Biochemical Engineering* **17**, 101–129.

Lee, Y-H., Fan, L. T., and Fan, L-S. (1980). Kinetics of hydrolysis of insoluble cellulose by cellulase. *Advances in Biochemical Engineering* **17**, 131–168.

Lieth, H. (1975). Some prospects beyond production measurement. *In* "Primary Productivity of the Biosphere", pp. 285–304. (Lieth, H. and Whittaker, R. H., eds.) Springer-Verlag, New York.

Linhares, L. F. and Martin, J. P. (1978). Decomposition in soil of the humic acid-type polymers (melanins) of *Eurotium echinulatum*, *Aspergillus glaucus* and other fungi. *Soil Science Society of America Journal* **42**, 738–743.

Loewenberg, J. R. and Chapman, C. M. (1977). Sophorose metabolism and cellulase induction in *Trichoderma*. *Archives of Microbiology* **113**, 61–64.

Lory, A. S. and Collier, R. J. (1978). Enzymic properties of *Pseudomonas aeruginosa* Exotoxin A. *Abstracts Annual Meeting of the American Society of Microbiology* B-92.

Lousier, J. D. and Parkinson, D. (1976). Litter decomposition in a cool temperate deciduous forest. *Canadian Journal of Botany* **54**, 419–436.

Lundquist, K. T., Kirk, T. K., and Connors, W. J. (1977). Fungal degradation of kraft lignin and lignin sulfonates prepared from synthetic ^{14}C-lignins. *Archives of Microbiology* **112**, 291–296.

Lynch, J. M. and Gunn, K. B. (1978). Use of a chemostat to study decomposition of wheat straw in soil slurries. *Journal of Soil Science* **29**, 551–556.

Lynch, J. M., Bennett, J. A., and Slater, J. H. (1980). The characterisation of a mixed microbial population degrading cellulose in straw. *In* "Colloque Cellulolyse Microbienne", pp. 187–190. CNRS, Marseilles.

Macfadyen, A. (1975). *Journal of Applied Ecology* **12**, 397, cited by Hayes (1979).

McLaren, A. D. and Peterson, G. H. (1967). Introduction to the biochemistry of terrestrial soils. *In* "Soil Biochemistry", vol. 1, pp. 1–18. (McLaren, A. D. and Peterson, G. H., eds.) Marcel Dekker, New York.

Mandels, M., Parrish, F. W., and Reese, E. T. (1962). Sophorose as an inducer of cellulase in *Trichoderma viride*. *Journal of Bacteriology* **83**, 400–408.

Manners, D. J. (1979). The enzymic degradation of starches. *In* "Polysaccharides in Foods", pp. 75–91. (Blanchard, J. M. V. and Mitchell, J. R., eds.) Butterworths, London.

Martin, J. P. and Haider, K. (1980). Microbial degradation and stabilization of ^{14}C-labeled lignins, phenols, and phenolic polymers in relation to soil humus formation. *In* "Lignin Biodegradation: Microbiology, Chemistry and Potential Applications", vol. I, pp. 77–100. (Kirk, T. K., Higuchi, T., and Chang, H-M., eds.) CRC Press, Boca Raton, Florida.

Martin, J. P., Ervin, J. O., and Shepherd, R. A. (1959). Decomposition and aggregating effect of fungus cell material in soil. *Soil Science Society of America Proceedings* **23**, 217–220.

Martin, J. P., Ervin, J. O., and Shepherd, R. A. (1966). Decomposition of the iron, aluminium, zinc and copper salts or complexes of some microbial and plant polysaccharides in soil. *Soil Science Society of America Proceedings* **30**, 196–200.

Martin, J. P., Haider, K., and Wolf, D. (1972). Synthesis of phenols and phenolic polymers by *Hendersonula toruloidea* in relation to humic acid formation. *Soil Science Society of America Proceedings* **36**, 311–315.

Martin, J. P., Parsa, A. A., and Haider, K. (1978). Influence of intimate association with humic polymers on the biodegradation of [^{14}C]-labelled organic substances in soil. *Soil Biology and Biochemistry* **10**, 483–486.

Matin, A. and Veldkamp, H. (1978). Physiological basis of the selective advantage of a *Spirillum* species in a carbon-limited environment. *Journal of General Microbiology* **105**, 187–197.

Mayaudon, J. (1966). Humification des complexes microbiens dans le sol. I. Le lipoprotein polysaccharide carbone—14 dux *Pseudomonas fluorescens*. *In* "The Use of Isotopes in Soil Organic Matter", pp. 259–269. Pergamon Press, London.

Mayaudon, J. and Simonart, P. (1963). Humification des microorganisms marqués par ^{14}C dans le sol. *Annals Institute Pasteur* **105**, 257–266.

Moo-Young, M., Chahal, D. S., and Vlach, D. (1978). SCP production from various chemically-pretreated substances using *Chaetomium cellulolyticum*. *Biotechnology and Bioengineering* **20**, 107–118.

Moo-Young, M., Daugulis, A. J., Chahal, D. S., and Macdonald, D. G. (1979). The Waterloo process for SCP production from waste biomass. *Process Biochemistry* **14**, 38–40.

Muzzarelli, R. A. A. (1977). "Chitin". Pergamon Press, Oxford.

Nakas, J. P. and Klein, J. P. (1980). Mineralization capacity of bacteria and fungi from the rhizosphere-rhizoplane of a semiarid grassland. *Applied and Environmental Microbiology* **39**, 113–117.

Nelson, D. W., Martin, J. P., and Ervin, J. O. (1979). Decomposition of microbial cells and components in soil and their stabilization through complexing with mode humic-acid type polymers. *Soil Science Society of America Journal* **43**, 84–88.

Nisazawa, T., Suzuki, H., and Nisazawa, K. (1972). Catabolite repression of cellulase formation in *Trichoderma viride*. *Journal of Biochemistry (Tokyo)* **71**, 999–1007.

Norkvans, B. (1967). Cellulose and cellulolysis. *Advances in Applied Microbiology* **9**, 91–130.

Oberkotter, L. V. and Rosenbergh, F. A. (1978). Extracellular β-endo-1,4-glucanases in *Cellvibrio vulgaris*. *Applied and Environmental Microbiology* **36**, 205–209.

Oberkotter, L. V. and Rosenberg, F. A. (1980). Aspects of the cellulase system of *Cellvibrio vulgaris*. *In* "Biodeterioration", pp. 149–158. (Oxley, T. A., Allsopp, D., and Becker, G., eds.) Pitman Publishers Ltd and The Biodeterioration Society, London.

O'Day, D. H. and Paterno, G. D. (1979). Intracellular and extracellular CM-cellulase and β-glucosidase activity during germination of *Polysphondylium pallidum* microcysts. *Archives of Microbiology* **121**, 231–234.

Olness, A. and Clapp, C. E. (1973). Occurrence of collapsed and expanded crystals in montmorillonite—dextran complexes. *Clays and Clay Minerals* **21**, 289–293.

Olness, A. and Clapp, C. E. (1975). Influence of polysaccharide structure on dextran adsorption by montmorillonite. *Soil Biology and Biochemistry* **7**, 113–118.

Olutiola, P. O. (1976). Cellulolytic enzymes in culture filtrates of *Rhizoctonia lamellifera*. *Journal of General Microbiology* **97**, 251–256.

Olutiola, P. O. and Ayers, P. G. (1973). A cellulase complex in culture filtrates of *Phynchosporium secalis*. *Transactions of the British Mycological Society* **60**, 273–282.

Osmundsvag, K. and Goksøyr, J. (1975). Cellulases from *Sporocytophaga myxococcoides* purification and properties. *European Journal of Biochemistry* **57**, 405–409.

Oxley, T. A., Allsopp, D., and Becker, G. (1980). "Biodeterioration". Pitman Publishing Ltd and The Biodeterioration Society, London.

Parfitt, R. L. (1972). Adsorption of charged sugars by montmorillonite. *Soil Science* **113**, 417–421.

Parfitt, R. L. and Greenland, D. J. (1970). Adsorption of polysaccharides by montmorillonite. *Soil Science Society of America Proceedings* **34**, 862–866.

Paul, E. A. and Voroney, R. P. (1980). Nutrient and energy flows through soil microbial biomass. *In* "Contemporary Microbial Ecology", pp. 215–237. (Ellwood, D. C., Hedger, J. N., Latham, M. J., Lynch, J. M., and Slater, J. H., eds.) Academic Press, London and New York.

Paulson, K. N. and Kurtz, L. T. (1969). Locus of urease activity in soil. *Soil Science Society of America Proceedings* **33**, 897–901.

Pavlovskis, O. R. and Wretlind, B. (1978). A virulence factor of *Pseudomonas aeruginosa*: Protease II. *Abstracts Annual Meeting American Society for Microbiology*, B-89.

Peitersen, N. (1975). Cellulase and protein production from mixed cultures of *Trichoderma viride* and a yeast. *Biotechnology and Bioengineering* **17**, 1291–1299.

Percival, E. and McDowell, R. H. (1967). "Chemistry and Enzymology of Marine Algal Polysaccharides". Academic Press, London and New York.

Pettersson, L. G. (1975). The mechanism of enzymatic hydrolysis of cellulose by *Trichoderma viride*. *In* "Symposium on Enzymatic Hydrolysis of Cellulose", pp. 255–262. (Bailey, M., Enari, T-M., and Linko, M., eds.) Sitra, Helsinki.

Pettit, M., Smith, A. R. J., Freedman, R. B., and Burns, R. G. (1976). Soil urease: activity, stability and kinetic properties. *Soil Biology and Biochemistry* **8**, 479–484.

Phelan, M. B., Crawford, D. L., and Pometto, A. L. (1979). Isolation of lignocellulose-decomposing actinomycetes and degradation of specifically ^{14}C-labeled lignocelluloses by six selected *Streptomyces* strains. *Canadian Journal of Microbiology* **25**, 1270–1276.

Pilnik, W. and Rombouts, F. M. (1979). Pectic enzymes. *In* "Polysaccharides in Foods", pp. 109–126. (Blanchard, J. M. V. and Mitchell, J. R., eds.) Butterworths, London.

Pollock, M. R. (1962). Exoenzymes. *In* "The Bacteria", Vol. IV, pp. 121–178. (Gunsalus, I. C. and Stanier, R. Y., eds.) Academic Press, New York and London.

Postgate, J. R. (1974). Evolution with nitrogen-fixing systems. *Symposium of the Society for General Microbiology* **24**, 263–292.

Powell, D. A. (1979). Structure, solution properties and biological interactions of some microbial extracellular polysaccharides. *In* "Microbial Polysaccharides and Polysaccharases", pp. 117–160. (Berkeley, R. C. W., Gooday, G. W., and Ellwood, D. C., eds.) Academic Press, London and New York.

Quastel, J. H. (1965). Soil metabolism. *Revue of Plant Physiology* **16**, 217–240.

Ramasamy, K. and Verachtert, H. (1980). Localization of cellulase components in *Pseudomonas* sp. isolated from activated sludge. *Journal of General Microbiology* **117**, 181–191.

Rayner, A. D. M. (1976). Dematiaceous hyphomycetes and narrow dark zones in decaying wood. *Transactions of the British Mycological Society* **67**, 546–549.

Rayner, A. D. M. (1978). Interactions between fungi colonizing hardwood stumps and their possible role in determining patterns of colonization and succession. *Annals of Applied Biology* **89**, 131–134.

Rayner, A. D. M. and Todd, N. K. (1979). Population and community structure and dynamics of fungi in decaying wood. *Advances in Botanical Research* **7**, 333–420.

Reese, E. T. (1977). Degradation of polymeric carbohydrates by microbial enzymes. *Recent Advances in Phytochemistry* **11**, 311–365.

Reese, E. T., Siu, R. G. H., and Levinson, H. S. (1950). The biological degradation of soluble cellulose derivatives and its relationship to the mechanism of cellulose hydrolysis. *Journal of Bacteriology* **59**, 485–497.

Rogers, H. J. (1974). Peptidoglycans (mucopeptides) structure, function and variations. *Annals of the New York Academy of Sciences* **235**, 29–51.

Rosenberg, S. L. (1978). Cellulose and lignocellulose degradation by thermophilic and thermotolerant fungi. *Mycologia* **70**, 1–13.

Rosenberg, S. L. (1979). Physiological studies of lignocellulose degradation by the thermotolerant mould *Chrysosporium pruinosum*. *Developments in Industrial Microbiology* **20**, 133–142.

Rowell, M. J., Ladd, J. N., and Paul, E. A. (1973). Enzymically active complexes of proteases and humic acid analogues. *Soil Biology and Biochemistry* **5**, 699–703.

Ruinen, J. (1961). The phyllosphere. I. An ecologically neglected milieu. *Plant and Soil* **15**, 81–109.

Saddler, J. N. and Khan, A. W. (1979). Cellulose degradation by a new isolate from sewage sludge, a member of the Bacteriodaceae family. *Canadian Journal of Microbiology* **25**, 1427–1432.

Saini, G. R. and Maclean, A. A. (1966). Adsorption-flocculation reaction of soil polysaccharides with kaolinite. *Soil Science Society of America Proceedings* **30**, 697–699.

Scharer, J. M. and Moo-Young, M. (1979). Methane generation by anaerobic digestion of cellulose-containing wastes. *Advances in Biochemical Engineering* **11**, 85–101.

Scheifinger, C. C. and Wolin, M. J. (1973). Propionate formation from cellulose and soluble sugars by combined cultures of *Bacteriodes succinogenes* and *Selenomonas ruminantium*. *Applied microbiology* **26**, 789–795.
Seifert, K. and Becker, G. (1965). Der Chemische Abbau von Laub-und Nadelholzarten durch verschiedene termiten. *Holzforschung* **19**, 105–111.
Selby, K. (1968). Mechanism of biodegradation of cellulose. *In* "Biodeterioration of Materials", pp. 62–78. (Walters, A. H. and Elphick, J. S., eds.) Elsevier Scientific Publishing Co., Amsterdam.
Selby, K. and Maitland, C. C. (1967). The cellulase of *Trichoderma viride*. Separation of the components involved in the solubilization of cotton. *Biochemical Journal* **104**, 716–724.
Shewale, J. G. and Sadana, J. C. (1978). Cellulase and β-glucosidase production by a basidiomycete species. *Canadian Journal of Microbiology* **24**, 1204–1216.
Shewale, J. G. and Sadana, J. C. (1979). Enzymic hydrolyses of cellulosic materials by *Sclerotium rolfsii* culture filtrate for sugar production. *Canadian Journal of Microbiology* **25**, 773–-783.
Shields, J. A., Paul, E. A., Lowe, W. E., and Parkinson, D. (1973). Turnover of microbial tissue in soil under field conditions. *Soil Biology and Biochemistry* **5**, 753–764.
Shimada, M. (1980). Stereobiochemical approach to lignin biodegradation: possible significance of nonstereospecific oxidation catalyzed by laccase for lignin decomposition by white-rot fungi. *In* "Lignin Biodegradation: Microbiology, Chemistry and Potential Applications", vol. I, pp. 195–213. (Kirk, T. K., Higuchi, T., and Chang, H-M., eds.) CRC Press, Boca Raton, Florida.
Sihtola, H. and Neimo, L. (1975). The structure and properties of cellulose. *In* "Symposium on Enzymatic Hydrolysis of Cellulose", pp. 9–21. (Bailey, M., Enari, T-M., and Linko, M., eds.) SITRA, Helsinki.
Sineriz, F. and Pirt, S. J. (1977). Methane production from glucose by a mixed culture of bacteria in the chemostat: the role of *Citrobacter*. *Journal of General Microbiology* **101**, 57–64.
Skujins, J. (1976). Extracellular enzymes in soil. *Critical Reviews of Microbiology* **4**, 383–421.
Slater, J. H. and Godwin, D. (1980). Microbial adaption and selection. *In* "Contemporary Microbial Ecology", pp. 137–160. (Ellwood, D. C., Hedger, J. N., Latham, M. J., Lynch, J. M., and Slater, J. H., eds.) Academic Press, London and New York.
Smith, M. H. and Gold, M. H. (1979). *Phanerochaete chrysosporium* β-glucosidases: Induction, cellular location, and physical characterization. *Applied and Environmental Microbiology* **37**, 938–942.
Smith, O. L. (1979a). An analytical model of the decomposition of soil organic matter. *Soil Biology and Biochemistry* **11**, 585–606.
Smith, O. L. (1979b). Application of a model of the decomposition of soil organic matter. *Soil Biology and Biochemistry* **11**, 607–618.
Sørensen, L. H. (1975). The influence of clay on the rate of decay of amino acid

metabolites synthesized in soils during the decomposition of cellulose. *Soil Biology and Biochemistry* **7**, 171–177.

Spaulding, B. P. (1977). Enzymatic activities related to the decomposition of coniferous leaf litter. *Soil Science Society of America Journal* **41**, 622–627.

Srinivasan, V. R. (1975). Production of bio-proteins from cellulose. *In* "Symposium on Enzymatic Hydrolysis of Cellulose", pp. 393–605. (Bailey, M., Enari, T-M., and Linko, M., eds.) SITRA, Helsinki.

Srinivasan, V. R. and Han, Y. W. (1969) Utilization of bagasse. *Advances in Chemistry Series* **95**, 447–460.

Sternberg, D. (1976). β-glucosidase of *Trichoderma*: its biosynthesis and role in saccharification of cellulose. *Applied and Environmental Microbiology* **31**, 648–654.

Sternberg, D. and Mandels, G. R. (1979). Induction of cellulolytic enzymes in *Trichoderma reesei* by sophorose. *Journal of Bacteriology* **139**, 761–769.

Sternberg, D., Vljayicumar, P., and Reese, E. T. (1977). β-glucosidase: Microbial production and effect on enzymatic hydrolysis of cellulose. *Canadian Journal of Microbiology* **23**, 139–147.

Stewart, B. J. and Leatherwood, J. M. (1976). Depressed synthesis of cellulase by *Cellulomonas*. *Journal of Bacteriology* **128**, 609–615.

Stradling, D. J. (1977). Food and feeding habits of ants. *In* "Production Ecology of Ants and Termites", pp. 81–106. (Brian, M. V., ed.) Cambridge University Press, Cambridge.

Stotzky, G. (1972). Activity, ecology and population dynamics of microorganisms in soil. *Critical Reviews in Microbiology* **2**, 59–137.

Sturgeon, R. J. (1979a). Microbial polysaccharides. *Carbohydrate Chemistry* **11**, 265–308.

Sturgeon, R. J. (1979b). Plant and algae polysaccharides. *Carbohydrate Chemistry* **11**, 246–264.

Stutzenberger, F. J. (1972). Cellulolytic activity of *Thermomonospora curvata*. Nutritional requirements for cellulase production. *Applied Microbiology* **24**, 77–82.

Su, T-M. and Paulavicius, I. (1975). Enzymic saccharification of cellulose by thermophilic actinomyces. *Applied Polymer Symposium* **28**, 221–236.

Suberkropp, K. and Klug, M. J. (1976). Fungi and bacteria associated with leaves during processing in a woodland stream. *Ecology* **57**, 707–719.

Suzuki, H., Yamane, K., Nisizawa, K. (1969). Extracellular and cell-bound cellulase components of bacteria. *In* "Cellulases and Their Applications", Advances in Chemistry Series 95, pp. 60–82. (Gould, R. F., ed.) American Chemical Society, Washington.

Swift, M. J. (1977). The ecology of wood decomposition. *Science Progress (Oxford)* **64**, 179–203.

Swift, M. J., Heal, O. W., and Anderson, J. M. (1979). "Decomposition in Terrestrial Ecosystems". Blackwell Scientific Publications, London.

Swincer, G. D., Oades, J. M., and Greenland, D. J. (1969). Extraction, characterization and significance of soil polysaccharides. *Advances in Agronomy* **21**, 195–235.

Thayer, D. W. (1978). Carboxymethylcellulase produced by facultative bacteria from the hind-gut of the termite *Reticulitermes hesperus*. *Journal of General Microbiology* **106**, 13–18.

Tracey, M. V. (1957). Chitin. *Reviews of Pure and Applied Chemistry* **7**, 1–14.

Trojanowski, J., Haider, K., and Sundman, V. (1977). Decomposition of ^{14}C-labelled lignin and phenols by a *Nocardia* sp. *Archives of Microbiology* **114**, 149–153.

Tsao, G. T. (1978). Cellulosic material as a renewable resource. *Process Biochemistry* **13**, 12–14.

Tsao, G. T.,, Ladisch, M., Ladisch, C., Hsu, T. A., Dale, B., and Chou, T. (1978). Fermentation substrates from cellulosic materials: production of fermentable sugars from cellulosic materials. *In* "Annual Reports on Fermentation Processes", vol. 2, pp. 1–21. (Perlman, D., ed.) Academic Press, London and New York.

Vaheri, M. (1980). Cited by Eriksson, K-E. and Johnsrud, S. C. (1982).

Vance, I., Topham, C. M., Blayden, S. L., and Tampion, J. (1980). Extracellular cellulase production by *Sporocytophaga myxococcoides* NCIB 8639. *Journal of General Microbiology* **117**, 235–241.

Varadi, J. (1972). The effect of aromatic compounds on cellulase and xylanase production of fungi *Shizophyllum commune* and *Chaetomium globosum*. *In* "Biodeterioration of Materials", vol. 2, pp. 129–135. (Walters, A. H. and Hueck-van der Plas, E. H., eds.) Science Publishers, London.

Verma, L. and Martin, J. P. (1976). Decomposition of algal cells and components and their stabilization through complexing with model humic acid-type phenolic polymers. *Soil Biology and Biochemistry* **8**, 85–90.

Visser, S. and Parkinson, D. (1975). Fungal succession on aspen popular leaf litter. *Canadian Journal of Botany* **53**, 1640–1651.

Vohra, R. M., Shirkot, C. K., Dhawan, S., and Gupta, K. G. (1980). Effect of lignin and some of its components on the production and activity of cellulase(s) by *Trichoderma reesei*. *Biotechnology and Bioengineering* **22**, 1497–1500.

Wagner, G. H. (1975). Microbial growth and carbon turnover. *In* "Soil Biochemistry", vol. 3, pp. 269–305. (Paul, E. A. and McLaren, A. D., eds.) Marcel Dekker, New York.

Walker, E. and Warren, F. L. (1938). Decomposition of cellulose by *Cytophaga*. *Biochemical Journal* **32**, 31–43.

Webley, D. M. and Jones, D. (1971). Biological transformations of microbial residues in soil. *In* "Soil Biochemistry", vol. 2, pp. 446–484. (McLaren, A. D. and Skujins, J., eds.) Marcel Dekker, New York.

Weimer, P. J. and Zeikus, J. G. (1977). Fermentation of cellulose and cellobiose by *Clostridium thermocellum* in the absence and presence of *Methanobacterium thermoautotrophicum*. *Applied and Environmental Microbiology* **33**, 289–297.

Westermark, V. and Eriksson, K-E. (1974). Carbohydrate-dependent enzymic quinone reduction during lignin degradation. *Acta Chemica Scandinavica* **B28**, 204–208.

Westermark, V. and Eriksson, K-E. (1975). Purification and properties of cellobiose: quinone oxidoreductase from *Sporotrichum pulverulentum*. *Acta Chemica Scandinavica* **B29**, 419–424.

Williams, S. T. and Gray, T. R. G. (1974). Decomposition of litter on the soil surface. *In* "Biology of Plant Litter Decomposition", vol. 2, pp. 611–632. (Dickinson, C. H. and Pugh, G. J. F., eds.) Academic Press, London and New York.

Wilson, R. W. and Niederpruem, D. J. (1967). Control of β-glucosidases in *Schizophyllum commune*. *Canadian Journal of Microbiology* **13**, 1009–1020.

Winter, J. and Wolfe, R. S. (1979). Complete degradation of carbohydrate to carbon dioxide and methane by syntrophic cultures of *Acetobacterium woodii* and *Methanosarcina barkeri*. *Archives of Microbiology* **121**, 97–102.

Wolfe, R. S. (1979). Microbial biochemistry of methane—a study of contrasts. Part I—Methanogenesis. *In* "Microbial Biochemistry", vol. 21, pp. 267–300. (Quayle, J. R., ed.) University Park Press, Baltimore.

Wood, T. G. (1974). Field investigations on the decomposition of leaves of *Eucalyptus delegatensis* in relation to environmental factors. *Pedobiologia* **14**, 343–371.

Wood, T. M. (1980). Cooperative action between enzymes involved in the degradation of crystalline cellulose. *In* "Colloque Cellulolyse Microbienne", pp. 167–176. CNRS, Marseilles.

Wood, T. M. and McCrae, S. I. (1972). The purification and properties of the C_1 component of *Trichoderma koningii* cellulase. *Biochemical Journal* **128**, 1183–1192.

Wood, T. M. and McCrae, S. I. (1977). Cellulase from *Fusarium solani*: purification and properties of the C_1 component. *Carbohydrate Research* **57**, 117–133.

Wood, T. M. and McCrae, S. I. (1978a). The cellulase of *Trichoderma koningii*. *Biochemical Journal* **171**, 61–72.

Wood, T. M. and McCrae, S. I. (1978b). The mechanism of cellulase action with particular reference to the C_1 component. *In* "Bioconversion of Cellulosic Substances into Energy Chemicals and Microbial Protein", pp. 111–141. (Ghose, T. K., ed.) Indian Institute of Technology and Swiss Federal Institute, New Delhi and Zurich.

Wood, T. M. and Phillips, D. R. (1969). Another source of cellulase. *Nature (London)* **222**, 986–987.

Wood, T. M. and McCrae, S. I., and Macfarlane, C. C. (1980). The isolation, purification and properties of the cellobiohydrolase component of *Penicillium funiculosum* cellulase. *Biochemical Journal* **189**, 51–65.

Yamane, K., Suzuki, H., Hirotani, M., Ozawa, H., and Nisizawa, K. (1970). Effect of nature and supply of carbon sources on cellulase formation in *Pseudomonas fluorescens* var *cellulosa*. *Journal of Biochemistry* **67**, 9–18.

Yamane, K., Yoshikawa, T., Suzuki, H., and Nisizawa, K. (1971). Localization of cellulase components in *Pseudomonas fluorescens* var *cellulosa*. *Journal of Biochemistry* **69**, 771–778.

Zantua, M. I. and Bremner, J. M. (1976). Production and persistence of urease activity in soils. *Soil Biology and Biochemistry* **8**, 369–374.

Zeikus, J. G. (1980). Fate of lignin and related aromatic substrates in anaerobic environments. *In* "Lignin Biodegradation: Microbiology, Chemistry and Potential Applications", vol. 1, pp. 101–109. (Kirk, T. K., Higuchi, T., and Chang, H-M., eds.) CRC Press, Boca Raton, Florida.

Index